Thermo-Mechanical Solar Power Plants

Solar Energy R&D in the European Community

Series B:

Thermo-Mechanical Solar Power Plants

Volume 1

Publication arrangements: D. NICOLAY

Solar Energy R&D
in the European Community

Series B Volume 1

Thermo-Mechanical Solar Power Plants

EURELIOS, The 1 MWel Experimental Solar Thermal Electric Power Plant in the European Community

Final report of the Construction of EURELIOS, assembled by the Industrial Consortium for the Implementation of the European Solar Power Plant

edited by

J. GRETZ, A. STRUB and W. PALZ
Commission of the European Communities,
Directorate-General Science, Research and Development

D. Reidel Publishing Company
A MEMBER OF THE KLUWER ACADEMIC PUBLISHERS GROUP
Dordrecht / Boston / Lancaster

for the Commission of the European Communities

Library of Congress Cataloging in Publication Data
Main entry under title:

Eurelios, the 1MWel experimental solar thermal electric power plant of the European Community.

(Solar energy R&D in the European community. Series B, Thermo-mechanical solar power plants: v. 1)
1. Solar power plants-Italy-Sicily. I. Strub, A. S. (Albert S.) II. Gretz, J., 1928- . III. Palz, W. (Wolfgang), 1937- IV. Commission of the European Communities. V. Series.
TK1056.E97 1984 621.31'213 83-26962
ISBN 90-277-1728-1

CIP

Organization of the Contractors meeting by
Commission of the European Communities
Directorate-General Science, Research and Development, Brussels

Publication arrangements by
Commission of the European Communities
Directorate-General Information Market and Innovation, Luxembourg

EUR 9127

Published by D. Reidel Publishing Company
P.O. Box 17, 3300 AA Dordrecht, Holland

Sold and distributed in the U.S.A. and Canada
by Kluwer Academic Publishers,
190 Old Derby Street, Hingham, MA 02043, U.S.A.

In all other countries, sold and distributed
by Kluwer Academic Publishers Group,
P.O. Box 322, 3300 AH Dordrecht, Holland

Preface

This book reports on the construction phase of the European solar power plant EURELIOS.

EURELIOS was the first of half a dozen solar power plants in the MW range around the world to enter operation.

It is of the central receiver type employing steam as the heat transfer fluid and has a nominal electric power of 1 MW. The Project was built in Sicily as part of the Solar Energy R + D Programme of the Commission of the European Communities.

Funding was provided by the European Communities, the Governments of France Germany and Italy and the contractors themselves, i.e. ENEL, ANSALDO, CETEL and MBB.

The active involvement of the national utility of Italy, ENEL, the Commission's host partner for EURELIOS, was of paramount importance for the successful completion of the project and should be stressed here.

Many individuals made contributions. The original concept was developed by Prof. FRANCIA who was a leading expert in the field. Dr. STRUB, Director for Non-Nuclear Energy R + D at the Commission of the European Communities (D.G. XII) had the general responsibility for the project, Prof. CHABBAL from France and Dr. KLEIN from Germany very much contributed to its initiation, Dr. CORVI from Italy was actively following the project all the way through its implementation phase.

Implantation was carried out with Dr. GRETZ as the Commission's project leader assisted by General Technology Systems, Ltd from United Kingdom and under the guidance of a very efficient management committee. In this committee the industrial consortium was represented by the following members : Messrs BORGESE, DINELLI (ENEL), TUARDICH, AMADEI (ANSALDO), TREIBER, HOFMANN, SCHOBER (MBB), TOTH, FAURE, DEPAUW, BLANCHARDIE (CETHEL).

It is important to note that EURELIOS is a purely experimental plant. The prime objective of its construction was the technical evaluation of this new and highly innovative conversion process. Its implementation at a relatively large scale - still being much below the full scale for commercial power plants of this type - was necessary because of the complexity of the involved technology.

At this time it has become clear that the construction of EURELIOS was worthwhile. Its value as well as its technical and economical limits could be assessed and much detailed experience has been gained. It is also interesting to note that EURELIOS has become an outstanding example of transnational cooperation among Administrations and among industry in Europe.

W. PALZ,
Head of the EC Solar energy R + D Programme.

Preface

This book reports on the construction phase of the European solar power plant EURELIOS.

EURELIOS was the first of half a dozen solar power plants in the MW range around the world to be [illegible].

It is of the central receiver type [illegible] mirrors [illegible] the heat transfer [illegible] and has a nominal [illegible] power of 1 MW. [illegible] in Italy [illegible] the Commission [illegible] programme.

The plant was [illegible] by [illegible] consortium [illegible] Germany, France and Italy [illegible] ENEL, ANSALDO [illegible].

The [illegible] Commission's [illegible] EURELIOS, [illegible] successful [illegible] and should be [illegible] here.

[illegible] EURELIOS [illegible] Dr. [illegible] Directorate General [illegible] of the European Communities [illegible] and the general [illegible] Project [illegible] France and [illegible] ENEL [illegible] the project all the way [illegible] phase.

[illegible]

It is important to note that EURELIOS [illegible] technology.

[illegible] EURELIOS was [illegible] the designed [illegible] assessment and much detailed experience has been [illegible]. It is interesting to note that EURELIOS has [illegible] components [illegible].

W. Palz,
Head of the EC Solar Energy R & D Programme

Contents

Preface v

1. INTRODUCTION 1

2. SYSTEM DESIGN CONSIDERATIONS 5

2.1 BASIC REQUIREMENTS 5
2.2 SYSTEM DESCRIPTION OF THE PLANT 6
2.2.1 Heliostats
2.2.2 Receiver
2.2.3 Steam cycle
2.2.4 Thermal storage system
2.2.5 Electrical systems (CETHEL and ANSALDO)
2.2.6 Control system
2.2.7 Central services
2.3 THE NOMINAL POWER BUDGET 11
2.4 SAFETY 11

3. SUBSYSTEM DESIGN INFORMATION 15

3.1 HELIOSTATS 15
3.1.1 Reflecting panels
3.1.2 Drive units, columns and foundations
3.1.3 Heliostat electronics
3.2 RECEIVER 34
3.3 STEAM CYCLE 46
3.3.1 Description
3.3.2 Emergency feedwater system
3.3.3 Steam cycle alarms
3.4 TURBINE 50
3.5 THE MOLTEN SALT THERMAL STORAGE SYSTEM 55
3.5.1 General description
3.5.2 Operation and control
3.6 ELECTRICAL SYSTEM 60
3.6.1 Introduction
3.6.2 CETHEL electrical system
3.6.3 ANSALDO electrical system
3.7 CONTROL AND COMPUTER ROOMS 71
3.7.1 Heliostat central controls
3.7.2 Master Control Centre (MCC)

3.8 CIVIL WORKS 95
3.8.1 Site selection
3.8.2 Survey and site preparation works
3.8.3 Description of works

4. OPERATION OF THE PLANT 101

4.1 INTRODUCTION 101
4.2 OPERATING PROCEDURES 101
4.2.1 Morning start-up
4.2.2 Cold start-up with full radiation
4.2.3 Hot start-up
4.2.4 Normal operation
4.2.5 Operation with cloud passing
4.2.6 Evening shut-down
4.2.7 Emergency operations

5. PHASE D 105

1. Introduction

A "power tower" solar-thermo-electric plant comprises a field of many mirrors controlled as heliostats which focus sunlight onto a receiver mounted on a tower. The resulting thermal energy is transferred by means of a working fluid, or fluids, to an electricity generator - in this project water and a steam turbine driven alternator, respectively. Optionally some means of short term energy storage may be included - in this project high pressure hot water and an additional superheating fluid, molten salt - to allow the plant to continue operation during, for example, the passage of clouds.

The Commission of the European Communities (CEC), DG XII - Research Science and Development, placed a contract for an initial feasibility study (Phase A) between May and October 1976. This contract resulted in the specification of a plant which would provide valuable design and operating experience on a number of novel problems inherent in achieving viable solar thermo-electric plants.

Phases B and C, which ran for three years from November 1977, comprised respectively detailed design and construction of the plant near Adrano in Sicily. Named EURELIOS, it was completed before Christmas 1980, commissioning operations having already started.

In January 1981, during the formal acceptance process of the Commission and ENEL (Ente Nazionale per l'Energia Elettrica, the operator of the plant) the project moved into its fourth phase, D, during which the output power is to be increased to the design figure and an experimental programme started.

The work was sponsored jointly by the CEC and the governments of Italy, France and F.R.Germany. The Commission's contract for design and construction provided for the payment by the Commission of half the expenditures incurred.

The organisation of the project is shown in Fig.1.1. The plant has been built by an industrial consortium consisting of ENEL and ANSALDO-AMN (now Ansaldo Impianti) - Italy, CETHEL - France, and MBB (Messerschmitt Bölkow-Blohm GmbH) - F.R.Germany. ENEL also acts as host-partner, providing free of charge the site and supporting site services such as security and having been responsible for ensuring the safety and legality of the complete plant. Within the Consortium, ENEL was responsible for civil works interfacing with the Sicilian electrical grid, the design of the control system and the definition of the operation procedures of the plant. ANSALDO supplied the tower and receiver, the steam cycle, hot water/steam storage and associated electrical and pneumatic controls. CETHEL's responsibility was to provide approximately half of the heliostat field and associated monitoring and controls, the molten salt thermal storage system and the internal electrical power system (alternator, main transformer and auxiliary power distribution). The other half of the heliostat field (rather more than half in terms of numbers of heliostats but, since its heliostats are smaller than those of CETHEL, less than half the total reflecting area) was supplied by MBB. Each company provides special electrical conversion and, where necessary, backup equipment, drawing power from Eurelios itself or from the grid. General Technology Systems Ltd., has been retained separately by the Commission, through its Belgian subsidiary, GTS SA, for assisting the Commission's project leader in the monitoring of the work and to act as the Commission's agent during acceptance testing at the end of Phase C.

A Helioelectric Plant Consultative Committee (HPCC), which consists of members proposed by the Commission (3) and the participating member states (3) and which has observers from the Consortium and from ENEL (as host), advises the Commission on overall policy and strategy of the project, on matters relating to the project needing a decision by the Council of the European Communities and on the passage from Phase B to Phase C and commented on the contractual reports of the Consortium.

This report first outlines the original requirements of the plant, and briefly describes it (Chapter 2) and then describes the plant in detail, subsystem by subsystem (Chapter 3).
Chapter 4 outlines the operating procedures and the report ends with a look forward into Phase D.

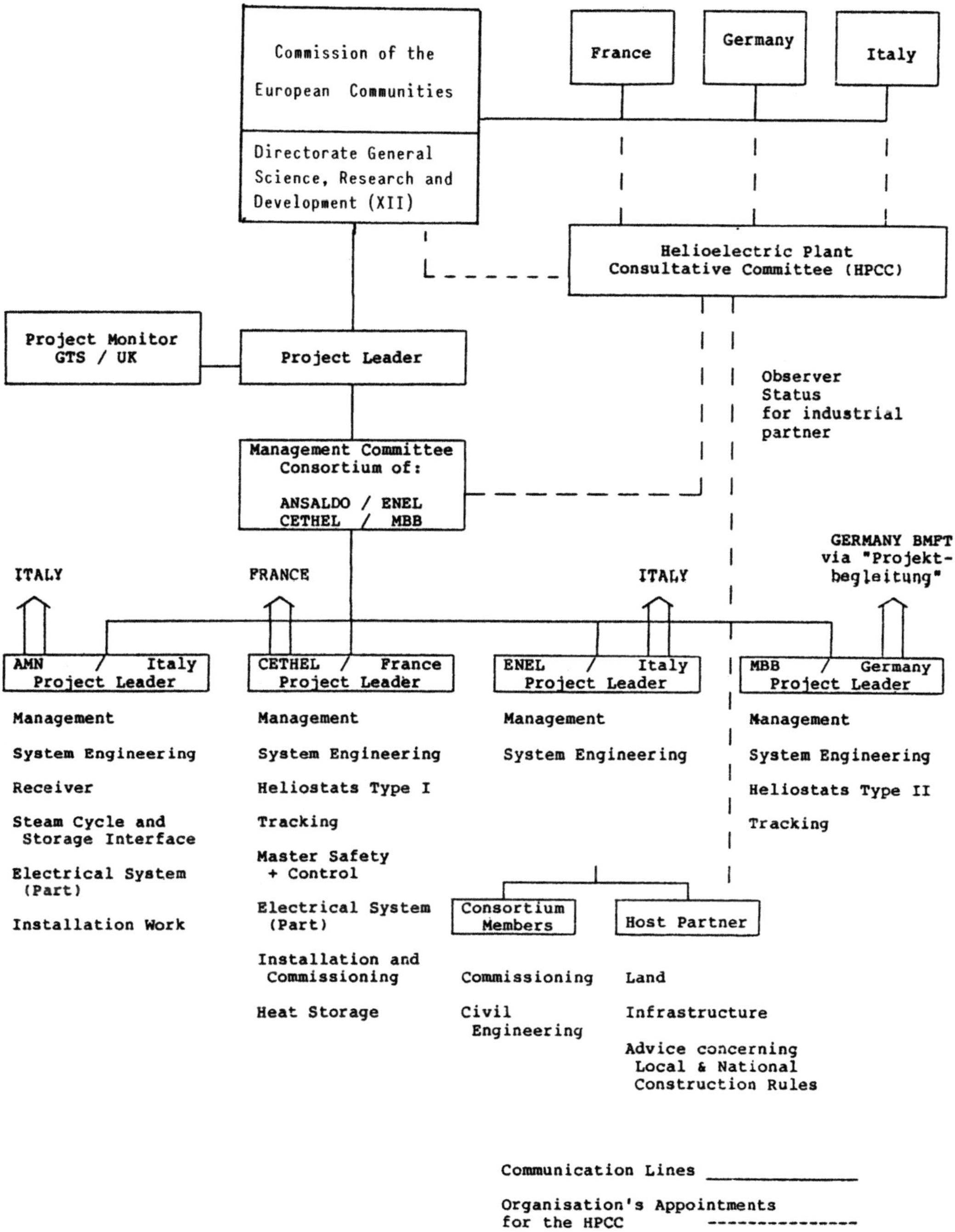

Figure 1-1 - EURELIOS Project Organisation (Phase B and C)

Fig. 1 - EURELIOS Project Organization (Phases B and C)

2. System Design Considerations

2.1 BASIC REQUIREMENTS

Ideally, all parameters of the plant could be considered to be adjustable to maximise cost-effectiveness according to system level requirements: e.g. the increase of temperature of the receiver could lead to a higher efficiency of the power conversion cycle and to a reduction of number of heliostats but requires an increase in pointing accuracy. In practice, however, the opportunities for parameter variation are limited by available or preferred technology.

For the EURELIOS plant, once the site had been selected, the following features were specified (although many were slightly altered during the design phase as a result of optimisation).

. Nominal power output: 1 MWel at an insolation of 1 kW/m^2 at equinox noon at the site of Adrano/Sicily with the following geographical coordinates:
    37.6466° north
    14.8034° east.
  The site slopes down about 4° from north to south and 2.5° from east to west. The elevation above sea level of the foot of the tower is 214 m. The heliostat field to be north of the receiver.
. Tolerable wind forces:
    - nominal output power to be available up to 18 km/h, the maximum tower deflection being ± 10 cm;
    - operation (reduced power) to be possible up to 50 km/h;
    - survival (no permanent deformation) up to 120 km/h (130 km/h for the tower, a figure which was, in the event, adopted for most of the plant).
. Utilisation of a steam cycle and cavity type boiler with anti-radiating structure (made of Pyrex glass, design of Prof. Francia) using water/steam directly as heat transfer and working medium with a flow rate of up to 5300 kg/h allowing growth over the then nominal conditions of 4860 kg/h. Nominal steam conditions are 510°C, 64 atm. Receiver aperture about 4.5 m diameter (amended as a result of early optimisation work from an initial 4.2 m x 4.2 m square aperture in a 5 m cube). Incoming radiation flux into the aperture is 4800 kW at the nominal design point.
. A steel tower to support the receiver, about 50 m high (centre of aperture) but this increased to 55 m, again resulting from optimisation during design.
. CETHEL heliostats each with 52 m^2 reflective surface and MBB heliostats with 23 m^2, each giving a reflected beam accurate to ± 4 mrad

(1 sigma). The CETHEL field section to give at least 57% of the total requirements; the MBB section at least 43%.
- Thermal storage to allow the plant to continue operating for 30 minutes without insolation.
- A conventional electrical system with a 1.1 MW alternator (minimum), connected to the national grid's Contrasto hydroelectric power station via a 1.0 MW output transformer (minimum). To have automatic emergency shut-down and appropriate emergency devices to ensure the safety of the plant despite equipment failures or loss of power.
- Centralised control.
- The whole plant to conform to the relevant Italian law and regulations.

Priority has been given to the experimental character of the plant. Hence although automatic start-up and shut-down are ultimate and important objectives, operation will involve considerable manual intervention. Similar comments apply to recovery from abnormal, including emergency, conditions subject, of course, to the need for automatic initiation of activities needed to ensure the safety of personnel and plant.

2.2 SYSTEM DESCRIPTION OF THE PLANT

Figure 2.1 is a simple schematic of the EURELIOS plant, the principal characteristics of which are summarised in Table 2.1.

The individual subsystems are described in detail in Chapter 3 of this volume. This section explains the overall configuration so that the individual sections of Chapter 3 can be read in isolation by those interested in only limited parts of the plant.

2.2.1 Heliostats

Sunlight is intercepted by the heliostat "field" and reflected into the receiver (boiler) just to its south. The field is split into two roughly equal "subfields", East and West, supplied by MBB and CETHEL, respectively. The subfields are each divided, for control purposes, into four independent "groups". Each heliostat has its own microprocessor-based local electronics unit which is controlled by, and reports its status to a "Central Unit" (one each for CETHEL and MBB). The Central Units can be used manually to control individual heliostats, groups or the corresponding subfield. The Central Units are also controllable by, and report to, the plant's Generator Control System (GCS) for routine and emergency control at group level and above only.

The CETHEL heliostats also contain emergency batteries with several times the amount of energy needed to move the heliostats to a safe position in the event of loss of the electrical supply.

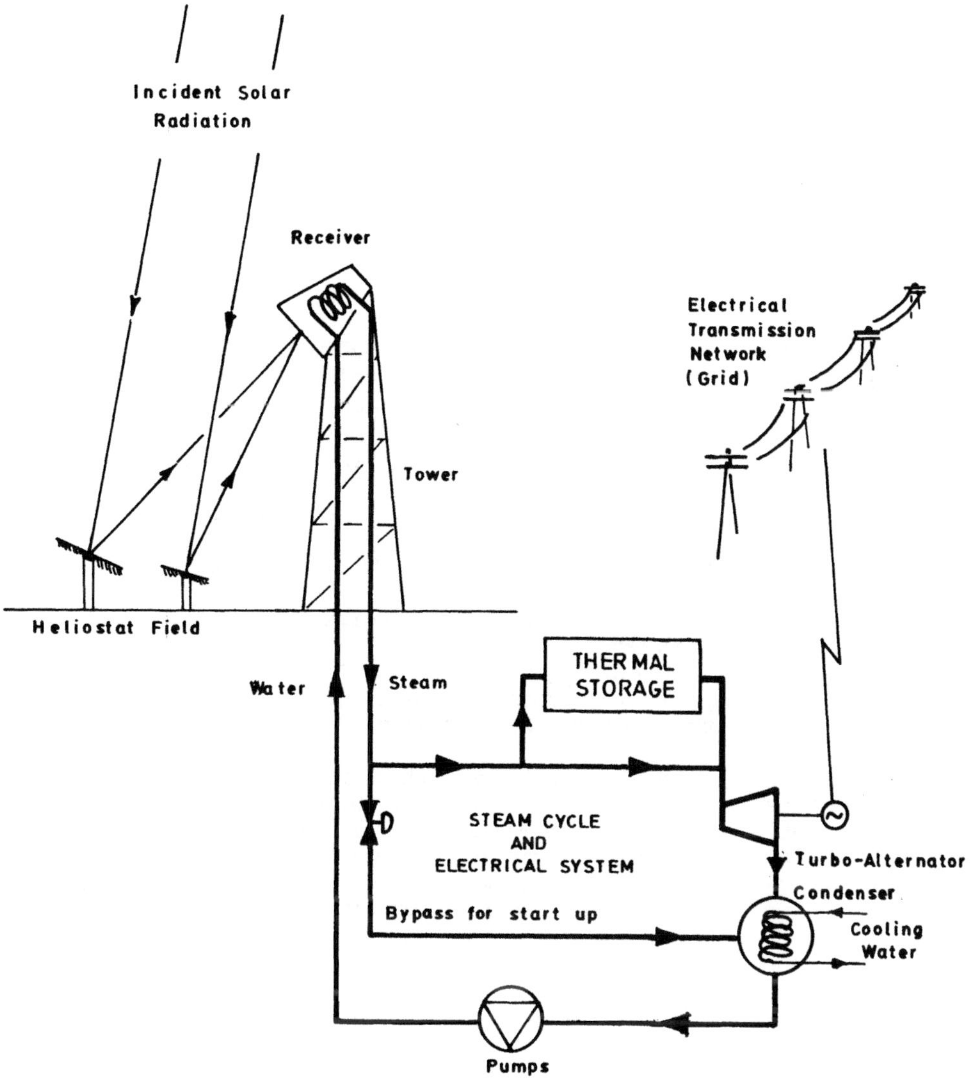

Figure 2-1 – EURELIOS Concept

1 MW_{el} SOLAR POWER PLANT OF THE EEC

Technical Parameters

General characteristics

Experimental plant of the central receiver multiheliostat type
Location : 37.6° North; 14.8° East (Adrano, Sicily, Italy)
Design point : equinox noon and assumed insolation of 1,000 W/m^2
Power generated : 1 MW_{el} into existing grid at design point

Design parameters (power figures at design point)

Heliostat field	: 4800 kW_{th} to receiver - 6202m^2
Heliostats	: two types, two axis controlled, focusing, overall inaccuracy $\pm$ 4 mrad (1 sigma)
Cethel type	: ca. 52 m^2, 8 modules, 70 heliostats
MBB type	: ca. 23 m^2, 16 square elements, 112 heliostats
Receiver/tower	: Cavity type receiver 4.5 m aperture at 55 m height, 110° inclination
Receiver outlet steam conditions	: 512°C, 64 atm, 4860 kg/h
Steam cycle	: Turbine connected to receiver (no intermediate heat exchanger)
Nominal power	: 1100 kW_{Mech} with steam of 510°C, 60 atm
Feed water temperature at receiver inlet	: 36°C
Cooling water temperature	: 25°C maximum
Thermal storage	: Reduced electrical output for ca. 30 min.
Capacities	: Vapour 300 kWh; Hitec : 60 kWh
Equipment	: Pressurised water reservoir for 4300 kg vapour (reducing from 19 to 7 bar) Two storage tanks, containing about 1250 kg Hitec salt (overall capacity) Heat exchangers for 64 bar, 480°C and 19 bar, 410°C steam
Electrical system	: Alternator 1500 KVA giving internal power and 1000 kW for external users
Transformers	
Emergency power supply	
Interface to grid	: equipment to connect transformers to public grid
Steam cycle control equipment	
Command, operation and monitoring is centralised in the control centre	

Table 2.1 - Technical Parameters, Summary

2.2.2 Receiver

The ANSALDO receiver is a stainless steel, dualchannel, once-through, cavity-type steam boiler located at the top of a steel tower. Output steam conditions are controlled by adjusting the feedwater flow rate and by the use of attemperator sprays within the boiler. It is equipped with pressure, temperature and thermal flux sensors which allow the control system to make the appropriate adjustments.

The design concept, due to the late Prof. Francia, includes the use of a Pyrex glass anti-radiating structure inside the receiver to reduce losses due to re-radiation and convection. The Pyrex is to be installed after experience without it has been obtained, in order to verify the effect of the anti-radiating structure.

The receiver is protected in various ways against damage due to possible electrical or pump failure or misalignment of heliostats (by ceramic bricks on the rim, by a protective casing, by an over-pressure release valve, by an emergency feedwater supply system, etc.).

2.2.3 Steam cycle

The steam cycle built by ANSALDO is made of stainless steel. Demineralised water is pumped at pressure into the receiver where it is boiled and superheated before being passed for use in one or more of three ways:

i) to drive the turbine for electrical power generation;
ii) to heat the salt of the thermal storage system;
iii) to heat the water for the thermal storage system.

During start-up, when the system is being pressurised and heated, or in emergency, steam is sent through a turbine bypass valve and desuperheater before being passed via a flash (sparge) tank to the condenser for recirculation. In normal circumstances the electrical generation load on the turbine is automatically adjusted to make full use of the available steam despite variable insolation, the control system of the steam cycle adjusting feedwater flow rate to maintain steam pressure and temperature.

The steam cycle control system, built by ANSALDO according to a functional design of ENEL, has as one vital role the protection of the receiver against damage by over-heating. In addition to having duplicated electrical supplies and duplicated pumps the steam cycle includes an emergency pressurised feedwater storage system which operates in the complete absence of the normal feedwater supply until the receiver is free of insolation.

2.2.4 Thermal storage system

This system stores heat in pressurised, saturated steam/water and molten salt, its purpose being to keep the plant operating for half an hour without insolation during cloud passing. The salt, contained in a system provided by CETHEL, is charged (heated) by pumping it from a "cold" to a "hot" tank through a heat exchanger taking heat from the main steam supply. The heavily insulated steam/water accumulator, part of ANSALDO's equipment, is charged by first partially filling it with cold feedwater and then bubbling (condensing) main steam into it until the desired working pressure is reached. During discharge the hot salt is pumped through a second heat exchanger, giving energy to superheat the saturated steam from the accumulator, this steam then being fed to the turbine.

2.2.5 Electrical systems (CETHEL and ANSALDO)

During normal operation, the plant generates electrical power in an alternator driven by the steam turbine. Part of this power is used (via a transformer) at 380 V to run the plant, the greater part being sent (via a second step-up transformer) to the nearby Contrasto hydroelectric power station where it contributes to the Sicilian grid. During start-up or maintenance, power comes normally by the same route from Contrasto. A second, independent, line brings in power to operate the auxiliaries if the normal supply is for any reason not available. Services essential for the emergency survival of the plant can also obtain electrical power from a bank of lead-acid batteries.

Each subsystem itself carries out any necessary further power conditioning on the 380 V three-phase supply distributed around the plant.

2.2.6 Control system

The plant is centrally controlled from a room overlooking the machine room. The control room contains the main control console for operating the plant, vertical boards, in front of the operator, displaying a synoptic of the process, the alarms, meteorological data, plant thermodynamic and electrical data.

Near the console are the desks and computers for detailed monitoring and control of the two heliostat subfields. Beneath the control room is located all the electrical switchgear whereas the control apparatus are installed in an adjacent equipment room.

2.2.7 Central services

The facilities for operating the plant are mostly housed in a building to the south west of the tower and receiver. There are in addition ancillary buildings housing spare parts, workshop facilities and exhibition, conference and office accommodation for visiting specialists not involved in the day-to-day operation of the plant. Meteorological monitoring is carried out at a number of locations around the site.

2.3 THE NOMINAL POWER BUDGET

		kW
Insolation intercepted by heliostat field, gross (1)		6200 (2)
<u>Less</u> shading, cosine, misalignment, reflectivity and blocking losses (3)		1400 (4)
Power into receiver (visible and thermal radiation)		4800
<u>Less</u> reflection, reradiation and convection losses (inc. pipework)		416 (5)
Net power absorbed by working fluid		4327 (6)
<u>Less</u> losses in steam cycle		3227 (6)
Shaft power from turbine		1100 (6)
<u>Less</u> alternator losses		50 (7)
Electrical output from alternator		1050
<u>Less</u> consumption and losses in auxiliaries (normal operation):		
steam cycle	32 kW	
CETHEL field	4 kW	
MBB field	4 kW	
central controls	5 kW	
other	5 kW	50
NET OUTPUT OF THE PLANT		1000

<u>Notes</u>:

1) Nominal, 1 kW/m^2 insolation but with mirrors normal to sun line, neglecting space between mirrors. All figures rounded.
2) Based on total mirror area: 70 x 51.8(CETHEL)+ 112 x 23 (MBB) = 6202 m^2.
3) Shading and blocking are zero at equinox noon.
4) Figure obtained by differencing from the one below.
5) Figure obtained from the one below, assuming 91.3% efficiency.
6) Figure obtained from AMN "nominal output" data for turbine. Net power absorbed is obtained by differencing enthalpies of steam-to-turbine and feedwater for the nominal flow rate.
7) Based on 95% efficiency of alternator for 1200 kW load.

2.4 SAFETY

Safety is particularly important in an experimental facility to which the public has access but which involves high voltages, high pressure steam and highly concentrated sunlight. The primary safety objective must be to avoid personal injury. Following this closely are damage outside the plant and damage to the plant itself.

The main unconventional risks to both personnel and equipment result from concentrated sunlight. It was established early in the project that light reflected from the receiver would not be a hazard but that care would be necessary to avoid risk from the focussed light from the heliostat mirrors. Light from a single heliostat is only a hazard, causing serious dazzling and possibly eye damage, to people near the focus of the heliostat. Since no heliostat has a focal distance of more than a nominal 200 m, such risk is limited to the immediate vicinity of the plant itself. To extend this limitation to the whole heliostat field rather than its applying simply to one heliostat, the concept of a Safety Trajectory was introduced. A vertical line is defined, adjacent and just to the east of the tower. The beam centre of every heliostat is constrained to follow this line wherever it is in transit between the ground and a Standby Position immediately east of the centre of the receiver aperture. Since heliostats are by this means focussed within the plant, no pair is ever likely to converge on any distant object and in addition the individual heliostat beams would even then be well out of focus themselves.

However, there still remains at risk the immediate vicinity of the plant, either from a single heliostat or from numbers of them. An analysis has been carried out to find the effect of any failure of a heliostat which causes it to stop or move out-of-control, so as to illuminate the surrounding area. The value of 3 kW/m^2 was selected as one which was very dangerous for the human eye.

To compute the iso-flux the following hypotheses were made:

a) The normal direct radiation intensity can be written as follows:

$$I = \exp(\sin E/\sin E_{1200} - 1)$$

where E is the angle between the solar radiation and the horizontal plane at the time and the day considered, E_{1200} is the value of the above angle at noon on the day in question.

b) The linear dimension of the heliostat's focus, due to astigmatism, increases as the inverse of the third power of the cosine angle between the incoming radiation and the normal to the heliostat ($\cos \phi/2$).

c) The single mirrors of the heliostat have the same focal length as the heliostat itself.

d) The reflected radiation is uniformly distributed, i.e. any orthogonal section of reflected beam has a constant power density.

e) The iso-flux lines are on a horizontal plane including the top of heliostat's frames.

The boundary obtained through the above conditions is approximate but is sufficient to give a general indication of the area within which care would have to be taken in the event to such a heliostat failure. Even in the case shown in the figure, a particularly unfavourable one, only a very limited area beyond the site boundary is affected.

Within the site, particularly at the foot of the Safety Trajectory, there would be risk whenever there were heliostats in transit and so a protective wall (shown in the figure) has been built to protect personnel near the foot, and to the east, of the tower. Further, personnel are not normally allowed on the tower or in the field when heliostats are in use.

Because of the Safety Trajectory concept the remaining danger is limited to the effects of the focussed spot near the receiver - either from poor focussing or misplacement of the spot as a whole. In the former case, which may occur whenever the sun's position is distant from that at which a heliostat was focussed, the beams spread, illuminating the sides of the receiver rim and possibly the structure of the tower. Structural illumination is minimal but an unprotected rim may be overheated. However, since the whole spot is formed from the full heliostat field, the individual members of which are individually controlled, failure only affects the whole field in the event of power failure, in which case the CETHEL heliostats still move out of the receiver under battery power. The MBB heliostats remain stationary. Provision has to be made to protect not only the receiver pipework (Eurelios includes an emergency feedwater system not requiring electrical power) but also to protect the eastern rim of the receiver while the spot moves eastward due to the motion of the sun. In fact the entire rim of the receiver is protected with ceramic bricks.

People are protected against the conventional risks in conventional ways: by excluding them from dangerous areas by using guards, fences, passes, closed cabinets, colour codings, warning notices, use of protective head gear. The plant is of course built to satisfy Italian national safety regulations and includes duplicate and safety-specific equipment in many places, to protect the operation of the plant itself from disruption.

3. Subsystem Design Information

3.1 HELIOSTATS

The heliostat field is required to provide a maximum of power into the receiver aperture. Fig. 3.1.1 illustrates the most important effects influencing the amount of radiation flux finally entering the receiver aperture. With the help of computer programs numerical calculations of the radiation power were performed for various arrangements of heliostats, leading finally to the field layout shown in Fig. 3.1.2.

Apart from the primary requirement, i.e. optimum power input into the receiver, there are several secondary requirements, which constrain the layout of the EURELIOS heliostat field.

For both types of heliostats, i.e. CETHEL and MBB, it was required to study the time variation of the image of the sun (alignment, quality of mirrors, etc.) and its effects on optical efficiency, both for heliostats close to the tower and more distant from the receiver. It was therefore decided to arrange the heliostats in such a way that there is a west-field (CETHEL) and an east-field (MBB). This gives similar elevation (and azimuth) angles in the west-field in the morning, as in the east-field in the afternoon. As CETHEL is to supply 57% of the nominal receiver input power, some of the CETHEL heliostats are placed east of the exact north-south line.

A dense positioning of heliostats is preferred, to minimise use of land and to gain as much radiation as possible: unavoidable imperfections in reflection lead to increased spillage of radiation for more distant heliostats. Minimum spacings of 11.7 m for the CETHEL heliostats and 7.8 m for the MBB's ensure that adjacent heliostats never touch.

To simplify the civil work, it was decided to arrange the necessary access roads within the heliostat field in the east-west direction and to have one additional road north-south, approximately at the separation line between the two types of heliostats. Accordingly, the heliostats have been arranged in east-west rows as shown in Fig. 3.1.2, between 25 m and 197.7 m north of the receiver. In case of future enlargement, additional heliostats might be put along the same east-west rows, just increasing the number of units per row.

Within the rows, i.e. in east-west direction, the minimum distance between heliostats is adopted. Heliostat rows more distant from the tower are further apart so as to minimise blocking of reflected light by heliostats

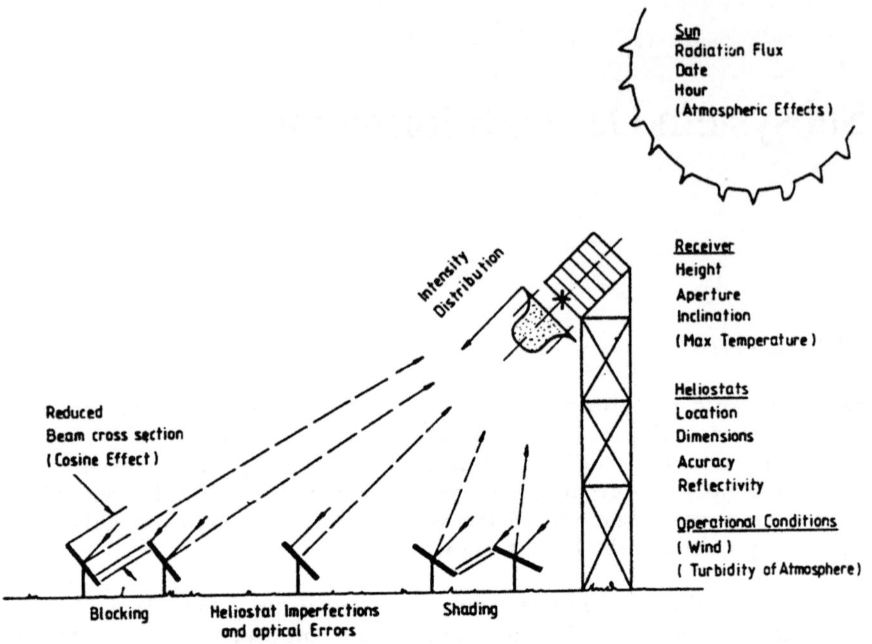

Figure 3.1.1 - Radiation Power Collected in a Central Receiver

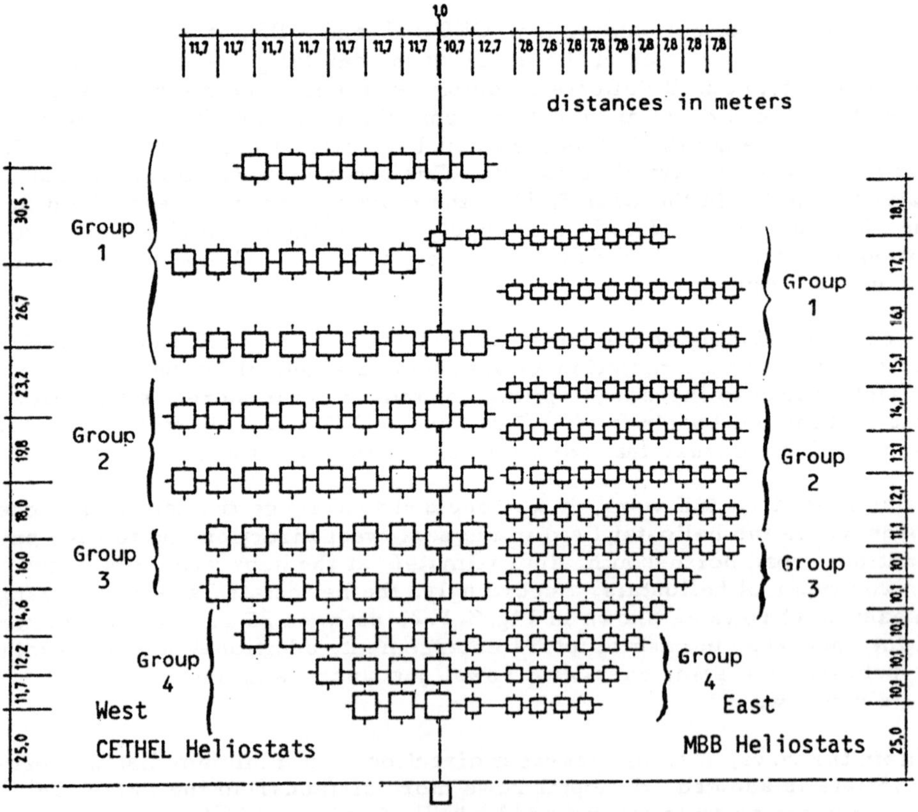

Figure 3.1.2 - Heliostat Field Arrangement

closer to the tower. Whereas the first rows of heliostats in the CETHEL section are as close as the above minimum distance (i.e. 11.7 m) the distance between the first rows of MBB heliostats is increased to 10.1 m (instead of 7.8 m) to allow easy passage for vehicles. Energy calculations have shown that this has an effect on the power output of the MBB field well below 1%.

The final arrangement has been evaluated by a mathematical model of the field with which the energy flux into the receiver was evaluated, taking into account the different geometrical effects indicated in Fig. 3.1.2 as well as the specified gaussian distribution of the direction of the reflected beam about the ideal direction of $\pm$ 4 mrad (1 sigma).

The result was that with the 70 CETHEL heliostats and the 112 MBB heliostats arranged as shown, the required radiation power of 4800 kW would be provided to the receiver aperture at nominal insolation conditions, viz. equinox noon, solar time, assuming a solar incident flux of 1 kW/m^2. This arrangement is optimum in the sense that slight variations of the distance of the heliostat rows do not change the power output dramatically but show the tendency that both larger row distances and shorter row distances lead to slightly reduced power outputs. There is neither blockage nor shadowing at equinox noon.

Averaged over the whole MBB field at equinox noon, the reflectivity is 85% and cosine effects and spillage due to imperfect reflection cause a loss of 5% (i.e. 95% "field performance") leading to an overall efficiency of 81%. With a total of 2576 m^2 of MBB reflective surface, this leads to a nominal contribution of 2086 kW, which corresponds to 43% of the required 4800 kW. Similarly for CETHEL with an 80% reflectivity, 3626 m^2 of reflective surface and 95% field performance, the contribution is 2756 kW, i.e. the balance of 57% required, with 1% margin. For single heliostats, the values differ according to their positions and they depend on time.

No system was installed during phase C that could allow any quantitative measure of the actual performance of the heliostats.

Only a 5 m square auxiliary target marked with a half-meter grid was located immediately below the receiver. This allows focussing and alignment of individual heliostats to proceed even when the receiver is being used for routine generation of power.

The east-west split of the field is largely irrelevant to the receiver and steam cycle. However, it is operationally desirable, during start-up and in emergencies, for the field to be subdivided. So, the MBB and CETHEL fields are each divided into four groups which are normally commanded as pairs (CETHEL group 1 with MBB group 1) by the General Control System (GCS).

Table 3.1.1 gives the basic requirements which had to be met by the heliostats. The following sections describe the approaches adopted by the two companies, Table 3.1.2 and Figs. 3.1.3 and 3.1.4 giving general information in comparative form.

Wind velocity	Full Power	18 km/h (5 m/s)
	Operational limit	50 km/h (14 m/s)
	Survival	130 km/h (36 m/s)
Time to reach stowage position (time allowed for 180° movement in elevation)		max. 30 min.
Environment	Temperature	0 to +40°C
	Qualification limits	-10° to +50°C
		Sandstorm Thunderstorm
Reflected beam accuracy at 18 km/h wind		± 4 mrad (1 sigma) Reflected beam

Table 3.1.1 - Heliostat Design Requirements

The most noticeable differences between the two designs of heliostat are their sizes and their back structures. MBB justified its layout as follows:

. development should be possible without too high a risk and within the given time;
. the heliostat should be optimum for relatively small, syy 1 MWel, central receiver power plants and not primarily designed for larger plants;
. the heliostat should be simple in terms of mounting (weight and dimensions) and transport. This might be particularly important for sites in developing countries;
. components should be available off the shelf. This was particularly critical with respect to gears: larger gears would have had longer supply time including additional development time;
. aberration losses can be kept smaller with smaller heliostats.

CETHEL's design was conceived in the context of the established French national programme, involving somewhat larger stations than Eurelios.

Both manufacturers adopted horizontal (elevation) and vertical (azimuth) tracking axes and a nominally horizontal, face down, stowage position for their heliostats. The heliostats can be driven at two speeds: fast to reach a desired position; slow during tracking of the sun. Sun tracking is open loop to the extent that it relies on computation of sun (and hence heliostat) directions as a function of time without direct measurement of the sun or of the reflected beams. There are, however, differences in the implementation of tracking and, from the safety point of view, they differ in that only CETHEL's heliostats have battery-powered emergency supplies available to drive them in the event of a power failure.

	CETHEL	MBB
REFLECTING PANELS		
Configuration (each heliostat)	8 modules of 6 mirrors of 1.8m x 0.6m = 51.8m^2	16 mirrors of 1.2 x 1.2m = 23m^2
Weight	2700 kg.	800 kg. (27 kg/mirror element)
Mirror elements	6mm backsilvered float glass	3mm backsilvered float glass in glass/glass or glass/metal foam-filled sandwich
Nominal reflectivity (clean)	0.8	0.85
Overall panel width	8.84m	5.63m
Overall panel height (top to bottom)	7.34m	5.01m
Height of elevation axis	4.20m	2.97m
Ground clearance	0.52m	0.47m
DRIVE UNITS		
Panel rotation speeds (azimuth)	12°/min fast; 0.6°/min slow	4.79°/min fast; 0.48°/min slow
Panel rotation speeds (elevation)	" " "	6.28°/min fast; 0.63°/min slow
Range of rotation	±120°Az; +75° -90° El.	± 120°Az; ± 90°El
Weight	1100 kg.	600 kg.
COLUMN		
Construction	Concrete Cylinder, integral with foundation	Steel tube, bolted to foundation
Dimensions	Ø 0.850 x 3.92m	Ø 0.406 x 2.62 m
Weight		180 kg.
FOUNDATION		
Construction	Circular concrete cylinder	Double, circular concrete cylinder
Dimensions	2.750 x 0.6m	0.800 x 0.32m on 2.50 x 0.5m base
NOS. OF H'STATS	70	112
TOTAL MIRROR AREA	3626 m^2	2576 m^2

Table 3.1.2 - Numerical Comparison of CETHEL and MBB Heliostats

Figure 3.1.3 - Heliostats Tracking into Receiver CETHEL (Upper) MBB (Lower)

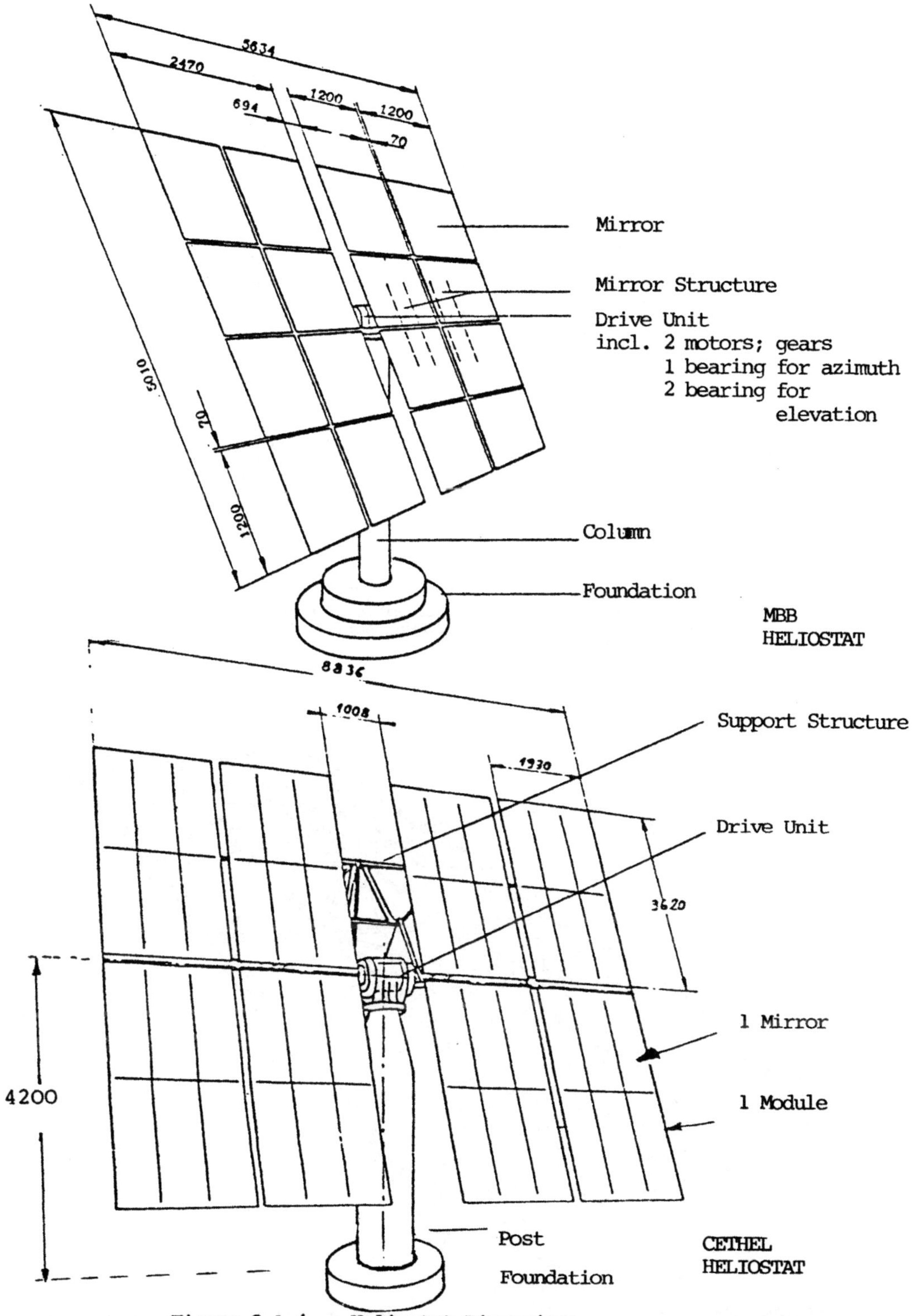

Figure 3.1.4. - Heliostat Dimensions

Heliostat Number	Focal length of module m	Focal length of panel m
1 - 2	80	57,5
3 - 4 - 5		63
6 - 8		69,1
7 - 9 - 10		73,6
11 - 14 - 15 - 16		80,8
12 - 17	100	84,5
18		87,8
13 - 19 - 22 - 23 - 24		93,5
20 - 25 - 26		100
21 - 27		107,5
30 - 31 - 32	140	108,9
28 - 33		110,9
29 - 34 - 35		118,5
36 - 39 - 40 - 41		126,5
37 - 42 - 43		131,1
38 - 44 - 45		139,6
46		148
47		151,9
48 - 49 - 50	200	148
51 - 52		151,9
53 - 54		159,2
55 - 56		169,5
57 - 58 - 59		176,6
60 - 61		182,9
62 - 63		191,9
64 - 65 - 66 - 67		203,6
68 - 69 - 70		210,8

The exact distance between heliostat and receiver aperture is never different from the adjusted focal length of the panel more than 3%.

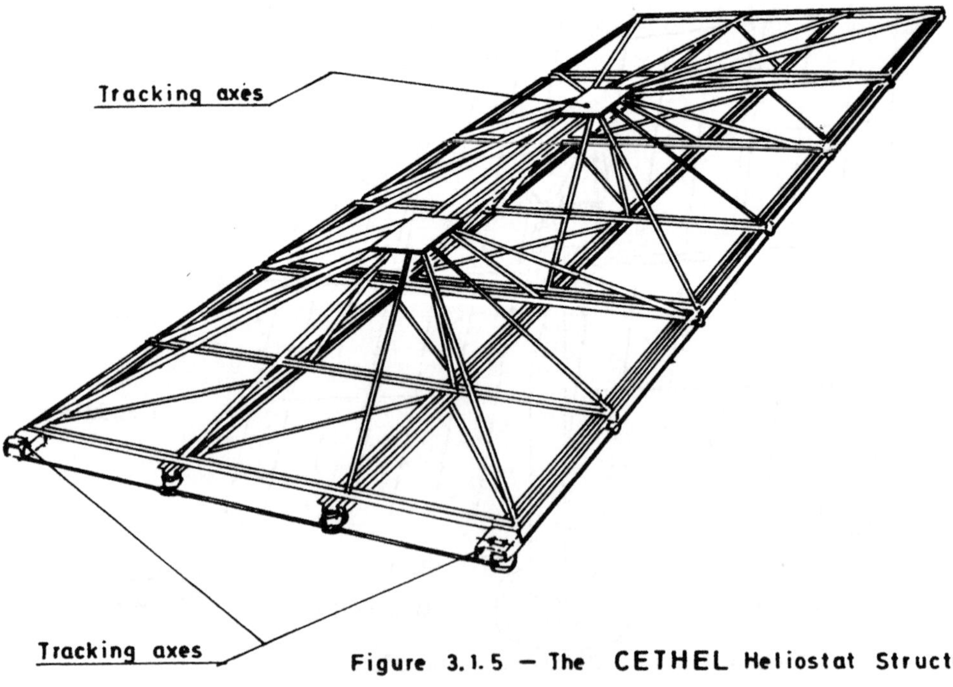

Figure 3.1.5 – The CETHEL Heliostat Structure

Figure 3.1.6 - CETHEL Heliostat Assembly

MBB Installation
Heliostat

MBB Focus on
Auxiliary Target

MBB
Mirror Element

Figure 3.1.7 - MBB Heliostats

3.1.1 Reflecting panels

CETHEL reflecting panels

The CETHEL mirrors are supported by an iron braced-framework structure (see Figs. 3.1.3 and 3.1.6) composed of:

- 8 modules, 3.62 x 1.93 m galvanised iron structure, each of them supporting 6 elementary mirrors 1.8 x 0.6 m;
- a rear structure, made of square section iron tubes, supporting the modules and fitted to the horizontal (elevation) axis of the drive unit.

The mirrors are clamped to the modules via rubber fittings whose elasticity allows for construction tolerances and expansion.

The CETHEL panel shape approximates a spherical mirror with a focal length equal to the distance between the panel centre and the receiver aperture.

First, the modules are shaped so that the 6 mirrors strips are bent cylindrically and canted to be tangent to the ideal sphere. Four types of modules are used in the whole field (sphere radius of 160, 200, 280 and 400 m corresponding to focal lengths of 80, 100, 140 and 200 m). Second the modules are oriented on the rear structure by 3 adjustable supporting bolts so that they focus at the receiver aperture. Again, to avoid too many adjustment parameters, the focussing is designed to be the same for a number of heliostats within a certain range (3%) of actual distance to the receiver (Table 3.1.3).

The CETHEL modules were completely manufactured, welded and galvanised in the workshop in France and transported to the site by train (560 modules, 24 in each wagon). The rear structures were prefabricated, the tri-dimensional structure being divided into flat elements, to be welded and painted outside. The shaping of the modules was also done at the site; aluminium blocks are first riveted to the structure and then cut in height with special tooling in order to give the spherical support to the mirror glasses.

The final erection work consisted of:

- fixing mirrors on modules;
- fixing rear structure to the drive unit, at ground level;
- fixing modules on rear structures, face down, horizontal;
- adjusting the focussing of the panel. The appropriate orientation is given to each module, using the horizontal reference, with water levels and wedges;
- transporting the complete heliostat and putting it on its post in the field with a mobile crane.

MBB reflecting panels

The basic mechanical configuration of the MBB heliostat is shown in Fig. 3.1.7 illustrating the structure before the mirrors are fixed to it. The supporting frames for the mirrors are attached to the drive unit by bolted flanges. All structural parts are finished with weather resisting paint.

The mirror elements are spherically shaped so that they form a sun image at a nominal 200 m distance, about the largest distance between heliostats and the receiver aperture. This use of one focal length may be contrasted with the four different values adopted for the larger CETHEL units.

Most of the MBB mirrors have a core of PVC-foam and an 0.6 mm galvanized steel backing. During the project a second type of mirror was developed, the most visible characteristics of which are that both back and front consist of glass and that the core material is not homogeneous but consists of two components. About 60 of those mirror elements have been mounted. This mirror design is fully compatible with the requirements for EURELIOS.

The mirrors are fixed to the support structure by means of 3 screws. This allows alignment of each mirror by canting it with respect to the structure and to its neighbour mirrors, using the sun. To find the optical axis of the heliostat, an auxiliary mirror is mounted on a bridge type fixture between the two supporting structures and aligned parallel to the elevation axis and to the support structure. Then the heliostat is oriented such that the auxiliary mirror generates a sun image at the auxiliary target of the tower. The 16 mirror elements can then be aligned such that their sun images coincide with that of the auxiliary mirror. After this activity the final readings of the encoder outputs are recorded. Fig. 3.1.7 also shows a focus of a complete heliostat at the auxiliary target.

MBB's calculated optical errors in an 18 km/hr wind are (1 sigma values):

mirror contour	1.47 m rad
mirror alignment	0.44 m rad
temperature effects	0.41 m rad
structural deformations	0.23 m rad
Total optical errors (1 sigma)	1.60 m rad

The total sun image angle including 95% of all reflected energy is 13 m rad.

3.1.2 Drive units, columns and foundations

CETHEL drive units, columns and foundations

The CETHEL mechanical unit, which supports and rotates the reflecting panel in azimuth and elevation, is located at the centre of the panel and fixed on top of the reinforced concrete post (Fig. 3.1.6). It includes all the necessary bearings, gears, motors, encoders, switches, etc.

It is a compact mechanism and so can be tightly sealed for transport, is very easy to link with the reflector and with the post, and can be completely sealed to protect the internal elements of the mechanism from dust, sand and humidity. Since the lower part of the reflector is well clear of the ground, the presence of vegetation, stones, snow, etc., is not important.

CETHEL's elevation and azimuth motors are identical. Their main characteristics are:

- . D.C. motors with permanent magnets;
- . 2 directions of rotation;
- . operating power: 90 W at fast speed; 40 W at slow speed, in tracking;
- . rated speed: 1500 rpm ($\pm$ 5%);
- . constant torque of 0.6 Nm $\pm$ 10^{-3} Nm from 80 rpm to 1500 rpm;
- . input voltage: 240 V at 1500 rpm;
- . rated current: 6 amp. (about);
- . efficiency: 60%;
- . closed motors, drop proof;
- . protection: IP 44.

The 1:11,200 reduction gears, also identical for azimuth and elevation, consist of parallel trains of gear-wheels driven by a worm-gear, all carefully manufactured with special steel to minimise backlash or deformations. The total error, including the effects of wind and weight, is limited to 1 mrad (3.4 arc min).

The azimuth axis relies on a toothed orientation ring bolted on one side to the pillar and on the other to the cast support of the mechanism. The ring is driven by the slow speed pinion of the (azimuth) reduction gear, the gear ratio being 32/129.

The elevation axis is implemented as a tube connected to the two wings of the reflector by means of two flanges. It also uses a toothed orientation ring driven by the slow speed pinion of the (elevation) reduction gear with the same gear ratio as for azimuth movement.

The unit is factory assembled and checked and the encoders are set. Then the mechanism is transported to the site, fitted with the reflector and fixed on top of the post. No adjustment is necessary on site.

The CETHEL drive unit is supported on an iron seating crown placed on top of the reinforced concrete post (height $\pm$ 2 cm, level $\pm$ 0.5%, south-reference aligned $\pm$ 1°, verified by water level and theodolite).

The foundation and post and their reinforcements were designed to satisfy Italian and French regulations for wind pressures, etc., with the reflecting panel in the most unfavourable loading conditions.

After completion of heliostat installation the sun image given by each of the CETHEL units was checked on the auxiliary target placed under the receiver. The focussing of a few of them had to be improved because of a lack of accuracy or perturbations during adjustment.

MBB drive units, columns and foundations

The main characteristics of the MBB drive assembly are:

motors	24 V DC, 300 rpm (3000 rpm for fast manoeuvres)
gears	5 stage spur gear, 1 stage worm gear, final spur gear

reduction ratio | Az 225 380, El 171 900
drive speed | Az 0.48°/min and 4.79°/min; El 0.63°/min and 6.28°/min
max. allowable mirror torque: Az = 3330 Nm; El = 4170 Nm
total gear efficiency | 0.35
bearings | preloaded ball bearings
position feedback | 13 bit digital encoders
pointing errors in 18 km/hr wind (1 sigma):
tracking | 0. 6 m rad
adjustment | 0. 3 m rad
deformation of pedestal and gear drive: 0.25 m rad
Total error | 0.7 m rad

Since the determination of the actual attitude has a very large impact on the overall tracking accuracy, MBB fits 13-bit absolute value angular shaft encoders to the output axes of the gears thereby compensating for backlash, etc. The resolution of the 13 bit encoder is 0.8 mrad.

The drive unit itself (Fig. 3. 1. 8 in its final assembly stage) is covered by a glass fibre cover with access-holes for maintenance operations.

MBB's steel column, bolted to the concrete foundation (Fig. 3. 1. 9 shows the mould), supports the drive assembly consisting of the bearings, gears, motors and digital encoders (13-bit for 360° which give absolute angular information, derived directly from the moving axis).

The design of the MBB heliostat was experimentally verified in 1978 by means of a prototype and also by using the first heliostat manufactured according to the final manufacturing drawings. The latter, particularly, confirmed the performance characteristics, given above, including the accuracy of the reflected beam, which are well within specifications.

3. 1. 3 Heliostat electronics

CETHEL and MBB have both installed electronics in their heliostats but MBB has elected to give individual heliostats more autonomy - each "knowing" its own position relative to the tower and target and "knowing" the direction to the sun. CETHEL has a higher data-rate communication system, the Central Unit giving such data to the heliostats directly.

CETHEL heliostat electronics

A microprocessor-based local electronic system in each heliostat takes care of information exchange with the central unit and of the functions of heliostat command and control (Fig. 3. 1. 10). Particular emphasis has been put on failure-detection. The data transmission lines between the CETHEL central unit and heliostats are of multidrop serial, asynchronous, current loop type, with a modulation rate of 2400 bauds.

The control logic of each heliostat is able to receive from the central unit the tracking position values for azimuth and elevation every six seconds, compare these values with the current values of the encoders and decide

Figure 3.1.8 - MBB Drive Unit (Cover Removed)

Figure 3.1.9 - Mould for MBB Foundation

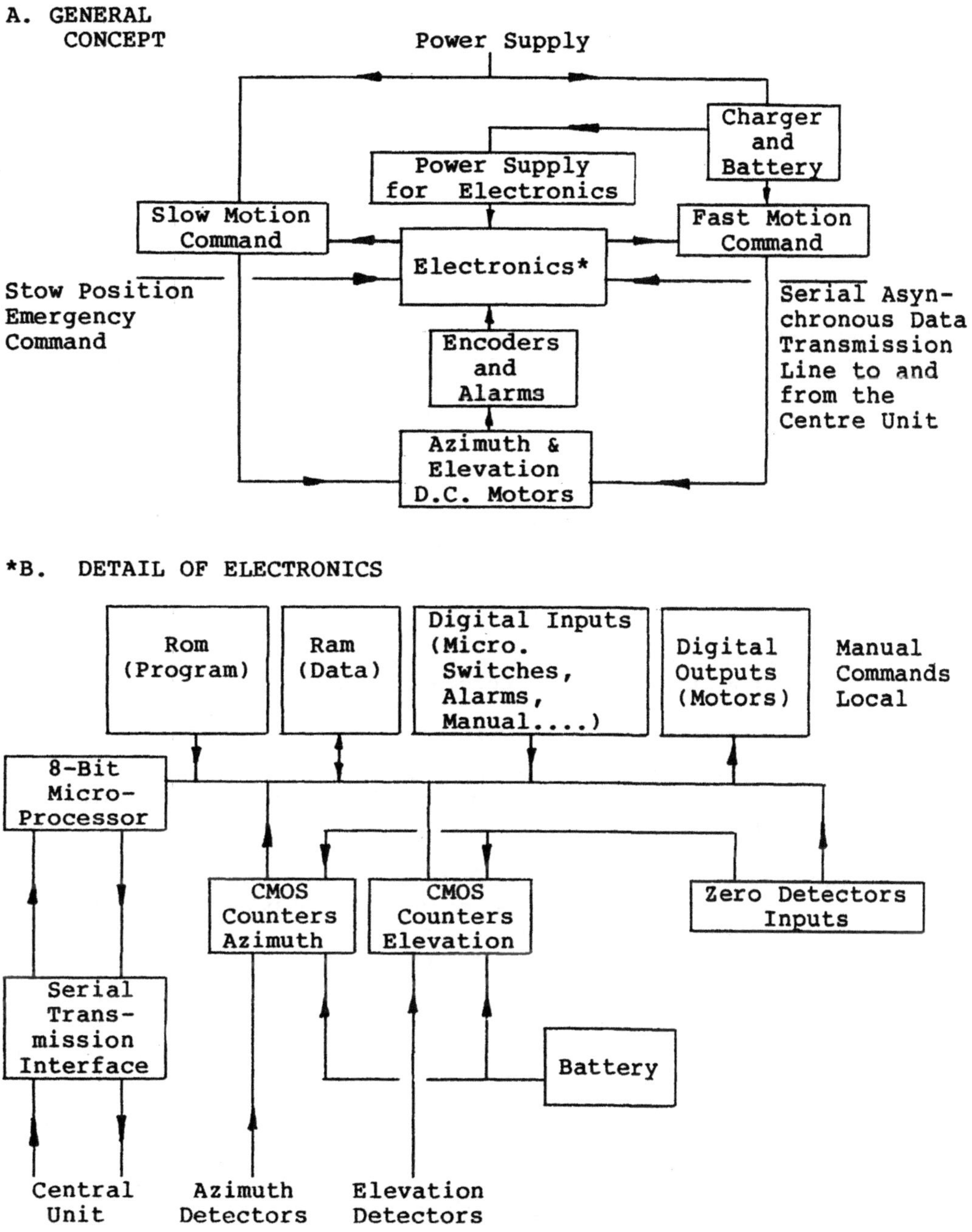

Figure 3.1.10 - The CETHEL Control System at Heliostat Level

at which speed and in which direction (sense) the heliostats must move. It also controls the alarms. The logic is primarily CMOS and TTL LS technology and, except for the DC motors and the encoders, the system is all mounted in one box.

Start-up

Once the current is established in the grid, the logic starts immediately. It checks whether the 24 V battery is charged. If not the logic informs the central unit and forbids fast speed.
If the battery is charged, the logic commands the closure of a relay, allowing operations with the battery.
The logic checks the state of the battery supplying power to the detectors, to inform the central unit of information loss. The counters can be reset to zero by a manual procedure.

Normal operation

The logic receives the commands from the central unit and commands the rotation of the motors to reach the assigned position. It also has a supervision role of heliostat operation.
The central unit may send an assigned position or ask for the current state or current position of the heliostat, automatic zero positioning and automatic command of speed. Other commands may be envisaged.
The logic compares the current position with the assigned positions in order to command both the sense and speed of rotation of the motors. A change of direction is always preceded by a delay to minimise mechanical stress.
The counters are forced to zero when the corresponding detectors are active.
When a movement takes place out of the normal range of operation, the logic cuts the power supply of the corresponding motor.

Emergency-command-to-stow-position transmission line

This line is shared among the heliostats of a group.
In case of failure of the field power supply, the operator can still order the heliostats to go to stow position, by depressing a button on the central console.
In this case the heliostats can move from the receiver position to stow position, following safety trajectories consisting of 12 points whose azimuth and elevation are computed for each heliostat, by the central unit, and sent every 72 seconds to the heliostats and memorized.

Central-unit-to-heliostat transmission line

This is a serial asynchronous current loop operating at 1200 baud. To provide failure detection current flows when there is no message. Data is transmitted as 8-bit words plus parity and stop bits:

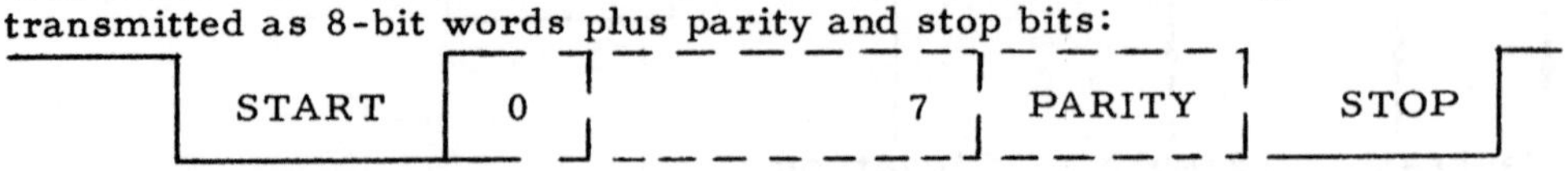

The line is organised as a bus, the failure of one heliostat not creating any disturbance to the rest of the line.
The line is optically isolated from the heliostat.

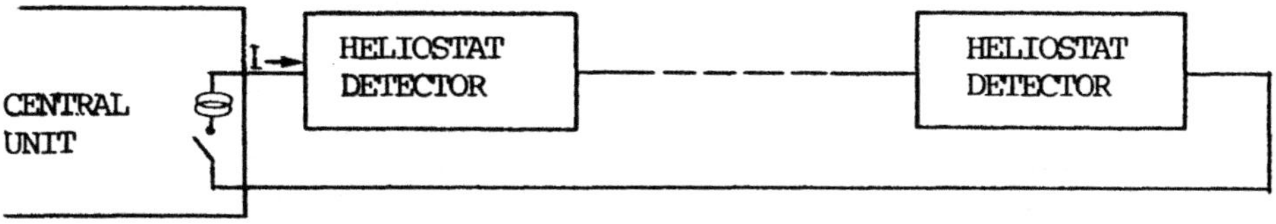

Heliostat-to-central unit transmission line

The general characteristics of this bus are similar to the above except for the parallel rather than series configuration.

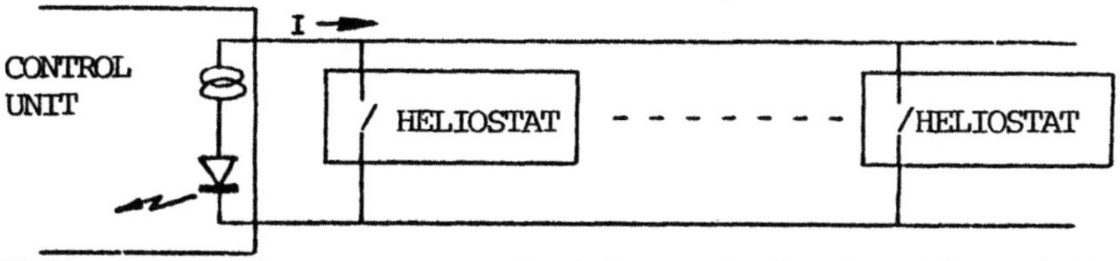

Heliostats send, upon request of the central unit, either their azimuth and elevation positions, or a word representative of their logical status. This status word is coded as one or two 8-bit words and contains information relative to:

. manual or automatic operation mode;
. alarms (general power supply);
. alarms (local power supply);
. alarms (batteries);
. micro-switches for extreme positions actuated or not;
. alarms (lack of rotation).

MBB heliostat electronics

The MBB tracking concept, which has been developed since 1978, provides somewhat more autonomy to the individual heliostats than does CETHEL's. Each heliostat can automatically track and maintain:

. stowage position (mirrors face down);
. standby position (near the receiver);
. receiver position;
. alignment position (to the auxiliary target).

The reflected beam is also controlled during the movement from one automatic position to another, to stay on the defined safety trajectories. Single heliostats and groups of heliostats can be commanded by the CU operator to manually defined positions for maintenance, washing and repair.

Figure 3.1.11 shows a simplified block diagram of the MBB tracking/servosystem. Fig. 3.1.12(a) illustrates the hardware related to every heliostat. Each local heliostat unit consists of an 8 bit microprocessor including program storage, battery buffered random access memory, timers, serial and parallel input/output circuits and in addition to a "simple" system, a separate arithmetic device. It performs at the same time input and decoding of commands, calculation of the sun position, determination of the real axis positions and servo-amplifier control and it sends status or position information to the Central Unit (C) upon request, by executing a multitasking program.

The heliostat control electronics receive serial information (commanded image position, date and time, data request) via the Command Data Bus from the MBB Central Unit in the plant Control Room. A single and a group address is dedicated to each heliostat in the field. The heliostat electronics decodes this information. It calculates the sun position and the related positions of the two axes more than 10 times every second. It compares the calculated positions with the actual positions retrieved directly from the angular encoders and sends drive signals to the servo amplifiers which power the motors until the right positions are reached. It sends, if requested, status and position informations back to the CU via the serial Response Data Bus.

The signals on the Command Data Bus and the Response Data Bus are in accordance with CCITT V. 24 (DIN 66 020) standard. The baudrate is 110 bit/s.

In addition two separate signal lines are wired from the CU to each group of heliostat electronics. These signal lines are used only in case of emergency (e. g. serial bus breakdown, emergency in the thermal system). They excite directly the two motor-drive electronics without involving the microcomputer. One signal (E1) turns the azimuth axis eastward with maximum speed. With the second (E2) the elevation axis is turned down to the stowage position. The motor drive electronics convert and combine the signals from the computer, the emergency lines, and the end position switches to power signals by providing to the DC-motors a high (fast) or a low (slow) voltage in the right direction.

The installed power supply converts the 220 V into the voltages needed by the microcomputer, the angular encoders and the drive electronics. It is a high efficiency primary-switching power converter. A small battery in each heliostat protects against loss of data in RAM in the event of a short loss of power.

Special purpose test units (Fig. 3. 1. 12(b)) have been provided to assist with the installation and maintenance of the MBB heliostats.

3.2 RECEIVER

The receiver is a once-through circular cavity-type steam generator based on the design concepts of the late Prof. Francia of Genoa. It is installed at the top of a tower at a height of 55 m, the centre line being inclined downwards 22° from horizontal (Fig. 3. 2. 1) towards the heliostat field.

The main waterway within the receiver consists of two virtually identical, parallel and independent branches, each 482 m long. This pipework is divided into four sections (Figs. 3. 2. 2 to 3. 2. 4).

i) an economiser (ECO), an open bottomless basket-like structure in the centre of the cavity designed to preheat the incoming feedwater to about 200 - 280°C;
ii) a truncated conical evaporator section (EVP), forming the greater part of the wall of the receiver and designed to boil the feedwater

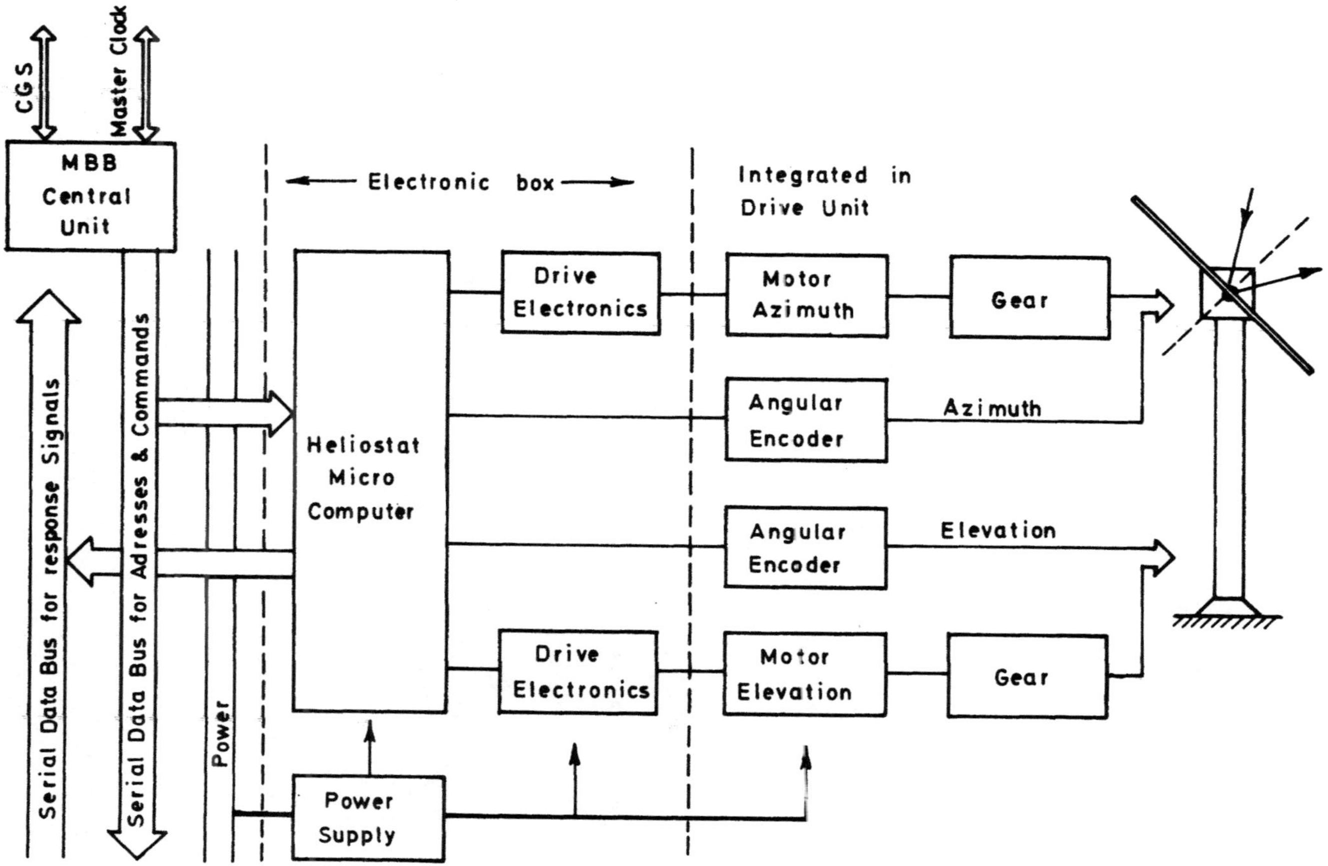

MBB Tracking System Heliostat for EURELIOS

Figure 3.1.11 – MBB Tracking Servo System

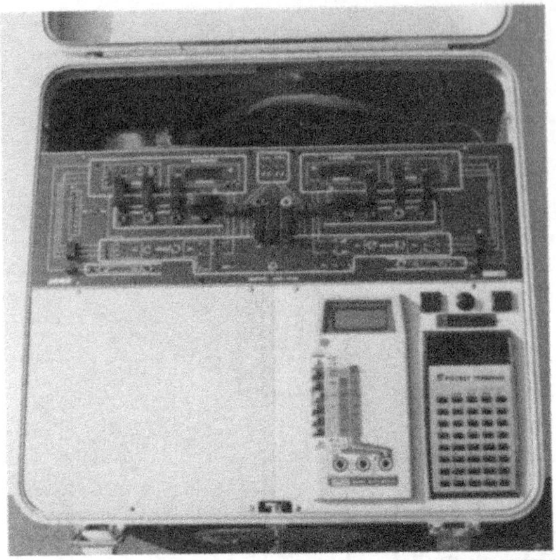

Heliostat Unit Tester for
Servo System (HUT)

MBB Electronic Box

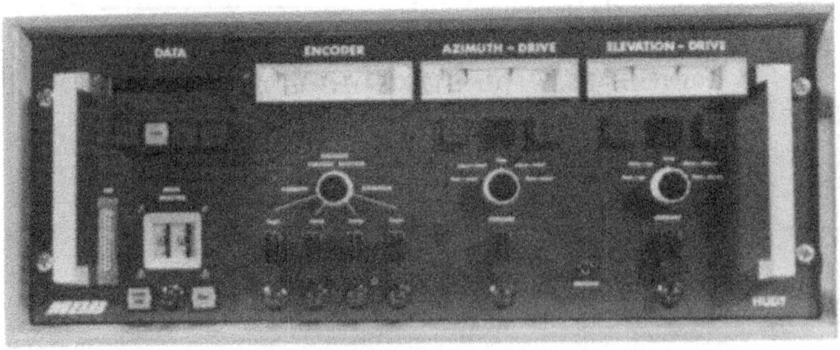

Heliostat Unit
Drive Tester
for Drive Unit
Testing (HUDT)

Figure 3.1.12 - MBB Heliostat Electronic Units

Figure 3.2.1 - The Receiver and Tower

with very little superheating;

iii) a shorter, wider angle truncated conical primary superheater (SH1), within which the steam is superheated to between about 350 and 400°C. It is mounted immediately above the evaporator;

iv) a cylindrical secondary superheater (SH2) consisting of interlaced circular and straight pipes forming an open square-cell honeycombe-like structure in which the steam is brought up to its final state (nominally 512°C at a pressure of 62 kg/cm^2).

The receiver pipework is of A213 TP321 stainless steel painted with matt black Pyromark 2500 paint to increase its absorbtivity. The pipe diameter is 44.5 mm in the economiser (where the fluid is liquid) and 57 mm thereafter, the total length being 835 m. The two branches are fed separately, the main feedwater pipe having bifurcated prior to the flow meters and control valves which are on the access gantry at the base of the tower. The branches come together again into a common "main steam" pipe at the end of the secondary superheater, immediately before leaving the receiver. The receiver, without auxiliaries (valves, support structure, etc.) and lagging, has a volume of 3.4 m^3 and weighs 11.57 tons.

The overall concept of Prof. Francia's receiver includes the use of Pyrex glass sheets to improve its efficiency.

For EURELIOS these are to be installed inside the cavity, as shown by the distorted M-shaped section in Fig. 3.2.2. There they will offer a "greenhouse effect" for the hotter parts of the pipework, leaving the cooler economiser (ECO) open to direct illumination by the heliostats. The Pyrex will not be installed until sufficient experience has been gained of routine operation without it to permit a detailed analysis of its quantitative benefits.

To minimise heat loss, the outside of the receiver is covered with a mineral wool blanket (?packaged in glass-fibre sacks?) attached to an open-mesh stainless steel net and enclosed within a riveted stainless steel casing which retains and protects the insulation from the weather. The individual elements of the casing are preformed to make fins (Fig. 3.2.5) designed to minimise the adverse effects of possibly maladjusted heliostats focussing on the outside of the receiver. For the same reason, and also to protect the receiver in the event of a total electrical power failure which might cause all or part of the spot from the heliostat field to drift slowly to the east as the sun moved west, the rim is protected by a circle of silicon thermal protection bricks (Fig. 3.2.6).

The receiver includes appropriate monitoring and control equipment connected into the control room. The flow conditions under typical circumstances with a steam flow rate of 4200 kg/h at 62 kg/cm^2 and 510°C, result in a transit time for water (from the ECO inlet to the superheater outlet) of about 16 minutes. This implies that careful design of monitoring and control systems is essential to maintain steady operating conditions despite the comparatively rapid variations of incoming flux which result from the frequent presence of clouds.

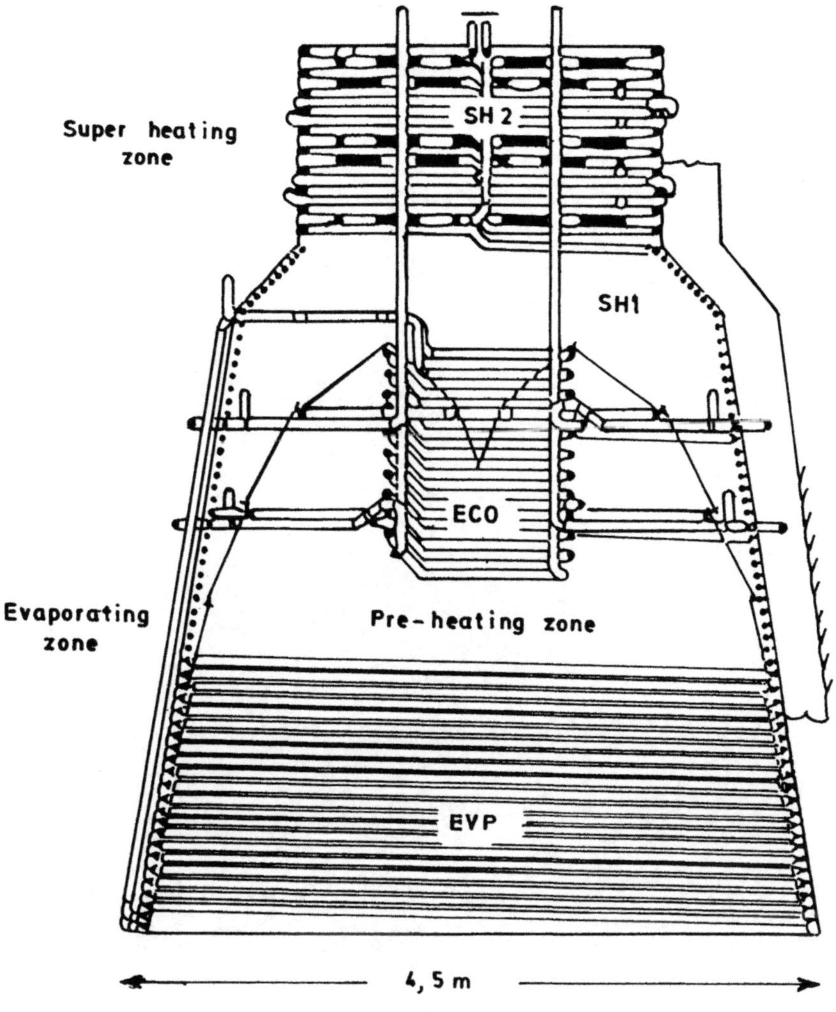

EVP = Evaporating section
ECO = Economizer
SH 1 = Primary superheater
SH 2 = Secondary superheater

Figure 3.2.2 - Receiver - A Sectional View

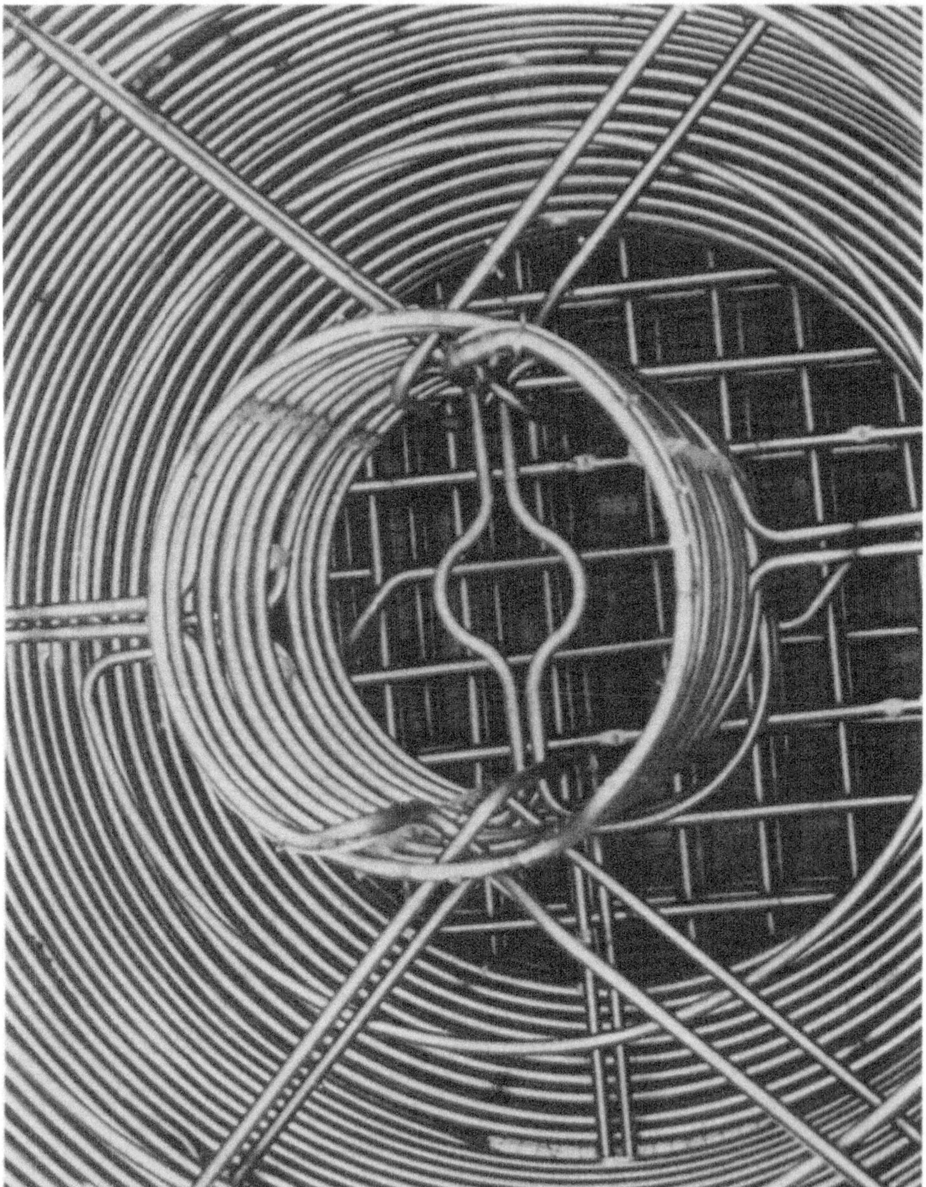

Figure 3.2.3 - View into the Mouth of the Receiver

Figure 3.2.4 - The Receiver and Its Supporting Structure (Without Auxiliaries or Lagging)

Figure 3.2.5 - The Receiver Prior to Hoisting

Figure 3.2.6 - The Receiver at the Top of the Tower (Showing Auxiliary Target and the Fitting of the Attemperator Access Gantry)

First, a safety valve at the receiver outlet, set at 64 kg/cm^2, protects against overpressures. Second, the receiver contains 80 thermocouples and 47 flux meters welded to the tubing. The average tube temperature and average heat flux into the tubes obtained from these sensors are used by the control system to regulate, instant by instant, the feedwater flow rate. The receiver also contains a number of water temperature and pressure sensors which are used by the regulation system, particularly to control relative to an operator-selectable secondary-superheater temperature set point.

Finally, the receiver is provided with a desuperheating/attemperating system to keep the output steam temperature as constant as possible as the solar flux varies, particularly to avoid temperature increases above the design figures. This system consists of four automatically controlled Yarway sleeve-type attemperator spray valves, two in each branch. The first pair, one at the end of each evaporator section, is controlled on the basis of superheater inlet temperature and can provide up to 8% of the maximum total flow. The other pair, at the inlet to the secondary superheater, controls superheater outlet temperature introducing up to 4% of the maximum total flow. The attemperators are all fed from a third branch of the feedwater piping. The pressure of this branch is controlled at a differential of about 12 kg/cm^2 over the pressure in the two main branches so as to guarantee that the water remains liquid and flows fast enough. The water temperature is raised to about 190°C in a heat exchanger situated between the economiser (ECO) and the evaporator, so as to minimise thermal shocks at the attemperators themselves.

The receiver is also protected by the monitoring of a number of other parameters. On the basis of these, the heliostat field is moved out of the receiver in the event of:

- a very low feedwater pump delivery pressure;
- a high differential between the flow rates in the two main branches;
- a low feedwater flowrate at the A/B branch inlet;
- the loss of either the 110 V or the 380 V power supply.

Further, in the event of loss of the 380 V AC supply and consequent loss of power to the feedwater pumps and the heliostat subsystems, when the basic feedwater pressure has dropped from the normal 101 kg/cm^2 to 86 kg/cm^2, a nitrogen-pressurised auxiliary water supply (Section 3. 3. 2) automatically continues to feed the receiver until the solar spot has moved clear of the receiver due to the east-to-west motion of the sun.

Structurally the main strength of the receiver is in the pipework itself. It is supported, along four generators of its conical shape, from a pair of substantial mild steel three-quarter-circular tubes (visible in the figures). The whole receiver is mounted on a trolley which can be winched to the top of the tower (Fig. 3. 2. 7) on integral rails. There the final connections are made to the steam cycle, control and monitoring systems.

Figure 3.2.7 - The Receiver Being Winched up the Tower

3.3 STEAM CYCLE

3.3.1 Description

The steam cycle (Fig. 3.3.1) is a conventional single superheating Rankine Cycle consisting of a boiler (the receiver, Section 3.2), a steam turbine (Section 3.4) and a condenser, with appropriate pumps, valves, controls, safety systems, etc. Associated with it is an auxiliary system designed to keep the plant operating for up to about 30 minutes in the event of cloud passing, using saturated steam from an accumulator (storage tank) superheated in a heat exchanger by a second working fluid, a hot molten salt (Section 3.5).

Typical operating parameters are given in Table 3.3.1. Starting from a temperature of 25°C it takes about 45 minutes to heat the receiver to 350°C, the temperature for turbine start-up, under conditions of full insolation - somewhat longer under normal early morning conditions. During this period the water or steam produced by the receiver is taken through a bypass valve (N11-F301) and attemperator/desuperheater into a Flash tank (N11-A002) before being passed into the Condenser (N61-3001) and pumped around the system again. The control system (Section 3.7.2) maintains the upstream pressure by adjusting the bypass valve. The control system also automatically adjusts the water level and pressure in the flash tank by means of the two valves N11-F309 and N11-F308 which, respectively, allow water and steam to flow to the condenser.

The condenser itself is an evacuated four-pass water cooled unit with a nominal operating pressure of 6 cm Hg (reduced to 3 cm Hg when using the salt storage system).

Once the steam temperature is high enough, steam can be admitted to the turbine, first via valve N11-F307 and then by the main steam control valve N11-F303 (the rotor having been turned over by the starter motor and the steam seals fed from the flash tank). This results in automatic restriction of the bypass valve to maintain the steam pressure. The (low pressure) exhaust of the turbine is passed directly to the condenser.

The 7-stage steam turbine runs at a speed of 8195 rpm, corresponding to 1500 rpm at the generator (alternator) after the appropriate reduction gearing, when synchronised with the 50 Hz main grid. The turbine is regulated by a Woodward regulator acting on the nozzle valves.

Condensate in the well of the condenser is extracted by two centrifugal "booster" pumps, N21-C001 A/B, each designed to extract 8800 l/hr when the condenser is evacuated to a pressure of 3 cm Hg. The water level in the condenser is controlled by the regulation system. This allows additional demineralised water to enter from (or causes excess to be pumped by the booster pumps back to) the cold water storage tank. This is particularly important during start-up of the plant from cold: water levels in the cycle tend to rise due to the conversion of water to steam in what was initially full of water. Air and uncondensable vapours are extracted from the condenser by two vacuum pumps, normally one in service, one in standby.

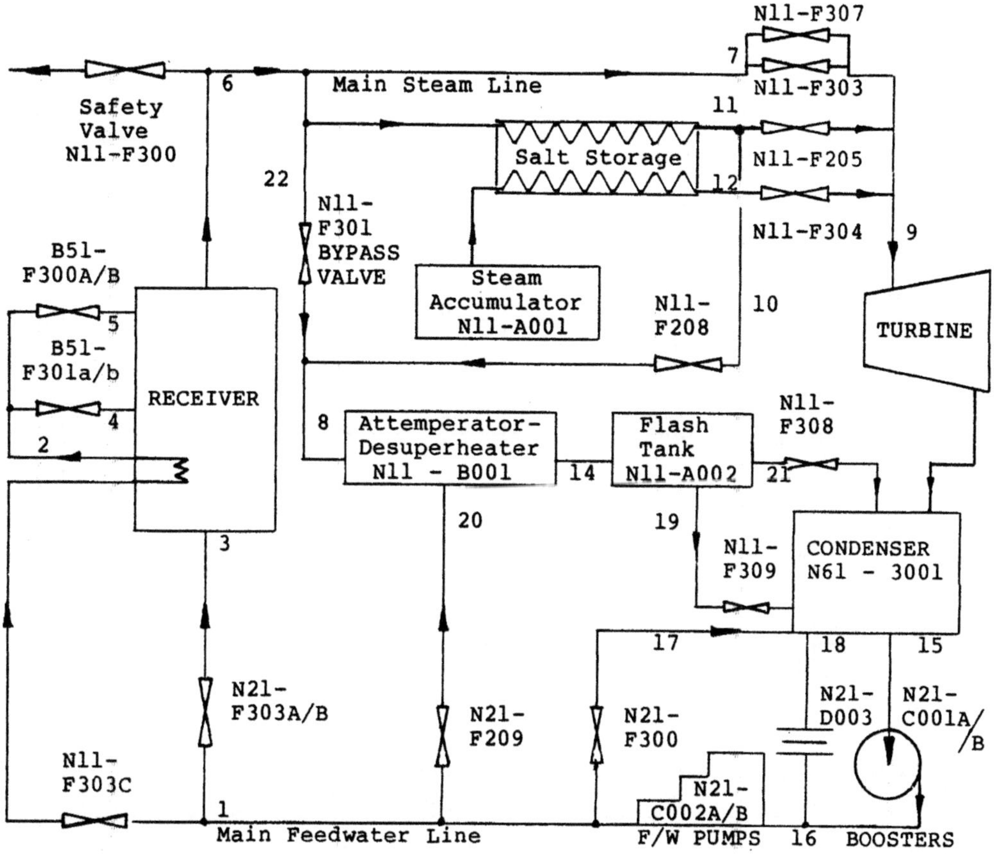

The numbered 'points' of Table 3.3-1 are shown on this sketch. Flow rates for A or B valves etc. are half the total flow rate in the common line.

DEFINITION OF MODES OF TABLE 3.3-1

- A Pressurising main feedwater line
- B Filling receiver
- C Receiver minimum load, on turbine
- D Receiver maximum load, on turbine
- E Receiver maximum load, 100% on bypass
- F Beginning of Salt Storage system intervention
- G End of storage system intervention (receiver isolated)
- H Receiver minimum load, heating salt system into bypass
- I Receiver maximum load, heating salt system into turbine.

Figure 3.3.1 - Steam cycle

MODE A

Point	1	2	3	4	5	6	7	8	9	10	11	12	13	14	15	16	17	18	19	20	21	22
Q=kg/h	0	-	0	-	-	0	-	-	-	-	-	-	-	0	8500	8500	8000	500	-	-	-	-
P=ata	-	-	-	-	-	1	-	-	-	-	-	-	-	1	1	2.2	1	1	-	-	-	-
T=°C	-	-	-	-	-	24	-	-	-	-	-	-	-	24	24	24	24	24	-	-	-	-

MODE B

Point	1	2	3	4	5	6	7	8	9	10	11	12	13	14	15	16	17	18	19	20	21	22
Q=kg/h	1000	-	1000	-	-	1000	-	-	-	-	-	-	-	1000	8500	8500	7000	500	1000	-	-	1000
P=ata	101	-	46	-	-	40	-	-	-	-	-	-	-	1	1	2.2	1	1.9	1	-	-	46
T=°C	25	-	25	-	-	25	-	-	-	-	-	-	-	25	25	25	25	25	25	-	-	25

MODE C

Point	1	2	3	4	5	6	7	8	9	10	11	12	13	14	15	16	17	18	19	20	21	22
Q=kg/h	1000	120	446	80	40	1000	-	-	1000	-	-	-	-	-	8500	8500	6900	600	-	-	-	1000
P=ata	101	75	72	66	65	64	-	-	12	-	-	-	-	-	.06	1.8	.06	.06	-	-	-	64
T=°C	25	150	25	150	150	510	-	-	500	-	-	-	-	-	25	25	25	25	-	-	-	510

MODE D

Point	1	2	3	4	5	6	7	8	9	10	11	12	13	14	15	16	17	18	19	20	21	22
Q=kg/h	5000	600	4400	400	200	5000	5000	-	5000	-	-	-	-	-	8500	8500	2900	600	-	-	-	-
P=ata	101	76	76	67	64	64	63	-	61	-	-	-	-	-	.06	1.2		.06	-	-	-	-
T=°C	36	180	36	180	180	510	510	-	510	-	-	-	-	-	36	36	36	36	-	-	-	-

MODE E

Point	1	2	3	4	5	6	7	8	9	10	11	12	13	14	15	16	17	18	19	20	21	22
Q=kg/h	5000	600	4400	400	200	5040	-	-	-	-	-	-	-	6400	8500	8500	1500	600	3200	1400	5000	5000
P=ata	101	76	76	67	64	64	-	-	-	-	-	-	-	50	.06	1.9	-	1.9	5	5	5	6.3
T=°C	40	180	40	180	180	510	-	-	-	-	-	-	-	160	40	40	40	40	150	40	150	510

MODE F

Point	1	2	3	4	5	6	7	8	9	10	11	12	13	14	15	16	17	18	19	20	21	22
Q=kg/h	1000	-	1000	-	-	1000	-	-	750	-	-	750	750	8400	8500	8500	6900	600	1200	1400	120	1000
P=ata	101	-	36	-	-	64	-	-	8	-	-	17.9	19	5	.06	1.9	.06	1.9	5	5	5	64
T=°C	25	-	25	-	-	510	-	-	440	-	-	440	210	160	25	25	25	25	150	25	150	510

MODE G

Point	1	2	3	4	5	6	7	8	9	10	11	12	13	14	15	16	17	18	19	20	21	22
Q=kg/h	-	-	-	-	-	-	-	-	300	-	-	300	300	-	8500	8500	7900	600	-	-	-	-
P=ata	-	-	-	-	-	-	-	-	7	-	-	8.6	9	-	.03	1.8	-	1.8	-	-	-	-
T=°C	-	-	-	-	-	-	-	-	400	-	-	440	175	-	25	25	25	25	-	-	-	-

MODE H

Point	1	2	3	4	5	6	7	8	9	10	11	12	13	14	15	16	17	18	19	20	21	22
Q=kg/h	2500	300	2200	200	100	2500	1000	1500	1000	1500	1500	-	-	-	8500	8500	4000	600	-	1400	-	-
P=ata	101	76	73	65	64	64	63	8	12	5	5	-	-	-	.06	1.8	.06	1.8	-	5	-	-
T=°C	28	150	28	150	150	480	480	430	480	310	330	-	-	-	28	28	28	28	-	28		

MODE I

Point	1	2	3	4	5	6	7	8	9	10	11	12	13	14	15	16	17	18	19	20	21	22
Q=kg/h	5000	600	4400	300	200	5000	3500	-	5000	-	1500	-	-	-	8500	8500	2900	600	-	-	-	-
P=ata	101	76	76	67	64	64	63	-	61	-	62	-	-	-	.06	1.9	.06	1.9	-	-	-	-
T=°C	36	150	36	150	150	480	480	-	450	-	430	-	-	-	36	36	36	36	-	-	-	-

NOTE : See figure 3.3.1 for Definition of modes and points

Table 3.3.1 - Steam Cycle Design Values

The output of the centrifugal booster pump(s) feeds either of the standby-redundant pair of reciprocating "feedwater" pumps N21-C002 A/B. Each of these delivers 8200 l/hr to the main feedwater line at the normal operating pressure of 101 kg/cm^2.

As mentioned in the description of the receiver, the flow of feedwater to the receiver is controlled (by valves N11-303 A/B/C) so as to regulate the temperatures achieved (subject always to the pressure control achieved by the steam cycle's turbine-bypass valve N11-F301). Any excess water, above that needed to maintain the normal operating feedwater pressure, is returned to the condenser through a "feedwater pressure control" valve N21-F300.

Water supply to the bypass desuperheater and to the storage steam accumulator is also taken directly from the main feedwater line. Water hammer in the feedwater line is damped by a bank of four tanks containing pre-pressurised (70 kg/cm^2) nitrogen bags and located just downstream of the reciprocating feedwater pumps.

The steam cycle's remotely operated valves are all electrically controlled and pneumatically operated. Compressed air is supplied by two two-stage reciprocating compressors controlled to give a working pressure of 75 kg/cm^2 in the instrument air circuit. The air from the compressors is cooled by the turbine oil cooler circuit: a closed cycle of demineralised cooling water driven by a pair of centrifugal pumps whose main purpose is to keep the turbine oil below 38°C at all times.

The last, but not least important, function of the steam cycle is to provide both superheated and saturated steam to the salt-storage system (Section 3.5) and to utilise the condensate or steam from it. To charge (i.e. heat) the salt, main steam (from the receiver) is admitted by valve N11-F203, the flow being controlled by the salt storage system's own controls (nominal flow rate 1010 kg/h) and finally passed either through N11-F208 to the bypass desuperheater or optionally through N11-F205 to be mixed with the main steam supply to the turbine. To charge the steam accumulator, N11-A001, it is first partially filled with water from the main feedwater, condensing and raising both pressure and temperature until the accumulator of saturated steam and water reaches its nominal 19 kg/cm^2, 210°C charged state.

Once the salt and the steam accumulator are charged and the system enabled, the steam cycle control will automatically transfer to use the stored steam and energy to keep the generator running for up to 30 minutes whenever a cloud interrupts insolation. This is implemented by starting the transfer sequence as soon as the pressure of the main steam falls below 8 kg/cm^2. Similarly, when the cloud has passed and the steam quality is acceptable the turbine will again automatically be fed from the main steam line. The system has been designed with flexibility in mind and the optimising of the detailed operating procedures will be an interesting part of the Phase D research programme.

3.3.2 Emergency feedwater system

This system (Fig. 3.3.2) is used only in case of emergency. It ensures that if the pressure in the feedwater line drops below 86 kg/cm^2 due to:

. loss of power to the booster or feedwater pumps, or

. electrical or mechanical failure in the feedwater system,

sufficient water is supplied to the receiver to protect it whilst it is still illuminated, namely for the roughly 20 minutes required for the motion of the sun to take the spot from the stationary MBB heliostat field out of the receiver.

The system consists of 6 one cubic metre bottles with a design pressure of 220 atm. Three of these bottles (N21-A003A/B/C) contain nitrogen at 200 atm., the others (N21-A002A/B/C) demineralised water for the receiver. A bank of six nitrogen storage bottles (N21-A004A/B/C/D/E/F) make up gas losses, if any. The compressor N21-C005 reinstates the system after use by transferring nitrogen back from the three water bottles to the main nitrogen bottles (N21-A003A/B/C) thereby allowing the water bottles to be refilled from the main feedwater line.

3.3.3 Steam cycle alarms

The steam cycle includes alarms to warn the operation of the following abnormalities:

. steam storage tank (N11-A001)

high condensate level	1600 mm
low condensate level	1100 mm

. condenser (N61 3001)

condensate low level	200 mm
condensate very low level	100 mm
condensate high level	600 mm
condensate very high level	720 mm

. flash tank (N11 A002)

high level	400 mm
low level	100 mm

3.4 TURBINE

The turbine is a 7-stage impulse turbine rigidly coupled to the reduction gear pinion. The reduction gear is connected to the alternator through a flexible joint. The combined turbo-generator unit, mounted on its chassis is illustrated in Fig. 3.4.1. General data are given in Table 3.4.1 and Table 3.4.2 gives nominal heat balance calculations for five sample turbine operating states.

The turbine casing is cast in two halves in a steel alloy. The top of the casing houses the nozzle diaphragms in grooves provided with drain holes. Leakage of steam between stages is minimised by metal labyrinth seals and there is a steam sealing box on the steam inlet and the exhaust side. The rotor is a special steel forging, heat treated to give the necessary high resistance to mechanical and thermal stress.

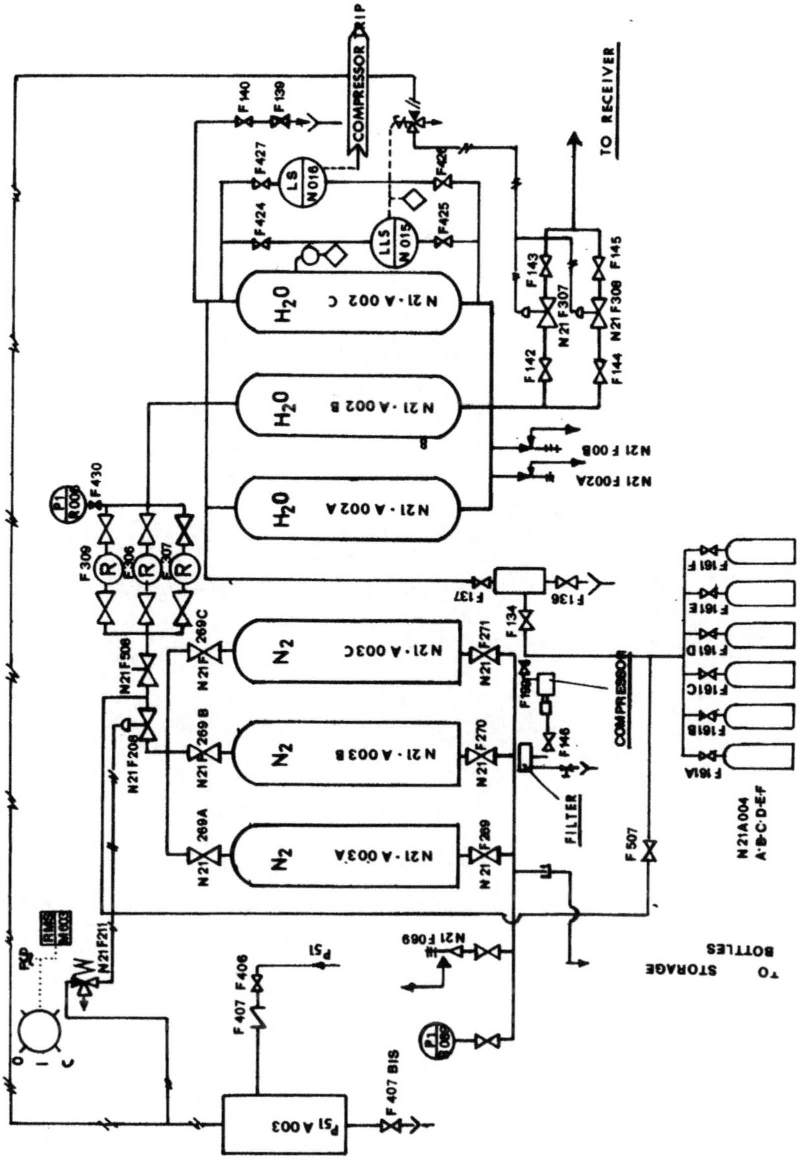

Figure 3.3.2 - Emergency Feed Water System

Figure 3.4.1 - The Combined Turbo-Generator Unit

TURBINE

- PART NUMBER	No. 1953
- MANUFACTURERS	ANSALDO - GENOA
- JOB	No. 18.477
- CONSTRUCTION YEAR	1980

STEAM CONDITIONS AT THE STEAM VALVE (Turbine design point)

- PRESSURE	60 ata
- TEMPERATURE	510°C
- FLOWRATE	5200 Kg/h

STEAM CONDITIONS AT THE TURBINE EXHAUST

- PRESSURE	0.06 ata
- TEMPERATURE	36°C

OPERATING DATA

- OUTPUT AT THE TURBINE/GENERATOR JOINT	1200 kW
- TURBINE SPEED	8195 rpm
- GENERATOR SPEED	1500 rpm
- OVERSPEED EMERGENCY TRIP	9250 rpm

OVERALL DIMENSIONS & WEIGHT OF THE COMPLETE TURBO-GENERATOR SET

- TOTAL LENGTH OF THE SET (TURBINE, REDUCTION GEAR, GENERATOR)	5.465 m
- MAXIMUM WIDTH	2.160 m
- MAXIMUM HEIGHT	2.400 m
- WEIGHT OF TURBINE AND ACCESSORIES	3746 kg
- WEIGHT OF ALTERNATOR (Power generator)	5000 kg
- TOTAL WEIGHT OF THE SET & ACCESSORIES	13500 kg

Table 3.4.1 - General Data on the Turbine

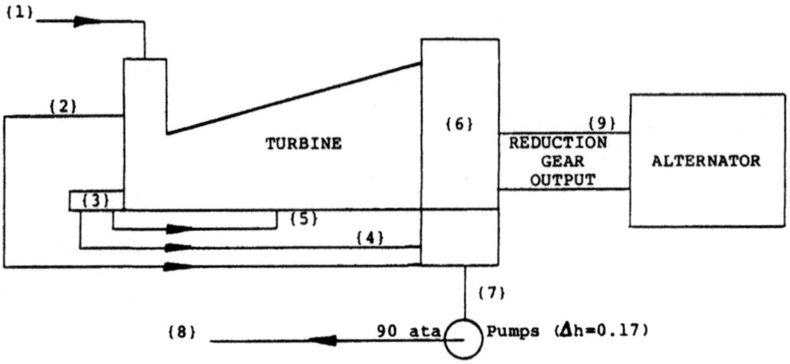

STATE		100% OUTPUT	** NOMINAL OUTPUT	75% OUTPUT	50% OUTPUT	AT START. OF SALT STORAGE INTER-VENTION
POINT*	PARAMETER					
1. Input to Turbine	Flow, kg/hr	5200	5200	3900	2600	1445
	Pressure, ata	60	60	60	60	16
	Temp. °C	510	450	510	510	410
	Enthalpy, kCal/kg	823.11	788.89	823.11	823.11	782.22
2. Input Valve Leakage	Flow, kg/hr	60	60	60	60	16
3. Gland Seal Steam	Enthalpy, kCal/kg	782.58	750.46	778.19	773.93	745.20
4. Gland Seal Loss	Flow, kg/hr	42	41	26	15	11
5. Gland Seal Re-use	Flow, kg/hr	136	137	109	85	40
6. Turbine Exhaust	Flow, kg/hr	5098	5099	3814	2525	1418
	Enthalpy, kCal/kg	603.57	587.89	602.57	604.12	601.40
	Pressure, ata	0.06	0.06	0.048	0.038	0.033
7. Condenser Output	Enthalpy, kCal/kg	35.83	35.83	31.82	27.75	25.35
8. Feedwater	Flow, kg/hr	5200	5200	3900	2600	1445
	Enthalpy, kCal/kg	36	36	31.99	27.92	25.52
9. Reduction Gear Output	Shaft power, kW	1200	1100	894	570	210
GHR	kCal/kWh	3410.8	3559.1	3451	3627	5206.8
GSR	kg/kWh	4.33	4.72	4.36	4.56	6.88

* Point numbers correspond to those on the sketch above.
** This is an example of one combination of steam conditions which would give the nominal power output.

Table 3.4.2 - Examples of Turbine Heat Balance

3.5 THE MOLTEN SALT THERMAL STORAGE SYSTEM

3.5.1 General description

The purpose of the thermal storage system is to keep the turbo-alternator and the auxiliaries running at reduced power for up to 30 minutes without solar input, e.g. during cloud passing. It can also be used to warm the turbine and/or pipework (or to keep them warm for limited periods) if required.

The system consists of two main parts (see Fig. 3.5.1):

- an accumulator (a pressurised water tank) provided by ANSALDO, able to deliver saturated steam at a pressure decreasing from 19 bar to 7 bar, by auto evaporation;
- a salt storage subsystem, designed and implemented by CETHEL, to superheat the steam using as working fluid a molten salt, such as HITEC (an eutectic mixture: 53% KNO; 40% $NaNO_2$; 7% $NaNO_3$), and including two steam/salt heat exchangers and two salt storage tanks.

Energy is stored using the latent heat of this salt between 430 and 275°C, making it possible to superheat the saturated steam from the accumulator to a high enough temperature for the turbine (410°C). The nominal 60 kWh of stored energy corresponds to only about 1100 kg of salt.

The salt cycle concept adopted, namely to have two storage tanks, one hot and one "cold" (although still hot enough for the salt to remain molten), allows salt to be pumped at an essentially constant temperature into the salt/steam heat exchangers: from the cold tank at the cold tank temperature through the charging exchanger to the hot tank and vice versa for discharge when the salt is giving up heat to superheat steam from the accumulator to drive the turbine.

During charging some of the steam from the receiver is diverted through the first heat exchanger (flow rate 1010 kg/h, up to 64 bar, 480°C/510°C) where it heats salt being pumped from the cold to the hot tank.

When the salt is at 430°C in the hot tank, discharge can start if the steam pressure at the turbine inlet drops below the accumulator pressure, 19 bar (although discharge normally waits for the pressure to drop to 8 bar). The saturated steam from the accumulator then passes through the second heat exchanger, where it is superheated by the salt to about 410°C and delivered to the turbine. The salt is then being pumped back from the hot to the cold tank. The mean cold temperature of the salt is about 275°C.

The main components of the salt system are therefore:

- for the charging circuit:
  - a cold tank SF1 where the salt is stored at a temperature of 275°C;
  - a circulating vertical pump P1 installed upon this tank;
  - connecting pipes between P1 and the charging heat-exchanger EC-1;
  - the charging heat-exchanger EC-1;
  - connecting pipes between EC-1 and the hot salt tank.
- for the discharging circuit:
  - a hot tank SC1 containing the salt at a temperature of 430°C;

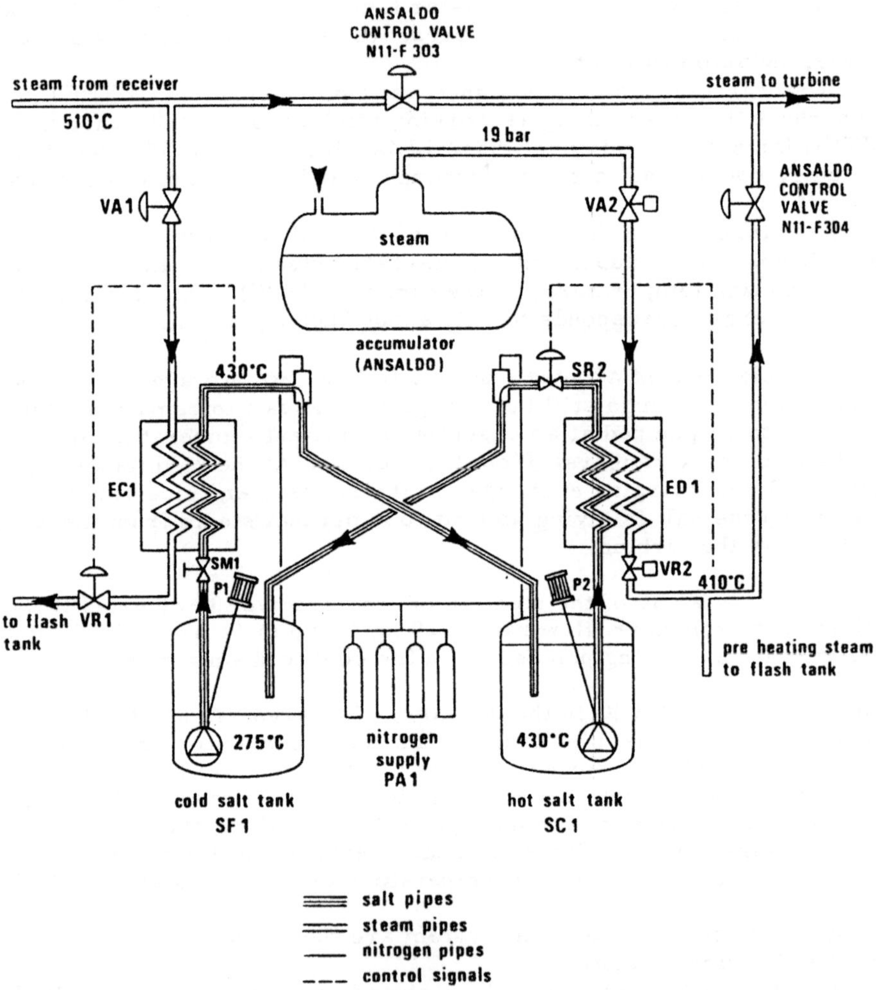

Figure 3.5.1 - Thermal Storage System (Main Components)

- a circulating vertical pump P2 installed upon this tank;
- connecting pipes between P2 and the discharging heat-exchanger ED-1;
- the discharge heat-exchanger ED-1 (a steam superheater);
- connecting pipes between ED-1 and the cold salt tank.

Each tank is large enough to hold the entire charge of salt, the ullage space being filled with nitrogen. All pipes, those for nitrogen as well as the salt circuit, are of stainless steel 304L, 1/2", schedule 10S.

The design of this system involved the solution of some interesting problems in Phases B and C but its operation on a day-to-day basis during the later phases is expected to provide even more useful insight into the strategy of using and/or keeping the stored energy in the design of larger thermal storage capacities.

The salt is of course solid at normal ambient temperatures (it freezes at about 160°C) and so the whole stainless steel circuit had to be designed to be self-draining into the tanks. The layout of the pumps and pipework in the tanks had to allow for electrical melting of the salt in such a way that its expansion prior to melting did not overstress the system and that there was no risk of starting a pump which still contained traces of solid salt. Further, to avoid any risk of salt freezing inside the pipes and heat exchangers and to prevent thermal shock when salt starts to pass through, these components have to be heated. They are kept at the correct temperature by means of electrical resistances, or tracing, and the whole inox steel salt circuit is self-draining (10% slope) into the tanks, in which the salt can freeze and be remelted by electrical heaters.

The design of heat-exchangers also had to fit with rather low heat-exchange rates and very small flows of both steam and salt, especially for the steam superheater. As a result, the heat-exchangers were made of concentric twin pipes throughout, the internal pipe being provided with external fins and partially filled with a core. In this way appropriately small cross-sections were achieved to provide acceptable heat transfer coefficients.

The compact layout of the installation and the overdesigned insulation are designed to result in reduced thermal losses. In addition two different electrical heating rates are available during day and night in order, by minimising it at night, to save auxiliary power. But during the day, if the system is to remain immediately operable at any moment, the temperature level of the salt and of the whole salt circuit, including empty sections, must be kept high enough to avoid excessive drops of temperature. Optimisation in this area will be one of the more valuable experimental activities with this subsystem during Phase D.
Additional numerical data are given in Table 3.5.1.

3.5.2 Operation and control

Charge

Valves VA1 and VR1 (refer to Fig. 3.5.1) are operated through electrically controlled pneumatic actuators.

HEAT EXCHANGERS	EC1. CHARGING HEAT EXCHANGER		ED1. DISCHARGE HEAT EXCHANGER (SUPERHEATER)	
Power	110.8 kW		190 kW (78 kW)*	
Charging (discharging) time	40 minutes		30 minutes	
Heating surface	16 m^2		37.7 m^2	
Internal tube diameter	52.5 / 60.3 mm		42.7 / 48.3 mm	
External tube diameter	82.8 / 88.9 mm		82.8 / 88.9 mm	
Core	42.2 mm		33.4 mm	
Height of the fins	7.11 mm		12.7 mm	
	Between the two concentric tubes	Inside the internal tube	Inside the internal tube	Between the two concentric tubes
Fluid	SALT	STEAM	SALT	STEAM
Pressure (abs. bars)	-	64	-	19 (9)
Flow (kg/s)	0.455	0.280	0.937 (0.365)	0.401 (0.151)
Entering temperature (°C)	275	480	430 (430)	210 (175)
Leaving temperature	430	330	300 (293)	410 (410)
Speed (m/s)	0.2	20.3	0.92 (0.4)	18.1 (10.8)
Pressure drop (bar)	0.1	1.4	1.1 (0.22)	1.1 (0.36)

MOLTEN SALT TANKS

- Overall volume (each) — 1m^3
- Diameter — 1.1m
- Overall height — 1.15m
- Salt capacity:
  - overall — 1,250 kg
  - for process — 1,100 kg
    (corresponding heat capacity 60 kWh)
- Temperature of the salt:
  - cold tank — 275°C
  - hot tank — 430°C
- Material — INOX 316L
- Insulation thickness — 200mm (Rockwool)

PIPING		ELECTRICAL HEATING	
Diameter	17.1/21.3 (½")	Installed Power	45 kW
Material	Stainless Steel 304L		

* Minimum working conditions in parentheses.

Table 3.5.1 - Molten Salt Storage System Numerical Data

During charging the salt flow is approximately constant and the temperature of the salt leaving the heat exchanger is maintained around 430°C by varying the steam flow through the steam control valve VR1.

The nominal required temperature of steam entering the heat-exchanger is 480°C. A secondary control loop acts to anticipate opening of VR1 when the steam temperature becomes lower than 480°C and vice versa. VA1 is an on-off valve.

Discharge

In the steam side of the discharge circuit, VA2 and VR2 are on-off pneumatic valves. SR2 is a pneumatic control valve on the salt side. During operation the steam temperature leaving the exchanger is regulated around 410°C by varying the salt flow through SR2.

At the very beginning of operation, during the first 30 seconds about, SR2 is forced wide open in order quickly to fill the heat exchanger with hot salt for operation.

The steam flow is not controlled by the salt storage system, possibly decreasing with the pressure in the hot water accumulator, although the steam cycle control normally restricts the flow to maintain a pressure of 8 bar at the turbine inlet.

Controls

The salt storage is operated using three turn-push-buttons with signal lamps (TPL-1, TPL-2, TPL-3) located in the vertical panel in the Control Room. Buttons TPL-1 and TPL-2 control the charging circuit; TPL-3 controls the discharging circuit.

Depending on the positions of these buttons, the storage can be set in the following states:

- stopped for night;
- warming-up of the connecting pipes of the charging circuit;
- charging operation;
- warming-up of the connecting pipes of the discharging circuit, enabling discharge.

Two other switches (MHS and MCS) allow control of the electrical heaters in the salt tanks.

The salt storage panel in the control room also displays the temperatures of the salt in the hot and cold tanks and has clocks which measure the running times of the charge and discharge circuits, respectively.

On "run", turn-push-button TPL-1 allows warming of the charging circuit pipes to 275°C for the pipes between the cold tank and the charging heat-exchanger, EC1, to 430°C for the pipes between the heat-exchanger and the hot tank and also for the heat-exchanger itself. This operation takes place early in the morning, lasting about one hour, whilst the boiler is being started-up. When the boiler produces steam at nominal characteristics, button TPL-2 can be set to "run", to start the charging sequence. TPL-2 controls the startup of pump P1 and the electrical supply to the control system for valve VR-1, which is operated according to the temperature of salt leaving the heat-exchanger. Start-up is automatically forbidden if the temperature, in any section of the pipework, or the level of salt in the cold tank, is too low.

The charging operation can be stopped voluntarily by positioning TPL-2 on "stop", or automatically by the safety system. In the event of over-pressure in the salt circuits (over 0.05 bar above atmospheric), which might be the result of a steam leakage into the salt, the safety system stops the pumps and closes the valves, isolating the steam circuits. The pressure measurement is performed in the nitrogen atmosphere maintained above the salt in the two tanks. This neutral atmosphere is in fact an extra precaution against oxidation of the salt, although it should already be stable enough at the envisaged temperature levels.

Turn-push-button TPL-3 on "run" causes warm-up of the discharging circuit pipework (to 430°C for the pipes between the hot tank and the heat-exchanger ED-1, to 275°C for the pipes between the heat-exchanger and the cold tank). After 1 hour the circuit is ready for discharge provided the salt is hot (430°C in the hot tank) and enough adequately-pressurised (about 19 bar) water is available from the steam accumulator. The start of discharge, i.e. starting pump P2, is ordered by a signal from the ANSALDO control system when the low pressure threshold in the main steam pipe is reached. This discharge order has to be confirmed by the complete opening of the steam valve VR2 within about 40 seconds from the start of P2. Without this verification, P2 would be stopped so as to avoid pointless circulation of the hot salt. Start-up is also automatically forbidden if the temperature, in any section of the pipes, or the level of the salt in the hot tank is too low.

Discharge can be stopped voluntarily using the CAJ switch, or automatically by the above mentioned safety system.

Stop for night

To stop the storage system for a short period, such as over-night, buttons TPL-1, TPL-2 and TPL-3 are turned to the "stop" position, which changes the temperature thresholds of the regulators controlling the electrical resistances: the 430°C threshold to 275°C and the 275°C to 0°C, effectively eliminating heating.

There are also switches: MHS to put on the electrical heaters to maintain the temperature of the salt at 430°C in the hot tank and at 275°C in the cold tank; MCS to put on the electrical heaters to re-melt the salt if it has been left to freeze in the tanks.

3.6 ELECTRICAL SYSTEM

3.6.1 Introduction

The function of the electrical system is to provide power, at 20 kV, to the Sicilian grid at the local, Contrasto, hydroelectric plant. This power is generated, at 5.5 kV, by the EURELIOS turbo-alternator. When the plant is operating normally, some of the 5.5 kV supply is transformed to 380 V and supplied to the plant auxiliaries. At other times the auxiliaries bus can be fed from the grid, either through the normal transformers or directly via a 20 kV/380 V emergency transformer. In addition, limited emergency operation is possible from batteries whose state of charge is maintained by the system during normal operation.

A general outline of the electrical system appears in Fig. 3. 6. 1. The plant is normally connected to the Italian grid at Contrasto, at 20 kV, through an underground cable, a circuit breaker, A, and a 20 kV/5.5 kV transformer, T1. The 5.5 kV line is normally supplied directly from the alternator through circuit breaker B but during start-up, etc., power comes instead via breaker A and T1. The auxiliary transformer T2 reduces the 5.5 kV to the 380 volts required by the plant's other systems and made available at several specific terminal points inside the power distribution cabinets. In emergency, i.e. any failure affecting the supply via T2, the 380 V auxiliaries can be supplied from a second auxiliary transformer T3 (20 kV/380 V) provided by ENEL as a back-up source and connected to another section of the Italian grid. The T2/T3 transition is automatic as soon as the voltage fails, as well as the reverse transition T3/T2 when it comes back. The emergency electrics also includes a battery charger, battery system and inverter to power priority (and DC) users.

In short, the normal operating procedure is first to start the auxiliaries using power from the grid through breaker A and transformers T1 and T2; then to operate the heliostat field, the feedwater system, receiver, turbine and alternator; and finally to perform automatic synchronisation and closure of breaker B, thus connecting the plant to the grid. The plant auxiliaries then consume a small fraction of the generated power and the balance is supplied via breaker A and T1 back to the 20 kV grid. If through any failure breaker A opens and the plant remains operating and so supplying only its own auxiliary power, automatic resynchronisation might also be performed in a slightly different way, synchronising and closing breaker A without opening B. This makes it possible to connect the plant again to the grid as soon as the cause of failure has been eliminated.

All the electrical equipment is rather classical. CETHEL was responsible for:

- the 1500 kVA alternator (procured from Ansaldo) with the automatic synchronisation and voltage control equipments;
- the main and auxiliary transformers T1 and T2, respectively 1250 and 250 kVA, with protection devices;
- the 5.5 kV cabinets, containing the medium voltage equipment;
- the low voltage 380 V power distribution cabinets with terminal points corresponding to the main following uses:
  steam cycle and control system
  MBB heliostat field (power and control)
  CETHEL heliostat field (power and control)
  CETHEL salt storage subsystem (power and control)
  utilities such as light, air conditioning ...
- the T2/T3 automatic transition contactor;
- electrical equipment and wiring system of the CETHEL heliostats and

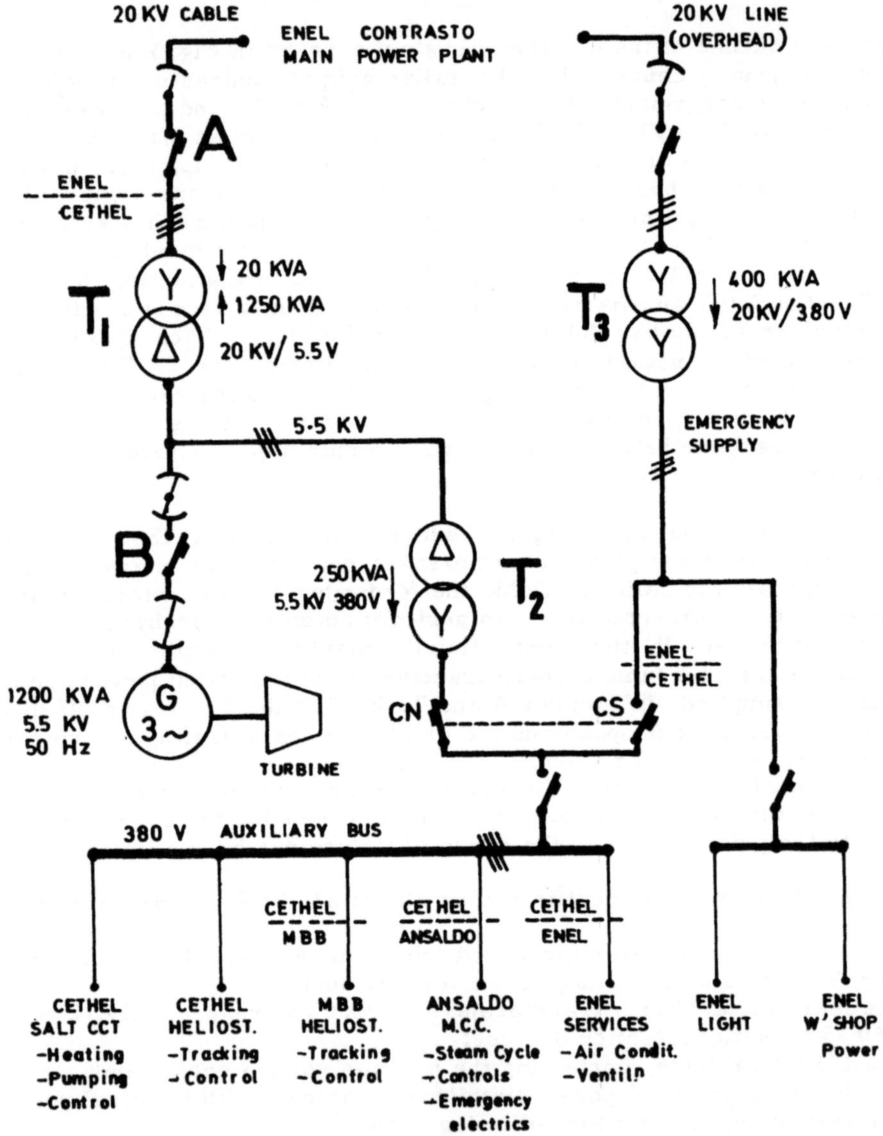

Figure 3.6.1 - ELECTRICAL SYSTEM OUTLINE

the salt storage, involving many heating resistances and protection devices.

ENEL took care of the connection to grid, providing breaker A and transformer T3.

Each Consortium member starts its own electrical installation from the CETHEL power distribution cabinets. ANSALDO in particular had many auxiliaries (pump, compressors, control systems, etc.) to be supplied and provided a specific back-up DC-current source with inverter, ensuring safe operation of the overall control system, and particularly of the protection devices and displays in the control room.

3.6.2 CETHEL electrical system

This section describes (with reference to Fig. 3.6.1) the electrical system primarily by describing its operation in some detail.

I) Preliminary conditions for operation

a) The plant is powered either by the (normal) closure of breaker A and use of transformers T1 and T2 or by the operation of the normal/emergency contactor (CN-CS) and use of transformer T3. The CN-CS contactor precludes any possibility of connecting T2 and T3 in parallel. The emergency supply allows the plant to be operated at any time except that it does not normally allow production of energy.
b) The auxiliaries necessary for the production of steam are started up.
c) The turbine is prepared for operation.
d) The turbine is started up.
e) Energy production.

II) Coupling of the turbo-alternator to the grid

Once the turbine speed has reached an 80% threshold the alternator (Table 3.6.1) can be pre-excited. When the parameters (speed, voltage, phase angle) are correct, it is possible to couple to the grid, either manually (using the ICB push-button in the control room) or, more normally, automatically.

a) Manual coupling
   . the selection switch is put to the "manual" position;
   . the synchronisation key is inserted and turned to "CETHEL";
   . the synchronisation rack at this point becomes live and indicates
     i) on the double-pointer frequency meter, the frequencies of the grid and of the alternator, respectively,
     ii) on the differential voltmeter, the discrepancy between the voltage of the alternator and that of the grid,
     iii) on the synchronisation display, the angular phase difference between the alternator and the grid.

   Coupling may be made when the frequencies are equal, the voltage difference is less than 5% and the synchronisation display is passing through zero phase difference (preferably advancing). These conditions can be achieved by pressing the "+ speed" and "- speed" push-buttons to equalize frequency and to synchronise in phase. Circuit

• Maker	ANSALDO
• Type	GSN 500 Y4
• Number of Phases	3
• Service Rating	continuous
• Power	1500kVA at 0.8 phase angle cosine.
• Voltage	5500V {+15%} {-10%} at 40°C
• Frequency	50 Hz
• Speed	1500 rpm
• Number of poles	4
• Efficiency (@ cosy = 0.8)	94.9% at full load 94.8% at 3/4 load 94.2% at ½ load
• Insulation	class F
• Overheating	class F - provision for 10% overload for 1 hour every 6 hours
• Protection	IP 23
• Bearings	Self lubricating bushes
• Total weight	5150 kg
• Exciter	type EDR 350/80, 0.31kW, 50V brushless, rotating diodes

Table 3.6.1 - Alternator Characteristics

breaker B is then closed by pressing the corresponding push-button.

b) Automatic coupling
- the selection switch is put to the "automatic" position;
- the synchronisation key is inserted and turned to "CETHEL";
- the voltage, frequency and phase parameters of the alternator and grid are in this case compared by the synchrocoupler which generates the "+ speed" and "- speed" commands for the turbine's speed controller until the parameters are equal and the "close of breaker B" order is automatically issued.
- the synchronisation equipment functions whenever coupling is being attempted and must be isolated during normal running by extraction of the synchronisation key. This is possible only when the key has been turned to "STOP" from the "ENEL" or "CETHEL" position.

III) Coupling of the plant to the grid

When the plant is in normal energy production and breaker A opens because of a fault in the grid, the turbine and transformers T1 and T2 remain in service but a command is sent to partially open the turbine bypass valve to maintain the turbine speed despite the reduction in load. After elimination of the fault coupling can again be achieved, either manually or automatically. This coupling, either manual or automatic, is carried out in exactly the same way as above except that the synchronisation key is turned to the "ENEL" position rather than "CETHEL" so that the final command is issued to close breaker A instead of B.

If breaker A opens because of a fault in the plant, breaker B will also open, the low tension (380 V) supply to the auxiliaries automatically switches (by the CN-CS switchgear) to the emergency supply (via T3) and a command is again sent to partially open the turbine bypass valve. After elimination of the fault the electrical supplies to the plant must be switched back to be through breaker A (using the CN-CS switchgear) after which the plant can be re-coupled as described in II above.

IV) Regulation of turbo-alternator

a) Speed regulator
A bi-directional servo-motor drives the control valve, for admission of steam to the turbine in response to "+ speed" or "- speed" orders received;

b) Voltage regulator
This regulator has three roles:
- voltage regulation when the turbo-alternator is isolated from the grid (breaker A open),
- cos ϕ (Phase) regulation when the turbo-alternator is connected to the grid; in this case voltage regulation is eliminated,
- voltage equalisation operates only before coupling to the grid so as eventually to make the turbo-alternator voltage equal to the grid voltage.

V) Protection of the 5.5 kV grid

All the safety devices issue an appropriate optical and audible alarm in the control room as well as the other reactions mentioned.

a) Alternator
   . mass stator (MS). Verifies stator insulation and causes, in the event of a fault, opening of circuit breaker B, ending of rotation (?) and removal of excitation from the alternator;
   . power return (RP). Protects against possible feedback of power into the alternator by causing, in response to feedback of about 1% for 5 seconds, opening of breaker B;
   . maximum voltage (MU). In response to a voltage increase of about 10% for 3 seconds this causes opening of breaker B, ending of rotation (?) and removal of excitation from the alternator;
   . minimum voltage (mU). In response to a voltage drop of about 20% for 3 seconds this causes opening of breaker B;
   . maximum current, first and second thresholds (MITA 1 and 2). In response to an increase of about 1.5 In = (normal operating current) for 3 seconds (MITA 1) or 4.2 In for 0.4 seconds (MITA 2) these devices cause opening of breaker B.

b) Protection of the T1 and T2 transformers (see Table 3. 6. 2).

	T1	T2	T3
Supplier	CETHEL	CETHEL	ENEL
Voltages	20kV/5.5kV dual use	5.5kV/400V step-down	20kV/400V step-down
Power	1250kVA as step-up 250kVA as step-down	250kVA	400kVa
Dielectric	Oil	Oil	
Coupling	star/delta	delta/star (neutral out)	star/star
Voltage Adjustment	± 5% (Adjustable off load)	± 5% (Adjustable off load)	

Table 3.6.2. - Transformer Characteristics

These transformers are protected by a pair of systems (DGTP2-type) which detect faults and actuate alarms and/or breakers.

. Temperature. This is monitored by dual-threshold thermostats. The first threshold triggers the usual alarm, at 100°C. The second, at 110°C, results in opening of both breakers, A and B, as well as the usual alarms;
. Release of gas. Gas release monitors cause only the issuing of the usual alarms;
. Overpressure. Two detectors respond to overpressure in the transformers by causing opening of both breakers A and B;
. Overload. Detectors respond to a 110% overload in nominal power of T1 by opening breaker B. In the case of T2 the protection responds to an overload of about 4 In in 0.2 seconds by opening both breakers A and B.

VI) Low voltage distribution

As mentioned above, low voltage (380 V) supply is assured by the use of two independent, alternative, supplies: via transformers T2 to T3, respectively. Normally the auxiliaries are supplied by T2, by the general safety equipment and by the CN contact of automatic transition equipment, CN-CS.

In the event of a fault in T2 the loss of voltage above contactor CN is detected and causes opening of CN and closure of CS. The short (less than 0.5 second) duration of the loss of voltage increases the security of the affected auxiliaries, particularly the 2 ANSALDO pumps supplying the receiver and the equipment for manoeuvring the heliostat field (especially MBB's which has no integral back-up power supply).

This 380 V three phase bus supplies the following users via appropriately calibrated circuit breakers:

. ENEL	60 kW	breaker	DDE,	3 x 250 A
. ANSALDO	75 kW	breaker	DDA,	3 x 250 A
. MBB	45 kW	breaker	DDMBB,	3 x 100 A
. CETHEL	30 kW	breaker	DDC,	3 x 100 A
. storage	50 kW	breaker	DST,	3 x 250 A
. regulation	20 kW	breaker	DC1,	3 x 100 A

The monitoring of the insulation of the 380 V system utilises a Merlin Gerin type TR2 controller which triggers the usual optical and audible alarms in the control room in case of a fault.

Two 380 V/220 V transformers are used to step down the 380 V supply for the regulation equipment and for the electrical tracing used to heat the salt storage pipework, respectively.

3.6.3 ANSALDO electrical system

Although each Consortium member was responsible for its own electrical conditioning, based on the 380 V three-phase supply, and these are described in the appropriate sections, Ansaldo had specific emergency responsibilities which make it appropriate to consider its electrical equipment as part of the overall electrical system.

The Ansaldo electrical equipment can be subdivided into three groups providing power, respectively, for the following purposes:

a) Main control centre (MCC), i.e. for distribution of power to equipment associated with the steam cycle, including the receiver. The MCC itself is described in Section 3.7.2. The distribution is powered from the normal 380 V, 3 phase, 50 Hz CETHEL supply, the functions of the individually switched and protected lines being as indicated in Tables 3.6.3 and Fig. 3.6.3.
b) Service board for ventilation and air conditioning. This board is powered from the normal 380 - 50 Hz supply and comprises 10 independently switched and protected lines as listed in Table 3.6.4.
c) Service board for power and light. This board provides lighting and general-purpose electrical services other than those required to operate the plant. It is fed directly from the T3 emergency supply and so

Line No.	Protection Rating (kVA)	Function
1	-	380v, 3 phase, 50Hz supply from CETHEL
2	0.25	Turbine starter motor
3	4.0	Demineralised water pump (A)
4	4.0	" " " (B)
5	5.0	Reserve
6	5.0	Reserve
7	6.0	Closed cycle cooling pump (A)
8	6.0	" " " " (B)
9	6.0	Nitrogen compressor for bottle recharge
10	6.5	Turbine lube. oil aux. pump
11	15.0	Condenser vacuum pump (A)
12	15.0	" " " (B)
13	15.0	Service air compressor (A)
14	15.0	" " " (B)
15	37.0	Receiver feedwater pump (A)
16	37.0	" " " (B)
17	3.5	Condensate extraction (booster) pump (A)
18	3.5	" " " " (B)
19	1.0	Vacuum break el./valve ? ?
20	15.0	Turbine lube. oil cleaner
21	10.0	Battery charger (see Figure 3.6-2)
22	1.0	Compressed air drier
23	65.0	Receiver and tower auxiliaries
24	1.0	Demineralised water control board
25	5.0	Inverter by-pass transformer (see Figure 3.6-2)

Table 3.6.3 - MCC Power Distribution

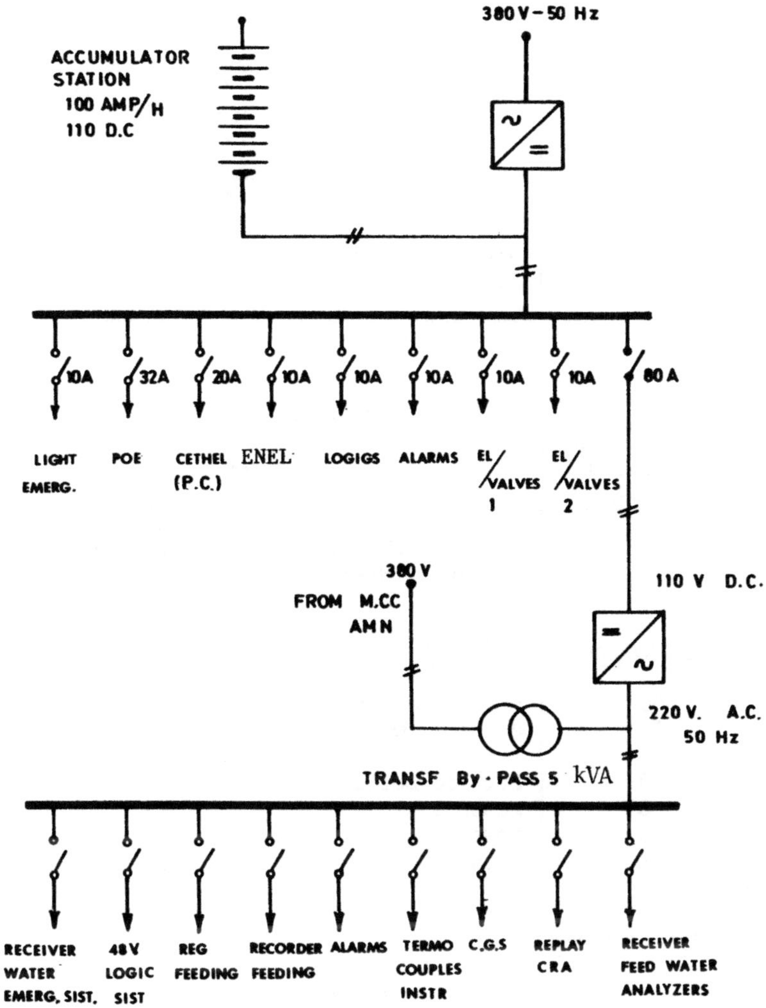

Figure 3. 6. 2. - Control Room Panel Supply

VENTILATION AND AIR CONDITIONING		
Line No.	Protection Rating (kVA)	Function
1	-	380 V, 3 phase, 50 Hz supply from CETHEL
2	0.5	Air extraction from battery room (A)
3	0.5	" " " " " (B)
4	3.0	" " " MCC " (A)
5	3.0	" " " " " (B)
6	3.5	" " " machine " (A)
7	3.5	" " " " " (B)
8	3.5	" " " " " (C)
9	50.0	Air conditioning
10	4.0	Reserve

POWER AND LIGHT		
Line No.	Protection Rating (kVA)	Function
1	-	380 V, 3 phase, 50 Hz supply from CETHEL (metered, direct from T3)
GENERAL ELECTRICAL SERVICES		
2	100	Mechanical workshop
3	70	Machine room power sockets
4	15	" " overhead crane
5	8	Service building
6	8	Reserve
7	8	Reserve
8	-	Future
9	-	"
LIGHTING (metered)		
1	5	Battery charger
2	24	Main building
3	10	Workshop and store-room
4	8	External zone
5	24	Reserve
6	-	Future

Table 3.6.4 - Electrical Service Boards

is not normally taken into account in the calculation of the energy efficiency of the plant. It comprises 15 independently switched and protected lines, metered as two groups, as listed in the second part of Table 3.6.4.

3.7 CONTROL AND COMPUTER ROOMS

3.7.1 Heliostat central controls

As indicated earlier there are differences between the control concepts of the two parts of the heliostat field. In summary these are:

- although both central units calculate the required angular positions for each heliostat, only in the case of CETHEL are these sent to the heliostats as commands. The individual MBB heliostats operate more autonomously, being given simple commands (e.g. track the receiver) and making their own calculations of the required angular positions. The MBB central control can periodically verify that the actual positions agree with its own calculations;
- the operator controls the CETHEL field via a teletype (TTY) machine and (for group or whole-field operations) push-buttons on the console. The MBB field is primarily controlled via a VDU and keyboard (normally using its TTY only for printing records of operations and faults). Of course these controls are in addition to the (identical) interfaces between the heliostat central units and the plant's general control system, GCS, used for normal plant operation;
- CETHEL heliostats can turn automatically away from the receiver, under battery power, in emergency. MBB's require power from the central unit and so do not move if such power is not available.

3.7.1.1 CETHEL heliostat central controls

Summary of CETHEL heliostat field operation

The CETHEL field is divided into four groups (Fig. 2.3) each controlled by a group controller (CDG). Automatic command and control of the field is managed by the field controller (CDC - Controleur de champ) which receives instructions from the CETHEL desk, the TTY or the general control system (GCS). As described below, each group can be commanded to any one of six operational positions; the group can be commanded in three different modes and, by using the TTY, a qualified operator can carry out a number of special manoeuvres, over-riding automatic operation and allowing maintenance, alignment tests, extraction and re-insertion of heliostats of a group, correcting time, etc.

Six operational positions

Stowage: The heliostats are fixed in a position in which they will best survive bad weather. Although initialised with the mirrors approximately face down (azimuth 0°, -84° elevation) the position is replaced by that indicated by the thumb controls on the desk as soon as transition to this mode is commanded in either of the "group" command modes. The operator should carefully select a position compatible with local weather conditions.

Safety: The heliostats are in a vertical position (elevation 0°) so that the reflected rays are directed towards the ground and no spot is formed. This state is initialised with zero azimuth but this value is replaced by that indicated by the thumb controls on the desk as soon as transition to "stowage" position is commanded in either of the group command modes.

Start: The heliostats are focussed on a point at the base of the safety trajectory, i.e. 20 m east of the base of the tower. Motion from "safety" to "start" is implemented first in azimuth and then in elevation, and vice versa.

Standby: The heliostat beams are focussed on a point at the top of the safety trajectory, i.e. 20 m east of the receiver. Motion between "start" and "standby" is constrained such that the beams from the heliostats all pass inside a 20 m diameter vertical cylinder whose axis passes through the "start" and "standby" positions.

Receiver: The spot formed by the heliostats is in the receiver.

Maintenance: This supplementary mode allows TTY (operator) control of groups and individual heliostats. The commands available to the trained operator are summarised in Table 3.7.1. If, whilst any heliostat(s) is in "maintenance" an emergency occurs which makes it necessary to move to the "stowage" position, the manoeuvre must be carried out locally, putting the heliostat(s) in manual mode by means of the switches in the heliostat electronics. In the case of strong winds, it is necessary to allow for the time required for this operation.

Three modes of command

Manual, group: Each group is commanded directly by "stowage", "safety", "start", "standby" and "receiver" push-buttons on the left side of the control desk (Fig. 3.7.1) and accepts no field commands from the right-hand side of the desk nor, except the E1 and E2 emergency alarms (E1 only when in "receiver"), from the GCS. The status of each group can be seen from the group signal lights to the left of the desk and operator and GCS messages are both available via the printer.

Manual, field: In this mode, movement of the groups is coordinated by the field controller (CDC) and is initiated by the "stowage", "standby" and "receiver" buttons on the right of the desk. The group signal lights, the printing of operator and GCS messages and the reactions to E1 and E2 alarms is as in the "manual, group" mode. The coordinates of the stowage position indicated by the thumb-wheels on the console are memorised when the manual, field button "stowage" is pressed to pass from the "standby" to "stowage" positions.

Auto, field: In this case the field is controlled by the GCS: except that the operator can intervene, using the TTY, by putting any group to "maintenance". Unless the operator has put a group into manual mode, using the TTY, the push-buttons on the console are ignored. The group signal lights operate as in the manual modes and the stowage-position thumb-wheel values are memorised when a "standby"-to-"stowage" command is received from the GCS. Transfer to a more automatic mode

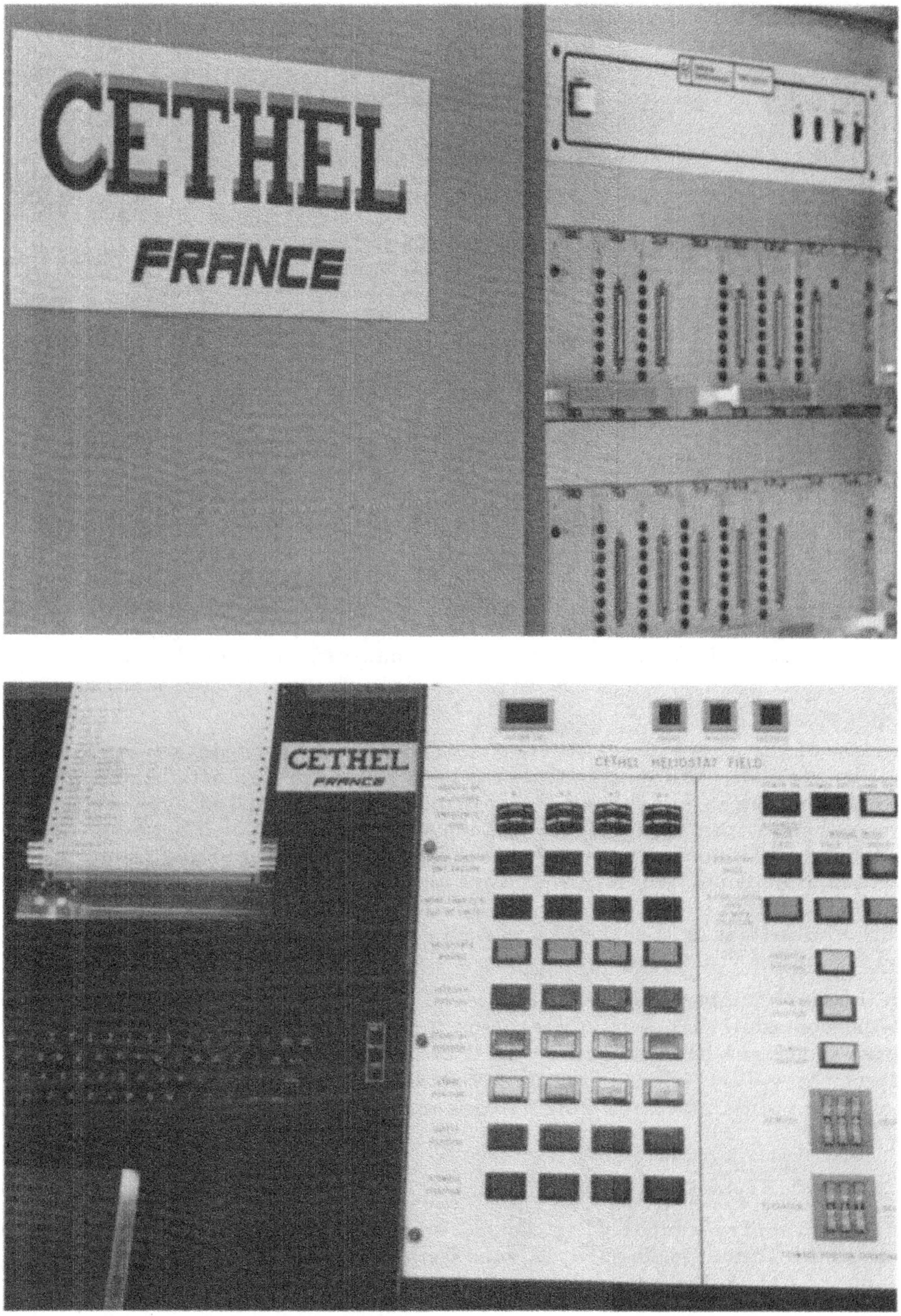

Figure 3.7.1 - CETHEL Control Desk

("manual, group" to "manual, field" or "manual, field" to "auto, field") is authorised only if all groups are in the same state ("stowage", "stand-by" or "maintenance"). Transfer "auto, field" to "manual, field" is always authorised. Transfer "manual, field" to "manual, group" is not authorised if the groups are in "receiver".

Logical status and other messages

A group status message is printed on the TTY in the form

"GROUP N, LOGICAL STATE" = .G.H.%.R.A (or M) .C.T."

with the following meaning:

- T is printed only if the group clock does not agree with that of the CDC
- C is printed only if the group arithmetic unit is faulty
- A (or M) is printed when the group is under Automatic (or Manual) control, respectively
- R is printed when an E1 alarm has tripped the group
- % is printed when more than two heliostats in the group are faulty
- H is printed when any heliostat in the group is faulty
- G is printed when an E2 alarm has tripped the group

There are also two similar heliostat status messages and a number of diagnostic printouts which can appear on the TTY, including those already indicated in Table 3.7.1. They are summarised in Table 3.7.2.

The CETHEL control console

In view of the above descriptions the illustration of the desk (Fig. 3.7.1) will be virtually self-explanatory. However, it may be noted that the time displayed is Greenwich Mean Time. The illuminated push-buttons are marked with cross, circle and square in the figure. Those on the "group" side light in green, when the group is in, or moving from, the indicated position, the yellow "heliostats moving" light distinguishing between the two cases. The red "10% out" light is illuminated when more than 2 heliostats in the group are faulty (malfunctioning or not responding) and not under "maintenance" by the TTY "HO" command.

On the field (right-hand) side the "lamp test" button tests the other lamps on the console. The green "operation mode" lamps indicate which operational mode has been selected whilst the "authorisation and op. mode selection" yellow-illuminated push buttons indicate which alternative modes are authorised for selection, according to Fig. 3.7.2.

3.7.1.2 MBB heliostat central controls

The MBB central unit (CU) is the main control device for the MBB part of the heliostat field. In turn, it sends status information to and receives commands from the general control system, GCS.

The MBB CU provides power to the heliostat field. It is connected to the heliostats via two serial data buses and two static signal lines. It commands the heliostats via the "command data bus" and receives requested status information from a selected heliostat via the "response data bus".

COMMAND*	RESULT
ROUTINE COMMANDS	
FT	Printout of Julian day and time of the central clock in the format: "MASTER CLOCK = xxxxx JD xx H xx MN xx S"
GL,N	Printout of logical status of group N in the format: "GROUP N, LOGICAL STATUS = X.X.X.X.X.X.X.X.X."
GT,N	Printout of Julian day and time of the group N clock in the format: "CLOCK, GROUP N : xxxx JD xx H xx MN xx S"
GA,N	Transfer of group N to 'Auto'
GM,N	Transfer of group N to 'Manual'
HL,N,NN	Printout of logical status of heliostat NN of group N in the format: "GROUP N, HELIO NN, STATE = .X.X.X.X.X.X.X.X A = .X.X.X.X S = .X.X.X.X"
HP,N,NN	Printout of position of heliostat NN of group N in the format: "GROUP N, HELIO NN, POSITION = A = xxx xxx S = xxx xxx"
HR,N,NN	Resetting heliostat NN of group N
COMMANDS FOR TRAINED OPERATORS	
FS,SS,MM,HH,JJL,JJH	Reset central clock time
GI,N	Re-insertion of group N
HI,N,NN	Re-insertion of heliostat NN of group N from maintenance or alignment
GO,N,±EEE,±AAA	Transfer group N to 'maintenance', position EEE, AAA degrees
VT,N,±XXXX,±ZZZZ	Correct target vector for 'receiver' by XXXX cm East, ZZZZ cm up
V1,N,±XXXX,±ZZZZ	" " " alternative target 1
V2,N,±XXXX,±ZZZZ	" " " " " 2
VA,N,±XXXX,±ZZZZ	" " " 'standby'
VD,N,±XXXX,±ZZZZ	" " " 'start'
HO,N,NN,±EEE,±AAA	Transfer heliostat NN of group N, to 'maintenance', position EEE, AAA degrees
HE,N,NN,±XXXX,±ZZZZ	Transfer heliostat NN of group N to 'alignment' on the given vector, $|X| \leq 9999$cm, $|Y| \leq 9999$cm

*Each command must be terminated by 'carriage return'.

Table 3.7.1 - CETHEL TTY Commands

1. MASTER CLOCK (response to command FT).

2. GROUP STATUS (response to command GL or to any change of group status).

.G	.H	.%	.R	.A (or M)	.C	.T
Emergency E2	Heliostat fault	More than 2 H'stat faults	Emergency E1	A-Auto M-Manual	Calcu- lator error	Clock error

3. CLOCK, GROUP N (response to command GT).

4. OPERATOR ERROR (response to operator error in TTY input).

5. GROUP N COMMAND ERROR (response to communication errror).

6. GROUP N OFF (results from lack of response to CDC commands).

7. HELIOSTAT STATUS (response to command HL or to any change of heliostat status).

HELIOSTAT STATUS 1

.A	.A	.R	.S	.B	.M
Emergency E2	Thermal Alarm	Moving	Secteur?? fault	Battery fault	Manual mode

HELIOSTAT STATUS 2

Azimuth				Elevation			
.O	.R	.H	.F	.O	.R	.H	.F
out of action	speed error	fast move- ment	position error	out of action	speed error	fast move- ment	position error

8. HELIOSTAT POSITION (response to command HP).

9. INCORRECT MASTER CLOCK, TIME SETTING (response, at CDC switch-on, if master clock battery has discharged).

10. FIELD CONTROLLER ON (response to CDC switch-on).

Table 3.7.2 - CETHEL TTY Messages

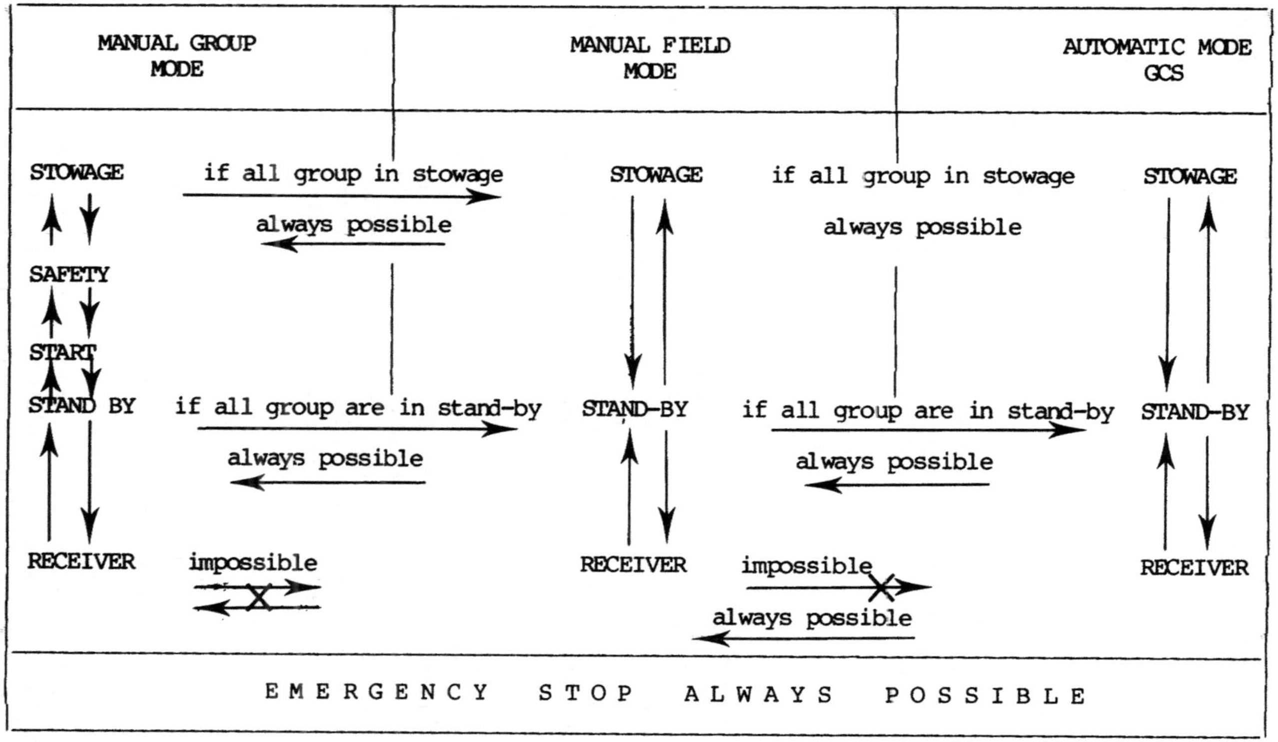

Figure 3.7.2 - CETHEL Mode Transitions

For each heliostat group two "emergency signals" are provided which control the movement of the heliostats in the azimuth (E1) and the elevation axis (E2) directly in case of a serious failure even if this occurs in the serial bus system.

Dependent upon automatic mode commands from the GCS or on manual commands initiated by the operator, the MBB CU sends corresponding commands to a single heliostat, to a group of heliostats or to all heliostats (date, time, position, status requests). The CU also continuously verifies the correct function of each heliostat by periodically requesting status information. As an additional control the actual positions about the two heliostat axes are requested from each heliostat after the heliostat indicates that the "receiver position" has been reached. The CU calculates for the heliostat the correct values and compares these with the actual values received from the heliostat. If the comparison indicates a difference, an error message will be produced. In any case all status information obtained from the heliostat field is assembled by the CU and indicated to the operator. The heliostat field status is also provided on specified data lines to the general control system.

The MBB central unit consists of two main units, the MBB console in the control room and a power cabinet in the "sottoquadro". A block diagram and photographs appear in Figs. 3.7.3 and 3.7.4.

The MBB console houses the electronic circuitry of the CU. One active electronic unit and a second one in cold standby redundancy are installed in the base of the console. The electronic units consist almost entirely of off-the-shelf microcomputer systems. The central unit processor (CUP) is an Intel 8080 processor-based unit containing the arithmetic sub-unit (which checks the heliostat position calculations), 16 kbytes of EPROM and 2 kbytes of battery-buffered CRAM, an interrupt unit, counters and interfaces.

The console itself contains a control desk, an Intercolour 8001 CRT plus keyboard and an HP 2635A line printer. The control desk contains (one per group) illuminated switches for power on, manual override and emergency deviation and also signal lights showing the status of each group (10% out of limits; moving; in receiver position; in standby position; in stowage position). It also has switches controlling power to the processor, to the interface sub-unit and to the whole field, initialisation and emergency reset and also signal lights showing the existence of GCS command sequences and of errors in the field. The CRT is normally used to display information to the operator for the dialogue between him and the central unit processor. Instead of the CRT and keyboard the line printer also may be used for this dialogue. It also provides a permanent record of error messages and other service information.

Other sub-units of the MBB CU are:

- Timing unit, a quartz-controlled adjustable oscillator and associated counters, supplied by an accumulator-buffered supply (in the AKKU, see below), to provide stability for about 7 days. It can be synchronised periodically to the CETHEL clock.
- Emergency unit. Able to operate independently of the central unit

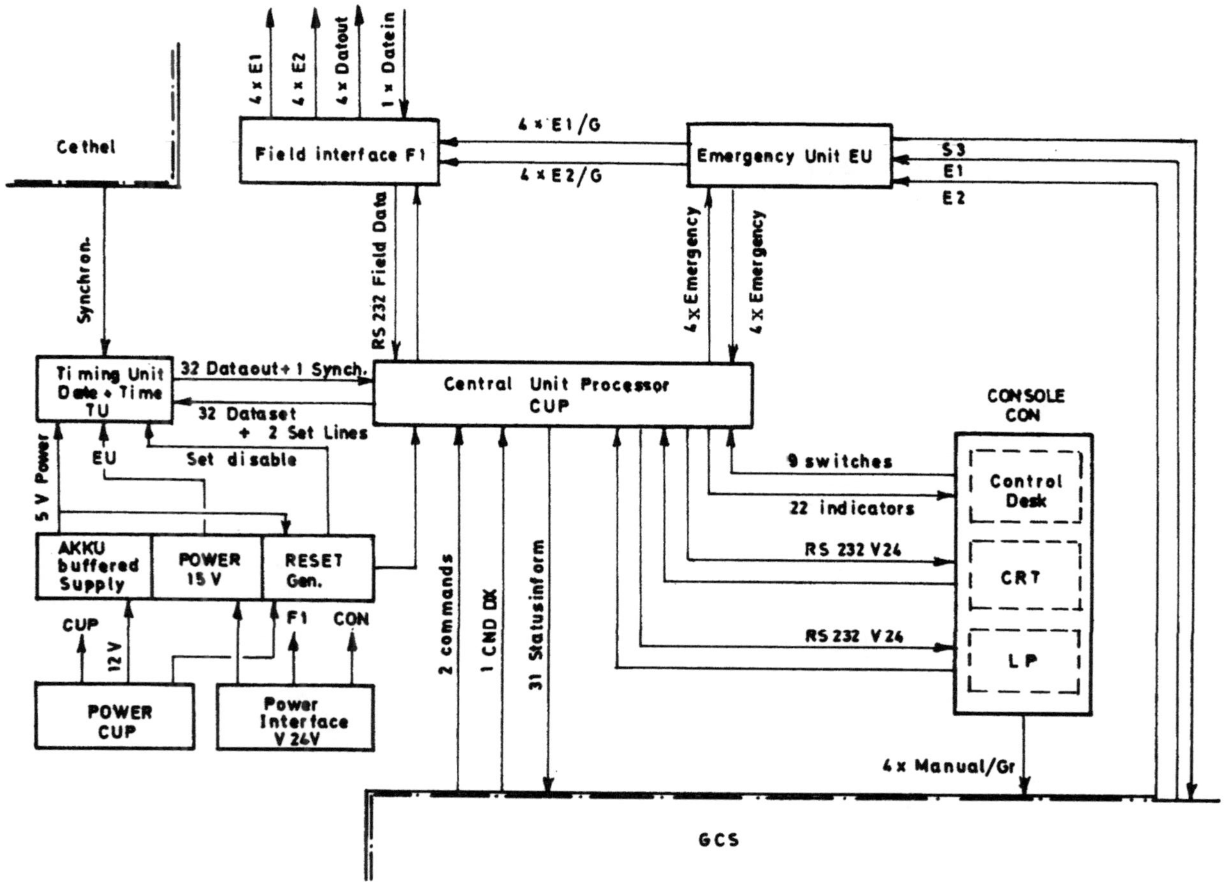

Figure 3.7.3 - BLOCK DIAGRAM OF MBB CENTRAL UNIT

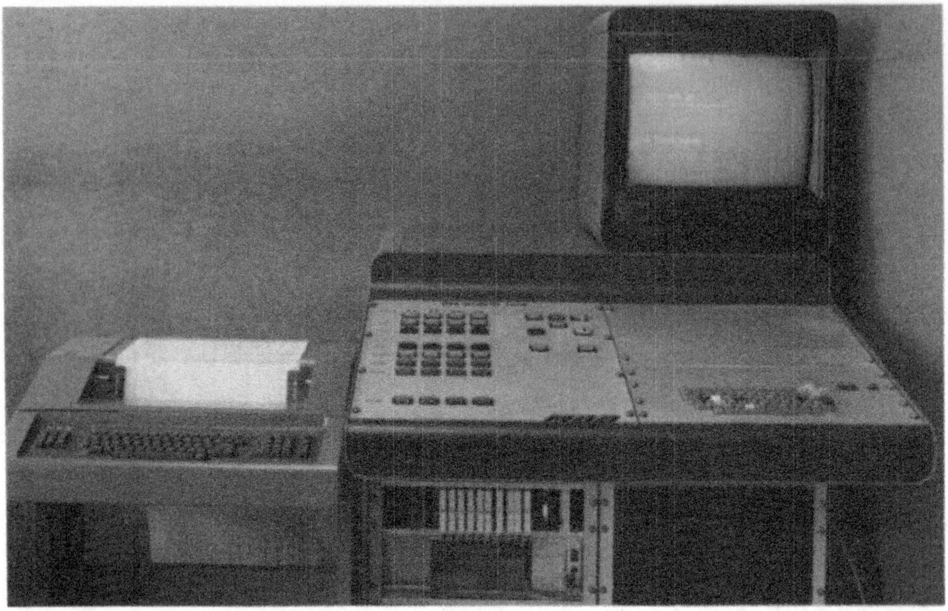

MBB Central Unit Console
(Control Room)

Power Cabinet for MBB Heliostats
(in "Sottoquadro")

Figure 3.7.4 - MBB Central Unit

processor, this unit handles the automatic emergency sequences required in response to E1/E2 commands from the GCS or to an emergency command from the CUP.

- Field interface unit which buffers all signals to and from the heliostat field and includes appropriate short circuit and over-voltage protections.
- AKKU. This contains an accumulator-buffered 5 V supply for the timing unit and for reset generation; a capacity-buffered 15 V supply for the emergency unit which avoids an "emergency" sequence if there is a short (about 1 second) power loss; a reset pulse generator to ensure correct status for the CU and the timing unit after power-up.

Detailed functions, operations and interfaces of the MBB central unit are specified in the "Operators Manual" but they may be summarised as follows.

In the absence of commands, the central unit continuously scans the entire MBB field, group-by-group, heliostat-by-heliostat, verifying that there are no discrepancies between expected values and those measured. If a discrepancy is detected the CU requests a "failure word message" from the affected heliostat. This results in the printing of a message on the line printer identifying it as a data handling failure; arithmetic logic unit failure; lack of power to drive servos; elevation or azimuth end positions reached or discharge of the small battery intended to preserve the contents of the heliostat control logics RAM. Alternatively, when a group is tracking the receiver, the position of the heliostat is printed out if it differs from that calculated by the CU processor's arithmetic unit.

Summary tables can also be displayed, in response to "0" commands (Table 3.7.3), which show the status of every heliostat in terms of whether or not it is under control; tracking the receiver; disabled from accepting group commands (to allow individual heliostat maintenance for example) or "cancelled" (i.e. automatically inhibited from responding on the response bus because of a short circuit which would otherwise block the entire bus).

If a "no bus answer" error is detected for a whole group, an E2 emergency message is issued (taking the group towards stowage position).

The "0" command has already been mentioned. As well as responding to group commands from the GCS, the MBB field can be commanded from the console, using the commands shown in Table 3.7.3. The GCS group commands include "sequence" commands as a result of which a group command is sent out to each group in turn over a period of time. Similarly the GCS emergency commands E1 and E2 activate the emergency lines regardless of the status of the CU processor.

Interface between the GCS and the MBB central unit

Via coded parallel data lines the CU receives commands from the GCS to move the heliostats into certain positions (stowage, standby, receiver) and sends heliostat field status information back to the GCS. The output and input circuits of the CU provide complete insulation by opto-couplers and, particularly, relays.

Command	Function
R (Adr)	Go to Receiver
T (Adr)	Go to Standby
S (Adr)	Go to Stowage
G(Adr)A(Grad) (Min)	Go to defined azimuth
G(Adr)E(Grad) (min)	Go to defined elevation
(moving starts after GE command)	
M (Adr) (xy)	Move with defined speed and direction
x/y	
0 Stop Azimuth/Elev'n	
1 Slow West/up	
2 Slow East/down	
3 Fast West/up	
4 Fast East/down	
A (Adrs)	Go to alignment
B (Adrs)	Send alignment table
P (Adrs)	Request actual position
Q (Adrs)	Request calculated position

Command	Function
W (Adrs)	Request status word
F (Adrs)	Request failure word
Y	Display date, time
O 1	Control-parameter table
O 2	Tracking-parameter table
O 3	Group-cmd-disable-param.table
O 4	Cancel-parameter table
E	Output status table
H	Output alignment table
K(Adrs)(Value 1,2 .. n)	Set alignment table
L (Adrs)	Enable group address
N (Adrs)	Disable group address
C (Adrs)	Cancel heliostat
J (Adrs)	Reset cancel command
V (hour,min)	Set time
D (day,month, year)	Set date

Note: (Adr) stands for either a single heliostat address, 1-112, or a group address 251, 252, 253 or 254 (operator's option). (Adrs) stands for a single heliostat address, 1-112.

Table 3.7.3 - MBB Central Unit Console Commands

The following commands can be initiated from the GCS via the MBB CU:

- a specific single group into stowage position, standby position or to illuminate the receiver;
- complete field to stowage, standby or receiver position in defined sequences, group by group.

Sequences to and from the stowage positions are introduced to avoid high peak power demands. The sequences for movement between standby and receiver position are related to the allowed temperature gradient of the receiver.

Two additional emergency commands GCS-E1 and GCS-E2 are generated by the GCS. The CU activates the corresponding field emergency lines independent of the internal CU status or any switch status on the console desk. After receiving the GCS-E1 signal a time-limited emergency signal E1 for the whole field will be generated by the CU causing a high-speed eastward movement about the azimuth axis. The TCS-E2 signal initiates in the CU to generate first the E1 field signal for a limited time as above and then the generation of a steady E2 field signal which lowers the mirrors by rotation about the elevation axis.

MBB power interface

The heliostat field is powered from the 380 V three phase AC-grid. The power cabinet of the CU converts this through a 45 kW, 380/380 V star/star transformer (which assures complete isolation from the CETHEL electrical supply) to provide the 220 V supply on which the CU operates and the 380 V power supply for the heliostat field. The power cabinet provides all necessary switching, fusing and distribution devices. The following values are specified as power demand for the heliostat field:

Input power:

. voltage	3 phase 380 V ± 10%
. power consumption:	
whole field during tracking, average	4 kW
whole field maximum (e.g. emergency)	35 kW
central unit max.	0.5 kW

Apart from small batteries to avoid reset or loss of data in the event of a brief interruption of power neither the heliostats in the field nor the CU include any significant power storage capability because of the inspection and service requirements, limited life time and need for additional charging circuits. The plant, therefore, normally provides emergency power for the periods when a power drop or switch-off occurs on the normal grid and can also tolerate the MBB field's remaining stationary until the sun has moved enough for the image to leave the receiver.

Interface MBB central unit/MBB heliostat field

The interface between the CU and the heliostat field consists of cables for power and for data. The data connections have two separate lines for the "emergency signals" and the two serial bus lines (command bus and response bus) using 110 bits/sec RS 232, CCITT V.24 (DIN 62 020) standard signals.

3.7.2 Master Control Centre (MCC)

General

The MCC provides control and supervision of the whole plant including receiver, steam cycle, heliostat field and storage. For the definition of its present configuration ENEL referred whenever possible to usual procedures adopted in thermal power stations.

3.7.2.1 Information and control layout

A) Foreword

Before starting to explain in detail the control and supervision modules it is important to understand the basic interconnections and relationship among them, shown in Fig. 3.7.6. In this figure boxes represent plant components or electronic equipments and lines mean flow of analog or digital signals. It is also appropriate to point out that practically all the relevant signals shown will (by the end of Phase D) be routinely collected and processed in real time by the supervision computer (called "DASS").

B) System outline

The SRES module (solar radiation elaboration system) receives signals, F1...F20, from 20 fluximeters welded on the boiler pipes and distributed (8/8/4) on 3 different layers, orthogonal to incoming radiation flux.
It also receives signals, P1... P4, from four sun tracking pyroheliometers (i.e. not incident radiation transducers). The SRES processes those signals and outputs into an analog signal, called Qer, the best available estimate of actual heat flux entering into the process fluid.
The SRES also drives a complete control panel (in vertical panel P611 in the control room) which allows the operator to visualize the most important variables and to eliminate faulty sensors.
The output signal Qer is monitored on the control desk and fed into the regulation subsystem where it plays a fundamental role as driver of feed-forward control actions during sun-radiation disturbances.
The meteo measurement system module measures and collects meteorological data, viz.

- wind speed and direction at 2, 10, 50 metres altitude;
- temperature, atmospheric pressure and humidity;
- global sun radiation,

and presents these data on pen recorders and digital meters fitted into the P611 vertical panel.
The 10 and 50 m wind speed signals are also processed by a special non-linear threshold circuit which closes pre-alarm contacts to warn the operator and to trigger the E2 emergency signal which in turn deviates the whole mirror field to the stowage position.
The "Command generation for MCUS's" module (or "Code") generates the digitally coded commands to drive the two mirror fields. The Code can:

- select MBB or CETHEL or both fields, under operator command;
- start or stop the deviation sequence (from receiver to stand-by) for selected field(s), under operator or regulation system command;
- start the focussing sequence (from standby to receiver), under operator command only.

Figure 3.7.5 - Master Control Centre (MCC)

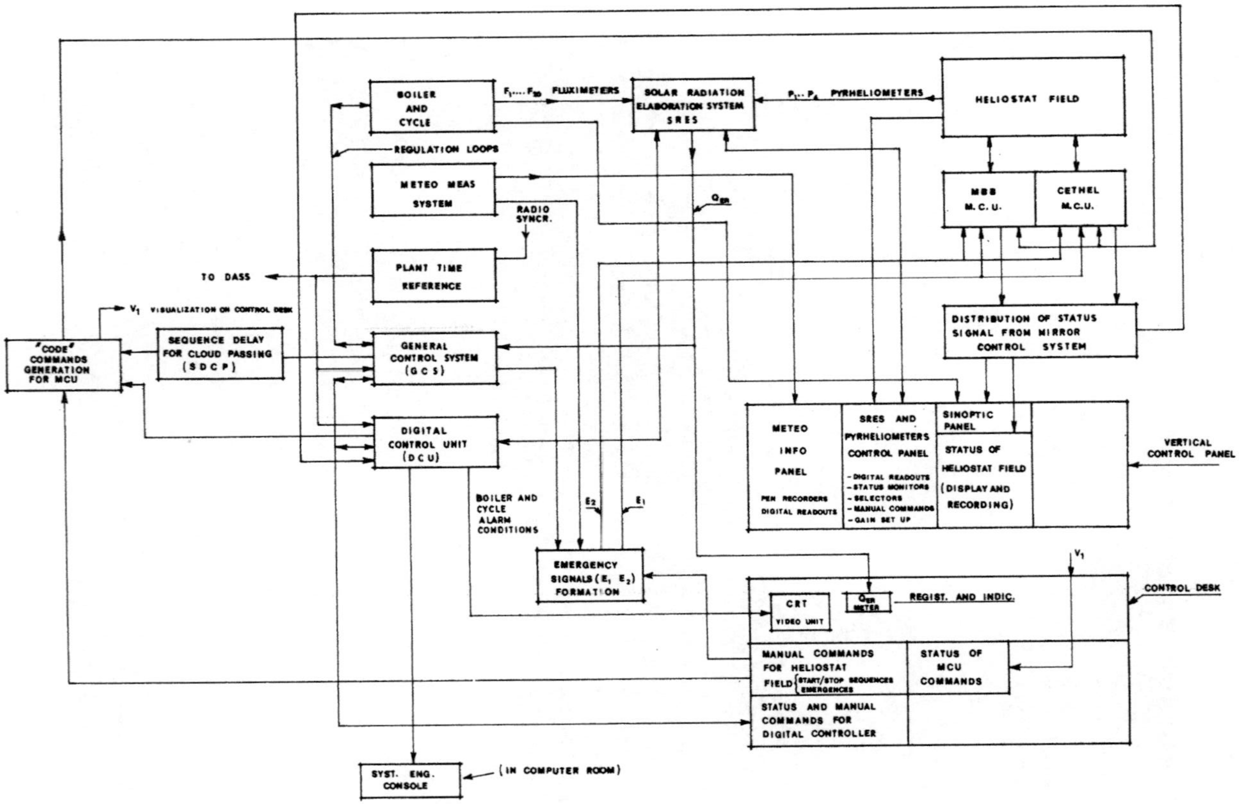

Figure 3.7.6 - MCC : Control & Supervision Block Diagram

Regulation commands come from the electronic module which detects a control system request for too low a feedwater flow rate, as happens if a cloud obscures the sun for more than 3 or 4 minutes.
If the sun returns before the deviation sequence has finished, the Code generates a "Stop sequence" command and allows the operator to restart the focussing sequence as soon as possible. The operator's commands come from push-buttons located on Control Desk.

The "distribution of MCU's Status" module receives the contact closures from the Mirror Control Units, giving the actual status of the mirror fields and (by a suitable set of relays) repeats and distributes them to the synoptic panel, for driving the lamps indicating the status of mirror groups, and to the event pen recorder for permanent storage of mirror group movements.

The General Control System (GCS) and the Digital Controller Unit (DCU) play fundamental roles as basic and more advanced control devices, respectively. Their functions are described in more detail in par. 3.7.2.2. Here it is sufficient to point out that they are connected to all other relevant systems and can be considered the "centroid" of the whole plant.
In particular, the E1 emergency signal is triggered by any of a set of boiler and steam cycle "danger" conditions.

The "Plant clock references" module:
. keeps the actual timing of the whole plant synchronised with the reference signals broadcast by the Italian network (RAI);
. drives the alfa numeric calendar and time displays located in the control and auxiliary rooms;
. distributes on a serial two-wire line (300 Baud, RS232 standard) a message containing actual time information and other relevant plant data of interest, in particular to the "DASS" system.

Vertical panels, control boards, CRTS' System Engineers Console, etc. implement the man-machine interface and allow supervision and control of the whole plant by the operator team. (Note that activation of the "DASS" system (foreseen for July 1982) will change in some way the role of control room instrumentation.) The instrumentation and the organisation of those panels is generally conventional. For some special functions like Qer formation, pyroheliometer control and monitoring, wind speed threshold, "CODE" control and monitoring and Mirror Control Unit status, special panels and monitor devices have been developed. Those devices are described in the following and in related documents.

C) Control console

This is in four sections, each housing the controls for one part of the plant:
i) Duplicates of the high level (group) heliostat controls. Once the heliostat controls are "ON", the MCC duplicates are sufficient for normal and emergency operation. The operator then needs to use the separate MBB and CETHEL control consoles only for non-routine activities.
ii) Controls for regulation of the receiver-feedwater flow rate and the attemperators/desuperheaters in the receiver.
iii) Controls for the thermal performance of the cycle and for the turbine.
iv) Controls for the auxiliaries such as the closed cycle cooling system,

compressors, vacuum pumps, and salt storage system.

The "central control" computers for the heliostat field and the control consoles used for experimental or maintenance operations with heliostats have already been described in Section 3.7.1. These consoles are alongside the MCC control console and can be reached very easily by the operator if necessary in case of non-standard (but not emergency) operation.

D) Vertical boards

The vertical boards are in six sections:

i) Meteorological signals.
ii) Controls and displays for the solar flux detectors (pyroheliometers) in the field.
iii) Controls, displays and alarms for the salt storage system (see Section 3.5.2) and for the alternator and its auxiliaries.
iv) A synoptic display of the status of the main components of the plant, together with alarm signals related to the receiver-feedwater, main steam, etc.
v) Temperature recording equipment for the receiver tubing and for the turbine, together with alarms for the turbine, the rest of the steam cycle and auxiliary machinery.
vi) Measuring, recording and alarm signals for the 200 kV and 5.5 kV power system including meters to record the energy produced by the alternator and that absorbed by the power plant's auxiliary systems.

3.7.2.2 Control concepts, overall description

The concept of the plant (extended) control system is described by the general scheme of Fig. 3.7.6. The role of the different components can be summarised as follows:

The Mirror Control Systems (MCS) supplied by the heliostat manufacturers ensure correct tracking of the mirror field and drive the heliostats along the prescribed trajectories. They supply the essential information about the field state both to the operators and to the automatic control devices. MCS commands can be issued both by the operator, through the mirror control consoles, and by the GCS (general control system), to which the MCS appears as a slave. The MCS consists of two centralised digital units (one of MBB and one of CETHEL) and of a distributed set of digital units, each unit devoted to the control of one heliostat. A more detailed description of MCS has been given earlier in this report.

The receiver and thermal cycle control system (RCS) consists of two separate parts: the receiver and thermal cycle regulation system (RRS) and the manoeuvre automation system (MAS). The RRS includes separate local regulators, of quite standard type, for the conventional components of the thermal cycle (turbine, condenser, flash-tank, thermal storage, etc.) and the integrated receiver regulation system, devoted to the control of pressure and temperatures in the receiver.

The essential functions of RRS components relevant to the receiver and steam cycle control are described in Fig. 3.7.7. In phase C, the RRS has been completely realised in analog technology. The MAS, though conceptu-

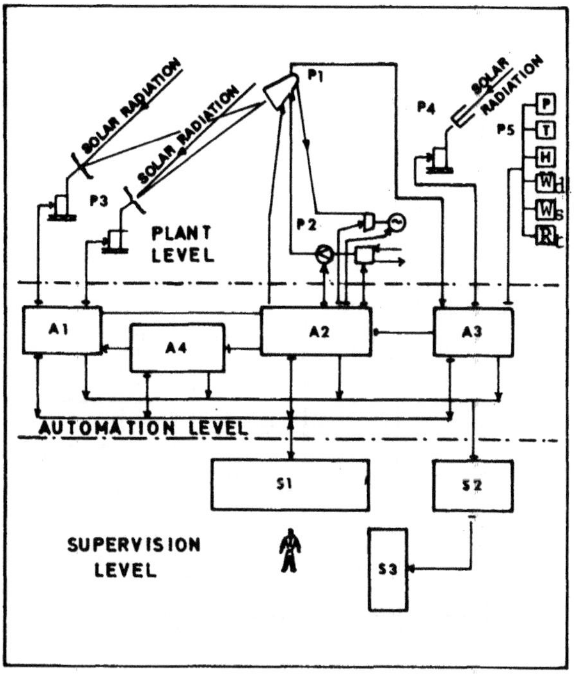

KEY

P1 Solar receiver, steam generator

P2 Thermal cycle

P3 Mirrors field

P4 Solar radiation measurements

P5 Meteo measurement devices

A1 Mirror control system

A2 Receiver and thermal cycle regulator

A3 Thermal flux and meteo measurement system

A4 Manoeuvres automation system

S1 Operator bench and vertical panels

S2 Supervision process computer

S3 Supervision consoles

W_d Wind direction

W_s Wind speed

R_t Total radiation

P Pressure

T Temperature

H Humidity

Figure 3.7.7 a

GENERAL CONTROL SYSTEM (GCS)

Structure

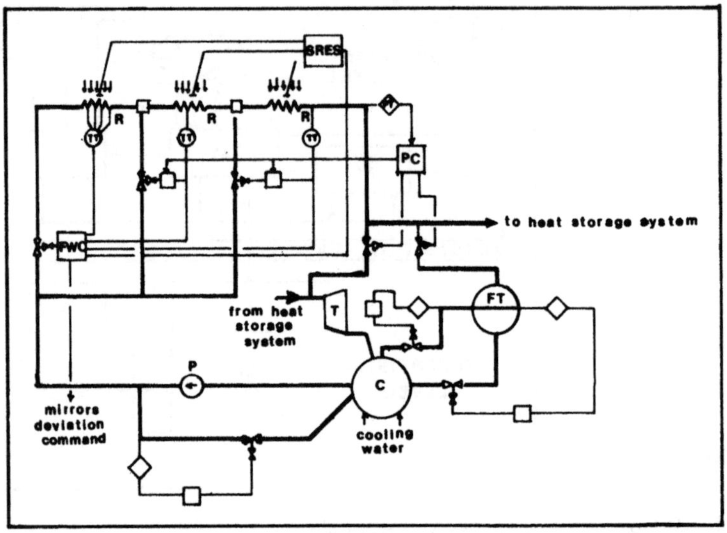

KEY

TT	Temperature Transducer
PT	Pressure Transducer
FWC	Feed Water Controller
ESC	Evaporator Spray Controller
SSC	Superheater Spray Controller
SRES	Solar Radiation Elaboration System
T	Turbine
C	Condensor
P	Feedwater Pump
R	Receiver
FT	Flash Tank

Figure 3.7.7 b

ally defined in Phase B, is to be implemented after Phase C but its interactions (as in Fig. 3. 7. 6) have been taken into account in the design of the RRS which includes the interfaces necessary for correct cooperation between the two control subsystems.

Special and peculiar features of the RCS are the interactions with the mirror field and with the thermal flux measurement system. These have been conceived to allow the RCS to command the MCS (e. g. for sequential and emergency mirror deviation) and to receive a continuously updated estimate of the total thermal power entering the receiver (from the SRES). Details on these special devices and their operation are given in sections 3. 7. 2. 3 and 3. 7. 2. 5.

A supervision process computer (SPC), and supervision consoles (SC), will be devoted to overall plant supervision, i. e. to plant data acquisition, monitoring and displaying, plant performance evaluation, operator alarming of plant malfunctions, plant data storing for post-mortem analysis, etc. Installation is planned for Phase D of the project.

The basic concepts of the RRS can be summarised as follows:

- Receiver outlet pressure is controlled, either by means of the turbine valve during normal operation or by means of the by pass valve during the plant start-up and shut-down, to avoid unwanted motion of the evaporating zone in the receiver.
- Receiver outlet temperature is controlled by modulating the feedwater mass flow rate in both receiver tubes and the attemperator spray valve positions, based on fluid temperature measurements at intermediate points in the receiver to anticipate control actions. As usual, feedwater flow is used for low-frequency regulation whereas spray flow is used for higher-frequency control.
- Receiver control in the event of large solar radiation disturbances (e. g. cloud passing) relies essentially on the feedforward control action obtained from the thermal power estimate computed by the SRES device.
- The feedwater control system (FWS) allows the temperature control system to be independent of the feedwater valve characteristics and of the pressure drop variations. In particular, to speed up the feedwater control, FWC has been designed to interact with the pump header pressure control.

An extensive and detailed description of RCS concepts and algorithms was described in March 1980 in a report by ENEL.

The RRS was tested at the manufacturer's facilities using a digital simulation of the plant. The structure of the simulation and the complete testing procedure have been documented by ENEL and the manufacturer and their essential features published.

3. 7. 2. 3 Special control features

Two aspects of the control system design are very peculiar to this application. The first is related to the need to deal with very long transport delays, which affect the receiver dynamics. This calls for the use of special feedback control actions which cannot be realised by traditional PID regulators. Therefore, the Manoeuvres Automation System (MAS), conceived in Phase B but not included in Phase C of the project, is intended to implement not only sequential automation of basic operating procedures but also some special digital control algorithms which are superimposed on the analog regulation system through suitable interfaces. In fact, the MAS is to be realised by Digital Control Unit (DCU) and the analog regulation has been designed to be compatible with the DCU outputs and inputs. The concept and performance of the digital control system and the results of simulations have also been reported in detail by ENEL.

The second special feature of the RCS is represented by its role as master control for the MCS. In this context, the RCS commands mirror sequence deviations when the thermal power to the receiver is not sufficient to keep the steam generator above the technical minimum. Since those events are normally due to cloud, which can pass more quickly than the several minutes required for depressurising the steam generator, the sequence deviation is commanded only if thermal power is lacking for a sufficiently long time.

This concept has been incorporated in the control system by means of two different devices connected in series:

- a "code device" (CD) which translates the simple logic command of the RCS to an MCS-compatible form by a suitable protocol;
- a "sequence delay device" (SDD) which transfers the actual deviation command to the MCS only if a thermal power lack is recognised by the RCS for a sufficiently long time interval (e.g. 5 min.).

The choice of the technical minimum, of the detection criterion for loss of power and of the sequence delay value have been based on receiver safety arguments and suitable dynamic simulations using the ENEL mathematical model.

3.7.2.4 Plant operation: manual -v- automatic options

The plant control status at the end of Phase C is essentially as follows:

- The receiver regulation system is kept in complete automatic operation after the start-up of the plant. Only receiver pressure regulation, pump header pressure regulation and feedwater control system are necessary and effective during the start-up. The temperature regulations can be put in automatic regulation only when the steam temperature at the receiver outlet is above about 400°C.
- The plant manoeuvres during start-up are performed manually by the operator at the control board except for a limited number of operations which should be done directly on the plant (seal steam system start-up, arming of the turbine and first opening of the turbine control valve, pre-excitation of the alternator, closing of the flash-tank regulating valves).
- Manual activation of the salt storage charging and of the steam storage filling.
- Possibility of commanding the mirror field from the main control board through the Code Device (CD). This enables the operator to activate standard sequence deviations of the mirror field without operating on the mirror control board.

A thorough description of the operating procedures, including specification of automatic and manual controls, can be found in the provisional operating manual (in English) compiled before the first plant start-up and in the updated operating manual (in Italian) which was developed from the earlier one after three months' operating experience. A summary is given in Chapter 4 of this report.

As a general remark, it can be said that the present semi-automatic operation of the plant during start-up is not a very significant drawback because the very slow dynamics of the solar receiver allow the operator adequate time for all the required manoeuvres. Full automation of the start-up manoeuvres would imply some reduction in operating personnel but little saving of time (perhaps 20 minutes).

3.7.2.5 Special control devices

The meteo measurement system consists of four pyrheliometers (P1... P4) installed on steel pedestals 5 meters high around the mirror field. Their operation and protection against bad weather by a coverage is controlled from the control room. Two wind speed measurements are taken from a mast at 10 and 50 meters height above ground. Besides the registration via pen recorders the device provides two logic signals which indicate that the adjustable threshold T1 and T2, corresponding to wind speed preset limits, have been overcome. Under automatic operation conditions (S1 on AUTO) the alarm results in heliostats going to stowage position as required for safety of equipment. Under manual conditions, the operator can over-ride under his own responsibility.

Elaboration of Qer by the Solar Radiation Elaboration System "SRES" is illustrated in Fig. 3.7.8. It calculates a sort of weighted sum of 20 individual fluximeters, taking into account the nominal flux values and giving the output signal as a percentile of the nominal value. It provides a high degree or redundancy because many fluximeters can be put out of service without affecting the value of the output signal. It calculates also a Qer estimation from pyrheliometer signals, which can be used for comparison and/or monitoring purposes, or, in case of total failure of fluximeters, as control signal for water flow rate regulation. It provides some manual and automatic monitoring possibilities which allow the system engineer to visualise the intermediate values of Qer's computation and which allow the DASS to digitally interrogate the "SRES" for having back the fluximeters values and the final computed Qer value.

Flux monitoring is achieved via a manual selector, mounted on vertical panel under the "SRES" device, which allows the operator to visualise on a 3 digit numerical display the actual value of each fluximeter output, up to a maximum of 40 independent signals. Up to now we have monitored by this device the 20 fluximeters also giving input to the "SRES" apparatus, in order to check the "SRES" input and to find wrong fluximeters. Because the above mentioned point by point manual monitoring does not give a synthetic view of a whole fluximeters set, we developed a special instrument (called "flux mapper device" (FMD)), which:

- accepts up to 48 fluximeter signals;
- converts them in logical pulses, whose duration is proportional to actual analog value;

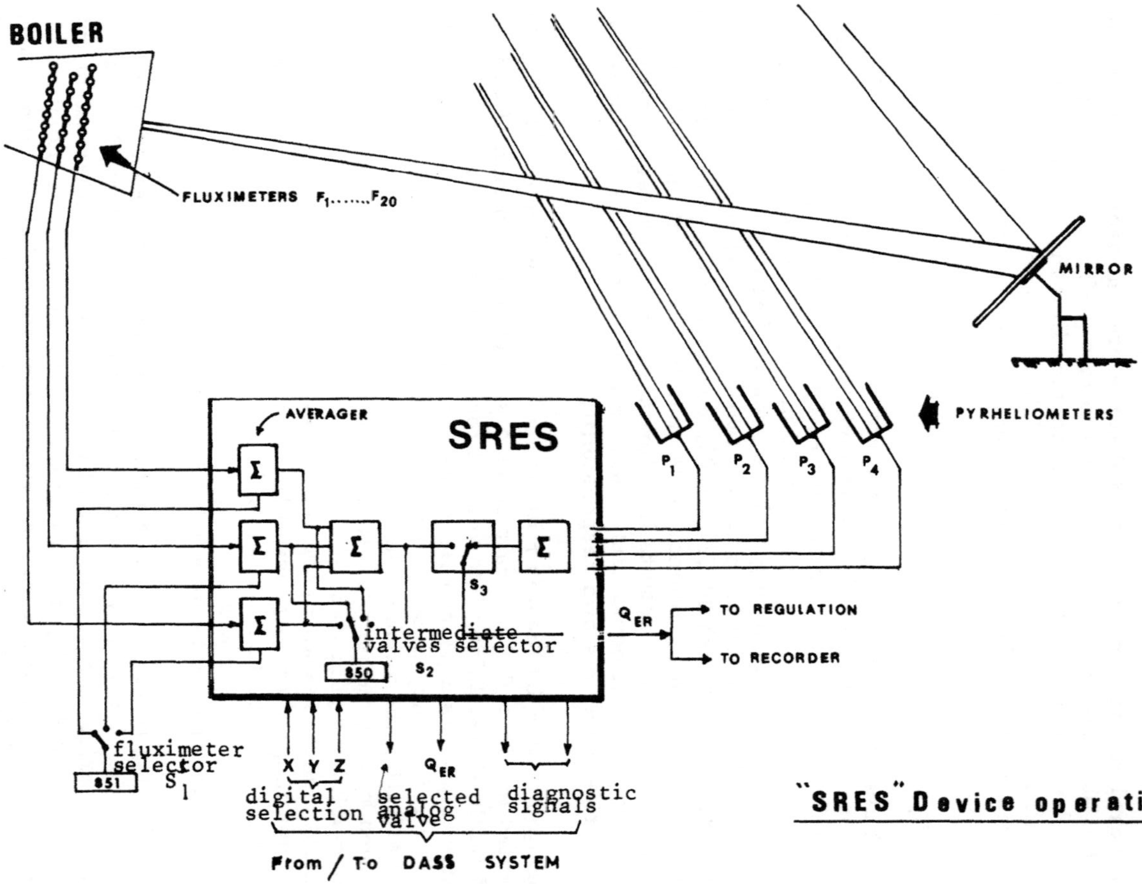

"SRES" Device operation

Figure 3.7.8

- presents them as a vertical "bar graph" diagram on a multipen event recorder;
- repeats automatically this kind of measure with a pre-defined time interval, adjustable from few minutes up to 6 hours; it drives also the paper advance and pen action of pen recorder.

The output diagram gives an immediate insight of actual flux distribution: furthermore, comparing a time series of those diagrams (for instance, during morning start-up or in case of cloud passing), people can gain a suitable knowledge for better understanding the dynamical behaviour of the receiver and related plant components.

3.8 CIVIL WORKS

3.8.1 Site selection

The site, selected by ENEL after a search covering different places in Southern Italy, is located in Contrasto, in the Commune of Adrano, Sicily, about 40 km north-west of Catania. The reasons for choosing this site are:

- good exposure
- easy access
- easy connection to the national grid
- availability of cooling water
- proximity to an existing ENEL hydroelectric station
- proximity to an international airport.

The overall area available is about 310,000 m^2 with a mean elevation of 220 m above sea level and a slight natural inclination north-east to south-west. The plant borders on regional road no. 16 (SS 575) to which it is directly connected, its position being shown in Fig. 3.8.1.

3.8.2 Survey and site preparation works

A soil investigation campaign performed in 1978 disclosed a basalt stratum beneath a subsoil and topsoil formation varying in thickness between zero and about 4 m. The subsoil consisted of sand and clay silt containing small rock fragments and was topped by a layer of humus. The land profile was quite irregular but was levelled off by removing all the topsoil (approximately 70 cm) and part of the subsoil. Final grading of the area where the mirror field stands was to about 5% north-south slope and 2% east-west whilst the buildings now stand on a flat area 214 m above the sea level.

After ENEL completed the acquisition of the site (15/10/78) the "Site preparation works" started with preparation of:

- access roads to the site
- provisional works and facilities area
- provisional works barracks (offices, canteen, etc.)
- provisional works connection to electricity and water networks.

An automatic meteorological station was installed at the site and started operation in January 1978.

Figure 3.8.1 - Location of EURELIOS

3.8.3 Description of works

The main civil works performed, outlined in Fig. 3.8.2, are the following:

- Mirror field. Ground preparation, mirror foundations, roads, draining system and fencing;
- Tower. Tower foundation, with an area of 100 m^2;
- Main building. The main building is composed of two rectangular blocks connected by a short corridor. Both buildings have steel frames, external tile curtain walls and metal deck roofing appropriately insulated and water-proofed. Offices, meeting room and a canteen are in the smaller, 250 m^2 block. After the corridor, the second block contains the machine room, battery room, air conditioning and boiler rooms, control and equipment rooms. This area is about 753 m^2 and height is between 4.70 and 9.50 m. In the machine room a travelling bridge crane is available to ease the handling of the heavier items of equipment. Adjoining the machine room on the south side, three external cells contain circuit breakers and two small transformers;
- Warehouse and workshop building. This building also has a rectangular plan and is constructed with steel frames, external tile curtain walls and roofing similar to that of thc main building. It contains the electrical and mechanical workshops and the warehouse and is also served by a travelling bridge crane. The total floor area for warehouse and workshops is about 480 m^2;
- Cooling water basin and pumps cabin. Near the buildings is a group of water-related tanks, etc., including an open reinforced concrete tank with a base area of about 30 m^2 and volume about 80 m^3 (for used cooling water), tanks to separate oil contamination from waste water, a fibre glass tank for demineralised water for the steam cycle and a small building containing water discharge pumps;
- Finishing. Roads, car parking and other open storage areas are asphalted. The heliostat field, between access roads, is finished with large chippings and the remaining areas have been appropriately landscaped and planted. The entire drainage system (buildings, roads, etc. mirror field and other sources of waste water, after treatment) feeds into a common drain for discharge to the Simeto river, nearby. A photograph showing a general view of the plant during testing is reproduced in Fig. 3.8.3.

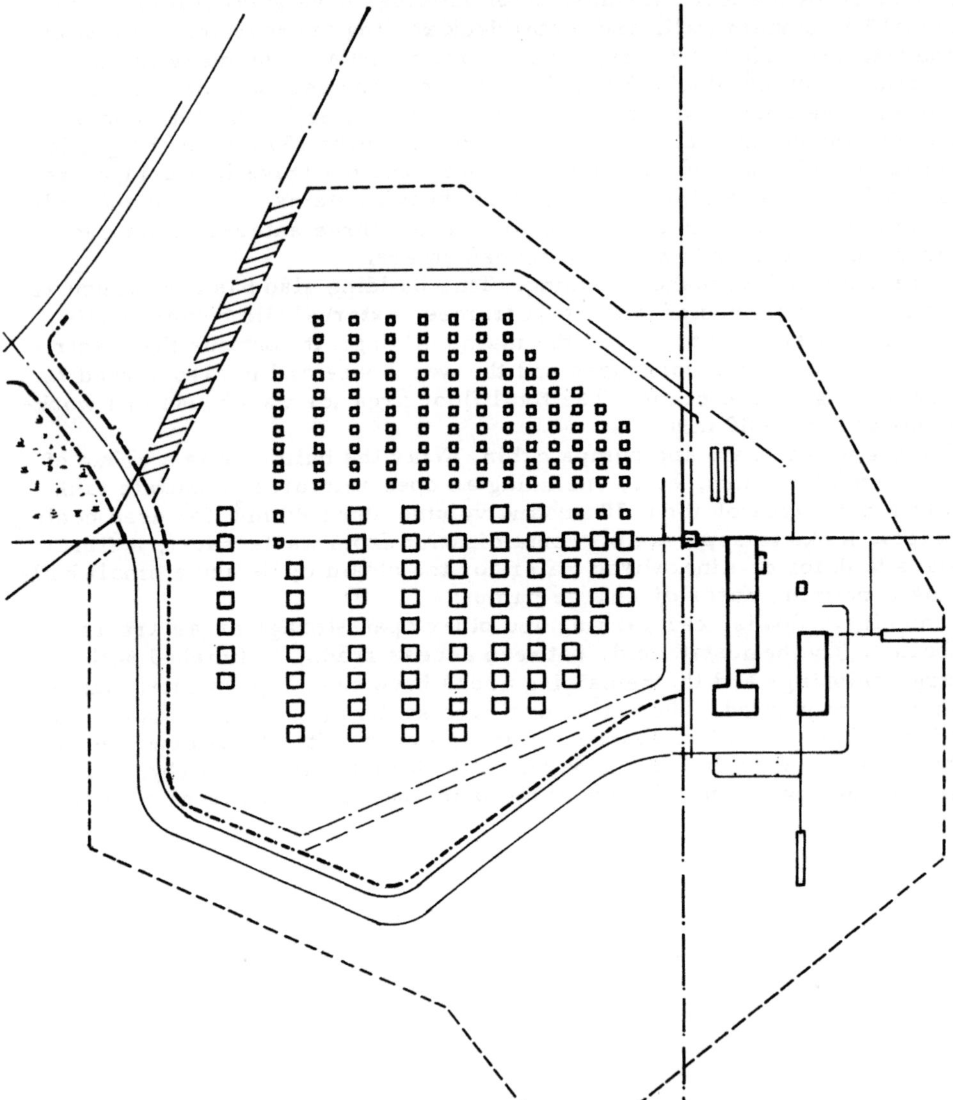

Figure 3.8.2 - General Lay-out

Figure 3.8.3 - General View during Testing

4. Operation of the Plant

4.1 INTRODUCTION

Detailed operation manuals have been produced by ENEL and ANSALDO and one important result of the future use of the plant will be the identification of amendments to the operation procedures, amendments which could improve the efficiency of such plants both thermoelectrically and in terms of manpower utilisation. This chapter, however, attempts to summarise the general operational principles and options as they appear in the Operation Manual as it now is.

4.2 OPERATING PROCEDURES

4.2.1 Morning start-up

Set-up the emergency feedwater system in start-up position and verify the correct positions of all relevant valves.
After verifying that the system is full of water and that the circulating pumps are in working order, start-up the cycle, pressurising the receiver and allowing water to flow through the receiver and the turbine bypass.
After flow is established and feedwater and receiver regulators and the emergency feedwater system have been put in automatic, the heliostats can be moved into the receiver (group by group at one minute intervals).
After about a half-hour of illumination, steam is produced and sent to the flash tank, the normal pressure of which is about 5 bars.
After a further half-hour, as the temperature at the receiver outlet passes 340°C, the turbine main header inlet can start being preheated. This takes about 20 minutes, after which the turbine is started and its speed increased under operator control at a standard rate of 200 rpm per minute until the synchronous speed of 8195 rpm is reached, at which point the alternator can be coupled to the grid. The total time from the pointing of the mirrors until the parallel is typically about two hours.
Once in parallel and with a receiver outlet temperature of about 470°C the plant can be put entirely under automatic control using a pressure set point appropriate to the time of year. The generator load is thereby adjusted to follow the fluctuations of solar flux.

4.2.2 Cold start-up with full radiation

The routine is virtually the same as for a morning start-up until the point at which the heliostats would have been moved from "standby" to "receiver". The required feedwater rate is first determined and set as a function of the observed and forecast pyroheliometer readings in the field and the helio-

stats are then moved one group every five minutes into the "receiver" position. The demanded feedwater flow rate is then adjusted according to the actual receiver fluxmeter measurements.
The procedure then reverts to that for morning start-up.

4.2.3 Hot start-up

This differs in concept from that of cold start-up in that, although the receiver is not initially illuminated, it is still pressurised with steam at at least 20 bar. The turbine may be shut-down or running on steam from the salt-storage system, the steam cycle pressure regulators generally having reverted to manual control.
Based on the observed insolation, appropriate setpoints are established and the bypass valve partially opened.
The mirror sequence "standby-to receiver" is initiated and the bypass valve put to automatic.
The receiver pressure is allowed to increase at about 1 bar/minute.
When the steam temperature is high enough, either

a) as soon as the steam pressure at the turbine inlet reaches 18 kg/cm^2 the automatic closure of salt storage steam valve is executed,
b) if the turbine is to be restarted, the hot start procedure is followed.

When the steam temperature reaches 470°C the whole system can be left in automatic mode.

4.2.4 Normal operation

After start-up the plant is automatically operated and the produced power is related to the daily insolation diagram. However, to maximise efficiency the receiver pressure set point may be adjusted periodically to keep the turbine inlet pressure valve about 70% open. Also it is possible, if necessary, to recharge the steam accumulator during the initial stage of normal operation.

4.2.5 Operation with cloud passing

When solar radiation drops below about 20% of nominal value for more than three minutes, the heliostat deviation sequence starts (receiver-to-standby) and if charged, the salt storage system is automatically put into operation. If the lack of sunlight is short enough for the heliostats not to reach standby, the plant will continue operating without operator involvement other than to send back to the "receiver" position any heliostat group which may have reached standby. The salt storage system will then not provide steam.
Steam starts to flow from the storage system as soon as the pressure to the turbine drops below the set point of the regulator, which should therefore be set as low as possible, commensurate with continued connection to the grid (typically 8 bar). Connection can then be maintained as long as sufficient steam is available from either source.
If all heliostat groups reach standby the operator prepares the plant, in manual mode, for the "hot start-up" procedure (Section 4.2.3).

4.2.6 Evening shut-down

Inhibit intervention of the salt storage system.
When the insolation drops below about 20% of nominal, the heliostat's

"receiver-to-standby" sequence is started automatically as above.
Put pressure regulation and receiver attemperator controls to manual and decrease the turbine load (by decreasing speed demand) as much as possible without breaking the coupling to the grid.
Decouple and reduce speed to a minimum.
Isolate the receiver and switch to"star t-up" position to inhibit operation of the Emergency Feedwater System.
Stop the feedwater and booster pumps and put the entire system to "manual".

4.2.7 Emergency operations

Immediate emergency reaction to loss of pressure from the feedwater pumps is automatic and causes intervention of the emergency feedwater system. If this loss is due to failure of the auxiliary electrical supply only battery-operated functions will be available (CETHEL heliostat deviation and those items powered from ANSALDO's batteries, section 3.6.3) plus those which can be carried out with the residual service air. Otherwise normal services will continue to be available.
The mirrors are commanded out of the receiver, the receiver bottling valve is closed and its pressure regulated by the relief valve, N11-F300. The regulation system continues to operate while service air is available, particularly to operate the feedwater regulating valves.
The operator:

- verifies that the emergency feedwater system operates correctly;
- verifies that the measured thermal input, Qer, to the receiver reduces correctly;
- when Qer is zero the operator closes the valves to avoid excessive thermal shock to the receiver.

The plant can then be normally shut-down.

5. Phase D

The contract for Phases B and C envisaged subsequent optimisation and improvement. A new two-year phase, Phase D, officially commenced on 5 January 1981.
Phase D is in two parts. The objective of Part I is to bring the plant up to full power and to establish a detailed work programme, DWP, for Part II.

A proposed DWP (dated June 14th, 1981) has, at the time of writing, been submitted by the Consortium. The main items of interest in the research programme planned for Part II, following the scheduled achievement of maximum power by the end of August, are:
1) plant dynamic behaviour and control system;
2) instrumentation and automation;
3) flux measurements in the receiver;
4) study of the mirror field behaviour;
5) analysis of the plant performance and optimisation of plant operation;
6) systems evaluation and final report.

These tasks are to be carried out by ENEL, CETHEL and MBB with assistance from Ansaldo should operational difficulties or other problems appear in those parts of the plant which Ansaldo supplied. Ansaldo, of course, still has its responsibility under the Phase C contract to install the Francia pyrex anti-radiation structure, deliberately delayed to allow quantitative experimental experience to be gained of the effects of such a structure on the performance of the receiver.

In addition to the work under the Consortium's Phase D contract, agreement has been reached between the Consortium, the Commission and Belgium for the latter to supply and install, by mid-1982, a Data Acquisition and Supervision System (DASS). A contract is being finalised between the Commission and Tractionel/Belgonucléaire. DASS will provide automatic and more comprehensive records of the operating plant than are at present available. GTS again has a direct contract from the Commission to provide advisory services in Phase D, particularly with respect to DASS, as it did in Phases B and C.

As for the future, provision was made in the layout of the heliostat field for slight enlargement. The receiver, turbine, generator, etc., were designed to accept the corresponding increase of power.

The possibility of some measure of integration between EURELIOS and proposed photovoltaic plant in an adjacent area has also been mooted - if only by co-ordination through an extended DASS.

Although other thermal cycles are not planned for the foreseeable future, the tower is compatible with the installation of different thermal cycles, making use of the existing heliostat field and other infrastructure of the plant. This type of development would, of course, require completely new negotiations with ENEL which would by that time presumably be full owners of the plant and using it as a routine, albeit small, contributor to the Sicilian electrical grid.

GPSR Compliance

The European Union's (EU) General Product Safety Regulation (GPSR) is a set of rules that requires consumer products to be safe and our obligations to ensure this.

If you have any concerns about our products, you can contact us on ProductSafety@springernature.com

In case Publisher is established outside the EU, the EU authorized representative is:

Springer Nature Customer Service Center GmbH
Europaplatz 3
69115 Heidelberg, Germany

Batch number: 10755292

Printed by Printforce, the Netherlands

Thermo-Mechanical Solar Power Plants

Solar Energy R&D in the European Community

Series B:

Thermo-Mechanical Solar Power Plants

Volume 2

Solar Energy R&D
in the European Community

Series B Volume 2

Thermo-Mechanical Solar Power Plants

Proceedings of the Second International Workshop on the Design, Construction and Operation of Solar Central Receiver Projects, Varese, Italy, 4-8 June, 1984

edited by

J. GRETZ
Commission of the European Communities, Ispra, Italy

A.S. STRUB
Commission of the European Communities, Brussels, Belgium

and

A. SKINROOD
Sandia National Laboratories, Livermore, California, U.S.A.

D. Reidel Publishing Company
A MEMBER OF THE KLUWER ACADEMIC PUBLISHERS GROUP

Dordrecht / Boston / Lancaster

for the Commission of the European Communities

Library of Congress Cataloging in Publication Data

CIP

International Workshop on the Design, Construction, and Operation of Solar Central Receiver Projects (2nd: 1984: Varese, Italy)
Proceedings of the Second International Workshop on the Design, Construction, and Operation of Solar Central Receiver Projects, Varese (Italy), 4–8 June 1984.

(Solar Energy R&D in the European Community. Series B, Thermo-mechanical solar power plants; v. 2)
Includes index.
1. Solar power plants–Congresses. I. Gretz, J. II. Strub, A. III. Skinrood, A. IV. Commission of the European Communities. V. Title. VI. Series: Solar energy R&D in the European Community. Series B, Thermo-mechanical solar power plants; v. 2.
TK1056.I59 1984 621.31'244 85–8356
ISBN 90–277–2049–5

Organization of the Workshop by
Commission of the European Communities
Directorate-General Science, Research and Development, Brussels

Publication arrangements by
Commission of the European Communities
Directorate-General Information Market and Innovation, Luxembourg

EUR 9766

Published by D. Reidel Publishing Company
P.O. Box 17, 3300 AA Dordrecht, Holland

Sold and distributed in the U.S.A. and Canada
by Kluwer Academic Publishers,
190 Old Derby Street, Hingham, MA 02043, U.S.A.

In all other countries, sold and distributed
by Kluwer Academic Publishers Group,
P.O. Box 322, 3300 AH Dordrecht, Holland

PREFACE

In this book, the Commission of the European Communities presents the proceedings of the Workshop on Solar Central Receiver Projects, held in Varese, Italy, in June 1984. This Workshop was supported by all operators of solar tower power plants around the world and, as a result, these proceedings provide a comprehensive overview of the technology in its current state of development.

The Workshop was organized by the Commission of the European Communities in the frame of the second solar energy R&D programme under the responsibility of its Directorate-General (XII) for Science, Research and Development in Brussels. The meeting place, Varese, in Italy, was selected because of its neighbourhood to the Ispra Establishment of the Commission's Joint Research Centre who cooperated in the organization of the Workshop.

Solar power plants of the central receiving type have two conflicting characteristics: they employ very simple and classical components but as a system they are of tremendous complexity. It was the hope for rapid progress by using available components that guided the decisions taken in the late seventies to build six large experimental plants: four in Europe, one in Japan and one in the United States. At that time, this technology enjoyed high priority in solar energy R&D around the world. Once the plants were completed, however, it became clear that the technical complexity combined with difficult meteorological conditions at most construction sites made the yields less favourable than anticipated. As a consequence the economic outlook became questionable, in particular for the smaller plants. The larger plants, on the other hand, still have the drawback that the required funding for one single plant is so high that it is difficult to raise from the capital market.

As a sponsor of the central receiver plant EURELIOS, the Commission of the European Communities took the initiative for this workshop in order to review the state of the art and enlarge the basis on which to build its own strategy. Before making further investments on EURELIOS, it was felt that a comprehensive appraisal of the technology was necessary taking into account the experience gained with other plants of this type.

Eventually, the Commission was very satisfied with the outcome of the Workshop. The experience put forward by the plant operators shows that the problems encountered are inherent to the technology in general and not to a particular plant. It is important to note that the engineering work was excellent for all plants and a comprehensive record of experience has been collected. This is true to the extent that most of the necessary knowledge to build such plants is now available and well recorded for the case where economic conditions may become more favourable.

It was remarkable to see the enthusiasm of all the people attending the workshop and the spirit of international cooperation. The success of the workshop is also due to the effort of its Chairman, Dr. Gretz, to whom I would like to express here my sincere appreciation.

Dr. W. Palz
Head of the Commission's
Solar Energy R&D Programme

C O N T E N T S

Preface v

OPENING SESSION

Workshop introduction
J. GRETZ, Commission of the European Communities, Joint Research Centre, Ispra, Italy 3

Welcome speech
G.R. BISHOP, Commission of the European Communities, Director of the Joint Research Centre, Ispra, Italy 4

Keynote speech
A. STRUB, Commission of the European Communities, Directorate General for Science, Research and Development, Brussels, Belgium 5

SESSION I - PART 1 : PRESENTATION ON THE GENERAL EXPERIENCE OF CENTRAL RECEIVER POWER PLANTS

Summary of the Session by the Rapporteur
F. PHARABOD, THEMIS, CNRS, France 8

Technical aspects of the EURELIOS plant operation
C. CORVI, ENEL-DSR, Roma; G. DINELLI, ENEL-CRTN, Pisa, Italy 10

Central receiver solar power plant of the IEA/SSPS project - Summary of results and experiences after three years of testing
W. GRASSE and F. RUIZ, DFVLR and Sevillana 21

10 MWe solar thermal central receiver pilot plant overview - Part I
A. SKINROOD, Sandia National Laboratories, Livermore, USA 35

SESSION I - PART 2 : PRESENTATION ON THE GENERAL EXPERIENCE OF CENTRAL RECEIVER POWER PLANTS

Summary of the Session by the Rapporteur
L. VANT-HULL, University of Houston, USA 48

A 1 MWe solar thermal electric power pilot plant (Sunshine project)
T. MUKAI, Agency of Industrial Science and Technology, MITI; T. HORIGOME, N. IKEDA and T. SAKAMOTO, Solar Energy Department, New Energy Development Organization, Japan 52

Themis plant operation progress report
V. MASCLE, EDF; F. PHARABOD, AFME and B. RIVOIRE, CNRS, France 62

CESA-1 project status report
J. AVELLANER, C. ORTIZ, F. MARTINEZ and F. SANCHEZ, Ministerio de Industria y Energia, Spain 73

SESSION II - PART 1 : HELIOSTAT FIELD

Summary of the Session by the Rapporteur
B. BONDUELLE, CNRS, Groupe d'Evaluation Scientifique Themis, France 84

Conceptional design of solar power plant with central receiver tower based on improved heliostats
H. YOSHIKAWA, Tokai University; N. IKEDA, New Energy Development Organization, Japan 86

SSPS-CRS Heliostat performance/history
P. WATTIEZ and J. RAMOS, ITET and Sevillana, IEA/SSPS Project 92

Heliostat field and receiver efficiency measurement on the IEA - 500 KWe - CRS at Almeria
W. SCHIEL, Institute for Technical Physics, DFVLR, Stuttgart, Federal Republic of Germany 100

Themis heliostat field
J.J. BEZIAN and B. BONDUELLE (CNRS), Themis Scientific Valuation Team, France 106

ASINEL heliostats for the GAST technological programme
J. DE MARCOS and L. CRESPO, Asociacion de investigacion industrial electrica, Madrid, Spain 112

CESA-1 heliostat field - evaluation status report
F. SANCHEZ, INITEC, Spain 120

10 MWe solar thermal central receiver pilot plant - heliostat evaluation
C.L. MAVIS, Sandia National Laboratories, Livermore, USA 124

SESSION II - PART 2 : RECEIVER

Summary of the Session by the Rapporteur
M.R. GENIER, EDF, Chatou, France 130

Operating experience and modifications in the solar receiver of the Eurelios plant
F. AIELLO, B. BELLAGAMBA and G. DINELLI, ENEL, Centro Ricerca Termica e Nucleare, Pisa, Italy 133

The SSPS advanced sodium receiver : transient response
R. CARMONA and J.G. MARTIN, IEA/SSPS ITET 140

Preliminary results on the performance of the Sulzer cavity receiver and the Franco-Tosi external receiver
C.S. SELVAGE and H. JACOBS, IEA/SSPS ITET 149

Relevant aspects in the design and construction of the advanced sodium receiver ASR for the IEA-SSPS central receiver system plant (Almeria - Spain)
P. CAVALLERI, V. BEDOGNI and A. DI MEGLIO, Research and Development Division, Franco Tosi Industriale SpA, Legnano, Italy 155

The Themis plant : start-up of the receiver - first results
B. RIVOIRE, CNRS-GEST, France 162

Stability of salt and corrosion resistance of circuit materials in the Themis power plant
P. SPITERI, Département Etude des Matériaux, EDF Les Renardières, Moret-sur-Loing, France 169

An originally scaled test unit of a modular gas cooled receiver of 50 MW thermal power
P. AGATONOVIC, H. FUHRMANN and D. ROOS, MAN Neue Technologie, Federal Republic of Germany 175

Ceramic as a material for large receiver constructions
H. GEHRKE, Dornier System GmbH, Friedrichshafen, Federal Republic of Germany 181

SESSION II - PART 3 : CYCLE, SYSTEM, CONTROL

Summary of the Session by the Rapporteur
R.L. GERVAIS, McDonnell Douglas, Huntington Beach, USA 190

Evaluation of sodium and steam system design for the solar power plant with central receiver tower
H. YOSHIKAWA, Tokai University; N. IKEDA, New Energy Development Organization, Japan 194

The advanced sodium receiver (ASR) for the IEA/SSPS central tower plant : operative conditions, control system design and performances
A. DE BENEDETTI and C. SALA, AGIP SpA, Milano, Italy 200

The front end processor of Eurelios supervision systems
M. MAINI, G. PANDINI, U. ZONI, ENEL - Italian Electricity Board, Research and Development Dept. - Automation and Computing Research Centre, Italy 209

The concept of the integrated design of process, control and operation in solar central receiver plants
C. MAFFEZZONI, ENEL - Italian Electricity Board, Research and Development Dept. - Automation and Computing Research Centre, Italy 215

CESA-1 - Control, system and cycle - Operation and status report
R. BALANZA, J. ROMAN, University of Sevilla, Spain 221

SESSION II - PART 4 : OPERATION, MAINTENANCE, STAFFING

Summary of the Session by the Rapporteur
L. CIVITANO, ENEL-CRTN, Pisa, Italy 228

Operation and performance status of the Eurelios power plant
A. CHILLE', L. CIVITANO, G. DINELLI, ENEL - Centro Ricerca Termica e Nucleare, Pisa; P. PERRONE, ENEL - Compartimento di Palermo, Centrale Eurelios, Adrano, Italy 231

CESA-1 staffing, operation and maintenance status report
F. MARTINEZ CORCOLES, CIA. Sevillana de Electricidad, S.A. 238

SESSION II - PART 5 : ECONOMICS

Summary of the Session by the Rapporteur
G. FANINGER, ASSA, Vienna, Austria 248

Solar radiation measurements in the Alps
W. DURISCH, P. KESSELRING, D. BARANDAO and M. BUEHLMANN, Swiss Federal Institute for Reactor Research, Würenlingen, Switzerland 253

Economic modelling of molten salt solar central receiver plants
T. DURAND, S. IBRAHIMI, Duc PHAM HI, Laboratoire LEIS, Ecole Centrale de Paris; C. ETIEVANT, Mission Maîtrise de l'Energie, Ecole Centrale de Paris, France 263

Comparison of solar thermal and photovoltaic electricity generation using experimental data from the IEA SSPS project
P. TOGGWEILER and R. MINDER, Electrowatt Engineering Services Ltd., Zurich, Switzerland 275

Development of an economic assessment model
H. KLAISS and M. GEYER, Institute for Technical Physics, DFVLR, Stuttgart, Federal Republic of Germany 281

10 MWe solar thermal central receiver pilot plant total capital cost
H.F. NORRIS, Sandia National Laboratories, Livermore, USA 293

Analysis of Themis experimental plant costs
C. ETIEVANT, Ecole Centrale de Paris, France 301

Central receiver costs for electric power generation
J.A. DIRKS, Battelle Pacific Northwest Laboratory, Sandia National Laboratories, Livermore, USA 307

SESSION III - LESSONS LEARNED AND ASSESSMENT OF THE TECHNOLOGY "HOW WOULD WE DESIGN TODAY A COMMERCIAL CRS SOLAR POWER PLANT?"

Summary of the Session by the Rapporteur
A.C. SKINROOD, Sandia National Laboratories, Livermore, USA 314

Lessons learned : considerations for new design
C.S. SELVAGE, IEA/SSPS - ITET, Almeria, Spain 317

SESSION IV - BASIC ASPECTS ON SOLAR CHEMISTRY AND STORAGE

Summary of the Session by the Rapporteur
M. FISCHER, Institut für Technische Physik, DFVLR, Stuttgart, Federal Republic of Germany 324

Some basic aspects on solar chemistry and storage
P. KESSELRING, Swiss Federal Institute for Reactor Research, Würenlingen, Switzerland 327

Solar central receiver costs for electric power generation
J.B. WRIGHT, Sandia National Laboratories, Livermore, USA 334

Chemical storage of solar energy and solar fuels
H. EICHER, T. WEISSKOPF, R. SCHLÖGEL and H.J. GÜNTHERODT, Institute of Physics, University of Basel, Switzerland 339

Application of solar energy for chemical processes
M. BECKER and R. KÖHNE, DFVLR, Köln and Stuttgart; R. HARTH, KFA Jülich, Federal Republic of Germany 344

A modularized sensible heat storage system for process heat generation
M. GEYER, Institute for Technical Physics, DFVLR, Stuttgart, Federal Republic of Germany 352

SESSION V - PART 1 : PROCESS HEAT GENERATION

Summary of the Session by the Rapporteur
E. BURCK, JRC, Ispra, Italy 364

ARCO central receiver solar thermal enhanced oil recovery project
F.A. BLAKE, D.N. GORMAN and J.H. McDOWELL, ARCO Power Systems, Littleton, USA 365

SESSION V - PART 2 : NEW TECHNOLOGIES

Summary of the Session by the Rapporteur
B.P. GUPTA, SERI, Golden, USA 386

Conceptional design of solar thermal electric power plant with optical fibers and channels
N. IKEDA, T. HORIGOME, NEDO, T. TANI, Electrotechnical Lab., Japan 390

Low tension stress design for a glass receiver concept
B. KARRAIS and R. KOCHENDÖRFER, DFVLR Stuttgart, Federal Republic of Germany 396

The university of Houston solar research program
L.L. VANT HULL, University of Houston, USA 404

SESSION VI - ROUND TABLE DISCUSSION ON THE "FUTURE OF HIGH TEMPERATURE SOLAR THERMAL CONVERSION"

Summary of the Session by the Rapporteur
F.A. BLAKE, ARCO Power Systems, Littleton, USA 410

Solid particles solar thermal loop for production of heat at 1000°C
C. ROYERE, Solar Energetics Laboratory, CNRS, Odeillo, France 412

Aspects of future high temperature solar thermal conversion
C.J. WINTER and M. BECKER, DFVLR Stuttgart, Federal Republic of Germany 417

SESSION VII - PRESENTATION OF NEW CRS PROJECTS

Summary of the Session by the Rapporteur
L.L. VANT-HULL, University of Houston, USA 430

The gas-cooled solar tower project "GAST"
P. WEHOWSKY, D. STAHL, INTERATOM, International Atomreaktorbau GmbH, Federal Republic of Germany; J. DE MARCOS, L. CRESPO, ASINEL, Asociacion de Investigacion Industrial Electrica, Spain 433

Central receiver research facility at the Weizmann Institute of Science
I. DOSTROVSKY and M. EPSTEIN, The Weizmann Institute of Science, Rehovot, Israel 439

Molten salt electric experiment
W.R. DELAMETER, Sandia National Laboratories, Livermore, USA 442

Solar thermal central receiver pilot plant overview - Part II : A utility perspective
J.N. REEVES, Southern California Edison Co, Rosemead, USA 449

WORKSHOP CONCLUSIONS
J. GRETZ, C. SELVAGE, A. SKINROOD 459

LIST OF PARTICIPANTS 463

INDEX OF AUTHORS 469

OPENING SESSION

Workshop introduction
J. GRETZ, Commission of the European Communities, Joint Research Centre, Ispra, Italy

Welcome speech
G.R. BISHOP, Commission of the European Communities, Director of the Joint Research Centre, Ispra, Italy

Keynote speech
A. STRUB, Commission of the European Communities, Directorate General Science, Research and Development, Brussels, Belgium

WORKSHOP INTRODUCTION

J. GRETZ
Commission of the European Communities
Joint Research Center
Ispra, Italy

Welcome to the 2nd International Workshop on the Design, Construction and Operation of Solar Central Receiver Projects.

This workshop is the follow-up of the 1st International Workshop on the subject, held in 1982 in Claremont, California, which most of us have in best rememberance.

At that workshop we established for the first time the potential of a new technology, the electricity generation with solar power tower plants, by discussing their operational results. These plants enjoyed by then a one year's operation at maximum.

During this second workshop we shall for two days deepen our look into the operational experiences and problems of the world's six solar power plants. Another two and a half days we shall spend investigating new aspects and applications of solar tower/heliostat field technology. Sufficient time is foreseen for rather extensive discussions, so that the workshop does not degenerate to a conference or congress. The final aim is to evaluate the technical and economic potential of the Central Receiver Technology and to situate its position in the ambitious of energy conversion.

Solar thermal power generation is feasible today : no major scientific breakthroughs are required, but rather the development and cost reduction of more or less available components, as well as optimisation of the systems linking together those components : receiver, heliostat, prime mover, storage, heat cycle.

We still have the chance of being a small group, the Solar Power Tower family with a hundred or so members, we all know each other, we are still colleagues and not yet competitors since we did not yet enter the industrial stage. We should use this chance and be frank with ourselves, avoiding euphorism and trying to establish a sound evaluation of the solar tower technology.

Before starting the actual workshop session Prof. Bishop, General Director of the J.R.C. Ispra, will welcome you here in Varese and then Dr. Strub shall give a keynote speech.

WELCOME SPEECH

G.R. BISHOP
Director of the Ispra Establishment, Joint Research Centre

It is an honour and a pleasure to welcome the participants in this Workshop on the performance of Solar Central Receiver Projects. I do so on behalf on my colleagues in the JRC Ispra Establishment, some of whom are engaged in various aspects of the EURELIOS project, and many more in the programme of studies of the thermal and photovoltaic conversion of solar energy which is executed with a set of facilities unique in the European scene. I hope that as many of you as possible will visit the Establishment in the course of your stay in this region, although I must admit that it abounds with alternative potential demands on your time.

Since the beginning of 1984 the JRC's four Establishments are embarked on a new four-year programme, in which the fraction of resources devoted to non-nuclear energies is about ten per cent. The programme is oriented towards a vocational interest in Safety, Environmental Protection and Standards and Norms. Safety in all aspects of the fission fuel cycle has long been a main stay of JRC activities but the European Community is faced with a host of other problems requiring the attention of scientists and technologists, so that the European Commission can put squarely and fairly before the Member States the issues involved in their policy making. Thus the rigorous methodologies developed in the fission field can be applied with success to the safety problems of other large scale industries, existing ones such as the chemical industry or embryonic ones such as fusion power generation. The objectives pursued are either to measure the environmental impact of present technologies or to anticipate that of future ones, so it is natural that we should participate in the work on Central Receiverrs and host the present workshop.

Performing this broad spectrum of programme activities in the same Establishment has its advantages, but also as a consequence we have to maintain a vivid awareness of the trends in research and the changes of emphasis required of us by our political paymasters. Thus although we could imagine, once its original purpose was fulfilled, several alternative uses in the future of EURELIOS in the areas of solar chemistry or fusion technology, these uses would be useful additions to our projects and not the basic reason for a Commission interest continued beyond the assessment phase.

It does appear that European conditions are not the best suited to operation of the first generation central receivers at least for electricity production. For a physicist that is a disappointment because solar heat is at the highest thermodynamic potential available to us without the immense technical difficulties that accompany fission, fossile and fusion fuel applications. There are certain to be other uses for concentrated solar spectrum energy where the mismatch between source and application temperatures is reduced. Perhaps a fresh impetus to European enterprise in this direction can be hoped from the entry of Spain and Portugal to the Communities. If we can put astronomical observatories elsewhere in the world where the atmospheric conditions are better or satellite launching stations with advantages for final orbits then we can do the same for central receiver projects, provided the collective will exists.

KEYNOTE SPEECH

Albert Strub
Commission of the European Communities, Brussels

It is a great pleasure for me to address this distinguished audience. First of all, I should like to welcome you all on behalf of the Commission of the European Communities. As you know, the European Communities are involved in energy research, development and demonstration in many ways and in many fields and you will be told during this meeting of the Communities' in-house research activities as well as its contractual research programmes in the solar field.

It certainly would not be appropriate to recall all the well known facts and historical developments which have led to energy matters and energy technology becoming a very important issue of our daily life. I think that most of you are in the energy business for a long while, and that you therefore remember - as I do - that the energy situation, and in particular the development of the oil price, has given us quite some headaches in the past 10 years and, I think, the subject is good for further surprises. But the very high energy price (and I mean by that not only the price of oil and gas, but alsoo of all other energy carriers which have followed the oil price development), has not prevented our society from adapting itself to the situation and from behaving like almost all systems we know as engineers: it only reacts to changes and adapts as long as these changes are not happening by huge step functions. This fact has led many of our energy policy makers to loose interest in long-term energy technology development, just because there is an oil glut and because the oil price did not increase very much in the last two years. But you have but to listen to the radio news every day and you know how dangerous it is, to rely on such a stability.

But let us recognize that people who are busy in the field of renewable energy technologies, do have a hard time at present. Quite some reduction in public funding has made their future much less brilliant than expected a few years ago.

But as an optimist, I am willing to consider this slowed down interest in renewable technologies as a challenge. We are thus forced to assess our actions, to carefully evaluate the results and the progress made in the past ten years and then, maybe, to reallocate our efforts.

This is one of the reasons why this Workshop is so timely and really appropriate. It is probably the most complete gathering of all the people involved in building and exploiting the various solar central receiver plants all over the world. I hope that you will be able to draw conclusions and come forward with the elements of guidance for the next step in this technology. I should like to underline that, in doing so, we should not be afraid of drawing also negative conclusions here and there. It is inherent to research and development that some of the results are less encouraging than others.

We also should keep in mind that the technology of central receiver plants and its prospects depend very strongly on the geographical location and the application one is aiming at. We, therefore, might not draw too general conclusions.

When assessing the role of solar energy in our future and long term energy scenarios, we must be aware that solar energy technologies will always remain relatively expensive when compared with more conventional energy technologies. From the point of view of an energy manager, I, therefore, would say that it is strategically sound to aim at transforming the sun's radiation into more noble and versatile forms of energy, such as electricity or hydrogen. This will increase its chances of becoming competitive. Thermodynamic conversion of solar energy into electricity or other solar fuels therefore has quite some more long term potential than, for example, solar heating and cooling of houses. Some of you will certainly make the point during this meeting that central receiver plants have an additional potential. They can even be used for quite a lot of non-energy applications, such as solar chemistry.

Let me conclude by stressing that the strategic importance of developing solar technologies is two-fold: on one hand, solar electricity might be the only long term energy option (in the event that thermonuclear fusion is late or even proves unfeasible). On the other hand, its development constitutes a challenge to our manufacturing industry. In that it might be one of the options for remaining competitive when faced with low labour cost in threshold countries.

And in addition, we may all say that the many efforts put into the development of renewable energy technologies in general, have certainly led to a reduction of our dependance on imported energies and have played a strategic role in the dialogue with the oil-producing countries which goes far beyond the barrels of oil it saved directly. And I think this will not change in the future.

I wish you a fruitful meeting.

SESSION I - PART 1

PRESENTATION ON THE GENERAL EXPERIENCE OF CENTRAL RECEIVER POWER PLANTS

Summary of the Session by the Rapporteur
F. PHARABOD, THEMIS, CNRS, France

Technical aspects of the EURELIOS plant operation

Central receiver solar power plant of the IEA/SSPS project - Summary of results and experiences after three years of testing

10 MWe solar thermal central receiver pilot plant overview - Part I

SUMMARY OF THE SESSION BY THE RAPPORTEUR

F. PHARABOD
Themis, CNRS, France

1. INTRODUCTION

In this session, three central receiver power plants were discussed:

EURELIOS (Italy), by G. DINELLI
SSPS-CRS (Spain), by W. GRASSE
SOLAR ONE (USA), by A. SKINROOD

2. POWER PLANTS

2.1 EURELIOS (1 MWe)

The EURELIOS project was presented by M. Dinelli (ENEL); it is sponsored by CEC with Germany, France, and Italy, and operated by ENEL, Italy. The first turbine roll occurred in June 1981.

In the project, the insolation measurements were less than expected. The reduced heliostat field had limited receiver performances and electric output (≈600 kWe). A receiver failure occurred in the transition zone between the boiler and superheater. Because of this failure, the tube supporting system had to be modified.

The start up time was near two hours. This time will be reduced using a by-pass. The conditions for start up are as follows:

irradiance greater than 450 W/m^2
heliostats available greater than 75%

During the experimental phase, the operating staff was 18 people.

The next phase of the project entails operation for next three summers (84 to 86) with shut down and conservation programs with reduced staff during next two winters.

2.2 SSPS-CRS (500 kWe)

The SSPS-CRS project was presented by W. Grasse (DFVLR); it is sponsored by IEA, managed by DFVLR, and operated by SEVILLANA. First electricity production occurred in May 1982.

In the project, the insolation was less than expected. There was no problem with the heliostat field; performance was reliable. But problems did arise because of lightning, despite substantial improvements on the protection. Slight corrosion on the backside silvering was observed.

Two sodium receivers were tested, 270°-530°C, operating when irradiance was greater than 300 W/m^2:

* semi-cavity with maximum flux of 600 kW/m^2;
* external panels with maximum flux of 1380 kW/m^2.

Sodium technology proved to be state-of-the-art. Sodium has excellent heat transfer and good thermal storage properties. Trace heating elements gave reason for many problems. Incidents with the cold vessel as well as with the steam motor were simply bad design.

In the power conversion system, a piston engine with very high thermal inertia is responsible for the waste of a large part of collected solar energy.

Design performances for the plant were experimentally verified. Due to its small size, the plant will never really be able to operate in an optimized power generating mode. The staff was 17 people. Still, it is an excellent test facility for solar thermal technologies in a complete system.

The next phase will be defined at the end of the present three year testing phase in December 1984.

2.3 SOLAR ONE (10 MWe)

The Solar One project was presented by Al Skinrood (SNL) and J. Reeves (SCE). SOLAR ONE is sponsored by DOE and Associates, managed by Sandia National Laboratories, and operated by Southern California Edison. The initial turbine roll occurred in April 1982.

In SOLAR ONE, system and component performances approximated the theoretical predictions:

* 10.4 MWe net receiver steam achieved October 1982;
* 33.6 hours on line in June 1983 (127 MWe-hr delivered);
* Maximum energy delivered in one day: 104.3 MWe-hr net;
* Minimum plant power level: 0.5 MWe.

Many of the original areas of concern have not been problems.

The reliability of the heliostats is excellent. Mirror corrosion was measured, and vents were installed in 10,000 modules. The receiver operates reliablty with stable flow; its efficiency is 0.76 and start up time is 1.5 hours with irradiance greater than 450 W/m^2. However, there were leaks in some of the tubes.

The thermal storage has met all the design requirements. It significantly increases operating flexibility. Due to thermal cycling though, several leaks have occurred in the heat exchangers. Thermal cycling of conventional as well as solar components must be thoroughly analyzed.

Digital displays and controls have many advantages. Ways to reduce operating and maintenance costs are needed. The operating staff is 34.

The next phase will see the end of 2 year experimental test and evaluation phase in July 1984 and the start of the 3 year power production phase from August 1984 to July 1987.

TECHNICAL ASPECTS OF THE EURELIOS PLANT OPERATION

C. CORVI*, G. DINELLI**

*ENEL-D.S.R. (Studies and Research Division), Roma
**ENEL-C.R.T.N. (Thermal and Nuclear Research Centre), Pisa

ABSTRACT

The experimental results of the operation of the Eurelios plant are presented and discussed with reference to the main factors influencing the performance of the plant.
Specific action to improve the operation of the plant are described, they concern the mirror field, the solar central receiver and the turbine.
Considerations on the anticipated size and performace of full scale heliothermoelectric plants are given.

1. FOREWORD

The Eurelios heliothermoelectric power plant is a project sponsored by the Commission of the European Communities, the plant was designed and built by a Consortium of European Industries: CETHEL (France) MBB (Fed. Rep. of Germany), Ansaldo Impianti and ENEL (Italy). The Commission of the European Communities and ENEL are co-proprietors of the plant that is operated by ENEL.

After completion of the construction of the plant on Dec. 1980 an experimental R&D programme started which at first concentrated on a stepwise run-up of the plant in order to reach its full power; afterwords and esperimental programme was performed ending on Dec. 1982 with the following main objectives; to gain experience of operation and improve the operation procedures, to evaluate the plant performance and to demostrate the feasibility of its operation in connection with the electrical grid.

From Jan. to Dec. 1983 the plant has been operated by ENEL under a provisional contract agreement with the C.E.C., while a new research programme has been considered for the period Jan. '84÷Dec. '86 in order to obtain an exhaustive information and experience after the required improvements and modifications. It is planned that the last 12-months period of the programme will consist on a demonstrative operation of the plant.

2. PLANT OPERATION

The organization of the operation of the plant was settled in similarity with the organization of a conventional thermoelectric power plant taking into account the specific requirements of the Eurelios plant. On the basis of two daily shifts and six working days per week the operation manpower was:

- Plant superintendent: responsable for the plant management from technical and administrative points of view;
- Mechanical maintenance group: composed of three people (1 assistant, 1 skilled workman, 1 workman) with the task of the mechanical maintenance of the plant subsystems;
- Regulation and instrumentation maintenance group: composed of three people (1 assistant, 1 skilled workman, 1 workman) with the task of the maintenance of the plant regulation and instrumentation (instruments set-up, repair etc.......);
- Operation group: composed of ten people (4 shift-supervisors, 4 bench operators, 2 workmen) with the task of plant operation and electrical maintenance; the normal shift is composed of three people;
- Administrative group: One clerk.

During the past period in addition to the normal operation of the plant several investigations and intervention at subsystem and component level were performed as a consequence of specific failures as well as in order to improve the plant net electrical power output.

With regard to the mirror field the main sources of failures were:

- for the MBB heliostats: power supply sets, arithmetic logic units, encoders, servo cards, eproms, reflecting pannels;
- for the Cethel heliostats: batteries, inductive sensors, electronic cards, reflecting pannels, power cables and connectors.

Concerning the thermal cycle repairs and/or replacements were necessary for same valves, drain pipes, feed water treatment instrumentation and control system, salt storage system.

The receiver requested three interventions for steam leakeges from several tube and welding failures: during summer 1982, on February-March 1983 and finally on September 1983.

A data acquisition and supervision system (DASS) was installed during 1983 to allow the acquisition of the data from the plant, display them on a suitable way by means of mimic diagrams and synoptic tables, and store the data for further analysis.

With regard to the performance of the plant the main factors limiting the electric power production were:

- reduced mirror field surface with respect to the design value (6200 m^2 instead of 8900 m^2);
- heliostat availability (less than 90%);
- mirror reflectivity (about 72% instead of 82% - design value);
- insolation (level and quality worse than expected);
- receiver and thermal cycle failures.

In addition the degradation on the performance of the mirror field at low sun elevation angles contributed considerably to the long start-up phase in the morning thus further penalizing the net power production.

A thermal cycle balance sheet is shown in Fig. 1. Global performance data for the operation period ending September 1983 are reported in Table I; besides the very small amount of generated electrical energy it is recognized the much higher value of the electric energy requested by auxiliaries.

For the operation period from April to September 1983 the incidence on the lack of parallel of meteo, heliostat and thermal cycle sources are shown in Fig. 2. Only in the last period of operation, September 1983, it has been possible to obtain a net energy balance of the plant as a consequence of a very favourable insolation pattern and of improved operating procedures that now imply that:

- the operation activities are started only when the weather conditions are extremely favourable (sky cover up to 2/8 and absence of haze);
- the heliostats and the thermal cycle are activated only if the insolation is 450 W/m^2 or higher and the heliostat out of service is less than 25% of the total reflecting surface;
- once in operation, the plant is put out of parallel when the generated power goes down to 100 kWe.

The utilization of the storage system has been dropped because of the plant efficiency and energy losses resulting from the thermal energy transfer from the receiver outlet to the storage system and then to the thermal cycle.

The objective of the heat storage system is to allow the plant to ramain in connection to the grid during cloud passing and after that to allow to restart the energy production with steam from the receiver. To obtain this the mirror field should be oversized so that the amount of energy necessary to charge the heat system does not penalize the electric energy produced.

For the Eurelios plant the utilization of the storage system would imply:

- a higher utilization factor of the solar energy equivalent to about 8.000 kWh-yr;

- a gross energy production with the storage system of about 1000 kWh/yr;
- an energy loss during the phase of thermal energy storage of about 6.000 kWh/yr;
- an energy consuption for tracing the molten salt system of about 48.000 kWh/yr.

Thus the owerall energy budget would be negative of about 45.000 kWh/yr.

3. PLANT IMPROVEMENT

3.1 Mirror field

As a consequence of financial limitations the actually installed mirror surface amounts to only 6200 m^2 instead of 8900 m^2 as foreseen in the original design. At the time it was decided to keep the receiver and steam cycle size at the original design value in order to have the plant running at nominal power, i.e. 1 MWe, at a later stage when financial conditions would allow to complete the mirror field. For this reason the achieved generated power level corresponds to its actual construction parameters.

A feasibility study has been performed in order to increase the mirror surface without increasing the number of heliostats. This solution would allow to improve the performance of the plant by increasing to thermal power input in the receiver and at the same time to test the performance of alternative material and mirror structures with higher reflectivity or less costly.

The heliostats which is most suitable for modification and surface enlargement is that one manufactured by Cethel; the modification would consist on the replacement of the lateral modules (1800x600) whith modules of increased width (850) and with the addition of a module (1000x3000) in the central upper side of the heliostat. By this means it would be possible to increase the mirror surface of the heliostat of about 3 m^2 with the central module and about 3.6 m^2 with the lateral modules thus increasing the total mirror surface of about 12%. In the most likely case that the added mirror surface would have a higher reflectivity, for instance 92%, the equivalent gain in the reflected solar power would be about 15% per heliostat. Since the Cethel heliostats represent about 60% of the total mirror surface, the overall gain in the peak power sent to the receiver from the mirror field would be in the order of 9% allowing a gain in the peak electric power of about 15%. Although this is a marginal gain this modification is under consideration since, as said previously, it could allow to test mirrors of different technology.

A much larger effort would be required to improve the availability of the mirror field by increasing the MTBF of the components thus reducing to a more acceptable level the interventions of the maintenance team; in fact the availability of the mirror field is at present in the range 80%÷90%.

3.2 Receiver

The operation of the receiver, which has a once-through circulation pattern of the water-steam working fluid, presents severe limits due to long start-up time and small flexibility to thermal input variation. In addition the mechanical design has evidenced weak points in the connections of the tubes of the receiver to the supporting frame which were not adequate to compensate the dilatation of the tubes at different thermal loads.

This led to exessive stress on weldings, failures on tubes at the connection to the rigid frame and steam leakege from several points in the upper cone of the receiver, that is in the transition zone between the evaporating the superheating phase of the working fluid.

3.2.1 Modified Circulation Pattern

A more stable thermal and fluid-dinamic pattern of the receiver will be achieved by modifying the present once-through system into a recirculation one, by adding a drum and a recirculation pump, as shown schematically in Fig. 3.

In addition it is expected that such configuration will shorten the start-up time of the receiver of about 30 minutes. During start-up water will be circulated through the receiver untill saturated steam is produced at 5 bar; then the circulation will be switched to normal operation and superheated steam will be produced. Under clear sky conditions it should be necessary about 50 minutes to reach the steam conditions required to warm-up and start the turbine.

3.2.2 Modified Mechanical Design

After two repair interventions which slightly modified the supporting system a detailed stress analysis study indicated that failures were the consequence of law frequency loads due to prevented dilatation. Consequently a new supporting system was designed in order to free the tubes of the superheating zone; in the present configuration these tubes are connected to the supporting structure by means of elastic suspensions. Such modifications are shown in Fig. 4.

3.3 Turbine

The turbine admission system, which consists on three ducts with two passages each, has been modified to match the actual steam flow conditions; in fact the average steam flow is in the range 1200÷2600 kg/h, the maximum value being below 3500 kg/h at 60 bar and 500°C instead of 5200 kg/h as in the original design. To avoid too high thermodynamic losses one of the two passages in the first and third duct where plugged thus reducing the passage area and associated lamination losses (Fig. 5).

4. REMARKS

The operation experience gained from the demostration heliotermoelectric plants already in operation has allowed to verify the feasibility to produce electric energy in connection with the grid and to operate the plant by standard procedures like those for the conventional termoelectric units. Furthermore the experimental campaigns have shown the limits of the plant and the possible intervention area for plant modification and optimization.

The question is whether it would be possible to reach the cost objectives for this technology, in fact the low density of the solar energy leads to plants which are highly capital intensive, and whether the large land requirement could be satisfied: for instance a plant having a size of 50 MWe would require a land area of about 80 ha.

A condition having a definitive incidence on the plant performance and then on the global economics is the aleatory and discontinuos availability of the solar energy. For instance under the meteo conditions met in Sicily almost 50% of the yearly gross energy production would be concentrated in the months of June, July and August.

Furthermore the maximum unit capacity attainable for heliothermoelectric plants, which on the other hand are characterized by a high technical sophistication, could give rise to a prohibitive incidence of operation and maintenance costs on the cost of the generated electric energy.

For these reasons it is believed that, at least in the European context, the capability of thermoelectric plants to become competitive will depend on a situation of conventional energy source prices and availability much more unfavourable than the present one.

REFERENCES

1 - EURELIOS - The 1 MWe Experimental Solar Thermal Electric Power Plant of the European Economic Community - Final Report for Phase D, June 1983.

2 - EURELIOS Plant Operation from April to September 1983 (in Italian), ENEL-CRTN Report n. 307.83.01, October 1983.

3 - New Energy Sources: State of the Art and Prospects for Development UNIPEDE CONGRESS Report n. 110.1, Brussels, June, 1982.

TABLE I - OPERATION DATA SUMMARY

		GROSS ENERGY PRODUCTION kWh	PLANT AUXILIARIES ENERGY kWh	PARALLEL HOURS h	MAXIMUM POWER kW	ELIOSTAT OUTAGE %
1981	MAY÷AGO	19404	118591	70	600	
1981	SEP.	2464	17768	7,5	533	10,2
1981	OCT.	7524	17179	25	550	6,0
1981	NOV.	11528	16510	33,15	566	6,9
1981	DEC.	2090	12838	9,32	500	6,1
1981	TOTAL 1981	43010	182888	145	600	7,3
1982	JAN.	3388	9454	9,73	500	3,6
1982	FEB.	1408	10740	4,54	500	6,5
1982	MAR.	4972	21894	14,93	550	8,5
1982	APR.	6094	21563	17,85	500	4,1
1982	MAY.	2354	19462	10,9	400	4,4
1982	JUN.	–	16021	–	–	–
1982	JUL.	–	11898	–	–	–
1982	AGO.	–	13562	–	–	–
1982	SEP.	–	12768	–	–	–
1982	OCT.	440	20468	2,95	350	12,6
1982	NOV.	2068	18118	10,53	300	13,1
1982	DEC.	3234	21663	13,18	450	15,8
1982	TOTAL 1982	23958	197611	85	550	8,6
1983	JAN.	3278	22255	12	535	12,2
1983	FEB.	286	19097	2	440	19,5
1983	MAR.	–	18705	–	–	11,8
1983	APR.	11088	24836	48	400	21,5
1983	MAY.	8866	21158	54	330	20,5
1983	JUN.	6028	16528	41	365	15,5
1983	JUL.	5082	24381	22	330	9,0
1983	AGO.	12496	28348	66	375	10,6
1983	SEP.	14332	14090	50	500	9
1983	TOTAL 1983	61456	189398	295	535	14,4
TOTAL (1981-82-83)		128424	569897	525	600	10,8

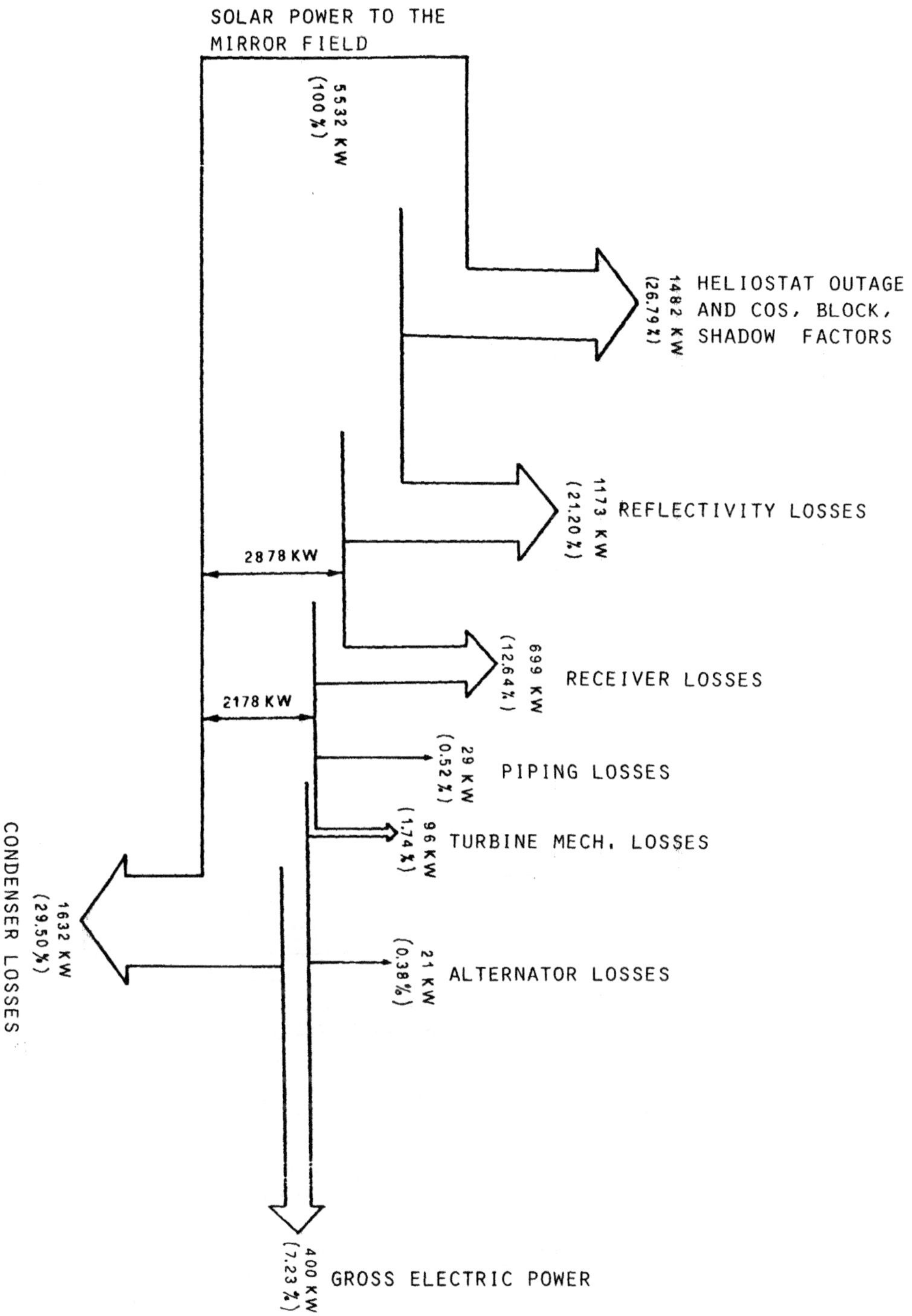

FIG. 1 - PLANT THERMAL BALANCE (DIRECT INSOLATION 890 W/M^2)

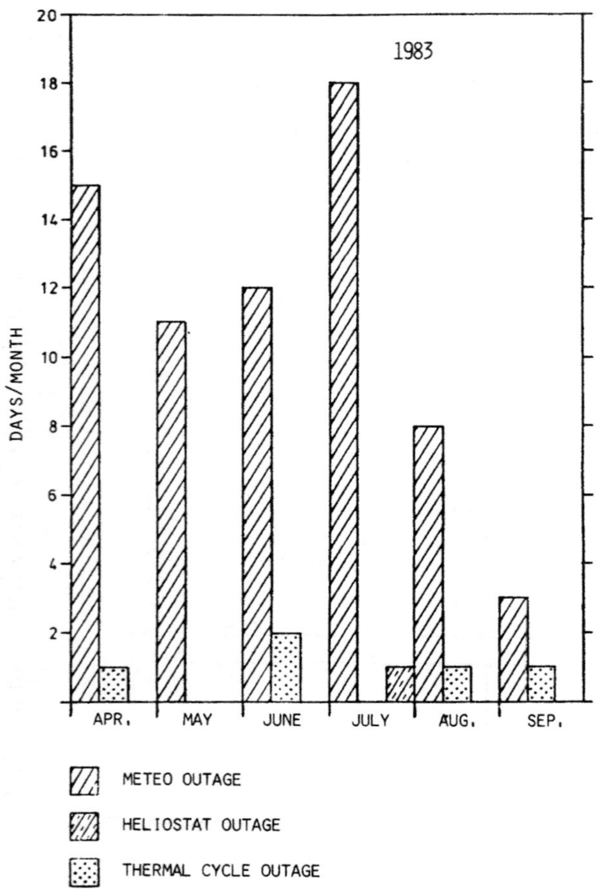

FIG. 2 - PLANT OUTAGE STATISTICS

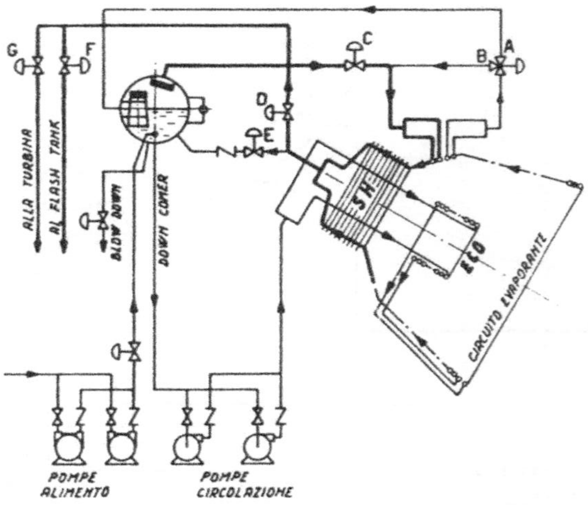

FIG. 3 - RECEIVER MODIFIED CIRCULATION SCHEME

FIG. 4 - RECEIVER MODIFIED SUSPENSION STRUCTURE

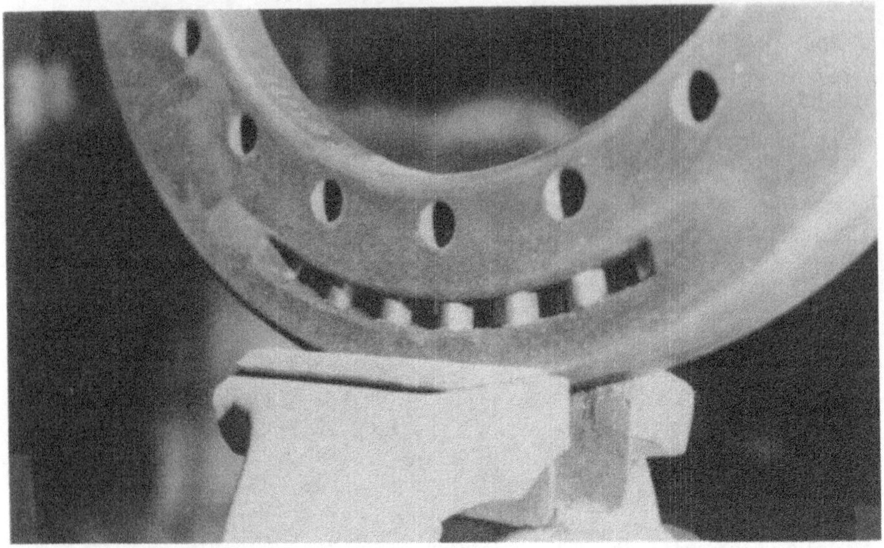

INITIAL DESIGN

MODIFIED DESIGN

FIG. 5 - TURBINE INLET DUCTS MODIFICATION

CENTRAL RECEIVER SOLAR POWER PLANT OF THE IEA/SSPS PROJECT
-Summary of results and experiences after three years of testing-

W. Grasse and F. Ruiz
DFVLR and Sevillana

Summary

Summary results are presented after three years of testing with the CRS plant were completed in June 1984. Whereas design performances were experimentally verified, longterm behavior did not meet expectations. Subsystem testing and evaluation was, therefore, emphasized rather than optimization of electrical power output. Through that program and with its widespread international cooperation, the SSPS Plants on the Spanish Plataforma Solar de Almería have become one of the most viable test facilities for solar thermal technologies worldwide.

1. INTRODUCTION

The Small Solar Power Systems Project (SSPS) conducted under the auspices of the International Energy Agency, consists of the design, construction, testing, operation, and evaluation of two different types of solar thermal power plants in Almería, Spain.

The project objectives are to demonstrate the viability of chosen designs for 500 kW_e net output at design conditions, to gain operational experiences in regard to reliability, to longterm performances and towards future technical and economical applications.

Design started in 1977, construction began in December 1979, and after an approximate six months period of functional and acceptance testing, the local utility Sevillana started routine operation of the plants in January 1982.

Nine member countries of the IEA (Austria, Belgium, Germany, Greece, Italy, Spain, Sweden, Switzerland, United States) participate in the Project and jointly provided the funds for investment and operation (approx. 45 MioDM). DFVLR, the largest German engineering research establishment, acts on behalf of the nine countries and is responsible for the overall technical and contractual management.

Already during the final design of the SSPS-CRS it became obvious that, due to its small size, the CRS plant was not well suited for delivering representative overall system data but rather technological know-how for forthcoming designs. In this sense, the most valuable contribution to the worldwide effort aiming at the development of commercial solar power plants concerns the SSPS heliostat field and sodium heat transfer system with the receiver and the steam generator in particular.

After three years of operation and with nine countries participating in the evaluation, the multitude of detailed results and experiences cannot be fully covered in a short summary report. Essential information is compiled in (1), (2), (3), (4), (5) and in further papers presented at this workshop (6), (7), (8), and (9). At the end of SSPS-CRS operation, which is presently scheduled for December 1984 as far as objectives established in 1977 are concerned, a concluding seminar will be held in Almería, where final results will be discussed with all international partners.

2. OPERATIONAL SUMMARY

Without extenuation, it is to be stated that SSPS-CRS is not a facility sufficiently mature for permitting utility like routine power production. With some surprise, this is mainly caused by conventional components and subsystems (i.e., electronics, vessels, power subsystem). The main system level weakness lies in the non-compatibility of the steam subsystem with the requirements of the "solar" front end.

Therefore, only in 1982 plant operation was accomplished according to utility-like standards on seven days with two shifts. Already in 1983, due to plant improvements and repairs, as well as replacement of the receiver by a more advanced design, the plant was in transition towards an international solar thermal test facility. Subject to such testing in 1984 is the heliostat field and the sodium heat transfer system with the receiver in particular.

Figs.1-4 record SSPS-CRS's operation history and clearly visualize the main three test periods separated by phases of technological plant improvements and test item installation, respectively.

It is, therefore, to be seen that recording of electrical gross or net outputs are only relevant for the basic understanding of the plant's performance. In no case, however, should such data be used for economic evaluation. This is valid for the manpower recording as well, which served more test purposes than routine operation.

In tabulated form, operation of the SSPS-CRS so far was as follows (in 1984 only from January to April):

	1982	1983	1984	TOTAL	%	%
Insolation I>300 W/m^2	2321	2574	766	5661	100	-
Tracking heliostats	2788	1508	687	4983	88	-
Energy collection	667	378	248	1293	23	100
Steam production	534	316	213	1063	19	82
Grid connection	58	53	43	154	2	12

Nonregarding outages due to improvement work, repairs, and preparation/implementation of tests, the operational potential of the SSPS-CRS can be drawn from Figs.3-4 where maximum values from a utility-like recording amount to 80% for the steam production factor, when related to time periods with an insolation of more than 300 W/m^2.

In the same plotting, the maximum electricity production factor reaches only 20% which, more than any other argument, indicates the need and the possibility of dramatic improvements to be made for better matching the two main subsystems of a solar power plant.

Availability of the SSPS-CRS is illustrated by the following table:

Outages due to	1982	1983	1984	TOTAL	%	%	%
-bad weather (I<300 W/m^2)	2127	1844	599	4570	48	42	44
-plant (when I>300 W/m^2)	1654	2196	518	4368	-	-	-

It is to be pointed out that only weather data are of practical relevance. Percentages of these data are related to the Almería daylight hours (4148 hours per year and 1365 hours in the first four months of 1984).

For reasons explained above, plant outage numbers have no statistical meaning whatsoever.

Maintenance work was well recorded over the whole operational period. Data are routinely distributed every month. From this, the following highlights can be summarized:

-Electronic components, mainly in the communication line from the heliostat array computer to the individual heliostats, and to the azimuth and elevation encoders showed an unexpected high failure rate. Components in the control systems for the receiver loop, the sodium heat transfer system, and the power system behaved well.

-Trace heating elements, one of the most critical parts in the sodium system, gave reason to many problems. They are difficult to replace, failures cause sodium plugs and spare part procurement turned out to be difficult.

-Lightning protection, even after substantial improvement, is not yet sufficient. This concerns in particular the shielding of the heliostat control and power box electronics. Severe damages were experienced after lightning strikes occurred in April 1982 and May 1984.

-Heliostat as well as data acquisition computers showed frequent failures caused mainly by disc drives and the CRT terminals.

-Incidents with the cold sodium vessel as well as with the steam motor were lamentable events because of their poor design and workmanship which caused many months of plant outages. From a solar specific standpoint, however, they are irrevelant experiences.

With its three years experience in high standard technology, the SSPS-CRS operation and maintenance team is one of the best trained solar engineering groups in the world. It is, therefore, well prepared to take over even more challenging tasks as foreseen in high temperature solar technologies on the Plataforma Solar de Almería.

3. METEOROLOGY DATA

After almost four years of continuous measurements, meteo data relative to insolation, wind, temperature, precipitation, and earthquakes allow a somewhat representative summary which is for the Plataforma Solar de Almería as follows:

Insolation

Irradiated energy is lower than was expected from a model used during the design of the SSPS Plants. 1730 kWh/m^2 were evaluated for 1982 and 2010 kWh/m^2 for 1983.

Figs.5-8 give some detailed information (4) regarding levels of insolation over a period of nine months and their seasonal characteristics. Average days have only 70% of insolation hours (in April) and 85% (in June) when compared to the relevant clear days.

During clear days, 98% of the insolation is direct. Slight haze reduces this to 92%, on overcast days figures of 65% until well below 50% have been determined by measurements made by R. Carmona and T. v. Steenberghe (4).

Wind

Wind velocities frequently exceed 50 km/h which is the upper limit for heliostat operation according to the manufacturer. With good experience, however, this limit was raised to 80 km/h at steady wind and to 65 km/h when high frequent gusts occur.

With this experience, the heliostat operation disturbance due to wind is only marginal.

Maximum wind velocity recorded exceeded 130 km/h which is very close to the specification for construction survival (144 km/h).

Temperature and precipitation measurements showed no unexpected values.

The strongest earthquake reached 4.2 on the Richter scale with its epicenter 70 km away.

4. SUBSYSTEM PERFORMANCES

It was already reported at the first International Workshop on CRS Plants (1) that design point performances were verified with almost all subsystems and components. The only exception is still the power conversion system which, however, is of no solar specific relevance as far as its nominal performance is concerned.

Test and operation of the SSPS-CRS in 1983 and 1984 was pursued mainly with the objective to learn more about design constraints, energy loss mechanisms, and actual subsystem performances in all relevant operational conditions.

Heliostat Field

The heliostat field of SSPS-CRS is the subsystem that has the longest record of successful operation. Design assumptions have been proven to be adequate in terms of performance and reliability.

Figs.9-10 summarize detailed recordings which started at the beginning of 1982. Operation, however, already took place since April 1981 for purposes of functional and acceptance testing.

In total, up to April 1984, the field operated for approximately 6,500 hours, 5,000 of them with careful recordings. Maximum operational time per year amounted to 2,788 hours in 1982 (seven days, two shifts). Non or only limited operation was possible during replacement of test receivers because of safety reasons (June to October 1983).

Availability of the heliostat field is well above 90% and frequently reaches 100%. Reasons for heliostat outages are mainly due to electronic problems (see above). Riveting of backside support plates reduced failures (which occurred before with only bonded plates) to zero. Other mechanical failures are marginal.

The image quality of individual heliostats is very high. Corrective measures were very limited.

With two German (HFD, HERMES) and one Swiss (FAS) measurement devices, several campaigns for determination of receiver power input were performed. It was confirmed that at design conditions and with clean mirrors, the field is able to deliver the specified 2840 kW into the receiver aperture.

Mirror reflectivity is monitored continuously (6). Losses of 0.2-0.3% per day are regular, peak values reach more than 0.5% per day. Washing frequency is about once a month.

Though only 0.009% of the total mirror area is affected (6), degradation in backside silvering was found to be an intriguing problem. Condensing water inside the honeycomb support structure is assumed to be the essential reason. Degradation rates are carefully observed (6).

Still not answered is the open problem of lightning strikes to the SSPS-CRS heliostat field (see above). Analysis was restarted after the second strike in May 1984.

Sodium Receivers

Within the three years of SSPS-CRS operation, already two differently designed receivers have been tested in conjunction with the total plant system. In no case, engineering or design problems were found. The receiver itself was never the cause for plant outages -apart from its assembling and functional testing period.

Both receivers, the more conventional design made by Sulzer/Interatom and the advanced design by Agip/Franco Tosi, were and still are the focal points in the evaluation of the SSPS-CRS's international team. Results of this work are reported at this workshop with papers (7), (8), and (9). Furthermore, M. Pescatore (10) analyzed performance and behavior of the two receivers with regard to efficiency, standby-losses, thermal losses during operation, start-up, and shut-down characteristics.

Fig.11 gives an overview of the key differences in design and efficiencies. Main design data are as follows:

	Cavity Receiver	External Receiver
Thermal Power (input/output)	2840/2508 kW	2840/2525 kW
Mass Flow	7.34 Kg/s	7.34 Kg/s
Inlet/Outlet Temperature	270/530 °C	270/530 °C
Heat Flux Peak/Average	0.6/0.16 MW/m^2	1.38/0.36 MW/m^2
Heliostat Aiming	single point	three points

The cavity receiver was operated on 250 days (with 1310 hours) and the external receiver on 80 days (with 373 hours). With some surprise, the external receiver proved considerably more efficient than the cavity receiver (7).

Sodium Heat Transfer System

In general, operating experiences with the sodium system confirmed its excellent heat transfer and good thermal storage properties.

The sodium technology proved to be state-of-the-art. Even with several sodium leaks, hazards of sodium fires could be ruled out by appropriate provisions. (The worst incident of that kind occurred at the cold sodium tank; it was, however, due to poor design and workmanship and not because of solar specific loads like e.g., unusual thermal cycling). The personnel of SSPS-CRS proved to be very well trained, and the equipment for handling sodium accidents was found to work satisfactorily.

From the various components of the sodium system, the bellows at the sodium valves seem to be weak points. Another area of special care is the steam generator where thermal loads and stresses due to transients during preheating and start-up procedures were tried to be optimized by modification of procedures.

At the occasion of the sodium tank repairs, two samples of sodium were taken (January 1983). Analysis showed them to be free of pollutants, and the sodium still had reactor quality. Sodium cleaning with the help of the cold trap turned out to be very effective. Expected sodium characteristics could always be reached after a short time.

Trace heating in SSPS-CRS is one of the less reliable components. In a small plant, it is also one of the main contributors to the parasitic consumption, in particular when no optimized plant operation routines are applied, like is the case in a test facility.

Power Conversion System

For principally financial reasons, an off-the-shelf offer with performance guarantee was accepted instead of the more expensive turbine competitive offer. The piston engine, though it finally and after very time consuming functional improvements was accepted because it proved the required rated output, is the least viable part in SSPS-CRS.

Its clumsy design, with which the power conversion efficiency was intended to reach 27.2% (compared to approximately 20% with a turbine) is mostly responsible for the waste of enormous amounts of collected solar energy. Figures as given above are the best illustration of what is called the thermal inertia of such a design with its high masses in steel and water.

System Performance

Representative and quantifiable system performance data are not--and probably will not--be available from SSPS-CRS operation. This is because, due to its small size, the plant will never be able to operate in an optimized power generating mode and, on the other hand, has qualified itself as an excellent test facility for solar thermal technologies in a complete system.

However, in a qualitative manner, system experiences deem to be valuable results for future designs:

- One should never try to plan routine operation with a solar thermal power plant without a sufficient solar multiple which only provides the inevitable operational flexibility.
- Electronic components in the heliostat field and all control systems must exceed usual standards. Computers, at least for control purposes, must be redundant.
- Decoupling of the heat transfer system and the power conversion system by means of thermal storage tanks is a worthwhile approach. However, the design engineer has to pay close attention to minimizing "thermal inertia" and at the same time have maximum flexibility for a clever "energy management" in all operational modes in order not to waste significant portions of energy collected with the cost intensive heliostats.
  From SSPS-CRS experiences, in (2) a very illustrative example is given of how minor cost savings (by procuring a ferritic steel tank instead of one made of stainless steel) penalized the operational optimization in a way which is orders of magnitude higher than the initial saving.
- As already outlined, a very typical system design sin was the attempt of suboptimizing the power conversion system by means of the most sophisticated thermodynamic tricks. Lacking system optimization finally turned out to be responsible for the non-viability of SSPS-CRS for routine electric power generation.

From Figs.12-15, some more SSPS-CRS specific lessons can be learned. They also give, for good operational days, indications for time constants involved in start-up, steady operation, and shut-down.

The two energy flow diagrams may initiate, on a broader basis, an exergetic analysis aiming at cost and energy effective design changes. Already existing simulation models can, furthermore, very well contribute to a better understanding of how to design an energetically optimized solar power plant for the intriguing conditions they are exposed to.

At the end of three years of testing, both from a personnel and technical point of view, SSPS-CRS became a test facility, well suited for advanced solar thermal research and development, rather than for routine operation in a power generating mode.

5. REFERENCES

1. GRASSE, W. and BECKER, M. (1982). Central Receiver System in the SSPS Project of the IEA (1st International Workshop on the Design, Construction, and Operation of Central Receiver Projects in Claremont, Cal).
2. SELVAGE, C.S. (1984). IEA/SSPS 500 kW_e Central Receiver (Annual Review of Solar Technology Development, San Diego, Cal).
3. BECKER, M. and STAHL, D. (1983). CRS Construction Report, SSPS-SR 2.

4. BECKER, M. (editor) (1983). SSPS Central Receiver System Midterm Workshop, SSPS-TR 4/83.
5. BECKER, M. and BUCHER, W. (1984). SSPS-CRS First Period of Operation, Preliminary Operation Results, SSPS-SR 4.
6. WATTIEZ, P. and RAMOS, J. (1984). SSPS-CRS Heliostat Performance/History (2nd International Workshop on the Design, Construction, and Operation of Central Receiver Projects in Varese, Italy).
7. JACOBS, H. and SELVAGE, C. S. (1984). Preliminary Results on the Performance of the Sulzer Cavity Receiver and the Franco-Tosi External Receiver (as above).
8. CARMONA, R. and MARTIN, J. G. (1984). The SSPS Advanced Sodium Receiver: Transient Response (as above).
9. deBENEDETTI, A. (1984). Design Considerations for the SSPS-ASR Advanced Sodium Receiver (ASR) (as above).
10. PESCATORE, M. Comparison of Cavity Receiver and External Receiver -private communication-.

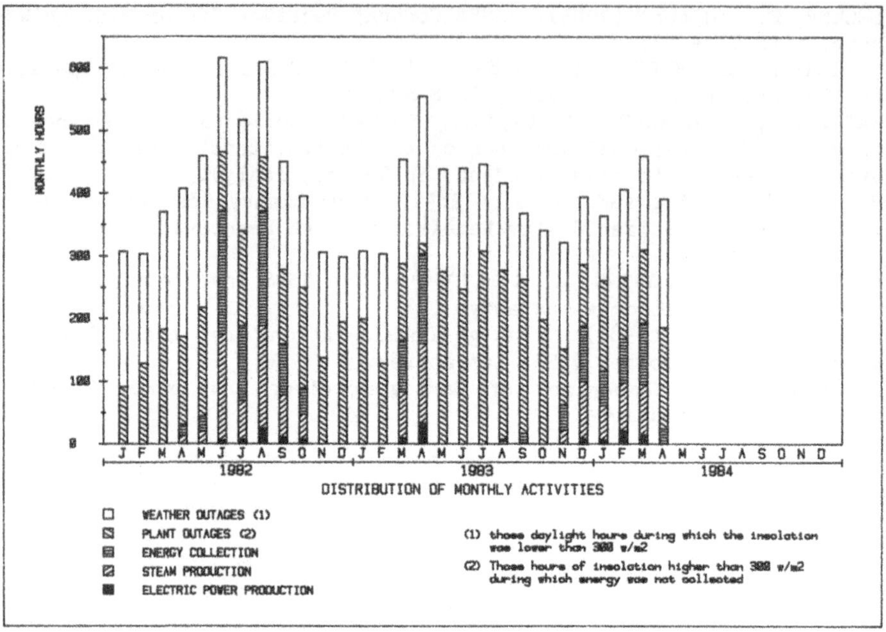

Fig. 1 SSPS-CRS Operation Statistic 1982-4/1984

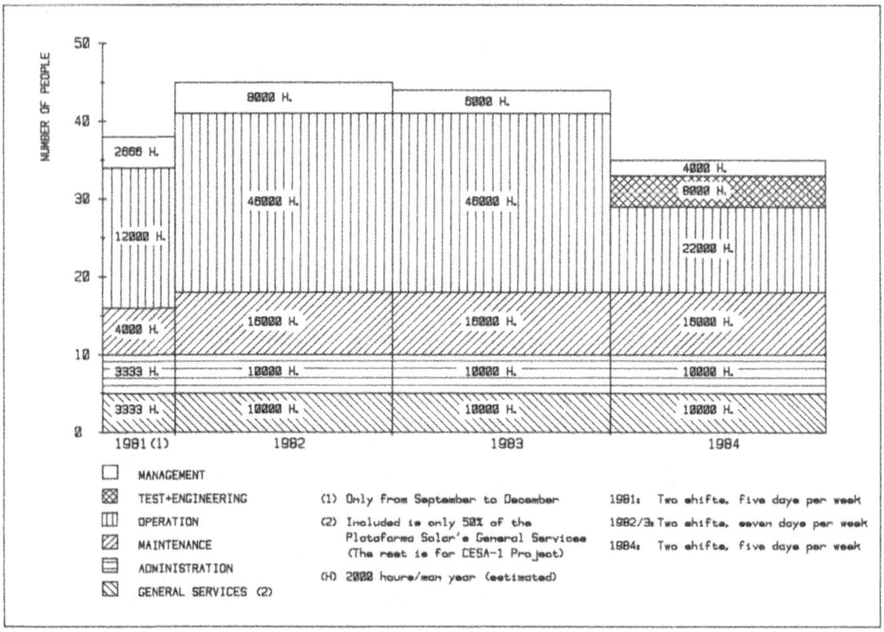

Fig. 2 Personnel of SEVILLANA (Plant Operation Authority) for the two SSPS Power Plants

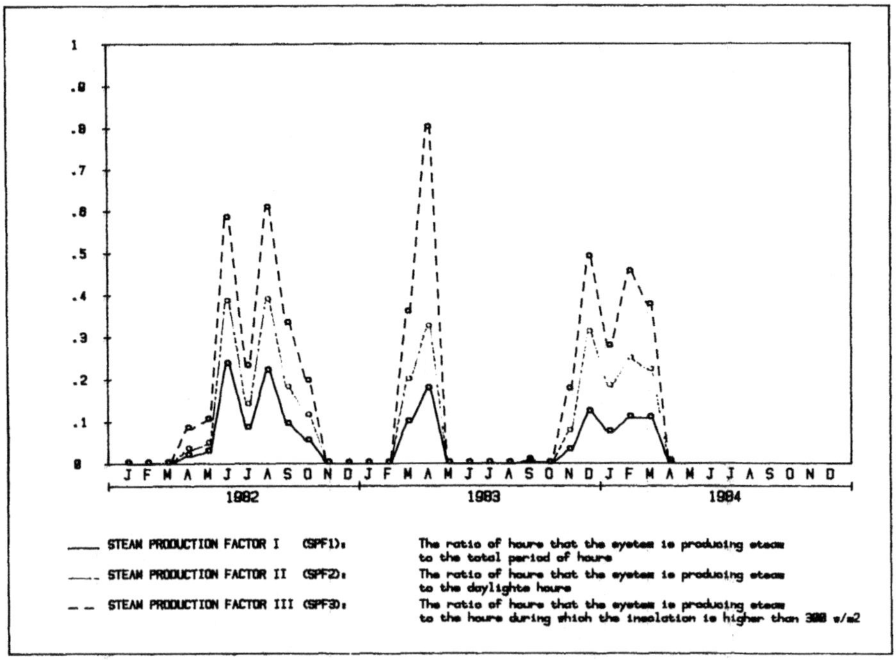

Fig. 3 Steam Production Factors for SSPS-CRS (1982-4/1984)

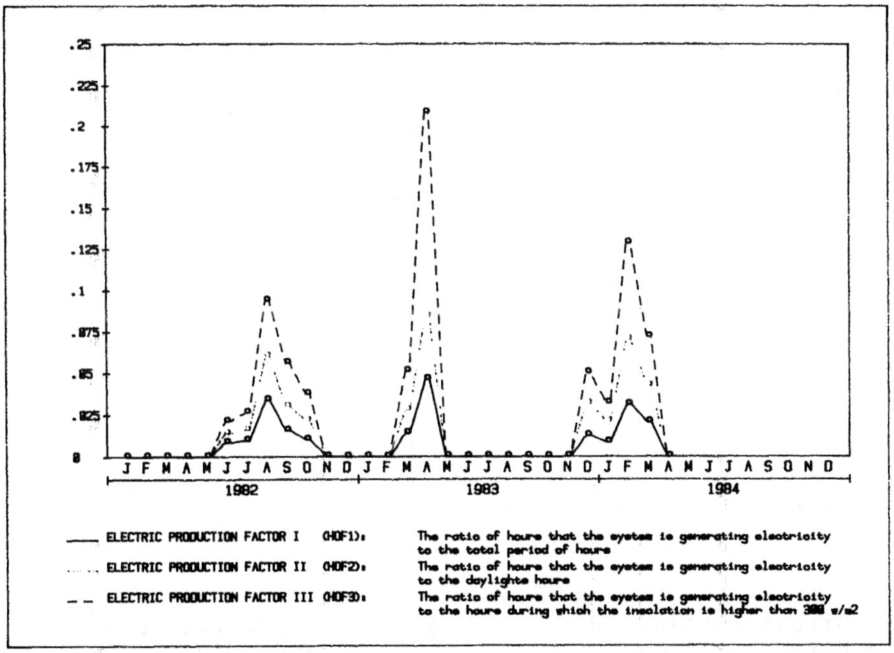

Fig. 4 Electric Production Factors for SSPS-CRS (1982-4/1984)

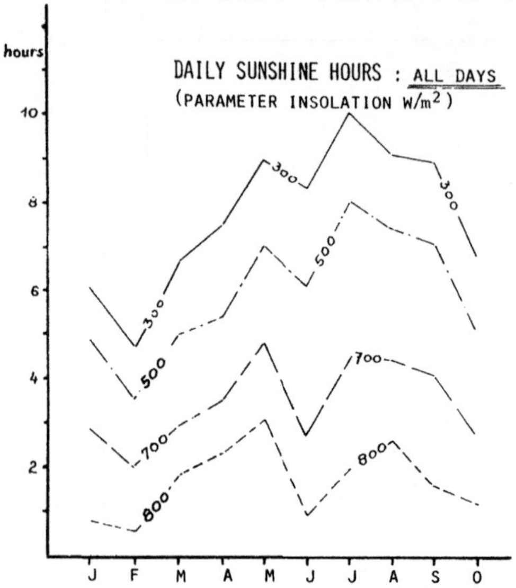

Fig. 5 Daily Sunshine Hours (Jan. - Oct. 1983)

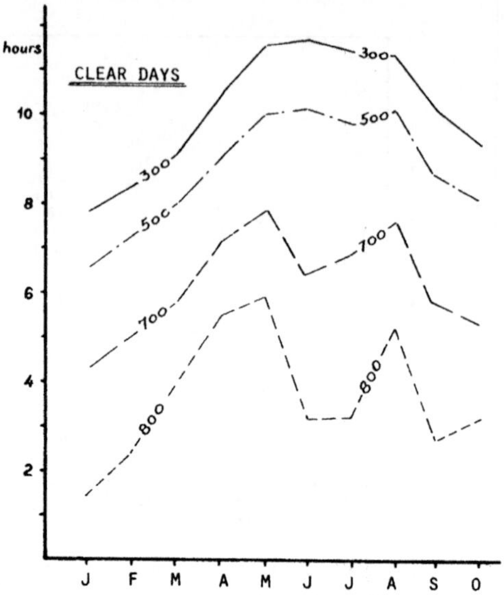

Fig. 6 Sunshine Hours (Jan. - Oct. 1983) Clear Days

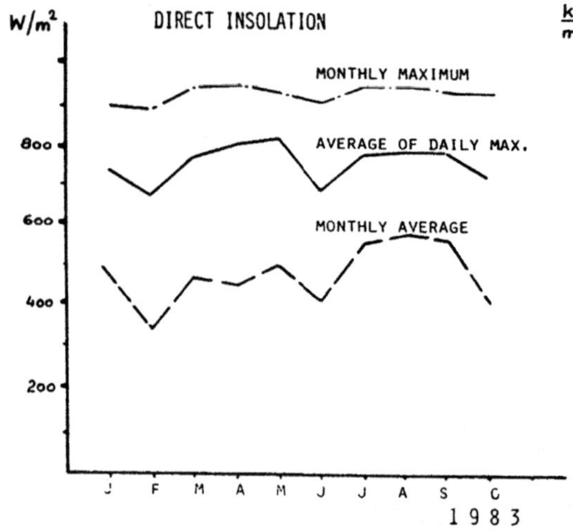

Fig. 7 Direct Insolation Data (Jan. - Oct. 1983)

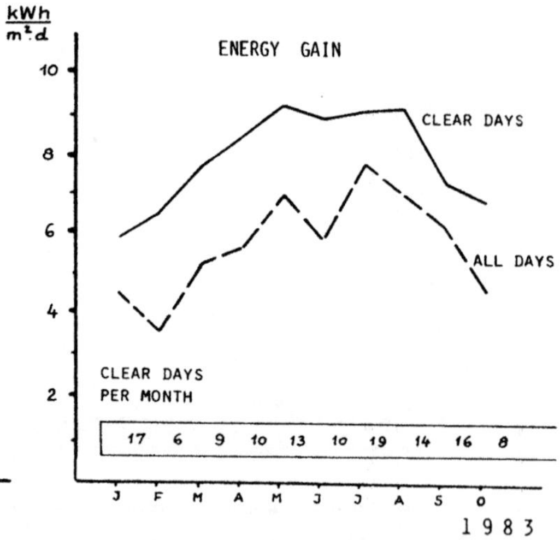

Fig. 8 Irradiated Solar Energy (Jan. - Oct. 1983) per Month

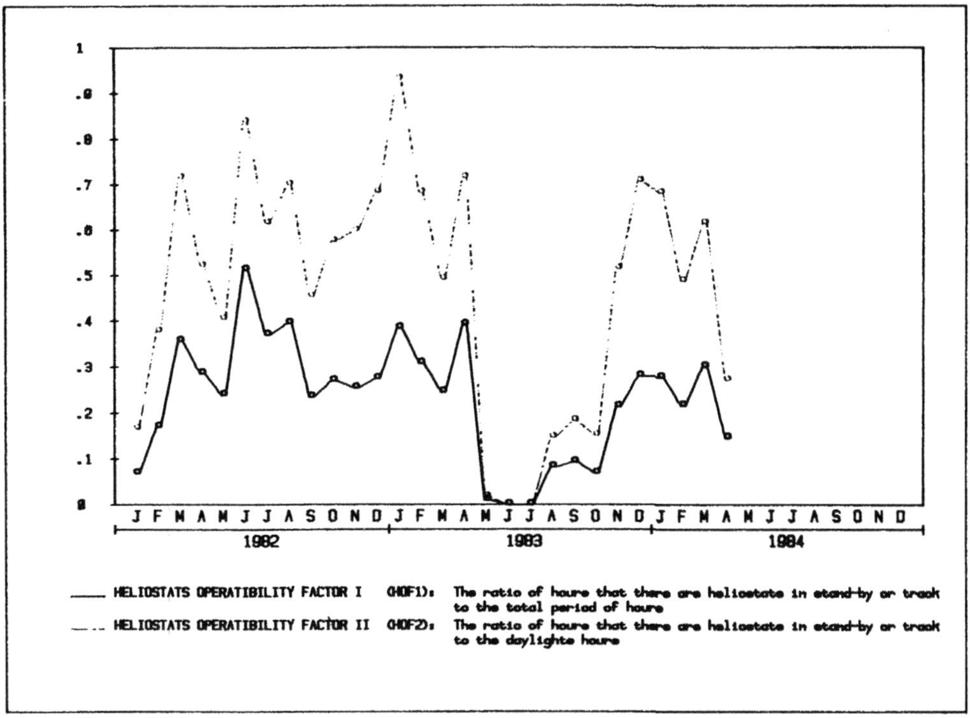

Fig. 9 Operability of SSPS-CRS Heliostat Field (incl. outages due to test preparation/installation)

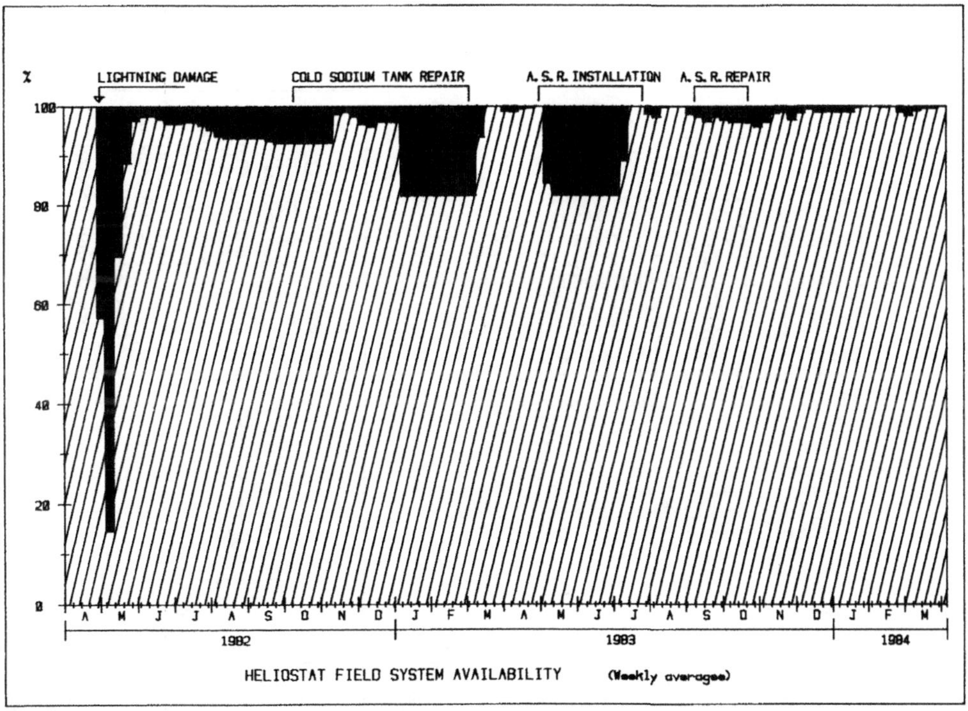

Fig.10 Availability of SSPS-CRS Heliostat Field

Fig. 11 SSPS-CRS Receiver Configurations and their Efficiencies

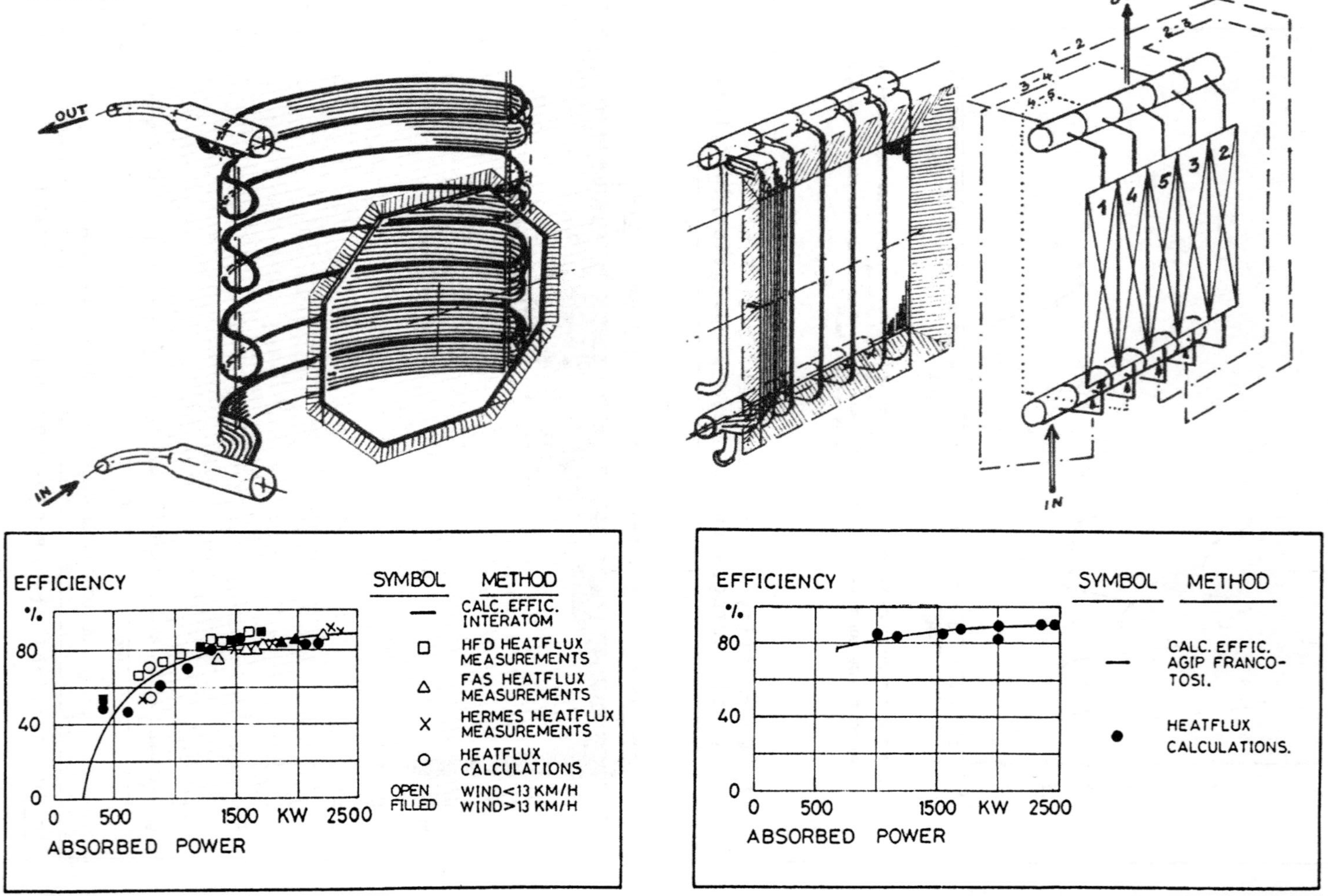

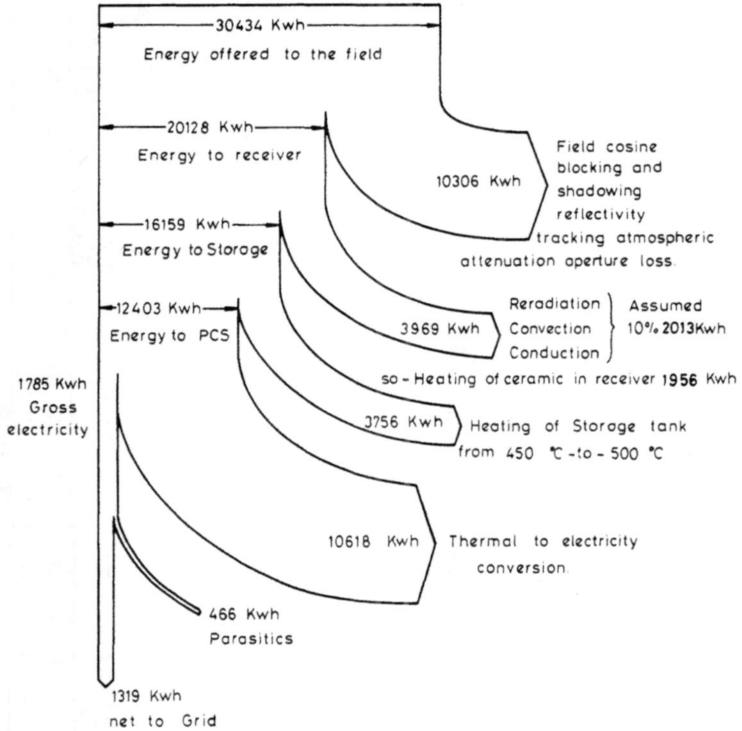

ENERGY **DISTRIBUTION** FOR CRS

April , 9 , 1983 (Paper 1,3)

Fig. 12 SSPS-CRS Energy Flow Diagram as measured on 9/4/1983

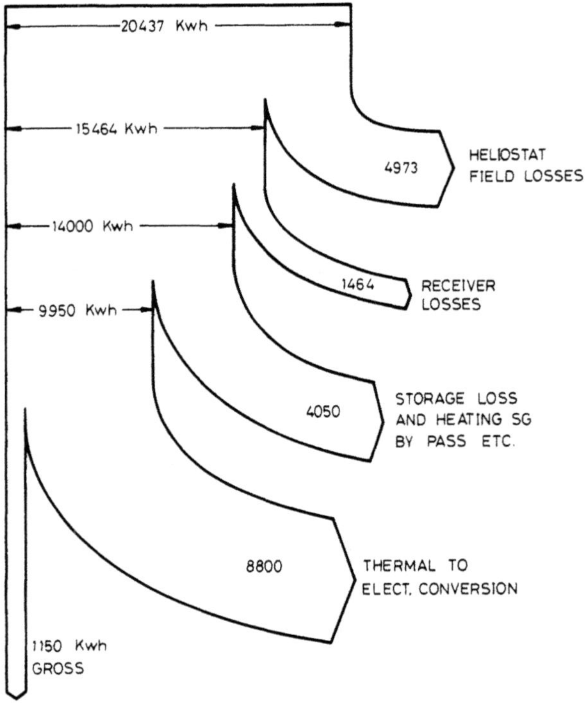

Fig. 13 SSPS-CRS Energy Flow Diagram as measured on23/12/1983

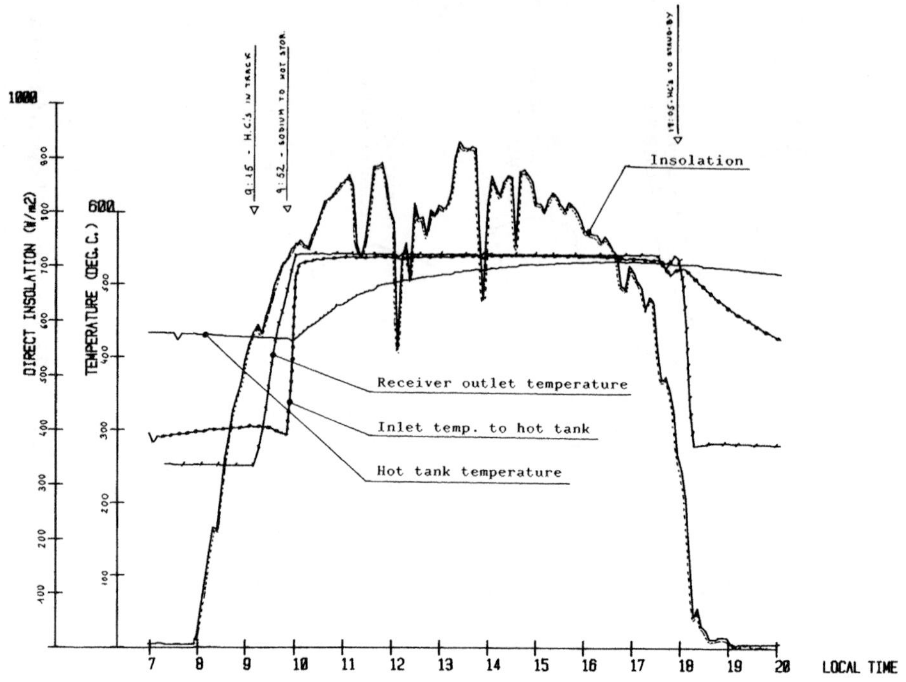

Fig. 14 DAILY OPERATION CHARACTERISTICS 21-FEBRUARY-1984

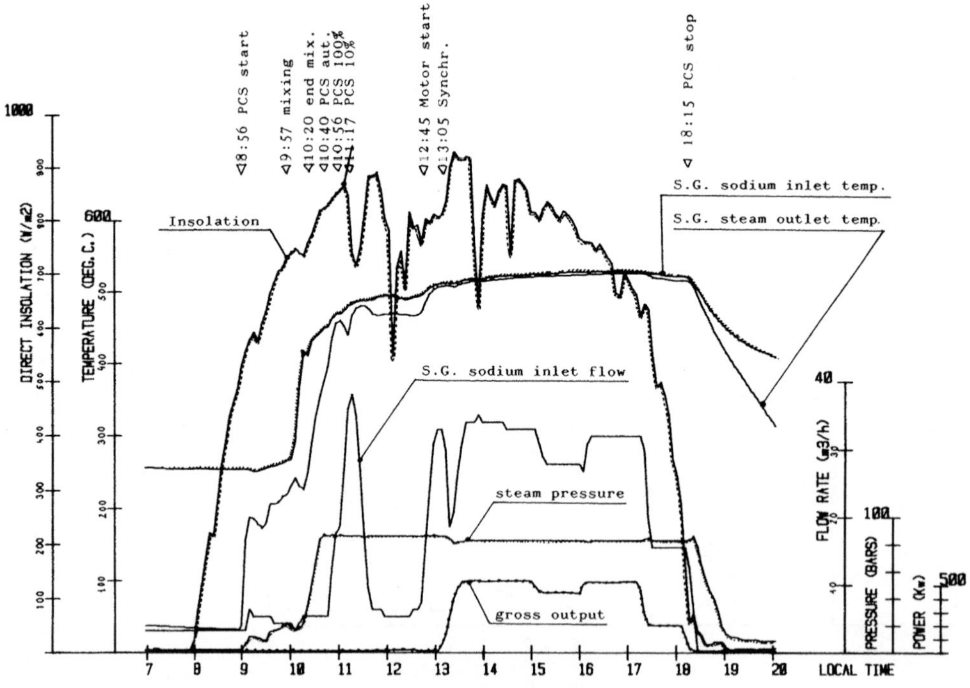

Fig. 15 DAILY OPERATION CHARACTERISTICS 21-FEBRUARY-1984

10 MWe SOLAR THERMAL CENTRAL RECEIVER PILOT PLANT OVERVIEW PART I

A.C. SKINROOD
Sandia National Laboratories
Livermore, California USA

Summary

The 10MWe Solar Thermal Central Receiver Pilot Plant located at Barstow, California, also known as Solar One, is a scale model of a 100 MWe electrical generating plant. Its primary purpose is to provide information for future central receiver power plants. Constructed at a cost of $141.5M, Solar One first delivered electricity to the commercial grid in April 1982. Solar One, consisting of a steam/water receiver surrounded by 1818 heliostats, has met many performance milestones including 10MWe net from receiver steam and 7 MWe net from storage steam. This paper presents an overview to the Solar One plant and assesses the plant's performance as well as the performance of the heliostats, the beam characterization system, the receiver, and the thermal storage subsystem.

1. INTRODUCTION

The 10 MWe Solar Thermal Central Receiver Pilot Plant (also known as Solar One) located at Barstow, California, (Figure 1) is a scale model of a 100 MWe electrical generating plant. Although the plant can supply electricity for 6000 people, its primary purpose is to provide information and data for future power plants of the central receiver type. Authorized by the U.S. Congress in 1976, Solar One is a cooperative project between the U.S. Department of Energy and a group headed by the Sothern California Edison Company.

The principal objectives of the plant are:

(1) To establish the technical feasibility of a solar thermal power plant of the central receiver type, and to identify areas where research and development may lead to significant performance improvements and increased capabilities;

(2) to obtain development, production, operating, and maintenance cost data to support private-sector decisions to invest in solar central receiver energy systems, and identify areas where research and development may most effectively be applied to reduce costs and areas of application of such systems; and

(3) to determine the environmental impacts of the construction, operation, and maintenance of solar thermal central receiver plants.

At Solar One, computer-controlled mirrors (heliostats) reflect and concentrate solar energy to a receiver (boiler) on top of a tall tower. Here, the solar energy heats water and generates steam which is piped to ground level where it is used to either operate a conventional steam turbine or to heat an oil/rock mixture which is part of an energy storage subsystem.

Figure 1. 10 MWe Solar Thermal Central Receiver Pilot Plant

The plant is designed to deliver 10 MW of electricity for a period of 8 hours on the longest clear day during summer when operated from direct receiver steam and 7 MW of electricity for a period of 4 hours when operated from steam generated in the storage subsystem.

The project was constructed at a cost $141.5M: $120M provided by the Department of Energy and $21.5 by the Southern California Edison team. Solar One first delivered electricity to the commercial grid in April 1982, and a Test and Evaluation Phase began in August 1982, and will be completed in August 1984. During this period, all of the operating modes are being checked out, engineering data gathered to evaluate performance, and software installed so that many of the plant functions can be automatically controlled. An overall schedule is shown in Figure 2.

Located on a 48.6 hectare site owned by the Southern California Edison Company, Solar One utilizes 1818 heliostats each with an area of 39 square meters. Each heliostat has 12 mirror modules with a reflectivity of about 91% when clean. The heliostats were designed and manufactured by the Martin Marietta Company. The receiver (Figure 3), designed and manufactured by the Rocketdyne Division of the Rockwell Corporation, can deliver up to 45 megawatts of thermal energy. Water is preheated in six panels located in the south side of the externally configured receiver. This water is then piped to the bottom of 18 boiler panels and is heated as it flows upward until it exits the receiver as superheated steam at a temperature of 515°C and a pressure of 10.4 MPA. The receiver is mounted on top of a 77 meter tall steel tower.

The storage subsystem, (Figure 4) also designed by the Rocketdyne Division, stores energy as sensible heat in 900,000 liters of Caloria oil mixed with 4,350 metric tons of rock and 2,200 metric tons of sand. The oil rock is contained in a single tank and utilizes the thermocline principle with the hot mixture in the upper portion of the tank and the cooler mixture at the bottom of the tank. A series of charging heat exchangers is used to heat the oil using steam from the receiver. The energy is extracted from the oil using a series of discharge heat exchangers. Steam is generated in these heat exchangers at 280°C at a pressure of 27.6 MPA.

2. PERFORMANCE

Generally, the plant is performing well. Many of the performance milestones has been met (Table 1). One factor which slowed the testing was the lower-than-normal insolation at Barstow during 1982 (Figure 5). During 1983, the insolation increased but still was substantially lower than the 25-year average. Insolation during the first few months of 1984 showed further improvement.

During the Test and Evaluation Phase, electricity has been produced whenever possible (consistent with the higher priority of completing the test program). More than 8000 MW-hr net have been delivered while the plant was connected to the grid (Figure 6).

Plant utilization during the hours when insolation is available has fluctuated, as shown in Figure 7. These fluctuations are not surprising in view of the fact that outages have occurred during checkout of the plant. During several months, the goal of 90% availability was achieved.

2.1 Evaluation Program

Sandia National Laboratories has conducted a program to evaluate the following aspects of the Solar One:

- Meteorological Conditions
- Plant Steady-State Performance
- Collector Performance
- Receiver Performance
- Thermal Storage Performance
- Plant Operations
- Plant Maintenance
- Safety
- Lessons Learned
- Plant Costs
- Environmental Assessment
- Automation

Table I. Performance Milestones of Solar One

REQUIREMENT	ACHIEVED	DATE
10.0 MWe NET RECEIVER STEAM	10.4 MW	OCTOBER 1982
7.0 NET STORAGE STEAM	7.3 MW	FEBRUARY 1983
10.0 MWe NET RECEIVER AND STORAGE STEAM	12.1 MW	JUNE 1983
28.0 MWe NET FROM STORAGE	43.4 MWh	MAY 1983
SYNCHRONIZED TO GRID FOR EXTENDED PERIODS	33.6 HOURS ON LINE 127 MWeH DELIVERED	JUNE 1983
90% AVAILABILITY	92%	FEBRUARY 1983
MAXIMUM ENERGY DELIVERED IN ONE DAY	104.3 MeH NET (15 HOURS)	JUNE 1983
HELIOSTATS SURVIVE 90 MPH WIND	60 MPH - NO DAMAGE	NOVEMBER 1983
RECEIVER ACCOMMODATES 0.3 MW/m	0.3 MW/m	APRIL 1982
MINIMUM PLANT POWER LEVEL 2.0 MW	0.5 MW	JANUARY 1983
PLANT SHALL AUTOMATICALLY SEQUENCE TO SAFE CONDITION	DEMONSTRATED	FEBRUARY 1982 TO PRESENT

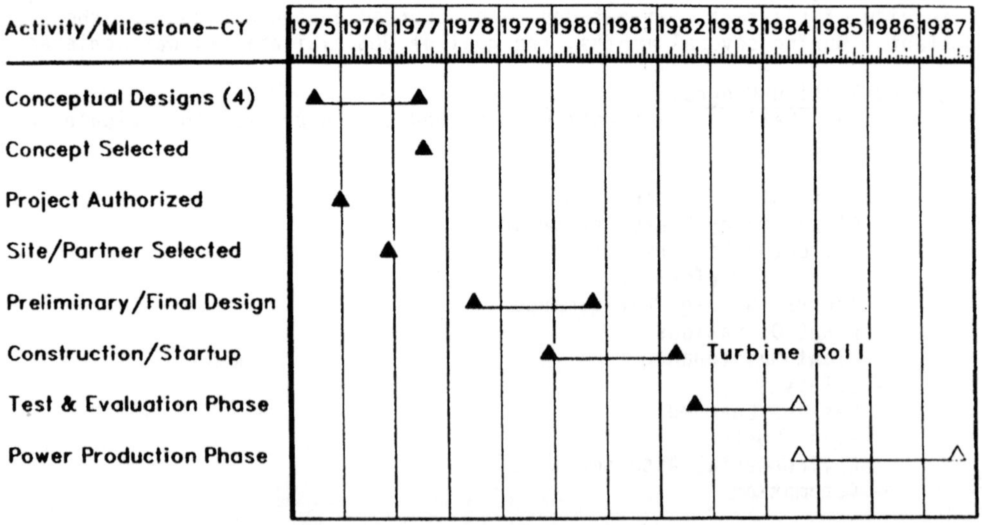

Figure 2. Solar One Schedule

Figure 3. Solar One Central Receiver

Figure 4. Solar One Storage System

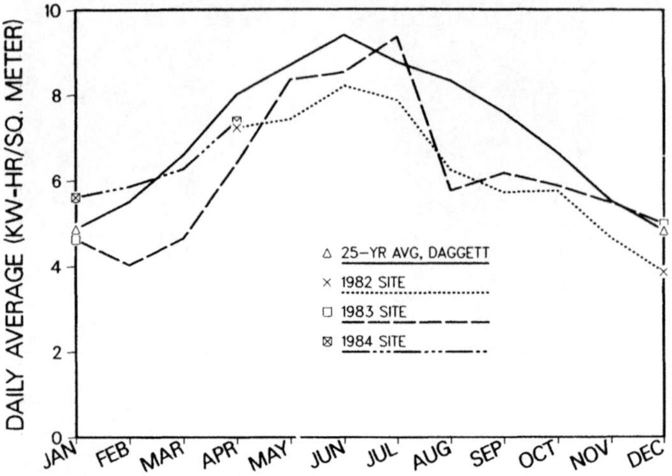

Figure 5. Direct Insolation Comparison at Solar One

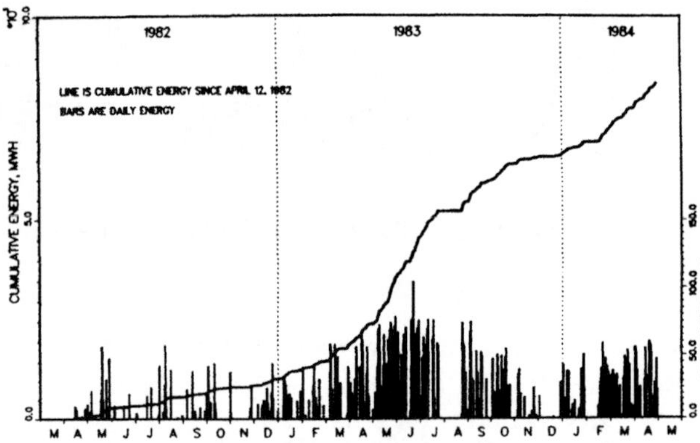

Figure 6. Solar One Net Electrical Production When Grid Connected

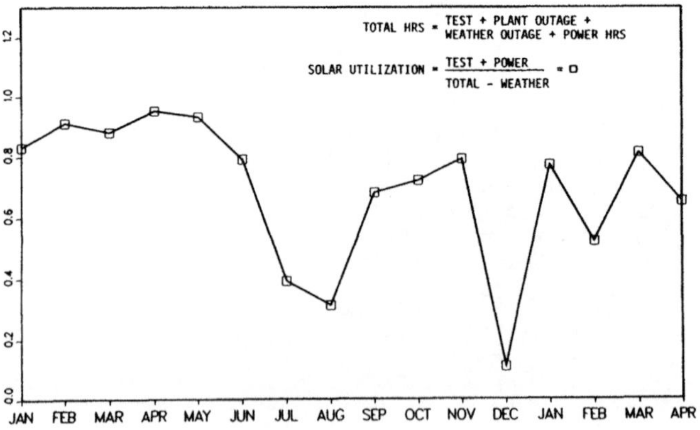

Figure 7. Plant Utilization at Solar One (1983, 1984)

During the Test and Evaluation Phase, measured characteristics of the plant are being compared to the principal objectives stated in the DOE Project Plan, expected performance, and the requirements defined by McDonnell Douglas (the overall design integrator for the plant). Many of the evaluations are well along, and Table 2 lists the reports which have already been published. Final reports on each major subsystem will be published in 1985. Southern California Edison is evaluating the peformance of the plant from the viewpoint of a commercial utility.

2.2 System Performance

An example of the evaluation being done is shown in Figure 8, "Pilot Plant Power Flow." Using a measured insolation value, and the total glass area of the field, a calculation is made of total insolation which would be incident on the field if the heliostats were pointing at the sun (64.33 MW). Calculations are then made of losses due to the fact that 21 of the 1818 heliosts were not operating at the time of the test (heliostat outage). The heliostats are bisecting the angle between the sun and the receiver rather than pointing at the sun; this orientation results in an 18% cosine loss. The measured reflectivity at the time of the test of the mirrors was about 84%, thus causing a further loss. Losses are also due to atmospheric attenuation (3%) and the fact that about 2% of the reflected energy misses the receiver. Thus the calculated solar power incident on the receiver is 40.42 MW. The next value of power, 29.79 MW is the measured energy in the steam the base of the tower. The difference between these two values, 10.63 MW is attributable to reflected solar power and thermal losses in the receiver and downcomer.

Of course, there is some uncertainty in the magnitude of the loss because of uncertainties in the optical performance calculations. These uncertainties arise because factors such as the actual aiming accuracy of the heliostats is not precisely known. The gross electrical output of the generator is measured and the measured parasitic loss subtracted to arrive at the net electrical output of 8.75 MW. These energy balances are quite helpful in understanding the performance of the plant and will be used in projecting the annual energy to be delivered from the plant. A daily energy calculation and measurement is shown in Figure 9.

2.3 Heliostat Evaluation

Performance of the heliostat field is critical to the successful operation of a central receiver plant since all optical losses directly reduce the electrical energy delivered. The overall reliability of the heliostat field at Solar One has been excellent, and maintenance costs are lower than expected. Corrosion has occured on many of the mirrors, but the corrosion area is only .016% of the total area of the field. A thorough investigation was conducted by Sandia as to the cause of the corrosion, and it has been concluded that a manufacturing process change occurred during production of the mirror modules causing some of the mirror module edge seals to leak. This change allowed water to enter the modules and become trapped behind the surface of the mirrors. Vents have been installed in about 10,000 of the 21,816 mirror modules to allow the mirror modules to dry out. The rate of corrosion will be monitored closely, but mirror corrosion is not expected to significantly reduce the electricity delivered in the next few years. One hundred and

Table II. Solar One Evaluation Reports*

EVALUATION CATEGORY	TOPIC/REPORT	DATE
Meteorlogical Summaries	SOLAR ONE: Solar-Thermal Central-Receiver Pilot Plant 1982 Meteorlogical Data Report (SAND83-8216)	5/83
	SOLAR ONE: Solar-Thermal Central-Receiver Pilot Plant 1983 Meteorlogical Data Report (SAND84-8180)	5/84
Plant Transient-State Performance	Analysis of 70-Tube Pilot Plant Solar Receiver Panel Test Data (SAND81-1220)	8/81
	Development of a Relap Model for the Barstow Thermal-Storage Subsystem (SAND81-1831)	10/81
Collector Performance	10 MWe Solar Thermal Central Receiver Pilot Plant Heliostat Experiences (SAND83-8220)	5/83
	10 MWe Solar Thermal Central Receiver Pilot Plant Mirror Module Corrosion Survey (SAND84-8214)	3/84
Receiver Performance	10 MWe Solar-Thermal Central-Receiver Pilot Plant: Receiver Cold Flow (TEST 1010) and the Receiver Steam Generation (TEST 1030) Test Report (DE83009236 - MDC H0141)	1/83
	10 MWe Solar-Thermal Central-Receiver Pilot-Plant Receiver Steam Generation (TEST 1030) Evaluation Report (SAND83-8005)	3/83

*Available From: 'NATIONAL TECHNICAL INFORMATION SERVICE (TIC)'
5285 PORT ROYAL ROAD
SPRINGFIELD, VA 22161
USA

Table II (cont). Solar One Evaluation Reports

EVALUATION CATEGORY	TOPIC/REPORT	DATE
Thermal Storage	10 MWe Solar Thermal Central Receiver Pilot Plant: Thermal Storage Subsystem Activation and Controls Testing Phase (SAND83-8015)	7/83
Lessons Learned	10 MWe Solar-Thermal Central Receiver Pilot Plant	
	Volume 1: Report on Lessons Learned	10/83
	Volume 2: Project Documentation (EPRI AP-3285)	3/84
Environmental Assessment	Wildlife Interactions at Solar One: Final Report (SCE 84-RD-5)	1/84
Other	10 MWe Central Receiver Solar Thermal Pilot Plant Data Evaluation (SAND82-8027)	12/82
	10 MWe Solar Central Receiver Pilot Plant Preoperational Readiness Review Meeting March 9-10, 1982, Barstow, CA (SAND82-2049)	1/83
	Overview of the Construction and Start-up of the 10 MWe Solar Thermal Central Receiver Pilot Plant (SAND83-8021)	6/83
	10 MWe Solar Thermal Central Receiver Pilot Plant: 1982 Operational Test Report (SAND83-8027)	11/83
	10 MWe Solar Thermal Central Receiver Pilot Plant Mid-Term Test and Evaluation Review July 20-21, 1983 (SAND83-8038)	1/84
	Accuracy of Barstow Solar-Thermal Pilot Plant Cost Estimates (SAND83-8217)	6/83

forty-six laminated glass mirror modules manufactured by Solar Kinetics Incorporated will be installed on heliostats to evaluate this alternate method of mirror module construction.

Mirror cleanliness is also important, and measurements at the Barstow location indicate that soiling reduces mirror reflectivity at the rate of 3 to 5 percent per month. A spray rinse, either artificial or due to rainfall, initially restored mirror reflectivity to 95% of its original value. Recent measurements, however, indicate that a film has formed on the mirrors which cannot be removed without mechanical washing action. An experimental wash truck, which has water spray and reciprocating brushes, is now being checked out.

2.4 Beam Characterization System

The Beam Characterization System measures heliostat performance parameters including accuracy of tracking, image shape, and power within the reflected image. Single heliostats are pointed at one of the four reflecting targets located on the tower below the receiver. Measurements are also made of the energy distribution within the sun's disk so that heliostat performance can be compared to theoretical predictions. The measurements are used to realign heliostats whenever necessary. A simple method has been used to identify heliostats which are considerably misaligned. The heliostats are programmed to direct the moon's image at aim points just above the Beam Characterization System Targets, and observations are made from the aim points looking back at the heliostats. If the image of the moon is not visible in the mirror facets, then that heliostat or individual mirror modules is misaligned.

2.5 Receiver Evaluation

The receiver has operated reliably and its efficiency approximates predictions. Outlet steam conditions are surprisingly insensitive to the passage of moderate size clouds. Since the receiver is based on the once-through-to-superheat principal--that is, water enters the receiver in the liquid phase and exits as superheated steam--there was considerable concern during the design phase as to whether the flow would be stable under low flow conditions. In fact, flow in the receiver has been quite stable under a wide range of operating conditions. In addition, there has occurred no plugging of the orifices which were incorporated in the tubes to increase flow stability.

The receiver is quite easy to control and the Southern California Edison operating staff has been pleased with the flexibility of receiver operation. Superheated steam can be introduced into the downcomer as early as 75 minutes after sunrise. Much of the start-up sequence has been automated. The receiver efficiency has decreased slightly since the absorptivity of the painted external surface of the tubes has decreased from .94 to .91. Periodic measurements of absorptivity are made, and the receiver will be repainted with Pyromark paint if substantial further reductions in absorptivity occur.

One area of receiver performance has been somewhat unexpected. Leaks have occured in 7 of the 1680 tubes near the top of the receiver panels. Some of the leaks occur in the first bend outside the zone where concentrated flux is incident on the receiver (Figure 10). These cracks start on the inside surface of a tube and propagate through the wall thickness. The cracks are thought to be caused by rapid cooling of the inner surface of the tubes during shutdown of the receiver. Shutdown procedures have been revised to reduce the rate of cooling, and a laboratory test program has been begun to attempt to duplicate the failures. The second type of crack starts at outer surface of some of the tube-to-tube welds. This type of failure is thought to be caused by differential thermal expansion between the receiver panels and the panel supports. A modification was made to the receiver; it consisted of cutting the welds in the region where the cracks were occurring. Since this modification was done, additional leaks have not occurred.

These tube failures point up the need to build experimental components and subject them to lengthy tests. The first tube leak did not occur until the receiver had been operated for 17 months. An extensive structural and thermal analysis was done on the receiver, but it is not practical or cost effective to analyze every aspect of design. Final performance verification must be done experimentally. Results of another receiver evaluation are shown in Figure 11. Receiver efficiencies were computed using calculated power input and measured thermal outputs. Receiver efficiency decreases rather gradually with increasing ambient wind speed. Although there is some scatter in the data, the overall trend seems to be less reduction efficiency than some of the predictions. Again, this demonstrates the value of experimentally evaluating performance.

2.6 Thermal Storage Subsystem Evaluation

The thermal storage subsystem has been very reliable and has met all of its design requirements. The loss of oil due to degradation is about 6 percent per year, a value which is close to the predicted value. Thermal energy storage has greatly increased the flexibility of operation at Solar One and is routinely used to generate auxilliary steam for sealing the turbine whenever the turbine is not being operated. Although the auxilliary steam generated is not included as electrical energy delivered to the grid, it does substantially decrease the parisitic power used since an electrical steam boiler would have to be used to provide sealing steam. Several leaks have occurred in the charging and discharging of heat exchangers due to frequent rapid thermal cycling. Repairs have greatly reduced flange leakage. Tube-to-tube sheet repairs were done on one of the heat exchangers in an attempt to eliminate leakage in this region, but the heat exchanger soon began to leak again. In January 1984, Sandia began an analytical and measurement program to understand the causes of the leaks. It is clear that standard heat exchanger design practices will not be adequate for solar plants because of frequent rapid thermal transients which are inherent in the technology.

3. CONCLUSION

In summary, the plant has performed well and the feasibility of the central receiver concept on the 10 MWe scale has been demonstrated. Much has been learned, but considerable work remains to be done. Evaluation of plant performance during the next phase--when the plant is operated for the primary purpose of power production--will be completed during the next 3 years.

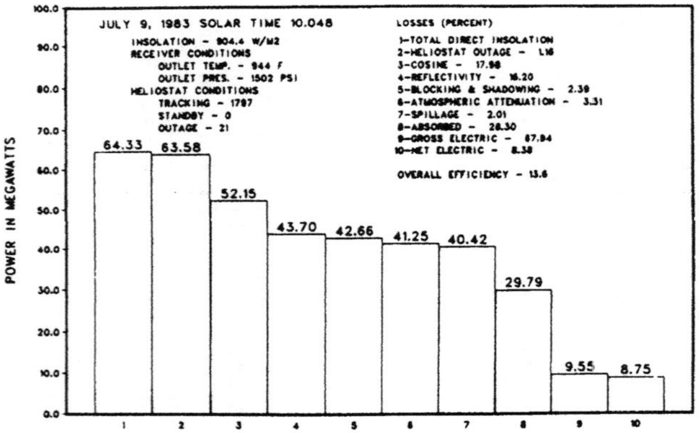

Figure 8. Pilot Plant Power Flow

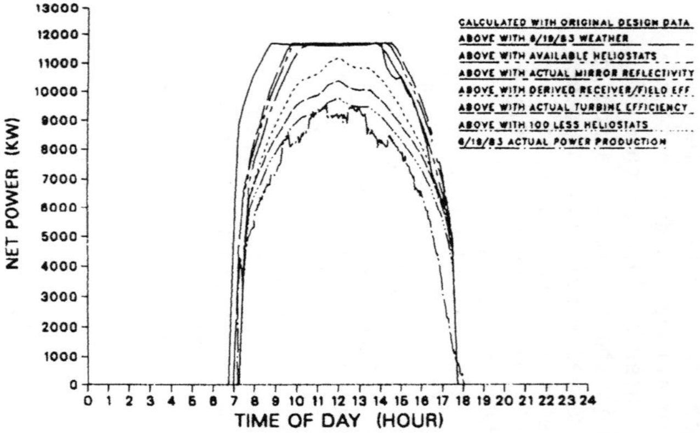

Figure 9. Summer Solstice Power Production for Solar One

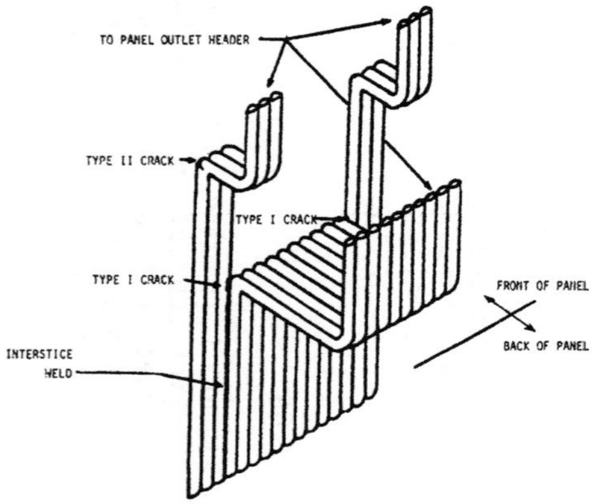

Figure 10. Schematic of Receiver Tube Panel Showing Location of Failure

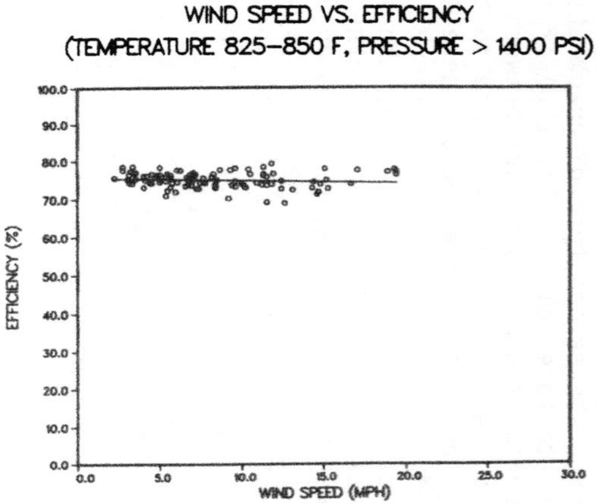

Figure 11. Receiver Evaluation of Solar One

SESSION I - PART 2

PRESENTATION ON THE GENERAL EXPERIENCE OF CENTRAL RECEIVER POWER PLANTS

Summary of the Session by the Rapporteur
L. VANT-HULL, University of Houston, USA

A 1 MWe solar thermal electric power pilot plant (Sunshine project)

Themis plant operation progress report

CESA-1 project status report

SUMMARY OF THE SESSION BY THE RAPPORTEUR

L. VANT-HULL
University of Houston, USA

1. INTRODUCTION

The following projects were presented in this session:
NIO Project (Japan) by N. Ikeda,
THEMIS Project (France) by F. Pharabod,
CESA I Project (Spain) by C. Ortiz.

2. PROJECTS

2.1 NIO Project

The plant design of NIO is a surround field with a 60m tower. The cavity receiver has a downward face of 8.5 m Ø. In the field, there are 807 close-packed 16 m^2 geometrically linked heliostats. The inlet receiver temperature is 187°C, and the outlet temperature is 249°C. Saturated steam is the working fluid and storage.

The first operation of the plant was September 1981. In August 1981, the plant met its 1 MWe outlet design goal.

Quasi sun following mode (daytime production of electricity), minimizing utilization of thermal storage, produced 2-3 times the electrical output compared to a quasi load following mode (evening operation).

The heliostat surface error was greater than 1° up to 10 m/sec wind velocity but grows rapidly above 10 m/sec. Electric production was 0 - 2.5 KW hr/m^2 day. The "2.5" depends on operating mode, season, and number of storage tanks in service.

Initial five steam accumulators were excessive; so if collected energy was all put into accumulators for a day, almost no electricity could be produced due to thermal losses. Reduction to one accumulator, or direct turbine operation, improved this situation.

Most significant problem has been peeling of the selective receiver surface (observed at annual inspection in 1982 and 1983).

These are the recommendations for the next system:

* reduce costs
* increase availability
  - resolve thermal stress problems
  - reduce heat losses
* consider system using optical fibers or light channels for conduction of collected energy.

The following question-answer period followed the presentation:

J. Gretz

Q - Why the low steam condition, 249°C?

A - To increase receiver efficiency and reduce heat loss at night.

Wattiez

Q - Can NIO change to a Photovoltaic plant?

A - Not appropriate for me to answer. Japan is changing direction of development.

W. Grasse

Q - NIO is very near the sea. Do you have any trouble with mirror degradation?

A - NO degradation of mirrors at NIO; they are laminated glass. But parabolic plant has bad trouble.

2.2 THEMIS Project

Operation of THEMIS has shown value of design concept: salt flow to receiver to storage tank; salt flow in steam generator to cold storage tank. This mode provides isolation of the steam generator and turbine from solar transients. From February to April 1984, field was built up to 199 operating heliostats--2.3 MW thermal, e.g. 100%.

THEMIS operation is very flexible; it can start operation at irradiance greater than 100 w/m^2. In 1984, the staff has 40 men. In 1985, the staff will be down to 25.

For the 1983-84 year, the following achievements were made:

* insolation matched nominal! (in contrast to reduced insolation at Barstow);
* mirrors are now stowed face up and locked in position to withstand high winds;
* pointing accuracy error equals 2.7 mr, sigma equals 1.8 mr and beam distribution approximates theoretical value for individual heliostats.

The lessons learned are as follows:

* receiver tubes are performing well;
* small pipes have plugged; vents and drains have failed;
* 18 flux probes in wall used to measure flux--excessive flux on;
* achieved desired (predicted and design value) of flux distribution with focus about 1 meter in front of aperature and fine control;
* salt flow, ending with passes to floor and roof of receiver, is satisfactory and protects hot salt from overheating;
* receiver heat up is very fast--about 10 minutes;
* heliostat field - receiver efficiency is about 75% at noon;
* annual average values are not yet available.

THEMIS will follow a 2 year program to achieve good thermal balance and closure of losses, then operate for one year to test out subsystems, etc.

Luis

Q Why change to face up stow?

A Back structure (trusses) catch wind when up.

Q Lightning?

A This was a problem in the first year; now have three horizontal "shield" wires and one on tower, no problems recently.

C. J. Winter

Q We appreciate fast start up and high efficiency. Why is receiver inclined downward?

A The cubic cavity requires radiation on all surfaces, including the floor.

Q What are the parabolic dishes for?

A To prevent freezing of HITEC Salt (340°C), so need to preheat piping with pressurized water from dishes. Receiver tubes preheated electrically.

Q Does face up stow affect mirror soiling?

A Probably would in dusty area, but THEMIS has clean air and rain.

M. Becker

Q Receiver flux shows surplus on lower part, excess on higher part. Why not equalize by adjusting aim points in aperture plane?

A Tried this but it resulted in some hot spots.

Q Reason for requested profile?

A Design condition.

H. Kleinrath

Q 21% efficiency from sunlight to electricty is higher than others have presented.

A 21% is at design point and good conditions.

Q Why start up so fast?

A Operation is independent of sun because of sizable salt storage.

C. S. Selvage

Q How are flux measurements accomplished?

A Scanning bar is not yet in operation. IR camera is used.

Q Discuss parasitics.

A Positive power output is achieved on single day basis, but not monthly.

R. Gervais

Q Can you quantify auxillary power required for trace heating?

A Trace heating is used rarely since primary loop is kept hot at night by circulating salt from cold tank through receiver. Cold start up requires twelve hours.

Q How long can this "standby" mode last?

A Several weeks.

A. Baker

Q SSPS showed long times to heat storage media.

A The THEMIS turbine is more "forgiving" than the Spilling engine used at SSPS. We can operate turbine at 350°C while SSPS requires 520°C. Receiver outlet reaches 420°C in 15 minutes. Also, turbine operation depends only on availability of hot salt in tank.

C. J. Winter

Q Solar multiple?

A 1.0 (0.8 for SSPS-Selvage).

2.3 CESA-I

The CESA-I plants has a fasttrack schedule: 2 years from start of basic design to start of construction; 3.5 years construction process. In 1985, the plant will test GAST components. As far as cost, the 1983 running cost equalled 5% of R&D and construction cost.

The system design of CESA-I is as follows: north field steam generating cavity receiver; 300 heliostats, each 38 m^2; 16 MW-hr of storage in HITEC, 4 hrs at 4 MWt-hr. CESA-I began running its turbine on October 23, 1983--synchronized at sunset. Charging loop accepted December 1983. Discharge loop accepted around July 1984.

The commissioning of CESA-I is as follows:

* heliostat field--no problems;
* receiver--10 weeks of delays due to pump seal failure, vent tube cracking (bad manufacturing procedure); also, superheater water traps required due to large thermal inertia;
* turbine--failures and modification (22 weeks delay), thrust bearing failure, electrohydraulic control problems, distortion of turbine casing by steam flow, exhaust pipe modification.

CESA-I requires a 33 man crew for 5 day/week operation plus 2 day/week "standby." In 1983, irradiation was greater than 950 w/m^2 nearly every clear day. The heliostat availability was 80-90%.

During commissioning, these are the following nominal values (do not project):

* 3500 sun hours
* 2500 insolation hours
* 2100 field hours
* 865 receiver hours
* 96 turbine hours
* 60 storage hours.

The collector field - receiver were working well, responsive. The receiver was sensitive to heat flux maldistribution. Thermal storage has excessive inertia. In mirror cleaning, it was found out that wiper is 5% better than spray, that dust builds up about 0.1% per day, and that about 70% of mirrors started to show corrosion.

In cold start, the steam drum pressure reaches "nominal" value in 2.5 hours. Receiver efficiency is about 80% (highest supposed inlet values seem to give lower efficiency, perhaps because inlet solar energy overestimated).

J. N. Reeves

Q Am I the only one that kills birds? (At Solar One)

A Bugs and flies are being burnt here. One sparrow was burned, one hawk blinded and killed.

A 1MWe SOLAR THERMAL ELECTRIC POWER PILOT PLANT (SUNSHINE PROJECT)

T. MUKAI
Agency of Industrial Science and Technology, MITI,
T. HORIGOME, N. IKEDA, and T. SAKAMOTO
Solar Energy Department,New Energy Development Organization

Summary

The results and experiences of a 1 MWe solar thermal electric power pilot plant with central receiver tower at Nio are reported. The main objectives of this project are to verify the technical feasibility of solar thermal electric power generation plant and to search the economical feasibility of it as one of the national "SUNSHINE PROJECT".
The pilot plant was constructed in 1981 and it succeeded in the operation of 1,000 kWe rated electric power output at August, 1981. Its stable operation has been ascertained. Several kinds of operation were studied to find the optimal operation method and to make the operational standard. It was found that operation mode that solar energy collected in a day was consumed completely in that day for electric power generation is most suitable (we call this mode quasi-standard operation mode). The operation of this pilot plant brought us various data and information. Especially, it brought us a prospect of one more renewable electric energy source, which has yet difficult problems to the practical solar thermal plant with economical efficiency.

1. INTRODUCTION

This project of a 1 MWe solar thermal electric power pilot plant at Nio is one of SUNSHINE PROJECT. The objectives of this project are to search for technical and economical feasibilities of solar thermal electric power generation plants with a central receiver tower. More specifically, they are to verify the technical feasibility, to obtain various data and information for construction, to accumulate plant operation experiences, to compare with another 1,000 kWe class solar thermal electric power plant with plane-parabolic system, and to clarify the problems to be solved prior to the practical use of solar systems.

The funding organization of the project is Agency of Industrial Science and Technology (AIST), MITI. New Energy Development Organization (NEDO) is entrusted with the project and administer it. The major supplier of this 1 MWe pilot plant is Mitsubishi Heavey Industries, Ltd. (MHI), and Electric Power Development Co.,Ltd. (EPDC) is responsible for the technical management and operation of the plant.

2. OUTLINE OF THE PILOT PLANT

In solar thermal electric power system of the pilot plant at Nio, solar energy is concentrated on the receiver at the top of the tower by heliostat mirror system, and the high pressure saturated steam is generated in the receiver. The steam is taken out from a drum at the top of the receiver and then charged into the steam accumulators. An impulse turbine for electric power generation is driven by the steam from the accumulators.This plant generates the rated electricity of 1 MWe (gross). Conceptual diagram of the system of the plant is shown in Fig. 1. The specifications of the plant are presented in Table 1.

Each heliostat has mirror area of 16 m^2 (4 m x 4 m) and is so-called G-L (geometrical linked) mounting which has two supporting poles with a rotating axis directed to the North Star. The receiver is a semi-cavity type with a cone-shaped and a cylindrical membraneous walls. The inner surface of the cone and the outer surface of the cylindrical pole are the heat receiving surfaces. The receiver generates saturated steam by natural circulation.

Thermal storage system is composed of 5 steam accumulators. Its capacity is 3 hrs of rated output. We already reported the detailed discription of this plant at the first workshop.

3. OPERATION OF THE PILOT PLANT

3.1 Operation methods

Four modes of operation were carried out during the operational studies. The modes are as follows (Fig. 2) :

1) Standard load pattern having a rated output of 2 hrs (1 hr in winter) in the evening and reduced output in the daytime. The standard load pattern was designed to follow the local load pattern.
2) Quasi-standard load pattern having a rated output of several hours in the daytime and reduced output in the evening.
3) Following insolation pattern with high and low, two constant load levels.
4) Mode with a constant load at rated output.

3.2 Research procedure of the plant

The fundamental design of this plant was completed in 1977, and construction was finished in 1981. Scince the approval by the government on August 31, 1981, operational studies, including electric power generation, had been conducted. The operation came to an end at the end of March, 1984. The research procedure of the plant operation studies is shown in Fig. 3.

4. GENERAL RESULTS OF THE OPERATIONAL RESEARCH

1) The pilot plant succeeded in the generation of 1,000 kWe rated electric power output in August , 1981, and its stable operation has been ascertained.
2) In the initial stage, the pilot plant was operated according to the similar operation standard and method of conventional thermal power plants.
3) In order to find the optimal operation method and to make the operational standard, four modes of operation were studied during two and a half years. It was found from the operational studies that the operation mode with a quasi-standard load pattern was the most suitable to get a maximum generated electricity and 2 or 3 times of electricity by this operation mode was obtained compared with that by the standard load pattern.
4) Large amount of data in the operational studies were accumulated. Especially, the relation between the generated electricity and

solar insolation, the threshold value of the insolation necessary to generate electric power, and several efficiencies concerned the plant were clarified.

5) Simulation program for whole system of the plant was developed.
6) Many know-hows were gained during operational studies through the troubles such as cracking in the receiver and valves, out of control of heliostats by lightning and freezing.
7) It was found that the weather conditions such as clouds and haze in the morning, had great influence on the electric power generation.

5. CHARACTERISTICS OF SUBSYSTEMS AND COMPONENTS

5.1 Tracking accuracy of heliostats

Sixteen reflected light sensors were distributed in the heliostats field. Tracking accuracy was measured by the sensors.

Average error patterns of the tracking are shown in Fig. 4.a. In this Figure, each diameter of the circle indicates the amount of the error at it's position of the heliostat. Tracking accuracies are almost within 1 degree except several points. For seasonal dependance of the tracking error, an example of a heliostat is shown in Fig. 4.b. It is found that the tracking error in all seasons have minimum values around noon.

5.2 Wind pressure load of heliostats

Heliostats of the pilot plant had been checked on the wind pressure load by wind tunnel. Displacements at the end of mirror panel of heliostat are shown in Fig. 5 in the relation of the mirror angles. There are some differences between the measured values at Nio and test values in the wind tunnel. The measured values are smaller than the test values. It is found from the measurement that the real displacement of the mirror panel is within 10 mm below 18 m/sec of wind velocity, and displacement is too small to be measured at the design value of wind velocity 3.5 m/sec.

5.3 Durability of absorbing surface of the receiver

Eight test pieces for durability study were installed at the surfaces of the receiver in four directions. An example of test results is shown in Fig. 6. Parameters in the figure are; "Fresh" is for initial, "Run 1" for exposed for 3 months, "Run 2" for 6 months, "Run 3" for a year, and "Run 4" for one and a half years, respectively.

It was found that the reflectance was not in general depend on the wave length, and it has increased only by a few percentages and there were no great drop in characteristics after one and a half years.

5.4 Characteristics of mirror reflectance

Outdoor test of mirror reflectance were conducted. There were no change of dependance on the wave length. Weather dependence on the mirror reflectance is shown in Fig. 7. It shows large decreases of it after windy days. The cleaning was very effective because the reflectance recovered to almost initial value. Small amount of rainfall (below 20 mm) did not effect its reflectance. Rainfall above 20 mm showed some effects in recovery of it.

6. RESULTS OF OPERATION

The objectives of this pilot plant was to confirm the electric power production from the solar energy, to make sure of some technical and operational methods and to have the first experience of solar thermal power generation system. It is expected from the operational studies that the plant can be operated continuously for a long time.

6.1 Electric power production

Generated electricity of each month is shown in Fig. 8. Total plant efficiency, normal direct insolation and expected values of generated electricity are also shown in the figure. It is found that the generated electricity is almost proportional to the insolation. This relation is shown in detail in Fig. 9.

6.2 Relation between generated electricity and normaldirect insolation

It is important for design and operation of the plant to know the relation between generated electricity and normal direct insolation. This relation is shown in Fig. 10. Many points were added to the figure since last workshop at Los Angeles. There are a little seasonal differences in threshold values of normal direct insolation for electricity generation. Their values are 2.5 kWh/m^2 day in spring, 1.93 kWh/m^2day in summer, 2.44 kWh/ m^2day in autumn and 3.60kWh/m2day in winter. It is found that the threshold values depend on number of the accumulators and on the operation pattern.

6.3 Plant efficiency

We take here the definition of the plant efficiency (monthly mean):

$$\text{Plant efficiency} = \frac{\text{generated electricity}}{\text{total mirror area x normal direct insolation}}$$

In the denominator, total mirror area is a summation of the mirror area operated within the given periods. Plant efficiency of the Nio plant is about 4 % as already shown in Fig. 8.

7. EFFORTS TOWARDS OPTIMAL METHOD OF OPERATION

Only gathering the solar energy into accumulators and no electricity production was experienced in initial phase of operational research. The reason was too much volume of accumulators for collector system and unexpected small normal direct insolation. Number of accumulators was decreased from 5 to 2 or 1 in order to make short the time for starting up the plant. At the same time, thermal insulation of components and pipes were reexamined.

To improve the plant efficiency, operation at 1,000 kWe output were conducted as long as possible and low pressure operation were followed after rated operation. Collected solar energy in a day was consumed completely in that day for electric power generation to minimize thermal loss in the nights. We called the operation mode quasi-standard mode. The consumption of electricity in the plant was decreased by stopping cooling pump in nights.

Control program was also revised to meet this optimal operation. Solar energy must be collected as much as possible to make the synchronizing time earlier. The whole heliostats were operated from the beginning of collecting phase, because it was found that the temperature change rate of the receiver did not exceed the design limit (110°C/hr) of it.

8. MAINTENANCE

Quality of feed water and condenser water was monitored by measuring PH and electric conductivity(Fig.11). Annual regular inspectin were conducted 2 times during operation. Cleanings of mirrors were also done 5 times.

9. EXPERIENCED TROUBLES

We had a few troubles during the operation period. The troubles were cracking of drain tube of the receiver (August 1981) and inlet valve of an accumulator (August 1982), out of control of heliostats (August 1982) by lightning, exfoliation of plasma jet coating of the receiver surface (October 1982, annual regular inspection), out of drive of heliostats by rusting and freezing (February 1983), over-run of heliostats by lightning (August 1983), backward flow between accumulators by siphonage (March 1983) and exfoliation of the coatings of the receiver surface (October 1983, annual regular inspection). It is found that important troubles for durability such an exfoliation of the receiver surface were included in them.

10. SIMULATION

Simulation code was developed and measured results are now under comparison with corresponding calculated values. As example of comparison between measurements and simulation, pressures and water levels in the accumulator are shown in Fig. 12. They show good agreement with each other. Comparison of electric power outputs shows also good as shown in the figure. Expected values of electric power output already shown in Fig. 8 were calculated by this simulation code.

11. PROBLEMS TO BE SOLVED

There are many problems to be solved for commercial solar thermal electric power plants. Economical problem is the most important. Electricity cost of the plant has to be decreased to that produced by conventional commercial power plants. For the object of this, construction cost must be decreased below 10^6 yen /kW. Especially heliostat cost which is about 50 % of total cost must be reduced. Increase of availability is important factor to decrease the electricity cost. Solar multiple of the plants and the plant with auxiliary thermal source have to be examined in earnest. Technical problems are reliability and durability of components, efficiencies, heat loss and wind load.

Each component, especially receiver, in this plant suffers severe thermal stress every day. Such a severe condition has never been experienced in conventional power plants. It is necessary for reliability and durability to develop withstanding components. Heat loss is essential problem in the plants. New technology should be researched from the point of long distance heat transfer.

12. FUTURE PLAN

Nio plant is evaluated thoroughly in technical and economical points and the gathered data during operation is put in order and the problems to be solved is clarified in this year. In addition, researches for the new systems and development of components will be continued for the practical use of solar thermal energy. System analysis about electric and thermal cogenerating system,central receiver solar power plant with compact sodium loop and solar thermal power system with optical fibers and channels etc. will be conducted. However,Japan now has no plan of construction of a large solar thermal electric power plant in near future.

ACKOWLEDGEMENTS The authors wish to thanks to Electric Power Development Co. and Mitsubishi Heavy Industries Ltd.

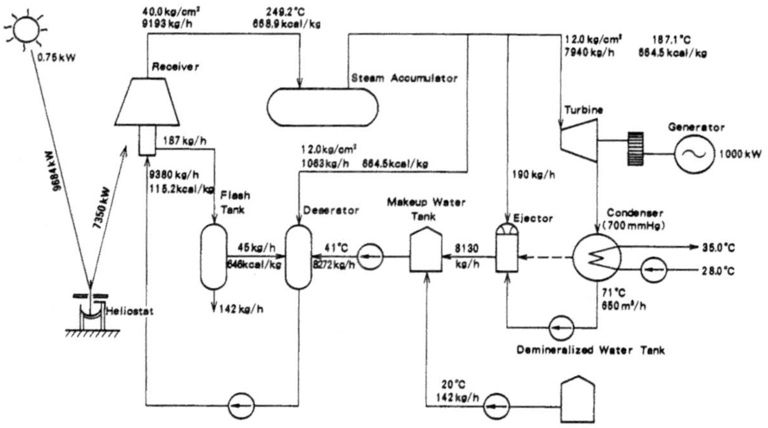

Fig. 1 Conceptual Diagram of the Pilot Plant

Table 1 Specifications of the Plant

Heliostat		
	reflector area	16 m² (4 m × 4 m)
	number of heliostats	807
	total reflective surface area	12,912 m²
Receiver		
	cone-shaped cavity type	
	diameter of cavity	8.5 m
	steam at receiver outlet	40 kg/cm², 249°C
	steam flow	9,200 kg/h
	tower height	69 m
Thermal storage		
	steam accumulator	60 m³ × 5 units
	pressure range	13 ~ 40 kg/cm² abs.
Turbine		
	impulse turbine	
	inlet condition	12 kg/cm², 187°C

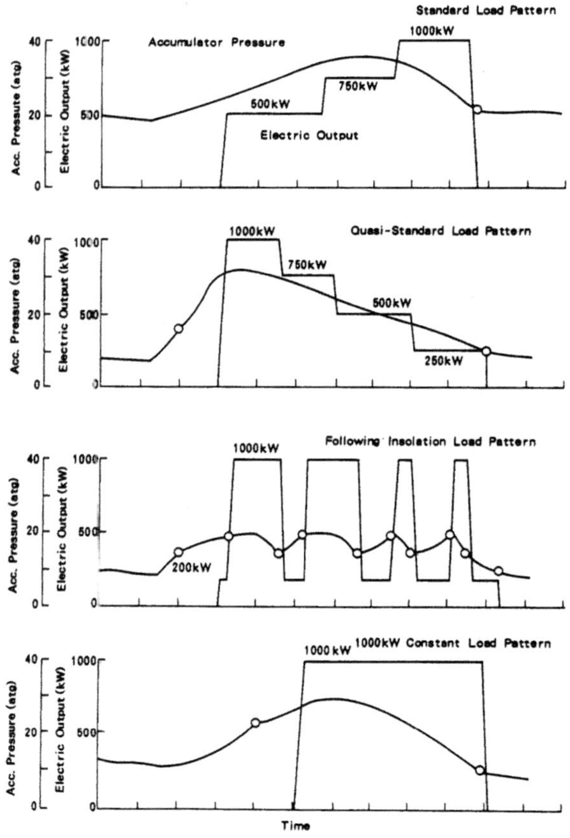

Fig. 2 Operation Mode of the Plant

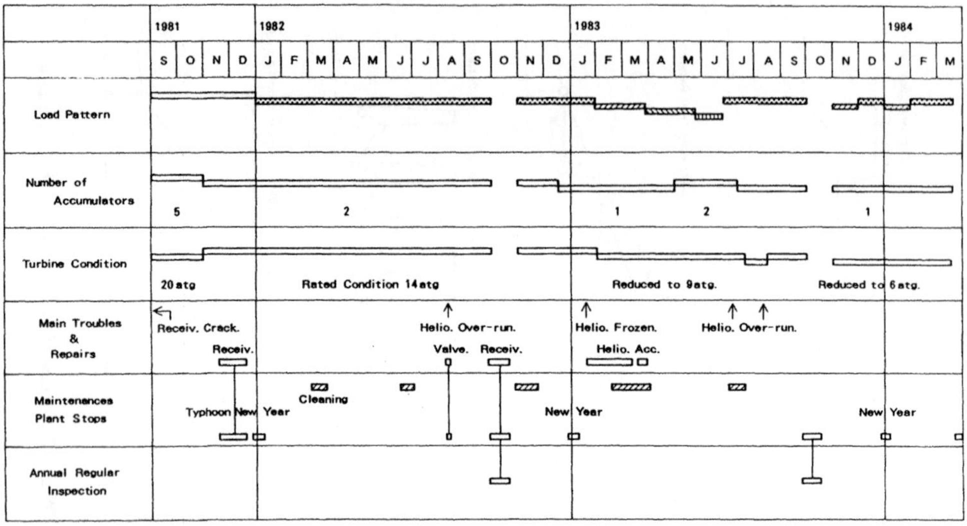

Load Pattern

Standard: Reduced output in day time and rated output in the evening leaving 1 MWeh worth storage.

Quasi-Standard: "Standard" with no thermal storage leaving.

Following Insolation: Load pattern with following insolation.

500kW Constant: 500kWe (1/2 rated) constant load pattern.

1,000kW Constant: 1,000kWe (rated output) constant load pattern.

Fig. 3 OPERATING SCHEDULE OF NIO PILOT PLANT (CRS)

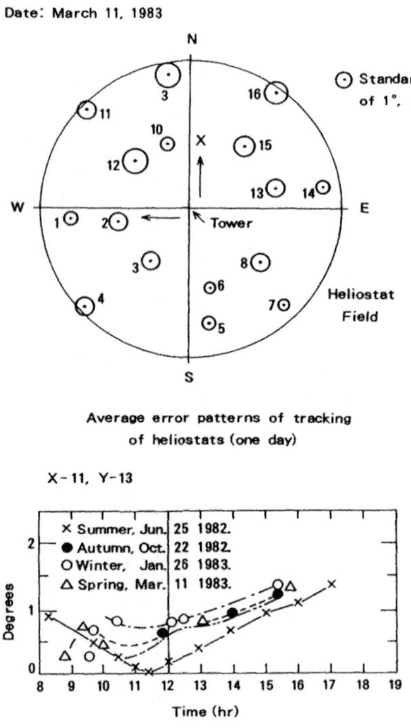

Average error patterns of tracking of heliostats (one day)

Seasonnal dependence of the tracking errors

Fig. 4 Tracking Characteristics

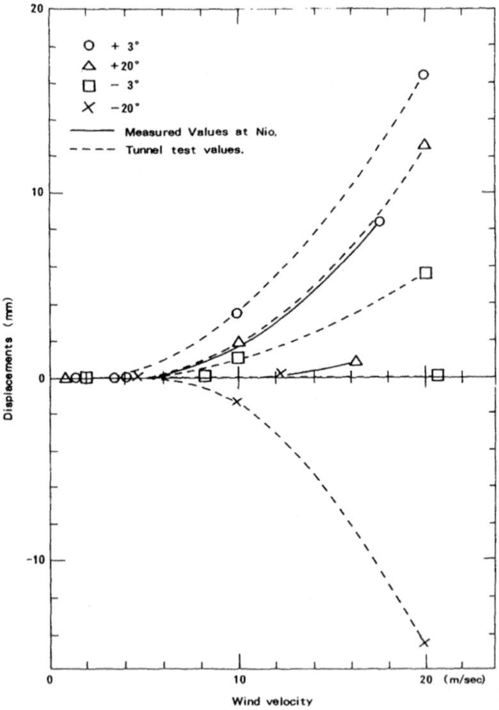

Displacements at the end of the mirror panel.

Fig. 5 Displacement of the Mirror

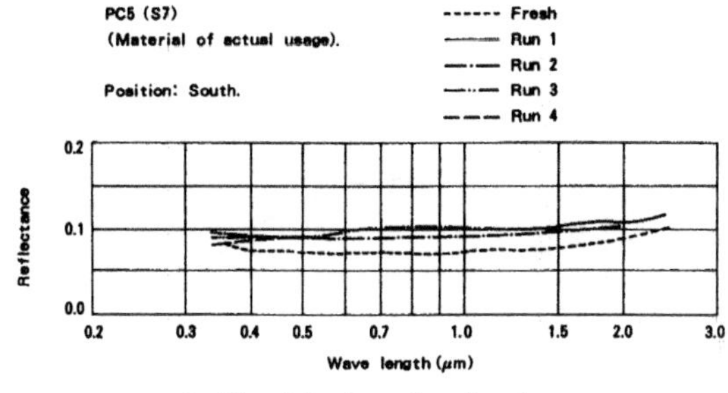

Durability of absorbing surface of receiver

Fig. 6 Durability of Receiver Surface

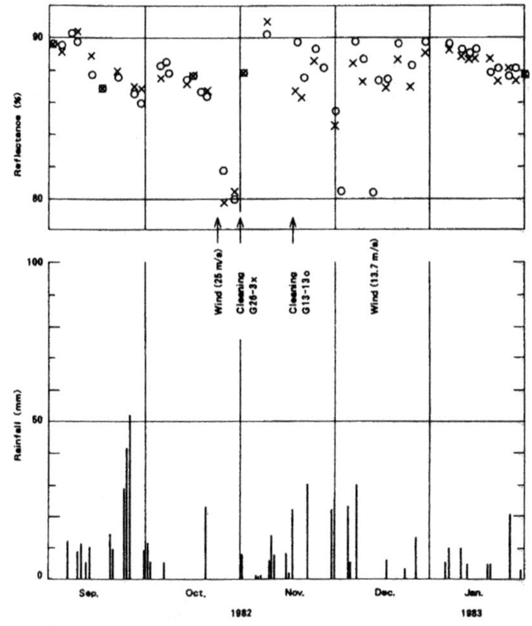

Fig. 7 Characteristics of Mirror Reflectance

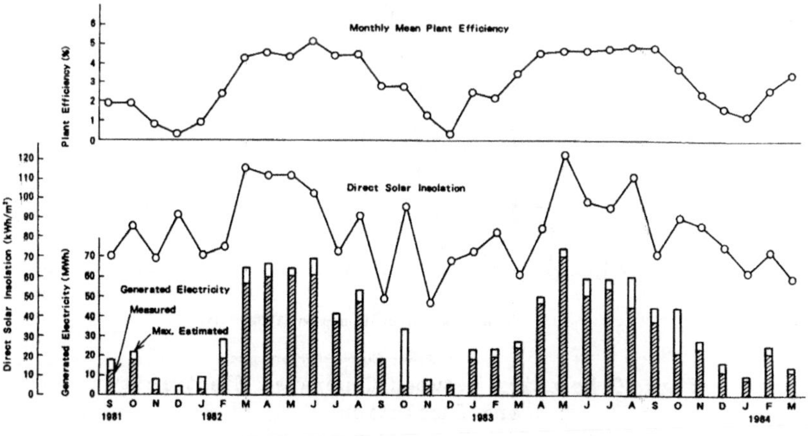

Fig. 8 Power Generation Performance

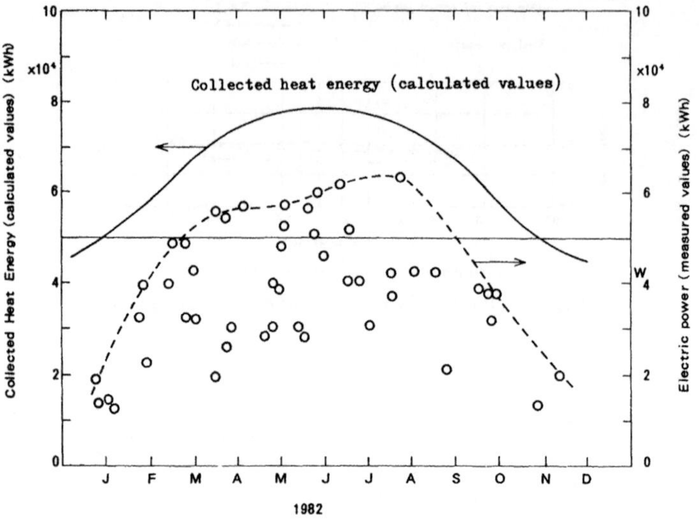

Fig. 9 Collected Heat Energy and Power Output

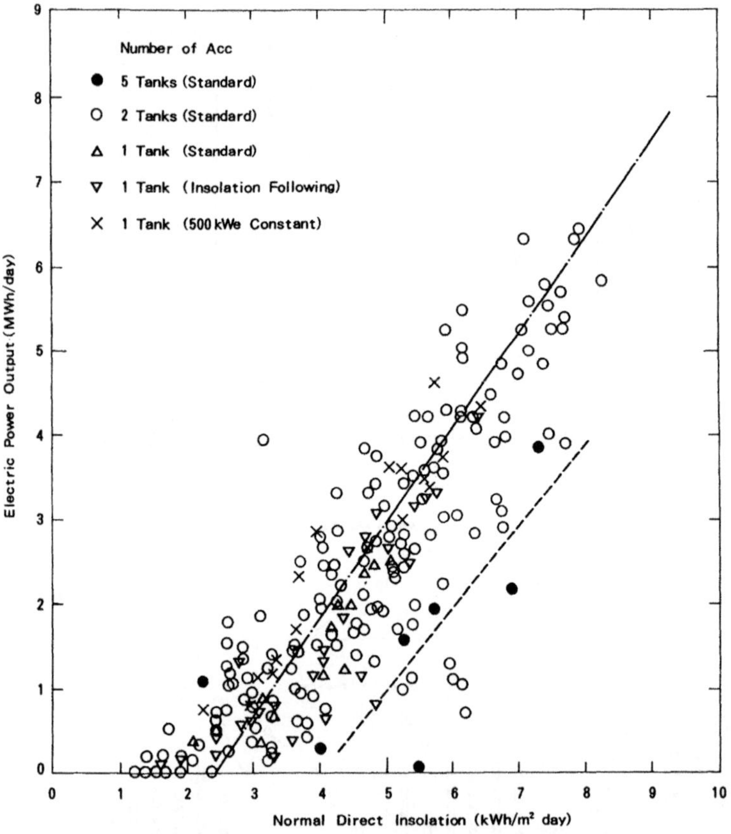

Power Output Performance

Fig. 10 Relation between Generated Electricity and Insolation

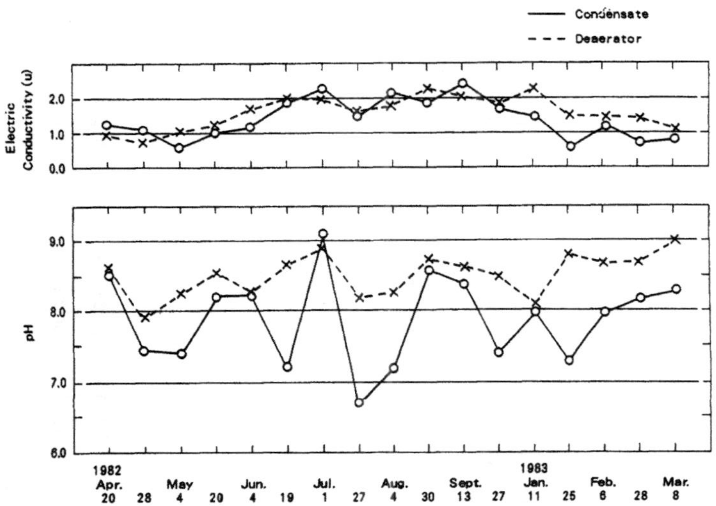

Fig. 11 Quality of Condensate and Deaerator

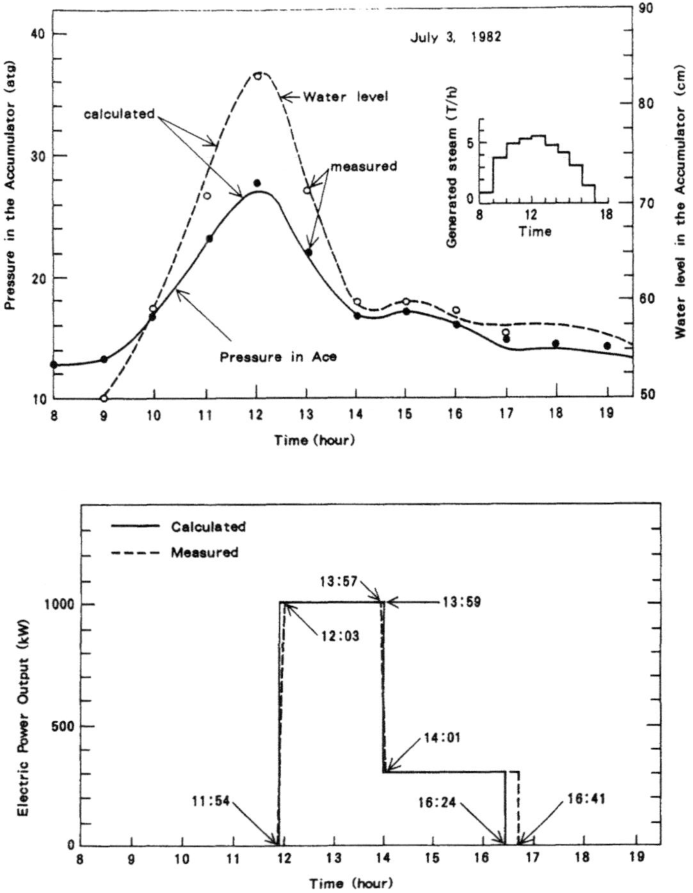

Fig. 12 Comparison between Simulation and Measurements

THEMIS PLANT OPERATION PROGRESS REPORT

V. MASCLE (EDF), F. PHARABOD (AFME) and B. RIVOIRE (CNRS)
AFME (Agence Française pour la Maîtrise de l'Energie)
CNRS (Centre National de la Recherche Scientifique)
EDF (Electricité de France)

Summary

The Themis Solar Power Plant was connected to grid in May 1983. The progressive power buildup that has occured since has demonstrated the validity of the combination of the salt receiver and intermediate thermal storage unit, which offers operating flexibility, rapid startup, and sustained efficiency with partial insolation.

The operating team has been running the installations continuously with the aim of maximum availability of the heliostat field receiver collector system. The main components display satisfactory behavior, whereas the secondary units incur long interruptions for maintenance.

The experimental team is involved in the definition of the operating conditions of 'solar' equipment, namely the heliostat field and receiver. Real time data processing serves to chart the fluxes in the receiver and to monitor the daily changes in output and energies in the different conversion stages.

The experimental phase will continue for two years, to allow an analysis of the power plant's operation in different conditions. This phase will be followed by a period of continuous operation and monitoring of developments in the solar equipment, for which international cooperation is sought.

1. INTRODUCTION

1.1. Review of technical characteristics

Without going into a detailed description of the main characteristics of Themis, already largely published and presented in the previous seminar at Claremont, it is important to stress the considerable value of the basic decisions. Experience has confirmed the soundness of the following alternatives selected in the design stage :

. salt flow in the solar receiver,

. intermediate thermal storage in two tanks.

This helps to collect the solar radiation from sunrise and to achieve a complete functional separation between energy reception and generation on the grid, without a drop in efficiency.

A number of specific construction features also gave full satisfaction during the startup period :

. locking of heliostats in the stow position,

. turbine with steam extraction ports and natural-draft dry air cooler,

. auxiliary parabolic dishes.

Figure 1 Themis Solar Power Plant in operation

1.2. Final touches during startup

Startup tests conducted in late 1982 and early 1983 helped to finalize the power plant operation systems, particularly monitoring and control and electronic data processing. Maintenance on equipment during salt filling and temperature buildup remained limited to the ancillaries, largely featuring 'non solar' technology : valves and fittings, electric heaters, secondary piping, conventional pumps.

The first receiver focusing operations were conducted in April 1983, and the plant was first connected to grid on 17 May 1983, with the receiver at one-third of its capacity.

1.3. Operation since connection to grid

The first year of operation was marked by long interruptions for finishing works and repairs to certain defective assemblies. Three periods may be distinguished :

- from June to September 1983, progressive power buildup of the solar receiver, illuminated by a partial field of 66 to 164 heliostats,
- from October 1983 to mid-February 1984, interruption of focusing for works :
  - absorbent coating of receiver
  - repairs to electrical heating system of receiver
  - modification of the support structure of the salt piping in the tower
  - maintenance on automatic control systems,
- from mid-February to April 1984, maximum collection of incident energy, buildup to power rating (199 heliostats, 2.3 MW).

Figure 2 shows the insolation in the month of March and the test periods with output to the grid.

1.4. Organization

Responsability for the installations was transferred on 1 January 1984 from the EDF Equipment Department, responsible for construction and startup, to the EDF Generation Department, responsible for operation. Experiments were conducted by the GEST team set up by CNRS and AFME, jointly with other EDF Research and Development and CNRS teams. The overall project is coordinated and funded by three organizations, AFME, CNRS and EDF, on an annual budget of around 20 MF.

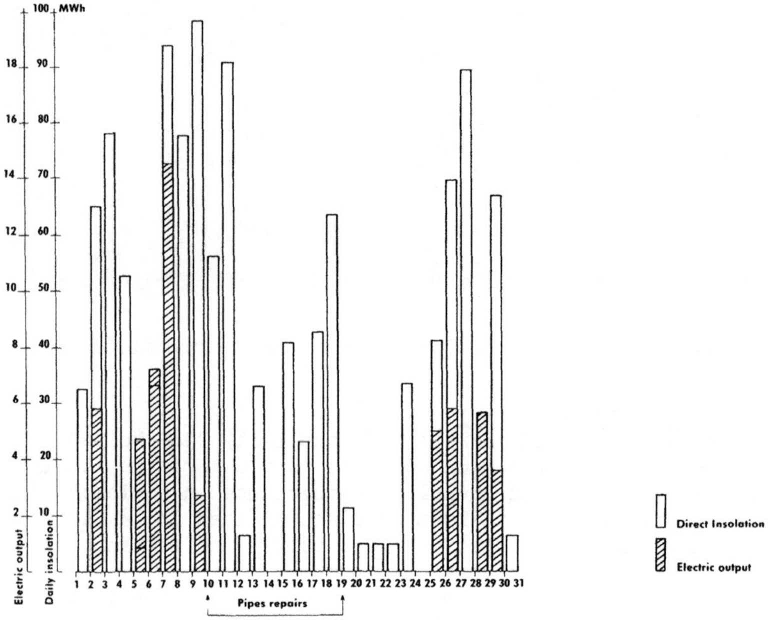

Figure 2 Insolation and electricity generation in march 1984

2. OPERATION AND MAINTENANCE

2.1. Normal operation

The operation of the power plant is remarkable for its great flexibility, which allows the optimal use of available solar radiation, thanks to the total separation of the primary and secondary circuits. As of today, three types of operation have been employed :

- thermal energy storage for deferred generation of electrical energy,
- generation of electrical energy in accordance with sunshine hours,
- discharge of thermal energy to generate electrical energy at a power level greater than that of the solar receiver.

The schedule of a typical day of operation can be described as follows:

- placing of heliostats in the standby position one hour before sunrise,
- opening of the solar receiver gate, circulation of salt, and placing of all available heliostats in the work position on the appearance of the sun ($P > 100$ W/m2 approximately),
- start of conditioning of the secondary circuit as soon as the quantity of hot salt produced is sufficient, or two to three hours after the start of focusing when starting with an empty stock,
- coupling of the turbine generator 30 minutes after the start of conditioning, and power rating possible 15 to 20 minutes later,
- interruption of focusing and placing of heliostats in the stow position when the sun disappears ($P < 100$ W/m2 approximately),
- shutdown of the turbine generator after exhaustion of the available hot salt (or earlier for fast startup on the following day),
- maintenance of the primary circuit temperature by periodic circulation of salt in the circuits until the following morning,
- maintenance of the secondary circuit temperature (175° C in the steam generator) by using thermal energy recovered by the auxiliary parabolic heliostats.

The advantages of the technique used in Themis are especially evident during cloudy periods. In fact, a temporary drop in the power of the primary circuit has no effect on the operation of the secondary circuit, because, since the salt temperature in the thermal storage unit remains constant, the steam characteristics can be kept at their nominal values.

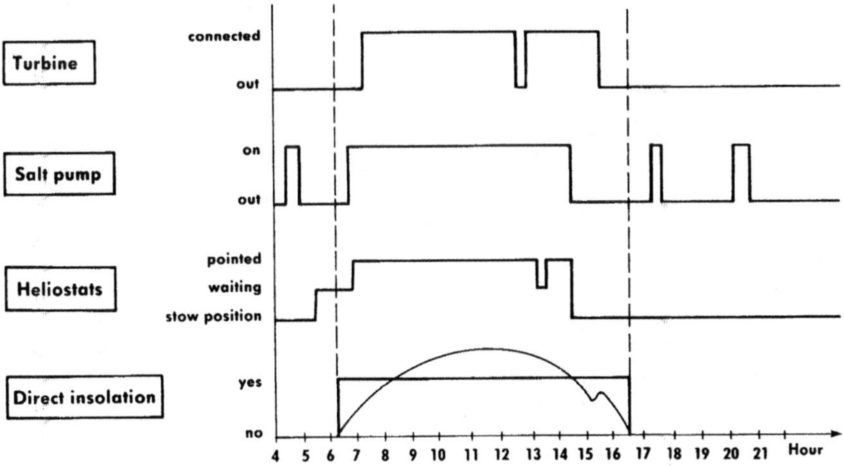

Figure 3 Operating diagram on 7 march 1984

Once the preliminary development phase is completed, the objective is to reduce the electricity consumption of the auxiliaries, by optimizing the use of the auxiliary heliostats. The latter were unavailable since the beginning of the year for alterations, and only re-entered operation in mid-April.

2.2. Operating teams

The power plant is operated by a team of 39 persons organized as follows.

Manager of Operations
Deputy Manager of Operations

MAINTENANCE	ADMINISTRATION	OPERATION
1 supervisor	1 office manager	1 supervisor
1 foreman	2 secretaries	1 foreman
5 control operators	1 driver	1 control operator
6 workers		6 teams including :
		. 1 watch engineer
		. 1 control operator
		. 1 plant equipment operator

To reduce this staff, it is planned from 1985 to install technical solutions that will serve to shut down the power plant completely at night, without delaying its morning startup, and thus to eliminate the night shift.

Moreover, the mix of qualifications will be developed so that the operating personnel can share in maintenance work on days of adverse weather conditions. The staff of the power plant will then amount to 26 employees.

Manager of Operations
Deputy Manager of Operations

MAINTENANCE	ADMINISTRATION	OPERATION
1 foreman	1 office manager	1 foreman
4 control operators	1 secretary	4 teams of 3 operators
4 workers		

3. EXPERIMENTS

3.1. Climatic mesurements

The six meteorological stations serve to measure the operating parameters, direct irradiance, wind speed and the values characterizing the site. The measurements taken during the year elapsed since connection to grid confirm the characterization of the site :

. annual direct irradiation : 1736 kWh/m2 (project 1700)
. maximum direct irradiance : 1020 W/m2 (project 1040).

The wind rarely exceeds 50 km/h in a sunny period, and is not a major cause of unavailability of the heliostat field. One or two annual storms occur in winter, with winds reaching the specified maximum (160 km/h in February 1984).

The analysis of recordings will help to determine the fine and adverse weather sequences, discontinuities in daily insolation, in relation to the management of the solar receiver (gate opening and closing) and its performance.

3.2. Heliostat field

3.2.1. Design, mechanical construction

The heliostats sucessfully withstood the violent storms that broke out in early February. The solutions adopted following the incidents of late 1981 are satisfactory, especially the lock, which is fully taken into account by the heliostat control system.

The mirrors do not exhibit any particular degradation. Substantial corrosion was observed on the bars of the back structure of some thirty heliostats. This problem, caused by unsatisfactory welds, has been partly resolved.

2.2.2. Monitoring and control

The centralized monitoring and control system was changed by the introduction of an interactive keyboard screen to facilitate operator-system dialog and to allow precise monitoring of the field. Automated pointing of the heliostat groups helps to maintain the balance of the high/low temperatures in the receiver.

At the heliostat, the most frequent defects résult from inductive encoders that perform position coding, requiring frequent zero resets. This problem has been resolved by using an encoder with modified characteristics.

3.2.3. Performances

In the past few months, the availability of the heliostat field has remained above 190 heliostats (95%) and the field does not raise any major problems. A campaign of 600 measurements on heliostat calibration shows that the average pointing error is 2.7 mrad (standard deviation 1.8 mrad).

3.3. Solar receiver

3.3.1. Receiver operation at reduced temperature

After 20 months of hydraulic operation of the solar receiver, the weak points encountered concerning electrical resistance heating and salt circulation have been corrected. The following lessons were drawn from this development phase :

- The 'high temperature' electrical trace heating or preheating applied does not adapt well to the design approximations and construction imperfections routinely accepted for antifreeze heating. For example, pipes were perforated by corrosion due to repeated overheating. The resistance heaters employed must be of top quality, design must be thoroughly detailed, and construction to the highest standards.

- The formation of several plugs in small pipes, drains and vents was observed when the salt was circulated. This defect, probably due to the accretion of oxide fines, disappeared by itself, as the salt flow thoroughly cleans the piping.

3.3.2. Power and temperature buildup

Since April 1983, focusing operations were performed at high salt throughput, without temperature regulation. They are intended to control the flux chart on the receiver walls. This control has proved difficult to achieve by adjusting the pointing of each heliostat group. However, satisfactory results were obtained by the following arrangements.

Figure 4 Solar receiver

- Broadening of the flux chart by slight defocusing. At present, the heliostat field is pointed 1 m in front of the theoretical focal plane. Losses at the receiver input are slightly increased, but the fluxes withstood by the walls are substantially reduced.
- Increase in the salt flow rate, causing a slight drop in the nominal temperature (430°C instead of 450°C), but which helps the receiver to accept higher fluxes.
- Use of an automatic pointing system that modifies the target point altitude nearly continuously, in order to achieve high/low symmetry of the receiver load.

These finishing touches made it possible to specify the thresholds to be programmed in the automatic safety system that monitors the maximum fluxes withstood by the circulation tube arrays, and the salt temperatures at different points of the receiver path. The exact algorithms of the automatic temperature control system were also determined. Today both systems are ready for routine operation.

3.3.3. Instrumentation

The receiver of Themis is provided with rather complete instrumentation. The strip flux meter at the focal plane and the infrared camera are in the course of starting up and have not yet yielded results.

On the other hand, the termocouples and fluxmeters installed in the receiver walls have been used intensively. The fluxmeters, which measure the thermal flux, proved to be difficult to use. Their accuracy is uncertain, particularly due to the unclear proportion of the convective flux affecting them. For better control of this particular measurement, they will be replaced shortly by optical fiber photon fluxmeters.

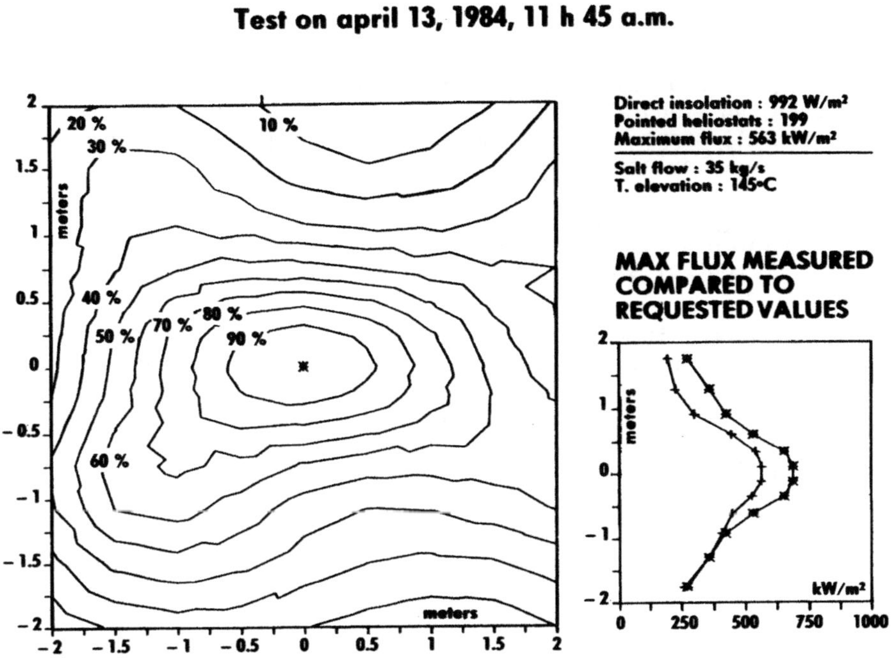

Figure 5 Flux chart of solar receiver

3.4. Thermal storage

The thermal storage system currently gives full satisfaction and plays an essential role of separating the solar energy collection and electricity generation functions. The experimental program on the storage unit will make it possible :

. to optimize the use of the available thermal stock,
. to point out variations in the salt decomposition rate with time, and corrosion problems that this could incur.

A number of measurements and tests are planned to quantify these problems and to assess their magnitude.

3.5. Steam production and turbine generator

Experiments on the secondary circuit are essentially aimed to optimize the use of solar energy by the steam generator, especially in the startup conditioning phases.

The specific consumption of the secondary circuit will also be checked to determine the operating procedures that should be followed to achieve the best possible operating balance.

3.6. Auxiliary system with parabolic dishes

The auxiliary heat production system by parabolic dishes re-entered operation in April, after changes were made to the receiver outlet temperature regulation system. A few minor changes are still likely, but the operation of these heliostats already provides full satisfaction.

3.7. Monitoring and control

The experimental monitoring and control system is being tested at Themis for possible use in future French thermal power plants. It consists of a data transmission bus materialized by a single coaxial cable.

that transmits all the information required for power plant operation (temperatures, pressures, valve positions etc) by multiplexing. These data are received by programmable automation systems which execute all the operating sequences of the equipment and also perform follow-up.

Moreover, an elaborate computer-aided operation system provides the operator with all supplementary data that he may require, either on a printer or on display screens.

3.8. Data acquisition and processing system for experiments

The experimental system has a small computer center distinct from the monitoring and control system used for operation. This computer performs three functions :

- Acquisition : it monitors all the data passing through the bus, can acquire 631 analog channels and 2277 binary data elements, representing the detailed status of the process. It can also directly acquire the data of specific instruments (target : 1024 cells, strip flux meter). The digitized infrared camera signal will soon be accessible to the computer.
- Processing : all the data (real time or off-line plot, filing, daily output, log) can be processed and the different programs adapted to each requirement (weather conditions, daily balance, monthly balance) are being developed.

Figure 6 Auxiliary parabolic dishes

- Modelling : models to calculate fluxes, thermal calculations of the receiver, and the balance of the installation, are being stored in the system and will be validated or modified according to the results of the experiment. Other programs will also be used in this system (field layout, miscellaneous optimizations).

3.9. Current performance

The availability of the heliostat field is now excellent with an unavailable heliostat rate under 5%. The receiver can collect solar energy from sunrise (as soon as the irradiance is around 100 W/m2). The entire heliostat field switches from the standby position to the working position in two minutes.

Temperature buildup of the solar receiver occurs in less than 10 minutes. The thermal storage unit is at 430°C. The turbine generator has reached the output level of 2.3 MW.

The curves in Figure 7 show the daily variations in the main characteristics : incident power on the heliostat field (irradiance x reflective area), thermal output of receiver, energy stored in the cold and hot tanks, electric power produced.

Curves 3 and 4 show the switch to the hot stock after the energy collected is transferred to the cold tank (the receiver outlet temperature regulation system was not operational on this day). Curve 5 offers an example of generation in two periods of the day : during sunshine hours and after sunset.

Preliminary performance results are favorable : overall plant gross efficiency reaches 21% at noon with an efficiency of 75% for the heliostat field-receiver subsystem.

4. CONCLUSION

The performance levels achieved after one year of operation demonstrate

- the rapidity of startup of the heliostat field-receiver system,
- the operating flexibility introduced by the thermal storage unit,
- the high efficiency of the conversion circuit.

The performance of the different subsystems met project anticipations.

Considering the importance of this salt receiver system for advanced studies in the thermodynamic conversion of solar energy, Themis is willing to welcome research workers from other projects to participate in a program under way and to contribute to its development. This program will serve to compile a precise balance at each step of energy conversion. The main effort is devoted to analyzing the heliostat field/solar receiver collection efficiency and the thermal behavior of the components of the salt circuit.

Continuous operation of the power plant during a sufficiently long period will help to improve monthly electricity generation by reducing the unavailabilities due to different causes, often 'non solar", as well as auxiliary consumption. The objective is to compile a productivity balance in about two years. After this date, Themis will be subjected to endurance trials in continuous operation, possibly simplified, to check the behavior of specific components, including the heliostats, solar receiver and salt system.

The Targassonne National Solar Test Center (CNESOL, Centre National d'Essais Solaires) is now ready for international joint projects. These projects should go beyond the state of exchanging information for the comparative assessment of central receiver power plants in the world. In fact, CNESOL hopes to become a forum for continued experiments on the salt system, which has demonstrated its excellent operation and flexibility.

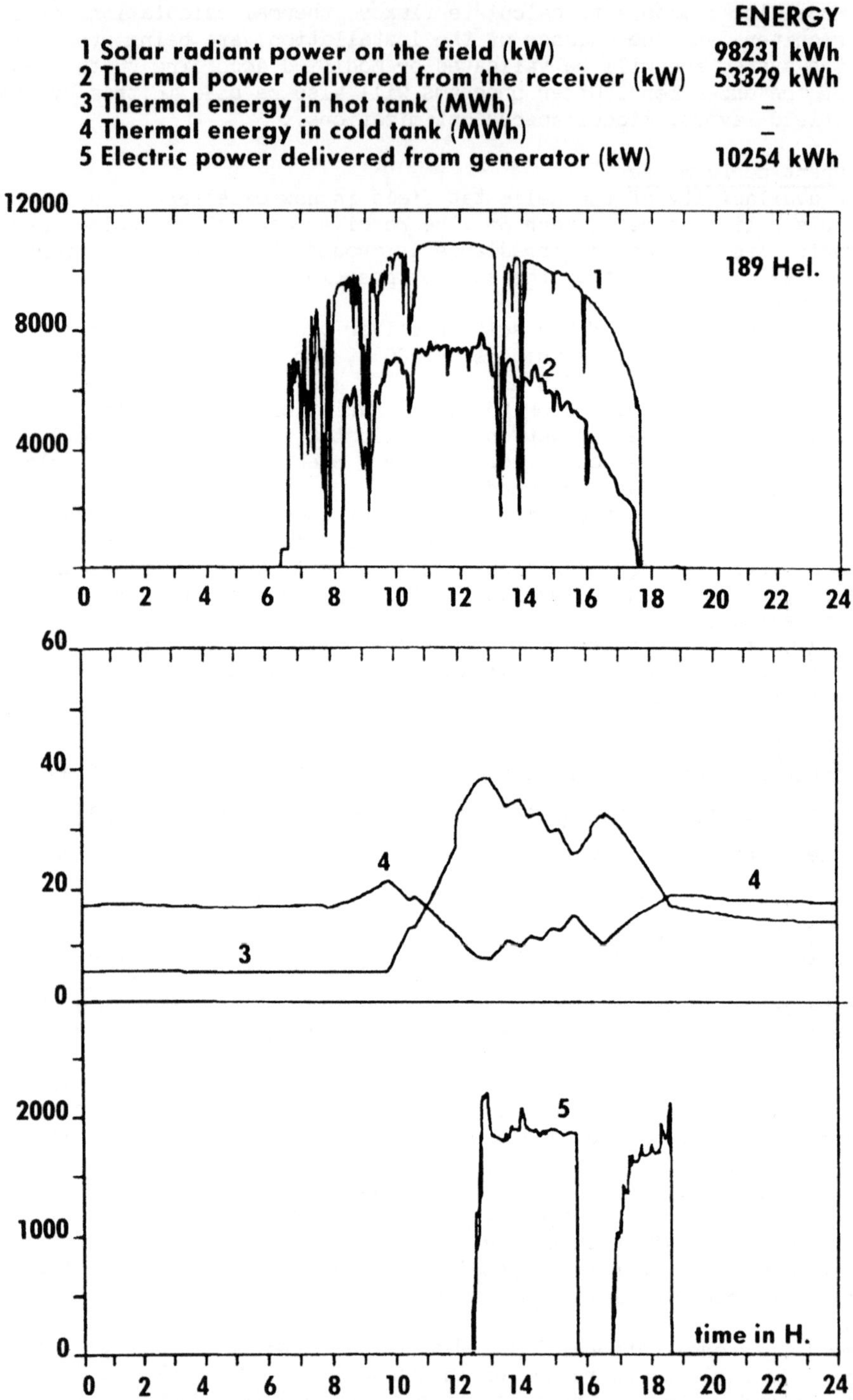

Figure 7 Daily operation: output and energy april 6, 1984

CESA-1 PROJECT STATUS REPORT

Ministerio de Industria y Energía
IER
J. Avellaner, C. Ortiz, F. Martínez, F. Sánchez

Summary

The CESA-1 Project, an enterprise of the Spain "Ministerio de Industria y Energía", consists of the design, construction, operation and evaluation of 1.2 Mwe Experimental Central Receiver Solar Power System. The CESA-1 plant, located within the boundaries of "Plataforma Solar" in Tabernas (Almeriá), includes collector field, water/steam cooled cavity type receiver, dual admission steam turbine, molten salt heat storage, as well as the associated controls and auxiliary subsystems.
The project,with a total cost of 2301 Mpts (1984), began in 1977 with a feasibility study and is deemed to be finish by the end of 1984, after 18 months of operation and evaluation.
This paper presents an overview of the project, summarize the organizatin, costs and schedule, main plant characteristics and the operation and evaluation activities performed to date.

1. INTRODUCTION

1.1.Project overview

Shortly after the 1973 oil crisis, the Spanish "Ministerio de Industria y Energía" set up an autononuous body called "Centro de Estudios de la Energía" (CEE), mainly devoted to the study and development of new energy sources. Among other activities, CEE began in 1977 a program, hereinafter named CESA-1 Project, which consists of the design, construction, operation and evaluation of a 1.2 Mwe Experimental Central Receiver Solar Power Plant. Figure 1 shows a simplified schedule of the project up to its completion in 1984.

1.2.Organization

In 1983, a new government organization, "Instituto de Energías Renovables" (IER), took over the responsibility of the project, stablishing the organization during the operation and evaluation phase indicated in figure 2. According to this, the evaluation activities are perfomed on site by the Evaluation Group. In parallel to this complementary studies are being performed.

1.3.Cost breakdown

The total installation cost of the project has been 2301 million pesetas (1984), 70% of the initial estimated cost. The distribution of this is as follows:

	Percent
Collector field	35%
Receiver subsystem	9%
Power conversion	15%
Heat storage	6%
Civil works	10%
Management and Engineering	10%
Research and Development program	15%
Total cost	100%

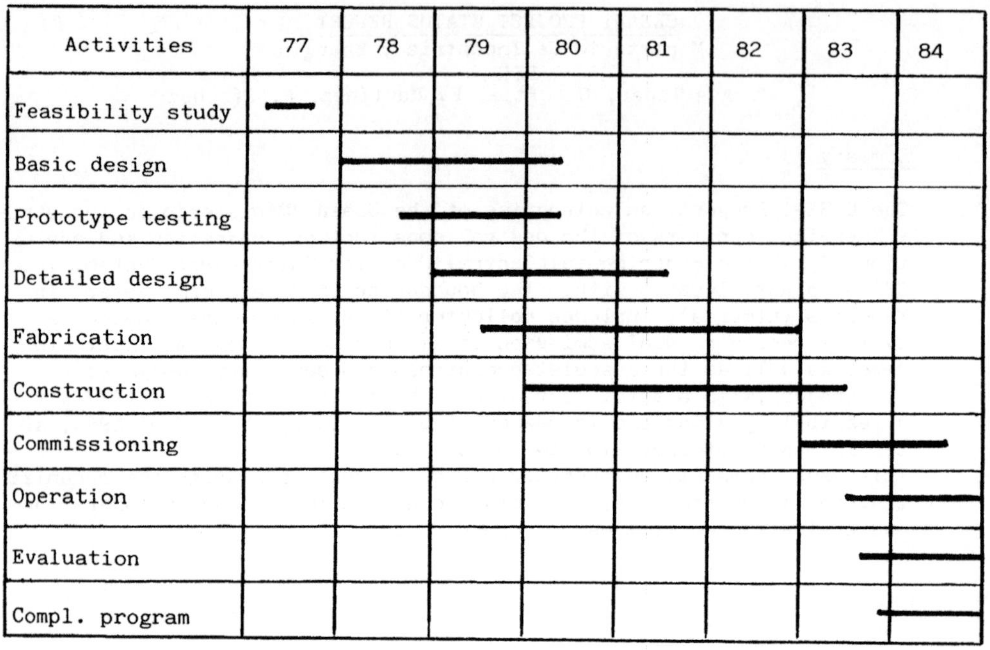

Figure 1. Project Schedule

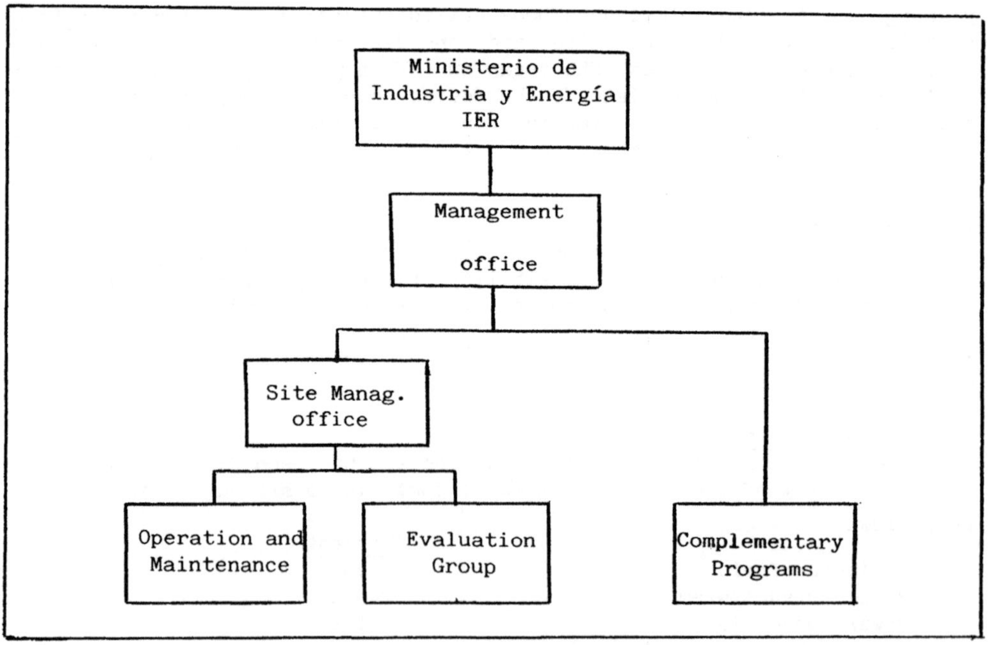

Figure 2. Project Organization

Project cost during Test and Operation phase has been budgeted in 251-million pesetas (1984), including O&M, evaluation and complementary activities.

1.4. Plant characteristics.

The technical concept selected for the CESA-1 project is a Central Receiver System with water/steam cooled receiver, steam Rankine cycle and molten salt heat storage, according to the basic scheme included in figure 3. The main characteristics of the plant are shown in table 1.

2. PLANT CONSTRUCTION.

On site activities began in January 1980 with the construction of the tower. The installation of equipment took place between March 1981 and March 1983, as shown in figure 1.

No significant problems were found during construction.

In total, roughly 140000 manhours were spent during the 24 months of erection, excluding civil works, distributed as follows:

	Man-hours	Percent
Collector field	19400	13.9%
Receiver	21260	15.2%
Power conversion	19670	14.1%
Heat storage	12160	8.6%
Controls and electric equipment	35200	25.1%
Auxiliary systems	32310	23.1%

Functional tests and checkouts of equipment and components were conducted between October 1982 and June 1983. Official inauguration took place on June 21st.

3. PLANT OPERATION STATUS.

3.1. General.

Since the beginning of 1983 CESA-1 plant systems and components have been put progresively in service, passing from start-up to routine operation situation. Through this procedure, a significant amount of experiences concerning operation and maintenance, plant performance and staffing requirements have been gathered.

At present, the plant, although close to it, is not yet in the routine operation phase, therefore, all statistics concerning production of electricity are not representative of the plant possibilities.

3.2. Operation and Maintenance team.

The CESA-1 O&M team is composed of 28 people under the responsibility of an O&M Manager. This team has the following composition:

Operation	Maintenance
1 Supervisor	1 Supervisor
1 Chemical engineer	1 Instrumentation technician
4 Shift chief operators	1 Electrical technician
6 Assistant operators	1 Mechanical technician
9 Watchers	3 Helpers

The plant is operated 5 days per week in 2 shifts of operation of 5 people (2 shift chief operator, 1 assistant operator and 2 watchers). During nights and weekends there is a watching shift of 2 people (1 assistant and 1 watcher). Maintenance team works only 5 days per week, 8 hours per day.

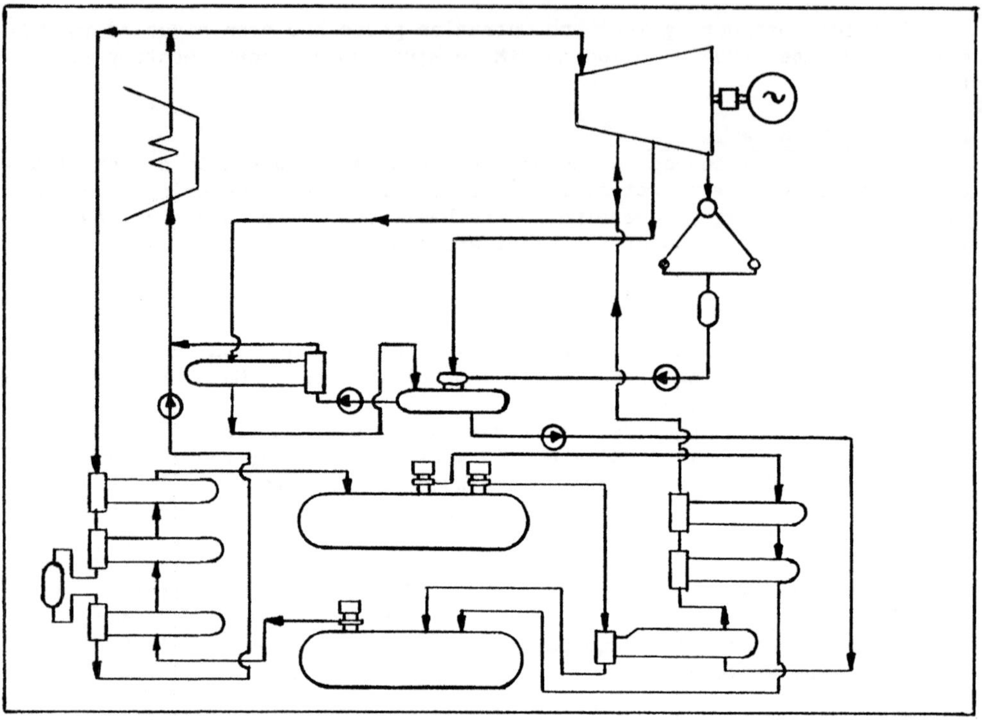

Figure 3. CESA-1 Cycle scheme

Figure 4. CESA-1 aerial picture

Table I CESA-1 Main Characteristics

GENERAL
- Location : Tabernas; 37º6'N, 2º21'W; 500 m. above sea level.
- Design point: 10 AM winter solstice, insolation ⩾700 W/m^2.
- Overall system efficiency: 15.5%

COLECTOR FIELD
- 300 heliostats distributed in 16 rows of north field.
- 39.6 m^2 reflective surface per heliostat.
- 12 mirror modules of 3 mm. float glass/70 mm. PVC foam/1 mm. steel sheet.
- 86.5% mirror clean reflectivity.
- Four focal lengths (90.140,200 and 250 m.).
- Open loop tracking with central computer PDP 11/34.
- Local controllers with μP INTEL 8085, 13 bit incremental encoders AC motors, 16 serial asynchronous full duplex communication lines.
- 69.5% design point efficiency.

RECEIVER
- Cavity type, water/steam cooled, forced circulation.
- 3.4 m. square aperture at 60 m. above ground.
- 525ºC/110 Kg/cm^2 steam conditions.
- 6110 Kg/h steam flow with 5500 Kwth solar input.
- 90.5% desing point efficiency.

POWER CONVERSION
- Regenerative water/steam Ranking cycle.
- Dual admission multistage condensing turbine.
- Operating conditions:

	HP steam	LP steam
- Nominal gross power	1200 Kwe	840 Kwe.
- Steam inlet	520ºC/100 Kg/cm^2	330ºC/15Kg/cm^2
- Efficiency	27.7%	21%

HEAT STORAGE
- Sensible heat storage, using HITEC, two tanks.
- Tank temperatures: 220ºC cold/340ºC hot.
- Storage capacity: 300 Tn of salt equivalent to 16 Mwht.
- Charging rate: 3984 Kwt.
- Discharging rate: 4165 Kwt.

3.3. Plant commissioning.

Start up activities began in January 1983 with the activation of the auxiliary subsystems (electrical supply, compressed air, cooling water, etc..). Up to now, all plant systems and components have been activated and operated on routine basis, with the exception of the heat storage discharging loop -- and the turbine in the low pressure steam mode, which are still under commissioning.

The most important start up milestones are listed as follows:

1983 February - Auxiliary systems fully operational.
- Heliostat Control system functional tests.

April - Collector field fully operational.
- Melting salt inventory in the cold tank.

May - First heliostats onto the receiver.

September - Activation of heat storage charging loop.

October - Receiver fully operational.
- Turbine roll and connection to the grid.

1984 February - Activation of heat storage discharging loop.

Plant commissioning period has been extended more than expected due to bad weather conditions and design modifications implemented to solve deficiencies, as is indicated hereinafter.

3.4. Statistic of Operation.

Gathering of operational data began in July 1983, as soon as solar related systems (collector field and receiver) became operational. Up to April 1984, the following statistics have been accumulated:

Theroretical sunshine hours....................	3527
Real sunshine hours	2528
Collector field tracking hours	2134
Receiver operating hours	865
Turbine operating hours	96
Synchronization hours	42
Gross energy production (Kwh)	17016
Gross peak power (Kw)	850
Heat storage charging hours	67
Heat storage discharging hours	54
Heliostats out of service (daily average)	50

3.5. Statistic of Maintenance.

Since July 1983, the maintenance team spent 18.2 million pesetas in -- scheduled as well as unscheduled maintenance of the plant, distributed among the diferent plant systems as follows:

	Labour		Materials	
	$x10^3$ pts	%	$x10^3$ pts	%
Collector field	2309	15.1	137	4.8%
Receiver	890	5.8	338	11.8%
Power conversion	2787	18.2	454	15.8%
Heat storage	1371	8.9	138	4.8%
Controls and electric equipment	1982	12.9	559	19.5%
Auxiliary systems	2755	18.0	1238	43.3%
Supervision	3231	21.1	-	-
	15325	100 %	2864	100 %

Material cost includes consumables used in plant operation.

3.6.Plant performance.

Although the whole plant is not yet fully operational in all modes of-operation and some of the systems have accumulated a relative small number-of operational hours, preliminary performance of the plant systems could be summarized as follows:

3.6.1.Collector field

The number of heliostats out of service is higher than expected, mainly due to electronic problems.

In total 26% of mirror modules shows mirror corrosion, ranging from very small spots up to 75% of its surface affected. Besides this, the mirror-modules present a noticeable distorsion of contour that leads to degradation of optical quality, not yet cuantified.

3.6.2.Receiver.

After the implementation of several design modifications, the receiver is performing, in general, within design conditions. Although overnight heat losses are higher than expected, cold start up time has been reduced down to 2½ hours.

The superheater section of the receiver is very sensitive to heat flux distribution and requires high attention from the operators.

3.6.3.Power conversion.

Up to now, the turbine has been operated only with high pressure steam coming from the receiver, during a limited period of time. Though the turbine is an "off the shelf" item, its performance is well below expectations. - Main turbine related problems are:

- Gland steam consumption higher than expected.
- Electrohydraulic control is not working properly.
- Turbine casing distortions of unknown origin.

3.6.4.Heat storage.

The main problem associated with the heat storage is its big thermal-inertia and slow response time to load changes. Due to this, it has been impossible up to now, to reach nominal temperature in the hot tank.

Electric heat tracing failures led into several pipe blockage due to - salt freezing, mainly in the valves.

3.7.Incidents.

From January 1983 to date, the following incidents were recorded:

1983	April	- Mechanical seal failure in receiver recirculation pump.
	June	- 1st. Receiver vent tube crack.
	October	- Pipe blockage due to salt freezing in charging loop.
	November	- Turbine thrust bearing failure.
	December	- Detected distorsion in turbine casing.
1984	January	- Central computer failure.
	March	- Pipe blockage due to salt freezing in discharging loop.

4. PLANT EVALUATION STATUS.

4.1.General.

Evaluation activities started shortly after the plant began to produce operation data. In June 1983 was stablished an evaluation plan to be implemented through second half of 1983 and 1984. Table II includes a list of -- evaluation activities to be performed. Up to date, all activities concerning

Table II List of evaluation activities

1100 COLLECTOR FIELD
- 1110 Overall efficiency
- 1120 Heliostat optical analysis. Beam quality. Reflectivity. Canting.
- 1130 Heliostat mechanical analysis. Backlash. Deflections.
- 1140 Control system. Rebiasing. Tracking accuracy.

1200 RECEIVER
- 1210 Receiver overall efficiency.
- 1220 Power distribution. Aperture plane/inside cavity
- 1230 Heat transfer coefficientes. Metal temperatures.
- 1240 Control system behaviour.
- 1250 Operative characteristics. Cold/hot start up. Transients.

1300 HEAT STORAGE
- 1310 Heat storage overall efficiency.
- 1320 Component behaviour. Heat exchanger/pumps.
- 1330 Molten salt evolution. Degradation/Induced corrosion.
- 1340 Control system behaviour.
- 1350 Operative characteristics. Cold/hot start-up. Transients.

1400 POWER CONVERSION
- 1410 Overall efficiency. HP/LP steam.
- 1420 Component behaviour. Turbine/Condenser/pumps.
- 1430 Control system behaviour.
- 1440 Operative characteristics. Cold/hot start-up. Transients.

2100 MODES OF OPERATION
- 2110-2170 Analysis of modes I to VII
- 2180 Analysis of transition between modes.
- 2190 Strategies of operation.

2200 PLANT OPERATION AND MAINTENANCE
- 2210 Operation experiences.
- 2220 Maintenance experiences.
- 2230 Economic analysis.

Collector field and Receiver, as well as some of Heat storage, Modes of -- Operation and Plant O&M were conducted, existing preliminary results of them. The remaining activities will be performed in second half of 1984.

In parallel with this evaluation plan, exist complementary studies which enlarges the assessment of Central Receiver technology, as indicated in table III.

4.2.Evaluation team.

On site evaluation activities are performed by a team of four engineers, leaded by an Evaluation Manager. In addition to this, personnel from the Complementary studies also performs, on part time basis, activities on site.

4.3.Instrumentation and tools for evaluation.

In addition to Data Acquisition System, used to gather operation data, the Evaluation team has the following equipment:

- PAIS, System based on a TV camera and a personal computer for optical analysis of heliostat images.
- SAIS, System based on radiometers, data logger and personal computer to analyze heliostat images and tracking accuracy.
- Portable reflectometer.
- Set of metrology instruments and a laser to perform mechanical tests
- Set of electronic instruments to perform control system tests.

4.4.Preliminary evaluation results.

In addition to what has been said in paragraph 3.6., the outcome of tests and evaluations performed up to date is as follows:

4.4.1.Collector field.

Beam quality of a sample of 17 heliostats tested with SAIS system shows a big dispersion of image sizes. Best heliostats images agree quite well -- with HELIOS calculations performed in the design of the receiver, which used 1.5 mrad beam quality error.

First test results show 0.7 mrad average tracking error of a sample of 17 heliostats, very close to expected performance (0.89 mrad).

Reflectivity measurements indicate clean mirror reflectivities slightly higher than expected (88.5%). Preliminary tests of mirror dust build - up show reflectivity degradation of 1.5% per week. Best cleaning method - seems to be the use of wiper, which allows to recover original clean mirror reflectivity.

The impact of mirror corrosion in reflected power, gear box backlash,- rebiasing and recanting requirements are under evaluation.

The analysis of failures in the electronics of heliostat local controllers shows the necessity of R-C cells and varistors in the power circuits- to protect solid state relays that control heliostat motors from overvoltages and high frecuency pulses.

4.4.2.Receiver.

Preliminary measured efficiencies shows values slightly below design - (85% instead of 90%).

The analysis of receiver tube metal temperatures indicate evaporator - behaviour as expected, while superheater zone seems to have lower heat transfer coefficients than expected.

Control loops of superheater attemperators have too slow response time to control steam temperatures. This regulation is performed using heliostat aiming strategies, as well as the number of heliostats tracking onto super - heater zones.

Heat and fluid losses of receiver during overnight standby are higher - than expected, being necessary to start-up the morning after almost from a cold condition. Nevertheless, cold start-up time has been reduced down to - 2½ hours.

Full cloud coverage or collector field tripout transients require 30 - minutes to 1 hour to recover original receiver conditions.

4.4.3. Other evaluation activities.

Activities concerning modes of operation and plant O&M has been already presented in paragraphs 3.4. to 3.6.

5. FUTURE PLANS.

Through second half of 1984, both, operation and evaluation, will be - finished. Considerable improvement of plant operating hours is expected,--- once commissioning is finished and more realistic statistics of operation - and maintenance will be obtained.

During 1985, CESA-1 plant will be used to test the gas cooled receiver of GAST project, therefore, only collector field will be evaluated in 1985.

Table III Complementary studies

1.- COMPUTER CODES

- Collector field optimization program
- Calculation of heat flux distribution in receiver aperture plane.
- Heat transfer model for cavity type receivers.
- Modification of thermodynamical model to real operational data.

2.- ECONOMIC STUDIES

- Dispatching electric networks with solar power plants.
- Evaluation of solar potential in Ebro region.
- Evaluation of solar potential in Andalucía region.
- Economic analysis of Central Receivers and Distributed Collector.

SESSION II - PART 1

HELIOSTAT FIELD

Summary of the Session by the Rapporteur
B. BONDUELLE, CNRS, Groupe d'Evaluation Scientifique Themis, France

Conceptional design of solar power plant with central receiver tower based on improved heliostats

SSPS-CRS Heliostat performance/history

Heliostat field and receiver efficiency measurement on the IEA - 500 KWe - CRS at Almeria

Themis heliostat field

ASINEL heliostats for the GAST technological programme

CESA-1 heliostat field - evaluation status report

10 MWe solar thermal central receiver pilot plant - heliostat evaluation

SUMMARY OF THE SESSION BY THE RAPPORTEUR

B. BONDUELLE
CNRS, Groupe d'Evaluation scientifique Themis, France

1. INTRODUCTION

This report will try to sum up the lessons learned by the different experiments described in the presented papers and give general orientations which can be pointed out from the new heliostat conceptions and developments.

2. TECHNICAL STATE OF ART

2.1 Availability

After a phase of initiation, the availability of heliostat fields can be maintained at over 95% without considerable difficulties. There are very few defects in the mechanisms themselves; therefore, most of the causes for unavailability stem from problems with the electronics (controllers, encoders, etc.).

2.2 Reflectivity

Already in some plants, reflectivity has reached 0.91 to 0.92 (for clean mirrors). The main causes of degradation are

* corrosion, which mostly depends on the conception of the rear structure of the heliostats, and
* atmospheric pollution, which depends on the site but is an important preoccupation. The washing strategy results from a balance between energetic losses and maintenance costs.

2.3 Tracking Accuracy

Tracking accuracy depends mostly on beam quality; there were no major disagreements on this point. The results were in accordance with expected specifications. Total budget error can easily be maintained in the range from 2 mrads to 4 mrads. Beam characterization systems were important in identifying heliostats and helping increase their efficiency.

2.4 Remaining Problems

The most obvious remaining problems were mirror corrosion, and lightning damages, for which there is no reliable protective method.

3. INCREASING THE COST EFFICIENCY OF HELIOSTAT FIELDS

First is given a non-exhaustive sum-up of different technical features which appear in the new heliostat concepts which are being developed. Then the cost figures of such development are given. Lastly, we give recommendations based on two considerations: the heliostat has to be site-adapted rather than world adapted; and secondly the fields are composed of great numbers of heliostats and a greater number of components, and the large amount of repetitive operations needs good attention paid to it--from the design conception to the maintenance operation of these components.

3.1 New Designs

The following is a technical orientation to new designs in the heliostat concepts:

* concrete (Themis): steady, reduced civil works, cheap corrosion protection cost;
* control and command system: tendency to simplify the central controller and give more intelligence and autonomy in the local heliostat controller;
* avoiding moisture traps in the mirror structure;
* binding laminated glass mirrors together;
* finding reliable lightning protection;
* investigating optical fibers for communications;
* stretched membrane heliostat.

3.2 Cost Figures

The installed price of $200/m^2 seems possible even without high production volumes. The goal remains at a level of $100/m^2 (or less).

3.3 General Recommendations

From the different experiments, the following are some general recommendations:

* make careful specifications according to realistic site conditions (wind temperature, humidity, etc.; overspecification needs to be avoided);
* investigate mixed fields;
* investigate wind loading in the field (edge versus center);
* insure careful selection of components (known reliability), thorough qualification, and strong manufacturing control;
* have effective maintenance equipment and procedure for handling of large fields (e.g. BCS, washing device,...);
* take photovoltaic applications into account; and
* last, but not least, develop a standard for heliostat comparison.

CONCEPTIONAL DESIGN OF SOLAR POWER PLANT WITH CENTRAL RECEIVER TOWER BASED ON IMPROVED HELIOSTATS

H.YOSHIKAWA and N.IKEDA*
Tokai University and New Energy Development Organization*

Summary

For technical and economical evaluations of solar power plant, the comparisons from technical points and electricity cost are presented on the solar plant with the heliostats based on new material and configurations because the collector system consisted of many heliostats, generally,occupies more than 50 % of total construction cost of the solar power plant. The improved types of heliostats provide smooth movement of sun tracking,strong resistance aginst wind and earthquake and good accuracy of mirror for reflection. According to the improved designs of solar power plants, each unit cost per kWe is examined.

(a) Single supporting column and double supporting column of heliostat mirror.

(b) New material and original material for supporting the mirror membranes.

The cost analisys with 1 MWe, 10 MWe, 100 MWe solar power plant with the improved heliostats was performed,extraporating from the current constructed plant. Arrangement and election of the improved heliostats on the down hill slop,expected for the decrease of total cost construction.

1.INTRODUCTION

Technical and economical evaluations for solar thermal power plant with improved heliostats are presented. The cost of heliostat system, in general, occupies about 50 % of total construction cost. It is necessary to develope the low cost heliostat with smooth movement with a good accuracy of sun tracking,the strong resistance against wind and earthquake and the small distortion of mirror pannel. The improved heliostats are examined in dependences of following items:

(a) Single supporting column or double supporting columns of heliostat mirror pannel.

(b) New material or ordinary material for the reinforcement of the mirror itself,

(c) Increase of thermal output by heliostat field.

The cost analysis for 1 MWe, 10 MWe and 100 MWe central receiver solar power plant with the improved heliostst are performed by an extraporation from 1MWe reference pilot plant at Nio.

2.FOUNDAMENTAL FUNCTION SURVEY OF THE HELIOSTAT

The most familiar heliostat is a"wing type" which has two mirror plates at the both sides about the central column like bird wings. The features of this type are very suitable to follow the sun tracking ,but have to support a heliostat weight and to handle a rotating and swing motions at the small part on the single column. Consequently,that part of heliostat has complex mechanisms and must be kept tough against strong winds and seismic forces. Three foundamental concepts of heliostat are examined from the points of the mechnism of weight supporting and rotating and swing motions.

Heliostat Type I shown in Fig.1,has a supporting column and its sun tracking mechanism which is separated the rotation and swing mechnism. The swing motion is operated by a linear displacement of oil pressure piston. The rotating motion is operated by moter-drive rotation of the column. The column is able to support its weight and wind pressure load and to become lower to near the ground level in the case of a strong wind.

Heliostat Type II shown in Fig.2, has two supporting columns for the pannel weight and the wind presure load. The swing motion is able to operate by up and down motion of the column height by oil pressure. The rotation is operated by the moter and gears which are located upper portion of the columns. Both columns,also, are able to become lower to near theground like the heliostat type I.

Heliostat Type III shown in Fig.3,has three supporting columns for its mirror pannel weight andthe load in severe conditions. The rotation and/or swing of a mirror pannel,and upand dwon motion of it are to be performed by only up and down motion of columns by oil pressure mechnism. One column among three columns has a fixed joint at the lower end and other ends have ball joints. The other tow columns have each caster at the lower end which is able to move linearly,and have each ball joint at another end. Three columns are able to become lower to the ground like other two types of heliostats.

The qualitative evaluation of three types of heliostats by 5 points method are performed and its results are shown in Tab.1. Only Type I and Type II of heliostats are selected as the objects of further analysis and evaluations from the result.

3.DESIGN OF IMPROVED HELIOSTATS

The improved heliostat design criteria in this paper are as follows:

1) Reflection accuracy of the mirror is within $\pm$ 1 degree.
2) Maximun distortion of mirror pannel is 0.6 mm.
3) Maximum wind velocity is 50 m/sec.
4) Mirror size is 16 m^2 (4 m x 4 m).
5) Total mirror area of the three types and the reference plant are identical.

Mirror pannels with the anti-distortion strength consist of thick plates as reinforcement of the mirror. A kind of truss structure for a light weight is selected in this design. A typical example is illustrated in Fig.4. Several materials for the strengthened membrane of the truss structure are examined in designed heliostats. These materials are a kind of stainless steel, aluminum alloy and high strengthened epoxy resin. The designs have done within the allowable stress for each material. The mechanical characteristics of the high strengthened epoxy resins are presented in Tab.2. This material consists of three components which are a kind of bisphenol epoxy resin as a base component,curing agent component and silica (SiO_2) as a filler component. They are mixed and cured at a certain high temperature and times in order to get a high strength by the chemical bridge method.

A driving power depends on the heliostat weight. The improved heliostats are designed by the defferent allowable stress for each anti distortion membrane material. Therefore, the mirror pannel weight is not linear function of the material density. The summary of the improved heliostat designs are presented in Tab.3. It is found that the driving powers decreases to be 1/3~1/4 of the reference plant,because of the light weight of each heliostat and reduction of number for them.

In arrangement of the many heliostats after grading on the tight site area, the effective collecting mirror area would be reduced by the shading and blocking. The mirror hight is adjusted by the column height in order to increase plant efficiency and efficient mirror collecting area. It is also better to arrange the movable column heliostats on the hill side for saving land grading cost and decreasing construction cost. The optimum slope of the hill is about 11 degrees for collecting efficiency of 100 % from calculation. According to this design, the plant power output has increased about twice as compared with the reference plant power output. The net electric output and the collecting efficiency of the solar energy for the improved heliostats and the reference plant heliostats,are presented in Tab.3.

4.COST EVALUATION

The estimated electricity cost is calculated by the following equatins:

$$L = P \times \left(\frac{\text{improved heliostats weight}}{\text{reference plant heliostats weight}} \right)^{0.6} + Q$$

$$R = L \, (1 - l) \frac{i(1 + i)^n}{(1 + i)^n - 1} \, i \, l \, L$$

where, P:total heliostat cost for reference plant,Q:balance of plant cost for reference plant,R:annual redemption,i:a rate of interest,n:lifes of redemption,l:a rate after redemption, and where, a) capital cost = R + 0.365 % L (property tax),b) general cost = operation man power fee (10 man) + 0.01 xL (maintenance fee), c) associated charge = 0.072 L,d) income tax = (a)+ b)+C))x1.523 %, e) electricity unit cost = (a)+b)+c)+d))x(net electrical output $)^{-1}$ (¥/kWh),f) estimated reference plant cost = 2.5 $\times 10^9$ ¥, g) annual generating electrical output: 1MWe plant = 0.48 $\times 10^6$ kWh/year, 10 MWe plant = 5.6 x 10^6 kWh/year, 100 MWe plant = 72 $\times 10^6$ kWh/year, h) heliostat cost = 53% of total construction cost including land cost etc., j) balance of plant cost estimates with the extraporation of the current cost.

The electrisity unit cost for 1 MWe, 10 MWe and 100 MWe plants with the improved heliostat using the new materials of reinforced structure,are shown in Fig.5 according to the plant redemption life.

5.CONCLUSION

As the results of this design studies with improved heliostats, the electricity unit cost of the central receiver solar power plant may be decreased about 1/4 compared with the reference plant about the 25 years redemption lifes. If the futher development of the heliostat mechanism, the applicable materials, the arrangement of heliostat by the movable column and the efficient use of the geological site figure, the net electricity unit cost would be able to decreased considerably the current unit cost of the solar thermal power plant.

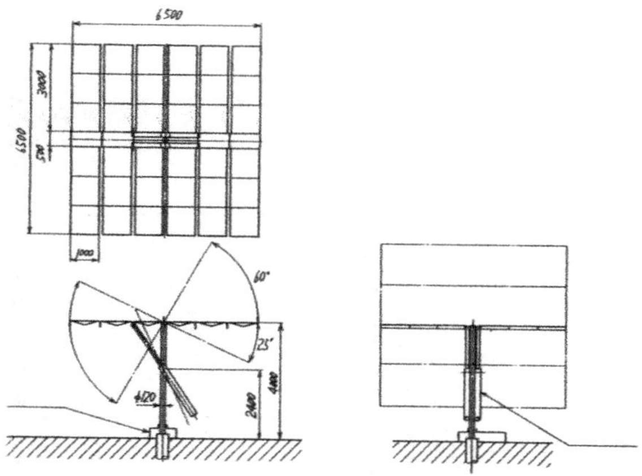

Fig. 1 Heliostat Type I

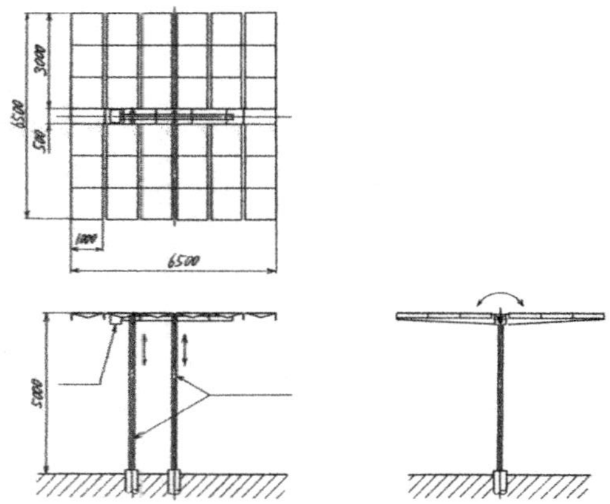

Fig. 2 Heliostat Type II

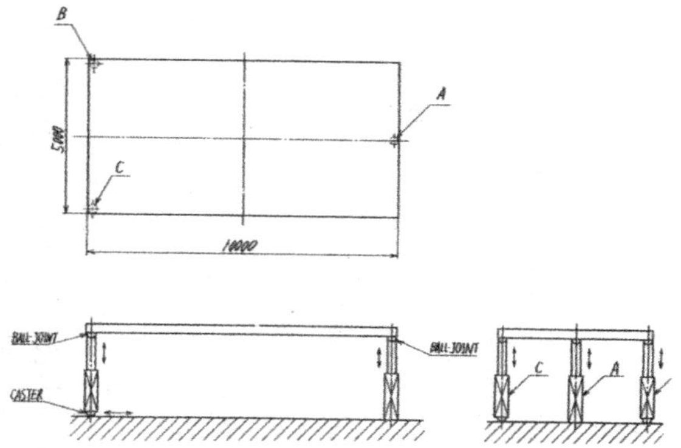

Fig. 3 Heliostat Type III

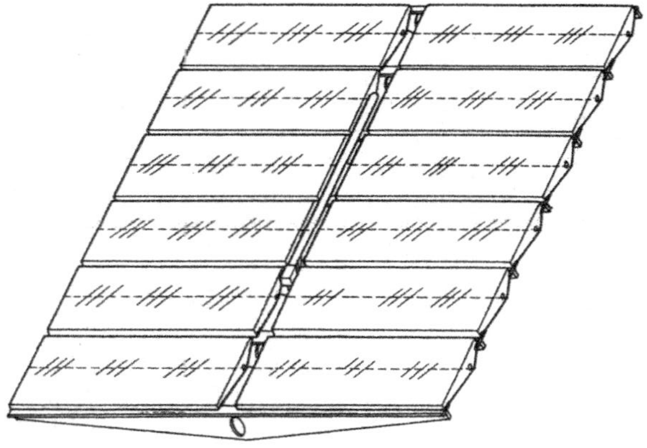

Fig. 4 Structure of Mirror Pannel

Tab.1 Qualitative Evaluation of Heliostats

	Type I	Type II	Type III
Flexibility	5	4	1
Easiness of Operation	5	4	3
Resistance for Wind etc.	3	4	5
Support of Its Weight	3	4	5
Reproducibility of Position	5	5	3
Total Evaluations	21	21	17

Tab.2 Typical Mechanical Characteristics of the High Strength Epoxy Resins

E807/MNA/2E4MZ/ SiO_2	Density (g/cm^3)	Elasticity (kg/mm^2) 296K	353K	383K	Compressive Stress (kg/mm^2) 296K	353K	383K
100 / 90/ 2 / 0	1.24	383	353	235	11.6	9.9	7.7
100 / 90/ 2 / 100	1.46	550	540	470	17.2	16.5	15.6
100 / 90/ 2 / 200	1.57	620	463	380	18.4	17.4	16.0
100 / 90/ 2 / 300	1.68	1000	875	583	22.2	20.8	20.2

E807: A kind of bisphenole epoxy resin,
MNA : Curing agent component,
2E4MZ:Accelerating curing component,
SiO_2: Filler.

Tab.3 Characteristics of Desined Plants with Improved Heliostats

	Types of Proposed Heliostats						Reference Plant Of Nio
	Type I	Type II	Type I	Type II	Type I	Type II	
Materials of Pannel	Stainless Steel		Aluminium Alloy		Epoxy Resin		Stainless Steel
Its Specific Weight (kgf/m^3)	750		2700		1500		750
Mirror Area (m^2/unit)	36		36		36		16
Total Weight of a Heliostat (kg)	1,560	1,760	1,070	1,320	970	1210	2,267
Driving Power of a Heliostat (W)	255	135	190	125	170	110	~ 200
No. of Heliostats	359		359		359		807
Total Area of Mirrors (m^2)	12,924		12,924		12,924		12,912
Total Weight of Heliostats (kg)	560,040	631,840	391,310	473,880	348,230	434,390	1,829,469
Total Driving Power of Heliostats (kW)	92	48	68	43	61	39	~ 160
Net Electric Output Power (kW)	1,785	1,829	1,809	1,834	1,816	1,838	840

Fig.5 ESTIMATED ELECTRICAL UNIT COST

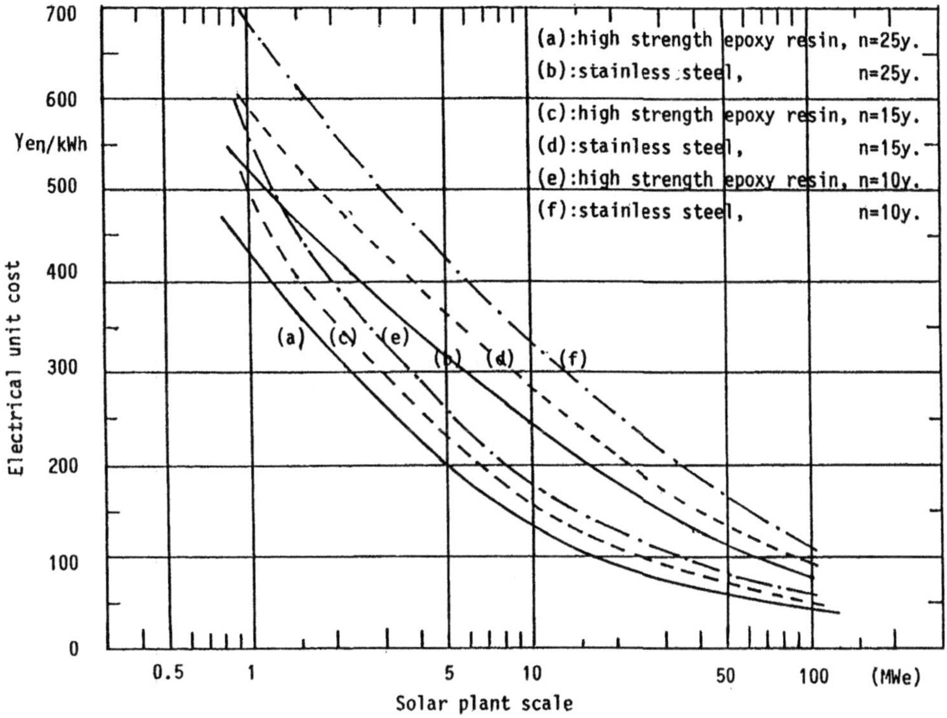

SSPS-CRS HELIOSTAT PERFORMANCE/HISTORY

P. Wattiez and J. Ramos
ITET and Sevillana,
IEA/SSPS Project

Summary

The heliostat operating experiences, tests, and evaluations of the IEA/Small Solar Power Systems - Central Receiver System (CRS), in Tabernas, Spain, are reported for the periods of 1982 and 1983.

Operation aspects of the heliostat field system, such as strategy and performance, are commented on, with regard to the main technical events which occurred during the time period under consideration.

Technical points are presented, such as reflectivity variation, mirror corrosion, and other technical problems. The availability and tracking attitude are analyzed from an operational point of view. The washing method is evaluated, including some recommendations. The operation strategy used for the year is explained.

Some further impacts on heliostat design are presented in the conclusion.

1. INTRODUCTION

During 1982-1983, the heliostat field system (HFS) of the IEA/SSPS-CRS has been in normal operation. Apart from some technical events which occurred in the Sodium Heat Transfer System (SHTS) or Power Conversion System (PCS). the behavior of the HFS has proved reliable, in spite of the meteo conditions observed on site. The HFS history can be summarized as follows:

* Heliostats installed	September 1980/October 1980
* Mirror installed	November 1980/January 1981
* HFS acceptance	April 1981
* Lightning	May 1982
* Mirror corrosion survey	January 1983 - 141 modules
* Vertical stow position	April 1983/July 1983
* Mirror corrosion survey	April 1984 - 280 modules
* Lightning	May 1984

2. OPERATION ASPECTS

2.1 Operation Strategy

The heliostat field has operated under several strategies since September 1981. During technical shut-downs of the SHTS, the applied strategy is to operate the HFS as much as possible, even if in standby position, to gain a maximum of operational experience from an operability and reliability point of view. The electronics in the HFS are powered 24 hours a day.

Besides these operational strategies, additional test and investigation have been conducted to observe HFS behavior under a vertical stow position.

The focusing point strategy on the receiver tube wall has been adopted to each receiver (Sulzer and ASR) in order to achieve more appropriate uniform heat flux distribution, depending on the Na flow circulation. The Sulzer receiver has a single aiming point, whereas the ASR used three aiming points. The field focusing procedure consisted in a row-by-row method, increasing approximately one row per minute, and did not change.

2.2 Weather Conditions

For two years, the irradiance time above 300 W/m^2 did not coincide with the expected value of 3,000 h/yr.

Nevertheless, the maximum daily irradiance observed has been above the expected value for winter, and below the expected value for summer. This is principally due to the lower sky transparency and dust suspended in the air.

The operation of the heliostat field has nonetheless been affected by the wind conditions due to the importance of the amount of gusty periods which for a short time exceed the operational safety limit (50 km/h) (1).

The current strategy is to stow the field when the wind speed exceeds 60 km/hr three times in less than 5 minutes, or when it reaches a peak of 80 km/hr. The whole field can be kept in track at wind speeds of up to 30 km/hr. For higher speeds, part of the field has to be defocused, starting with the last rows since the movement of the heliostats overheats the receiver frame.

2.3 Operation Data (Figure I)

Due to technical problems with the SHTS and PCS, the HFS was focused on the receiver only between April and October 1982. The completion of the cold storage tank repair in early 1983, has permitted tracking the sun with the HFS for two months more on the Sulzer receiver before the installation of the new Advanced Sodium Receiver (ASR).

During the installation period, the HFS was been in a new 'vertical' stow position in order to reduce the mirror silver corrosion and to investigate the reflectivity degradation in this position.

During this period. (May to July), the field remained in the new stow position day and night for over 1760 hours.

In August and September, the HFS was focused on the new ASR flat tube bundle before technical problems occurred.

Taking advantage of good days in the last two months of 1983, the HFS was again in track position.

After the first quarter of 1984, the HFS presented the following operation data:

		1982	1983	1984	To Date
Insolation	300 W/m^2	2323	2575	582	5480
Standby		2120	1130	358	3608
Track		668	379	224	1271

3. TECHNICAL BEHAVIOR

3.1 Reflectivity Variations

Over 750 days of mirror field reflectivity observations, the specular reflectivity curve of the heliostat field has been correlated with the heliostat soiling conditions (Figure II).

It has been noted that during the summer period the air contains the largest amount of dust, which has a direct influence on the cleaning frequency and the irradiance.

The reflectivity variation over the year gives evidence of the impact of natural cleaning forces such as rain, hail, occasional snow, and dew. This natural cleaning, in combination with low winter temperatures, maintained field reflectivity almost constant for a period of 50 days in 1982.

Effect of Washing

From experience gained from the washing activities, it appears that at the SSPS site, the cleaning procedure did not degrade the mirrors; instead, a layer of dirt remained on the glass after each washing.

The cleaning method used does not seem sufficient to remove all the deposited organic layers, and thus the reflectivity value after washing decreases slowly with time.

To remove a majority of these deposits requires mechanical or other cost-effective techniques. Through tests done on a dirty mirror, a mechanical cleaning procedure will restore the reflectivity value to within 1% of the original value. Over the last two years, the cleaning procedure has brought the field average reflectivity value to appoximately 94% of the nominal value.

The cleaning frequency of the heliostat field will vary over the year with reflectivity loss. Nevertheless, for economical reasons (2), the recommended minimum average reflectivity of the field should not be less than 85% of the nominal value.

3.2 Mirror Silver Corrosion

Beginning in October 1982, mirror module corosion, resulting from 2 years of heliostat exposure to the atmospheric conditions of the SSPS site, was first observed. In January 1983, a complete field survey was made. Of the 1116 modules, 141 modules were affected to different degrees by corrosion of the mirror silver layer.

. a total of 280 facets (25% of the total)

Of the 93 heliostat which make up the CRS field, only 4 have no indi cation of corrosion on their mirror modules. On the other hand, 6 affected heliostats were found with at least one very corroded panel, (± 30%). In general, it can be concluded that the corrosion tends to be near the edges of the mirrors and occurs more frequently at the bottom of the modules, rather than at the top.

CORRODED AREA ESTIMATION		
January 1983	Corroded Area Status	April 1984
124 x 0.5 cm^2 (x30)	low stage	228 x 2 cm^2
17 x 88 cm^2 (x300)	medium stage	39 x 30 cm^2
————	high stage	10 x 100 cm^2 3 x 200 cm^2
1553 cm^2	total area or	3226 cm^2
0.004%	% of total heliostat field mirror surface	0.009%

Based on corrosion impact observed at Solar-One (3), Figure III shows the SSPS-CRS mirror corrosion growth versus that observed at Barstow.

3.3 Other Problems

Lightning appears as a major threat to the SSPS heliostat field. Two lightning strikes have hit the plant since the beginning of operation; one on April 29, 1982, and the other on May 3, 1984. Most of the damage caused by the strikes have been in the heliostat field. The first one left 30% of the field out of service (28 heliostats). while the second affected 77% of the field (72 heliostats). Efforts should concentrate on determining the causes for such extensive damage, as well as finding a remedy to reduce the effects of lightning. Even in such a small plant, the repair is costly and time consuming. In a large plant with thousands of heliostats the effect could be devastating.

4. HFS AVAILABILITY AND MAINTENANCE

The availability of the plant reflects the actual need for full heliostat field operation. There have been periods when the receiver loop was out of operation, where the HFS maintenance was relaxed. Two such periods have been during the cold storage tank repair (October 8, 1982 - March 7, 1983), and the Advanced Sodium Receiver installation (April 29, 1983 - August 5, 1983). The low availability (close to 80%) during part of those periods is due to the fact that one HFC load was removed for use in the design of the test system.

Figure IV gives the availability of the heliostat field from April 1982 until April 1984. Table I shows the daily number of heliostat failures and the cumulative number of failures.

Main milestones in this figure are:

1. April 29, 1982. Lightning strike causes damages in the Heliostat Array Controller Communication Interface and in the electronics of 28 heliostats.
2. October 8, 1982. Start of Cold Storage Tank repair. The receiver loop is out of service during repair.
3. January 3, 1983. One HFC card is dismounted to help in the design of a test system. This renders 17 heliostats out of service.
4. March 7, 1983. End of Cold Storage Tank repair.
5. April 15, 1983. Design of test System for the boards is finished.
6. April 29, 1983. Start of ASR installation. HFC board dismounted again to help in the extension of the test system, also to cover the HFC boards.
7. August 5, 1983. ASR is ready to accept heliostats in track.
8. September 7, 1983. ASR tubes damaged due to improper filling.
9. October 21, 1983. ASR tubes repaired. Receive ready to start functional tests.
10. December 15, 1983. ASR acceptance tests are finished. Normal operation of the receiver begins.
11. January 30, 1984. On-site capability for test and repair of heliostat electronic controls.

The heliostat field has proved that it can have a very high level of availability. the main causes for failures have been the communication interfaces and the encoders (Table II). All other parts have proved to be quite reliable. To estimate the reliability of electronics it has to be kept in mind that the heliostat field has been continuously powered day and night since the start of operation.

A concept that has demonstrated its usefulness has been the on-site capability for repair of the heliostat electronics. This has minimized the stock of spare parts and the repair time. The SSPS experience is that when a system contains a large number of identical elements. the maintenance capability and facility for these elements should be directly available at the operation site.

5. CONCLUSIONS: IMPACT ON HELIOSTAT DESIGN

From heliostat behavior observed during operation of the CRS-HFS, several comments regarding the main components can be made:

- The mechanical part of the heliostat worked well under the operational conditions. Equipment design could be improved in certain areas such as the drive machanism and lock for stow position.
- Electronic equipment presented satisfactory behavior, taking into consideration such failures as:
  - ° connector plugs (weakness of the component itself)
  - ° high voltage protection
  - ° stability of the low voltage power
- The control component has shown satisfactory behavior. The main incidents have been HC communication loss and HFC command and communication loss. Those events have highlighted the need for the maximum flexibility of the software program from the point of view of operation and safety.
- A certain anxiety can be expressed with regard to the deterioration and degradation of the silver layer of the modules. The origin of those deteriorations can be found in the combination of:
  - ° delamination of the honeycomb
  - ° cracking of the protective painting under high tension (mechanical, thermal)
  - ° chemical attack of the silver under water condensation/residue
- Nevertheless, the abrasion of the front glass surface seems to be even less than expected. The nominal reflective value has a range of ± 1%.
- Some improvements which can be made in the next generation of the heliostats are:
  - ° Reflective surface and structure: Weight reducing per m^2 by using mirror sandwich and bigger reflective surface.
  - ° Mirror module: Use of very high reflective laminate mirror which will reduce the module weight and thus the parasitic consumption.
  - ° Electronic control equipment: For more reliable and resistant components. use of units are well distributed and, available locally.
  - ° Control and command system: Simplification of the logistic function to the essential positioning and security role which can be executed by radio communication.
  - ° Mechanical system: Reinforcement of the mechanical part corresponding to extended conditions of operation.

6. REFERENCES

1. WATTIEZ, P. and MARTINEZ, F. (1983). Performance, Experiences, Failures of the SSPS-CRS Heliostat Field. Central Receiver System, Midterm Workshop Proceedings, SSPS Technical Report Nº 4/83, paper 2.8.
2. WATTIEZ, P. (1982). SSPS Field Cleaning Frequency, SSPS Technical Report R-72/82.
3. NORING, J. E., MAVIN, C. L., DECKER, E.V., and SKVARNA, P. E. (1984). 10MWe Solar Thermal Central Receiver Pilot Plant-Mirror Module Corrosion Survey, Sandia National Laboratories Report, SAND84-8214.

MONTH	1	2	3	4	5	6	7	8	9	10	11	12	Year
Average Availability (%)				96.7	64.1	97.3	96.4	95.2	93.1	92.6	97.0	96.4	1982
Number of Failures				94	7	14	4	2	2	1	1	5	
Average Availability (%)	82.9	81.7	99.5	99.5	82.9	81.7	95.7	99.5	97.6	96.5	97.9	99.1	1983
Number of Failures	4	0	4	2	0	0	3	3	3	3	2	1	
Average Availability (%)	99.4	99.5	99.5										1984
Number of Failures	0	2	5										

Table I
Monthly Average Availability and Number of Failures (April 1982 - April 1984)

COMPONENT \ YEAR / QUARTER	1982			1983				1984	TOTAL
	2nd	3rd	4th	1st	2nd	3rd	4th	1st	
HC	27	--	1	5	2	3	4	1	43
HFC	--	--	--	1	--	1	--	--	2
Azimuth Encoder	2	1	1	2	--	--	--	1	7
Elevation Encoder	--	1	1	1	--	--	--	2	5
Azimuth Limit Switch	--	--	--	1	--	--	--	1	2
Elevation Limit Switch	1	--	--	--	--	1	--	--	2
Azimuth Motor	1	--	--	--	--	--	--	--	1
Elevation Motor	1	--	--	--	--	--	--	--	1
5V Power Supply	8	--	--	--	--	--	--	1	9
Mechanical Relay	--	1	--	--	--	--	--	--	1
Solid State Relay	--	1	--	--	--	--	--	--	1
Wiring Harness	2	1	--	--	--	--	--	--	3

Table II
Components Replacement (April 1982 - April 1984)

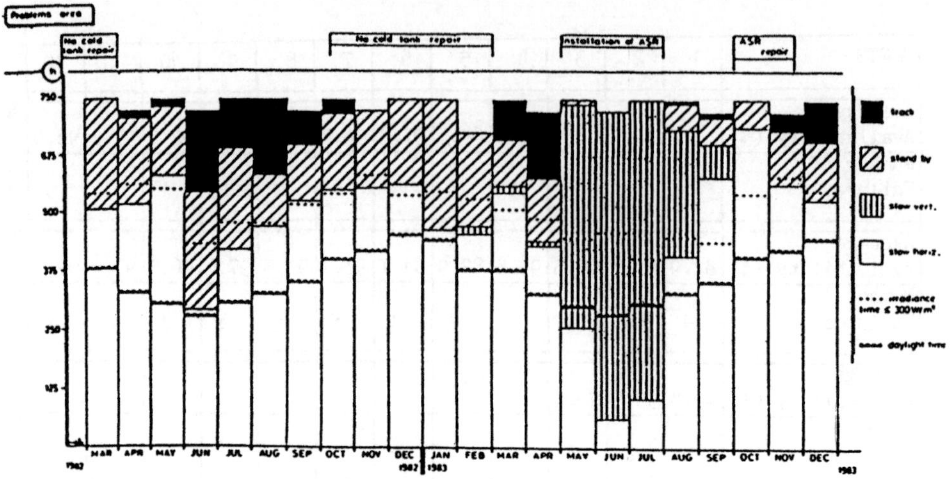

Fig. I   H.F.S. OPERATIONAL BEHAVIOUR

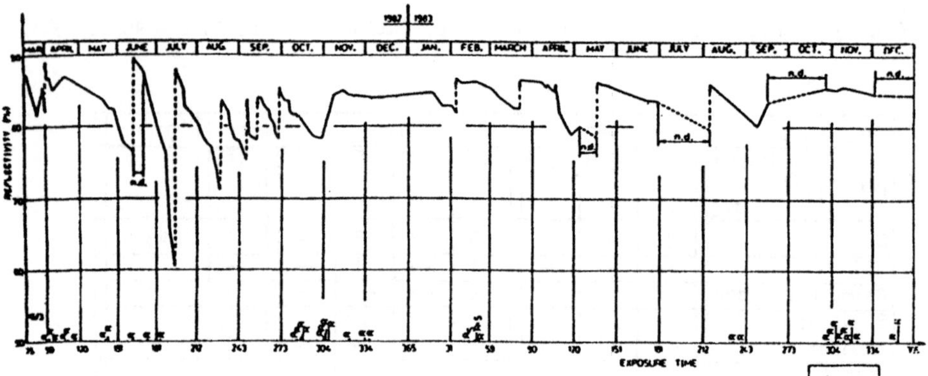

Fig. II   HELIOSTAT FIELD REFLECTIVITY CURVE

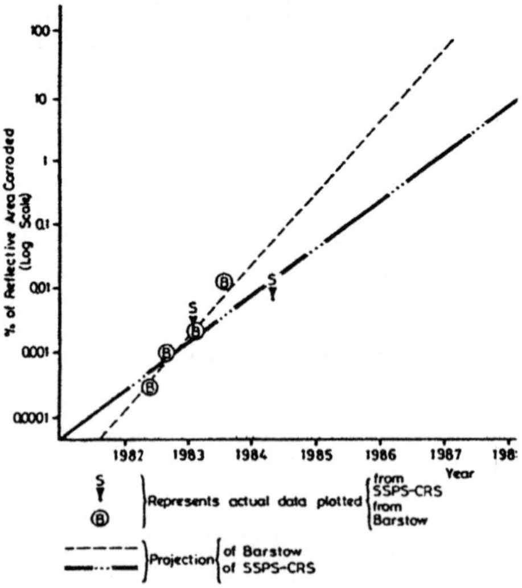

Fig. III   PROJECTED AMOUNT OF CORROSION AT PRESENT GROWTH RATE

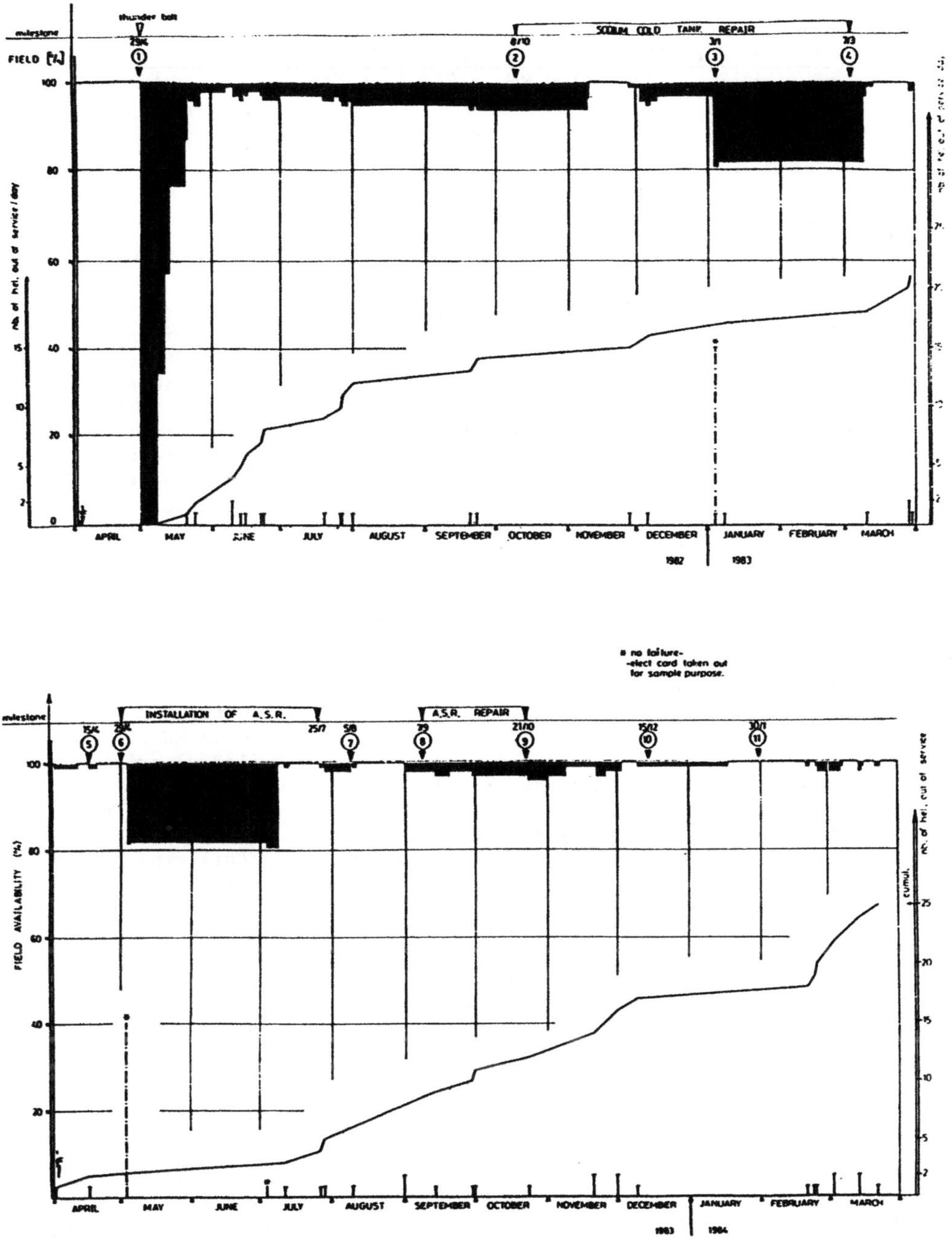

H.F.S. AVAILABILITY

FIG. IV

HELIOSTAT FIELD AND RECEIVER EFFICIENCY MEASUREMENT ON THE IEA - 500 kW_e - CRS AT ALMERIA

W. SCHIEL
Institute for Technical Physics
DFVLR - Stuttgart

Summary

The Heliostat and Receiver Measurement System (HERMES) is designed to evaluate the radiant flux distribution in front of a receiver aperture plane. The results of the Fall Measurement Campaign in 1982 on the SSPS-CRS power plant are presented in this paper. The incident power, its distribution, and by that the heliostat field and receiver efficiencies were measured. Besides the centroid of the flux distribution, energy versus radius, iso-flux contours, intensity cross sections and 3-D intensity diagrams were determined by the data map. The experimental results are compared with theoretical calculations and the HELIOS computer code.

1. Introduction

The central receiver system (CRS) at Almeria, Spain, is a small solar power station designed to produce 500 kW_e output at 0.92 kW/m² beam irradiance. The heliostat field consisting of 93 individually tracked mirrors with a total reflective area of 3655 m², redirects up to 2.8 MW of thermal energy into a cavity receiver on top of a tower. A detailed description of the complete system is found elsewhere (1).

The distribution of the radiant flux in the receiver aperture plane and the total power into cavity depends on tracking accuracy of each heliostat, alignment and reflectance of the mirrors and on actual windspeed and direction. In order to verify theoretical calculations and for better understanding of actual performances of the heliostat field as well as of the receiver, it is of interest to measure the incoming power and its distribution in the aperture plane.

The Heliostat and Receiver Measuring System (HERMES) has been developed to evaluate the heliostat field and the receiver and to improve both solar specific subsystems. HERMES measurements were carried out on the SSPS central receiver system during the Fall Measurement Campaign in 1982.

2. Experimental

During the years 1981 and 1982 the measuring system HERMES was developed by DFVLR-Stuttgart (2). It is principally similar to the American Beam Characterization System (3) and the Swiss Flux Analyzing System (4), however, allows measurements of complete fields and one line operation, respectively. The system is subdivided into three parts, the video camera system, the computer and the data acquisition system.

The measurements of the irradiance E at an aperture point (x,y) coming from the heliostat field is based on the reflectance of E at a Lambertian target screen which has to be brought directly into the beam. This was obtained by a coated moving target (4) of about 40 cm width traversing the receiver aperture plane in a very short time (5 to 8 sec). The motion of the coated target was steered by a radio signal and controlled by a computer system. The corners of the receiver aperture are identified via cursor interaction and stored in a geometric file for off-axis camera angle correction. Running from West to East the target activates 33 positioning sensors being mounted on the rail. Passing each sensor, a radio signal is transfered to the computer which triggers the camera to make a picture. The 33 single pictures are composed by a selection code into one total picture of the flux distribution. Before and after each measurement the computer asks for meteorological data needed for the evaluation to be done afterwards.

To measure the system constant k, a water cooled traversing bar driven by a stepping motor run through the aperture plane to the position with low, middle, and high radiant fluxes. An absolute self calibrating Kendall Radiometer was screwed at one side of this bar, surrounded by a screen coated with the same Lambertian material as the moving target. The absolute irradiance and simultaneously the response of the camera to the surrounded screen were measured. This delivers the conversion coefficient of the relative grey levels measured by the camera into absolute irradiance values. Taken into account the non-equability of $\rho(x,y)$ over the target area of 3 % (5), the inaccuracy of the absolute Kendall Radiometer of about 2 %, the camera signal $V(x,y)$ of 2 %, the accuracy of the flux measurements is estimated to be not larger than $\pm$ 5 %.

3. Results and Discussion

During August, 1982 the HERMES measuring device was transferred to the Central Receiver System (CRS) of the SSPS-project, and installed behind the last heliostat row in North-South direction about 200 m far from the tower. The main goals of this measurement campaign were the determination of the heliostat field tracking accuracy and the beam quality as far as the heliostat field and receiver efficiencies with respect to time of the day and different insolation conditions. All these informations can be got by the evaluation of the measured flux distribution in the aperture plane, and the sodium flow rate as far as the inlet and outlet temperatures. Thus, most of the measurements were concentrated on the flux determination.

An evaluation of one complete measurement set is presented in Fig. 1. At the top left part a 3-D intensity diagram of the corrected video picture is shown. Besides all meteorological data, the actual sun position and time are printed which were measured during the test time. The second graph shows the iso-contours of the incident power distribution, a theoretical aim point of the heliostat field, the gravity point of the measured distribution, and the aperture size projected in the measuring plane. Furthermore, it shows the intensity cross section through the centroid in vertical direction. The power versus radii graph was calculated by summing up the power inside concentric cycles around the centroid. The total power was computed by integration over the total measuring plane (5 x 5 m).

The deviations of the measured gravity point from the aim point in vertical and horizontal direction yielded the reflected beam tracking error. Fig. 2 shows this error with respect to local time at two different days. It can be seen that the center of gravity moves 10 cm East and about 15 cm down in the morning to 10 cm West and 5 to 10 cm up in the evening. This depends on the one hand on the position of the heliostats and on the other hand on actual wind loads. The dependence on weather conditions could not be found out because of too little data obtained during the measurement campaign. In the next two figures the measurements from October 6 and 7, 1982 (Figs. 3,4) are presented: Beam irradiance, total radiant power input to receiver, and thermal power output from the receiver through the day.

The measurement from October 6, 1982 shows a cloudy period around solar noon and how the systems responds. On this day the plant started at about 10:00 a.m. and stopped its operation at about 3:00 p.m. because of too high wind speeds. October 7, 1982 was a bright shiny day and the plant was operating from 8:00 a.m. to 5:00 p.m. With these results the heliostat field and receiver efficiency can be easily determined.

The measured heliostat field efficiency with respect to solar time is presented in Fig. 5, besides this the HELIOS-calculations dated from October 5, 1982 with a 100 % heliostat field reflectance and with a reflectance of 81.5 % measured at October 5. Note, however, that this reflectance value is measured punctually and that it is not yet cleared whether this value is representative of the total heliostat field. Besides this boundary condition the measurements are in good agreement with the theory.

The receiver efficiency depends on the radiant power input, the absorptance of the receiver tubes, the inlet and outlet temperature of the receiver (IR-reradiation), conduction and convection losses. In Fig. 6 the receiver efficiency and the product of the heliostat field and receiver efficiency are plotted as a function of the beam irradiance. At an insolation of 900 W/m² a receiver efficiency of about 87 % and a thermal efficiency of about 60 % were measured. The thermal efficiency measured over the time of operation (8 h:50 min) on a clear day (October 7, 1982) comes to 57 %. Converting this value to the total day from sunrise to sunset (11 h:25 min) by 8 h:50 min operation you get an efficiency of 50 %. The mean relative errors of the receiver and thermal efficiency measurements, resp., are estimated to be ± 9 % and ± 8 %, resp.

4. Conclusions

The HERMES measurements on SSPS-CRS and the evaluation by the plant's data acquisition system have shown that the solar specific subsystems, i.e. heliostat field and receiver fulfill the design criteria at steady state conditions. To find out the behaviour during transient conditions and simple mathematical expressions for simulation modelling measurements during longer periods are necessary.

References

1. BECKER, M., ELLGERING, H. and STAHL, D. (1983). CRS-Construction Report. DFVLR IEA SSPS-Operating Agent. Köln, West Germany.
2. SCHIEL, W. (1983). HERMES Measurements. SSPS Technical Report 4/83.
3. KING, D.L., ARVIZU, D.E. (1981). Heliostat Characterization at the Central Receiver Test Facility. Journal of Solar Energy Engineering, Vol. 103.
4. TOBEL, G.v., SCHELDERS, C.H. and REAL, M. (1982). Concentrated Solar Flux Measurements at the IEA-SSPS Solar Receiver Power Plant Tabernas/ Almeria. SSPS Technical Report 2/82.
5. GÖRLER, G.P. (1981). Messung zum spektralen Reflexionsgrad sowie zur Winkelabhängigkeit der Reflexion an einigen weißen Oberflächen. DFVLR-Interner Bericht IB 353-81/6.

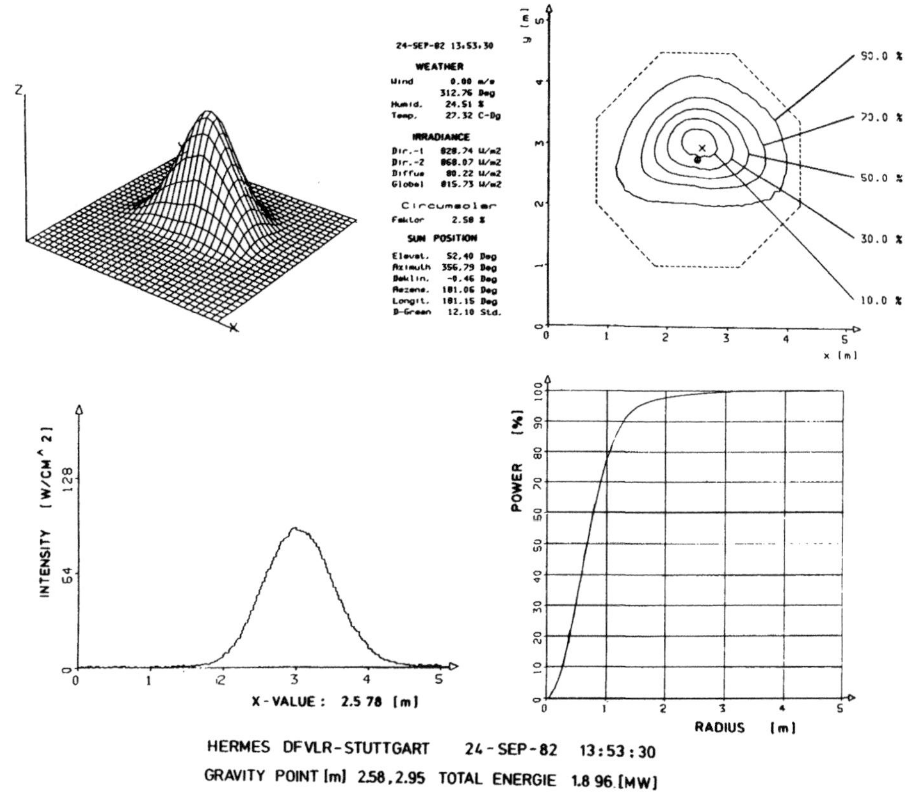

Fig. 1 One complete evaluation of the measured flux distribution 0.9 m infront of the SSPS-receiver aperture plane

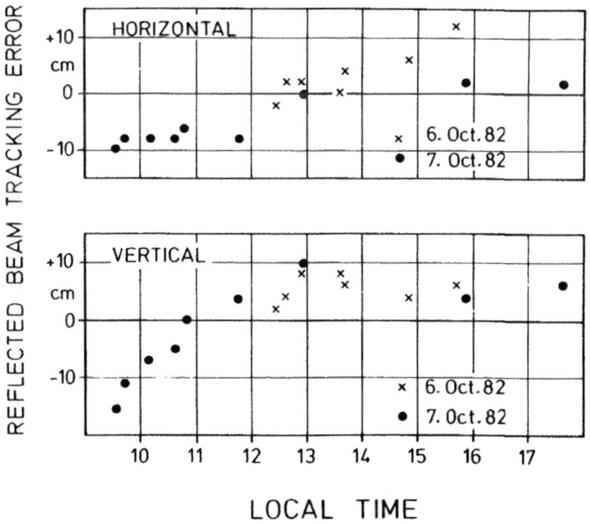

Fig. 2 Tracking accuracy of the SSPS-CRS heliostatfield

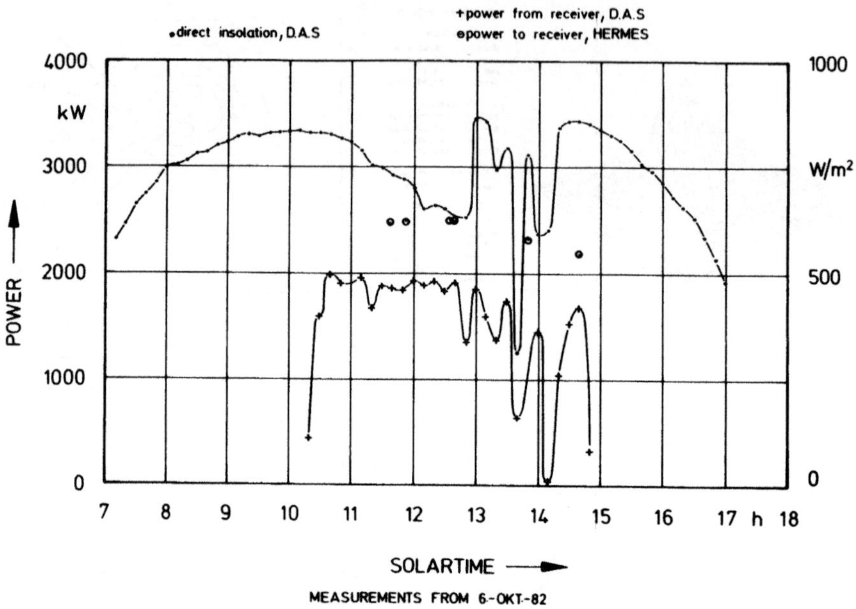

Fig. 3 Irradiance and power measurements from 6-Oct-1982

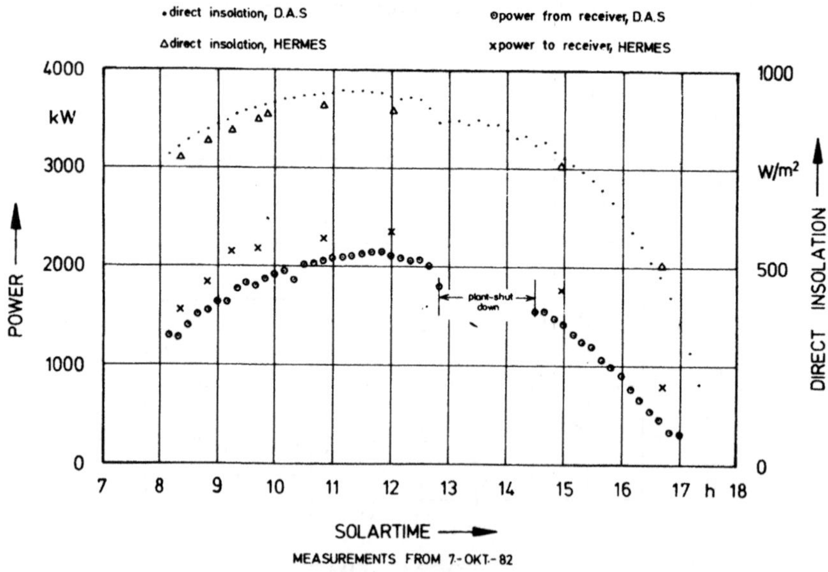

Fig. 4 Irradiance and power measurements from 7-Oct-1982

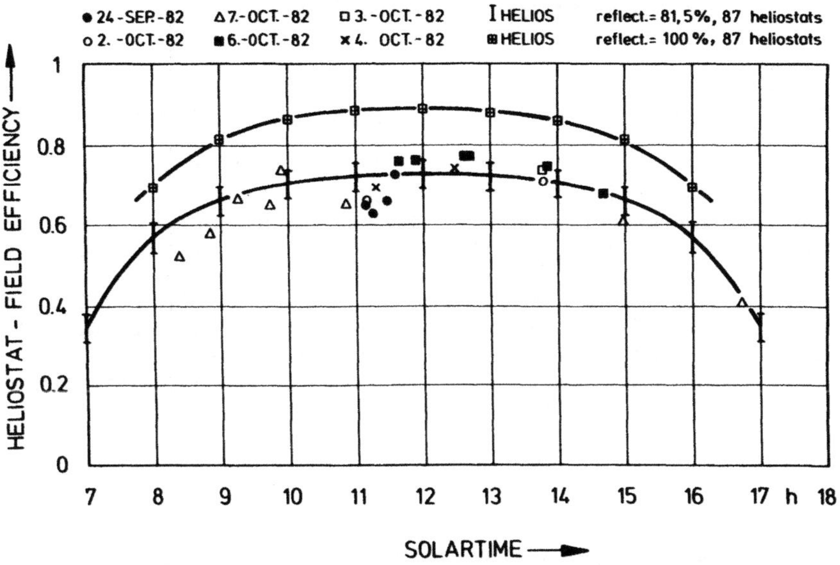

Fig. 5 SSPS-heliostatfield efficiency with respect to solar time in comparison with HELIOS computations

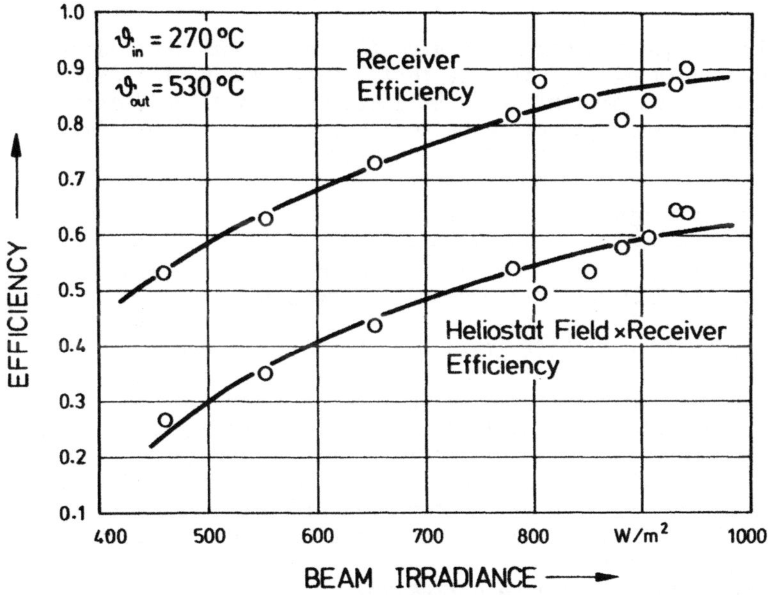

Fig. 6 Receiver and thermal efficiency measurements in dependance of the beam irradiance

THEMIS HELIOSTAT FIELD

J.J BEZIAN (A.F.M.E) and B. BONDUELLE (C.N.R.S)
Groupe d'Evaluation Scientifique Themis
Themis Scientific Valuation Team

Summary

This paper summarises the first results of operation of Themis heliostat field.
The optical tests show that the solar flux reflected by the heliostats fits satisfactorily the technical specifications.
The maintenance of the heliostats does not affect the solar plant performance. An heliostat avenlability of more than 95% is currently reached.
The evaluation of the reflective coefficient and the first estimation of the heat balance proove that the optical effiency of the heliostat field is in agreement with the design specifications.
Improvements have been brought to the central control program of the heliostat field, allowing more versatility and operation safety of the system.

INTRODUCTION

Themis heliostat field is composed of 200 heliostats of 53.95m² (1). The reflective surface is made of "sandwich" glass supported by a steel structure.
The heliostats are controlled by a three level processing system :
- the field central computer, nine group controllers, and 200 local heliostat controllers.

In the first section we give the individual heliostat tests. Then an history of the failures that occured during the last nine months is presented. We present also the first results of reflecting tests and field performance (geometric efficiency, field availability), and our conclusions on heliostat field management.

I RESULTS OF INDIVIDUAL HELIOSTATS TESTS

After the incident that occured in December 1981, Themis heliostats were modified. The new metallic pedestals introduced a change of the géographic position of the heliostats.
A new calibration of heliostats was therefor needed.
The tests concerned the 185 heliostats available in winter 1982/83. The fifteen others were replaced a few months later (2). 746 spots have been treated. But only 600 represent the state of the existing field. Data acquisition system, with 1024 cells on the active target, gives a very precise map of the reflected image. This image, compared with technical specifications, allowed us the check and sign for the heliostats (fig 1).
In order to analyse the image caracteristics, we introduced the three following criteria:

- the pointing accuracy criterium.

The energetic center of the image must be nearer than 6.6 mrad from the optical center (fig.2)

- the daily pointing accuracy variation criterium .

During the day, the variation of the position of the energetic center must *not be greater than 6.6 mrad*

- the shape criterium

we define an average radius of the spot, which is compared with the theoretical radius.

HELIOSTAT C 16

LE 25 1 1983 A 10 H 21

FIG. 1

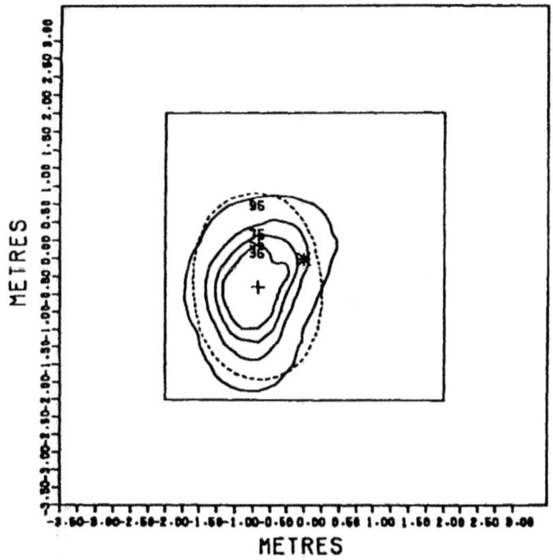

ORIENTATION 23.5

POINTAGE 6.74

FOCALISATION 0.63

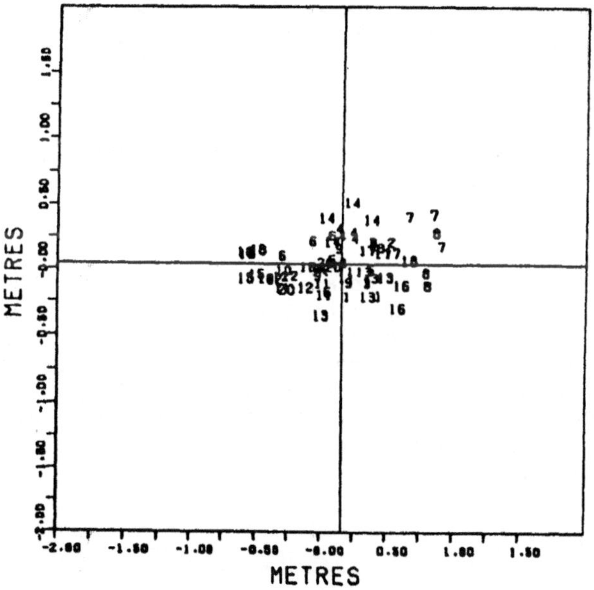

FIG. 2

XG = 16.5

YG = 4.6

GROUPE D

The results of all these tests prooved that most heliostats 85% met the designe optical requirements.

Energetical centers were always nearer than 3 cm from geometrical ones, given by the passive target. So this justify the use of the passive target to test heliostats. The abscissas and ordinates of energetical centers are fitting a gaussian distribution with a mean value of 2 cm . The average angular distance between the theoretical and real spot centers, over the 200 heliostats is 2.7 mrad (fig.3)

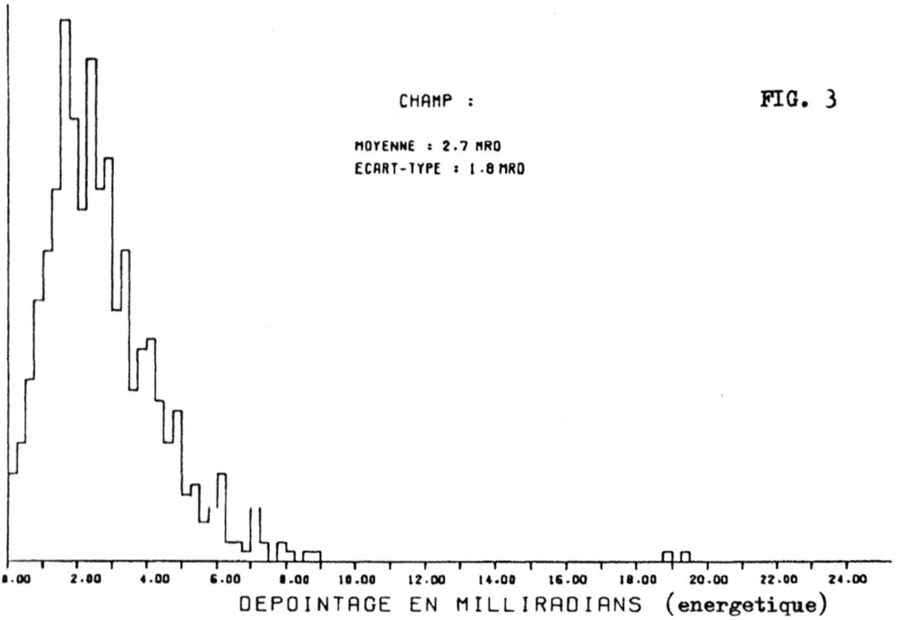

FIG. 3

We think it would be desirable to establish standardized criteria for the comparison of the heliostat fields.

II HELIOSTAT FIELD MAINTENANCE

Since the last mechanical modifications, no major failure occured. The heliostats behaved satisfactorily and stood without damage the tempest we endured on February 1984, with a 160 km/h wind.

About the mirrors we cannot notice any corrosion or any optical degradation. However, we have noticed, on the back structure of some thirty heliostats, some traces of corrosion, due to imperfect welding. Now, this problem is solved; about 195 heliostats are able to work : in April, we could work with 199 heliostats (fig.4)

Here is a table with the results of the nine last months exploitation, where we noticed the incidents that occured during this period.

INCIDENT CAUSES	Heliostat Concerned	Occurence	Unaivability	Remarks
Lightning	17	1	10 to 20 days	Only one storm
Battery	16	1	1 to 8 days	All along the period
Little electronic	157	1 to 10	1 minute	Initial adjustment needed
Serious electronic	40	1 to 3	1 to 3 days	Various Incidents
Glass shattered	3	1	10 days	Undefined cause

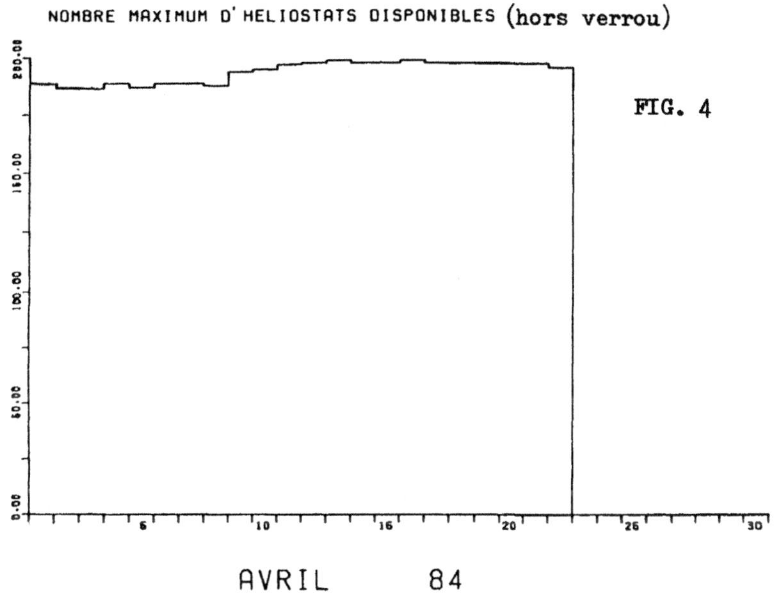

FIG. 4

More we have noted that 43 heliostats never felt in breackdown. The protection against the lightning needed some modifications. We do not know yet how it will resist.

Concerning the current maintenance of the heliostats, only one man is necessary to see about batteries, oil of motors...

It prooves that the heliostat field does not require much care to be efficient.

. 97

III HELIOSTAT FIELD PERFORMANCE

Field layout and height of the tower (the receiver is 86m high) explain the good geometrical efficiency (fig.5)

The other part of optical efficiency is the reflective coefficient.

The first tests we did gave us a coefficient about .9 . This is also otained by the first results of the heat balance of the plant. This good cofficient can be explained by the quality of the mirors, with a very thin glass cover. But we must take in account the fact that there is no atmospheric pollation in Themis, except when sirocco blous, followed by a little rain or snow.

The results show, at noon , on a shining day, a heat balance, between the sun and the enthalpy gap of the salt through the receiver, better than .7.

This result, associated with the availability, and an insolation which is often higher than 1000W/m², gives a good idea of the performances of the heliostat field.

IV HELIOSTAT MANAGEMENT

After the modifications, the introduction of a mechanical bolt, and the setting up of a new central computer, we develop a new central program to manage the heliostats.

The adding of an interactive system makes the operator system dialog easier and allows a continúous watching of the field.

More, after the first analysis of the receiver performance, we added in the program an automatic microvariation of the aim point to equilibrate the outlet temperature.

So, we can say that the heliostat field is precise, versatile, and performant.

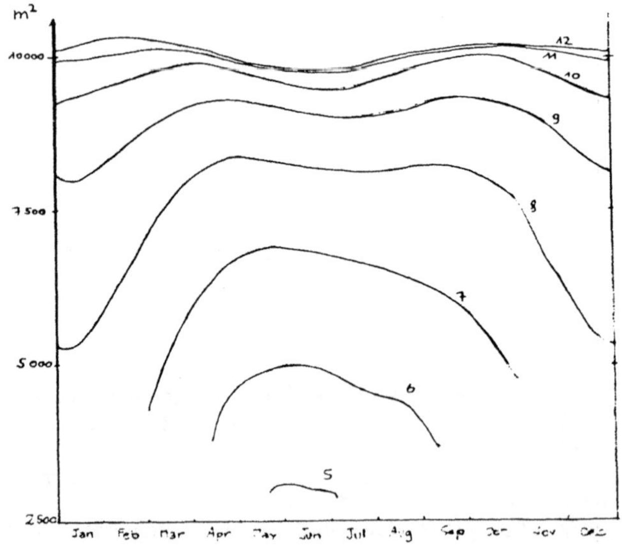

FIG. 5

EFFICIENT AREA

Morning hours

Include blocking
shading
cosine

REFERENCES:

(1) A.COLLON, A. LECLERC (1982) . *Le champ d'heliostats de Themis.* *Entropie N° 103 pp. 10 - 20*

(2) B. BONDUELLE (1984) . *Reception individuelle des heliostats Themis* *Rapport GEST 005.*

(3) J.J BEZIAN (1984) . *Surface efficace des heliostats.* *Rapport GEST 010*

ASINEL HELIOSTATS
FOR THE GAST TECHNOLOGICAL PROGRAM

J. de Marcos and L. Crespo
Asociación de Investigación Industrial Eléctrica

Summary

The ASINEL heliostat development process within the GAST Technological Program is described. The approach to the heliostat specification from the system point of view indicated that a field consisting of two heliostat types is worthwhile.
Two prototypes of aprox. 55 m^2 where developed with the aims of 3.9 mrad,total error budget and low cost. The main features of both heliostats are shown. The price of 150 $/$m^2$ for a 1000 unit production, has been achieved and a further reduction with the next development step is expected.

1. INTRODUCTION

The first approach in the GAST T.P. to the field lay-out optimization and correspondently heliostat characteristics was done on the bases of one single heliostat type for the complete field. In this way, a total error budget of 2.9 mrad (1σ) and a reflective surface of 52 m^2 were specified.

When the Spanish side joined the project, ASINEL started its own heliostat development activities by studying the parametric influence of the heliostat main characteristics in large fields and finally, whether for big Plants a field consisting of two different types of heliostats would make sense. As result of these studies, not only an optimized heliostat specification was obtained, but a very powerful computer program for heliostat field lay-out and single heliostat characteristics optimization (ASPOC - A Solar Plant Optimization Code) was created as well.

2. DEVELOPMENT GUIDELINE

The two different possibilities for field mixing were either the use of more accurate heliostats on the rear part or less accurate in the inner are.

Taking as a reference the existing German heliostat, 2.9 mrad (1σ) total error, investigations on annual receiver collected energy were performed, considering 1.9 mrad and 3.9 mrad respectively. The results can be seen in figure 1.

In order to realize the economical influence on the final energy price, different cost-error scenarios were supposed (Figure 2).

Scenario A corresponds to real quotation basis, while the others represent stronger or softer dependency of cost on total error. ($\pm$ 5% per mrad)

Figure 3 shows the final result.

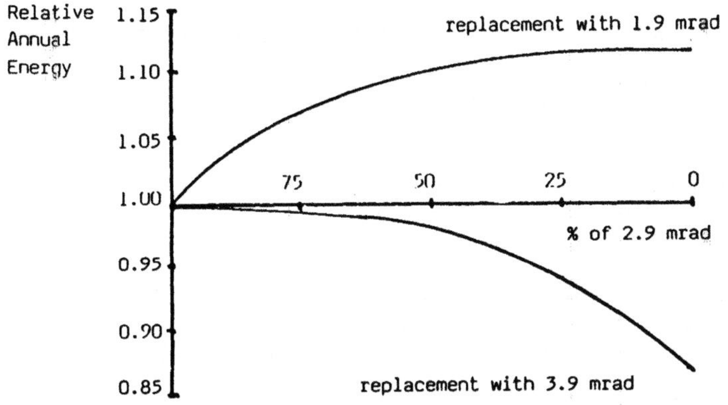

Figure 1

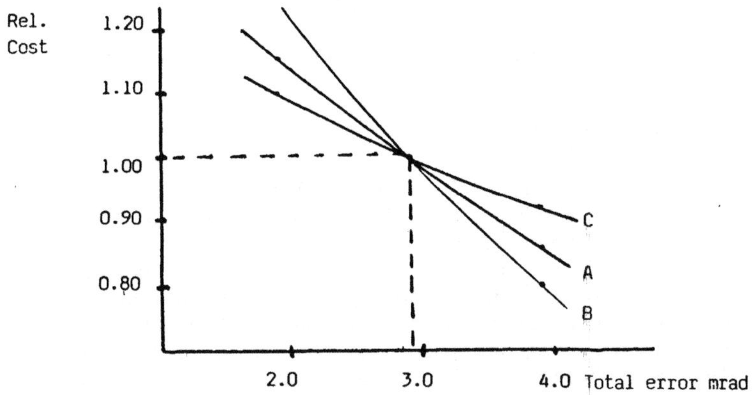

Figure 2

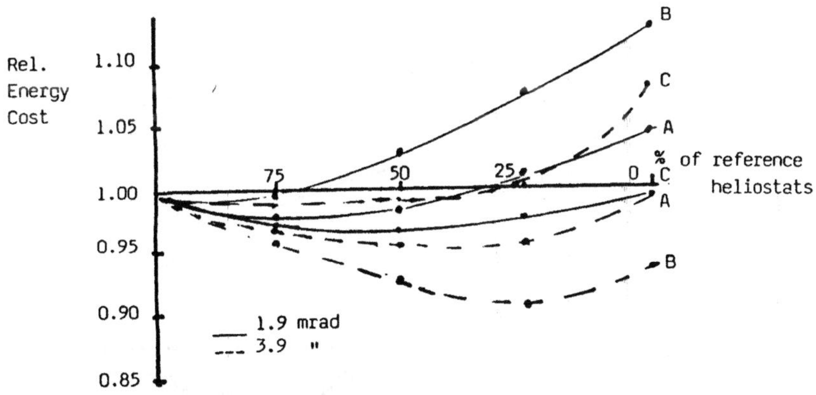

Figure 3

Taking these results into consideration, the most sensitive and attractive approach appeared to reduce the accuracy of the heliostat in the inner part.

The main requirements of the specification for the prototype were:

- Reflective surface 52 SQ-M ± 10 %
- Heliostat errors at normal operating conditions

Beam quality	3,6 mrads
Tracking accuracy	1,5 mrads
Overall	3,9 mrads

- Normal operation wind speed : 18 Km/h
- Survival wind speed : 140 Km/h
- Requirements contained in ASINEL's doc. AS-GAST-390/1-002

3. PROTOTYPE DESCRIPTION

The condensed sketches which are provided at the end, show the main features of both (CASA and INTECSA) prototypes.

The achieved cost for a 1000 units serie in the prototype stage was 150 $\$/m^2$. It could be even somewhat reduced by using the single mirror facet design.

4. ASINEL SERIES HELIOSTAT

After the succesfully conducted qualification test program, a compo - nent selection for the specification to be applied to the Spanish serie in the GAST T.P. is being done.

Additionaly, further system studies trying to increase the reflective surface per heliostat with the same gear box (that means by making the external geometry more rectangular) have been performed.

Figure 4 shows the dependence of the annual energy output of a 20 MW power plant with heliostats of 3.9 error budget on the width / height ratio and the single reflective surface. The curves have been plotted with the same total reflective area (i.e. 1980 hel. of 54.25 m^2 each) and considers a typical cloudy year.

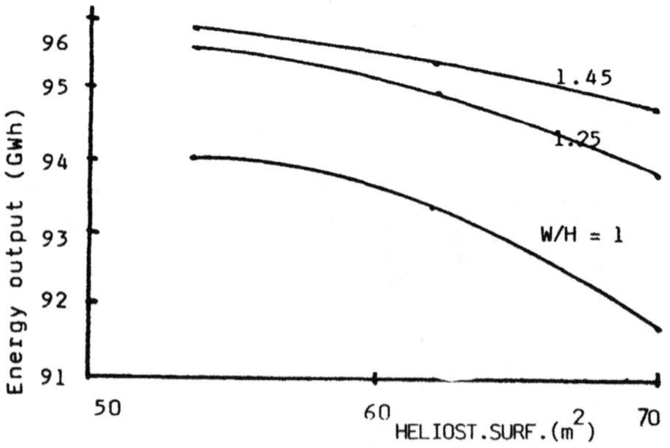

FIGURE 4

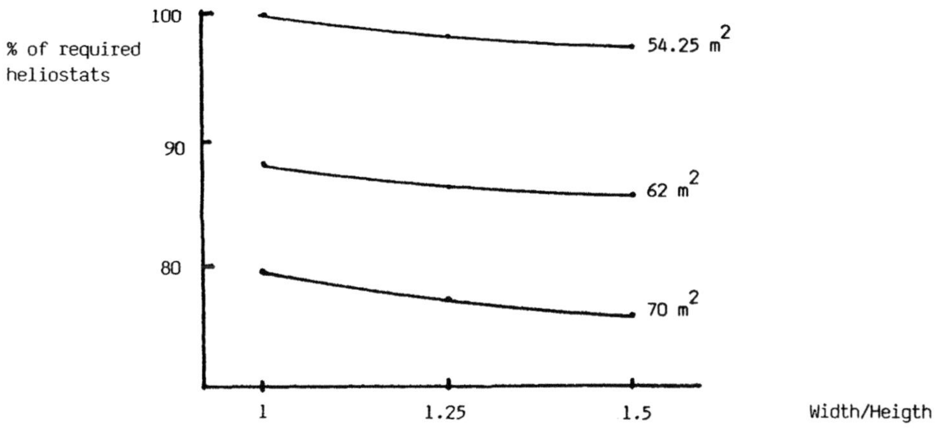

Figure 5

Taking into account this energy influence, one can figure out the number of heliostats which will be reduced. This is shown in fig. 5.

With 65 m^2 and 1.45 with ratio, one can save the 18 % of heliostats with respect to the present prototype design. Due to the fact that the reflective surface represents around 20 % of the heliostat cost, this surface increase of another 20 % will result on a relative cost increase per heliostat of aprox. 4 %. Therefore, a total field cost reduction of more than 15 % (considering cabling, civil work...) with this new heliostat is expected.

CASA HELIOSTAT PROTOTYPE I.

OVERALL DIMENSIONS

- HEIGHT ....8,4 M
- REFLECT.ASSEMBL.LENGTH...7,9 M
- REFLECT.ASSEMBL.WIDTH... 7,7 M
- CENTRAL SLOT .... 0,65 M
- ELEVATION AXIS HEIGHT... 4,2 M
- REFLECTIVE SURFACE...... 54,25 SQ-M
- WEIGHT (EXCL.FOUNDATION) 2,660 KG

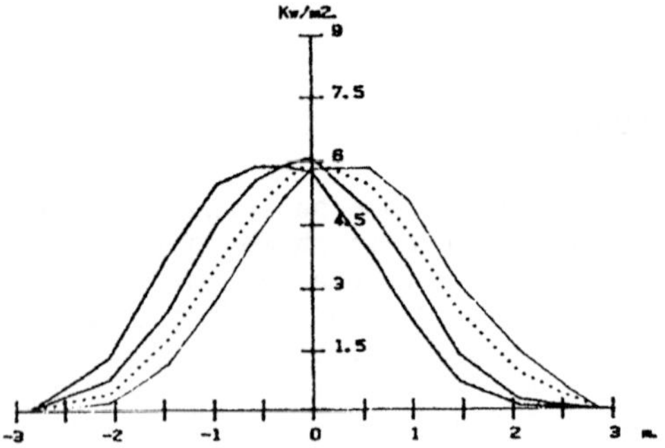

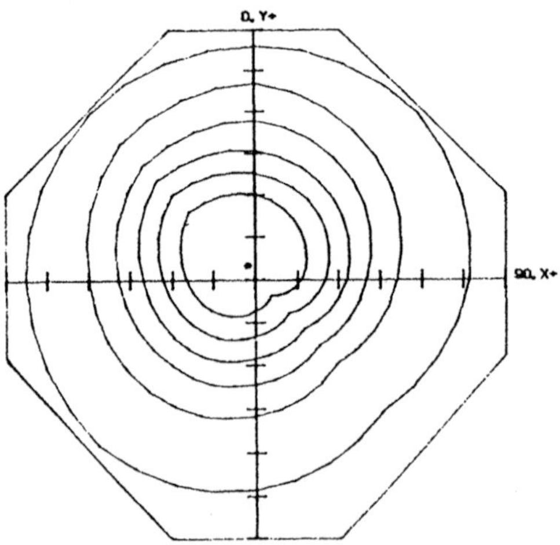

RADIACION DIRECTA	:	944 W/m^2
RADIACION DIFUSA	:	90 W/m^2
TEMPERAT.AMBIENTE	:	21°C
VELOC. VIENTO	:	7 Km/h
DISTANCIA H-B	:	302 m
ANGULO ABERRACION	:	39 grad
ANGULO INCIDENCIA	:	8 grad
CONC. MAXIMA	:	6.48
RADIO DEL 95 %	:	2.35 m

CASA HELIOSTAT PROTOTYPE UNIT

II.

MIRROR MODULE

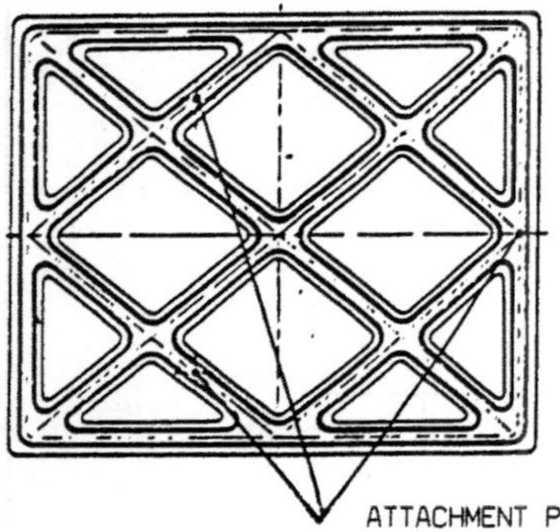

- LAMIN.MIRROR : 2.2+2.4 mm
- GALVAN.STEEL FRAME
- FRAME BONDED TO MIRROR ON A CURVED TOOL
- FACET WEIGHT : 46.5 KG
- REFLECTIVITY : 91 %

STRUCTURE

SIX TRUSSES MADE UP OF ROUND BAR STOCK AND ANGLE IRON.

EACH TRUSS BOLTED TO THE ELEVATION BEAM VIA AN INTERFACING PLATE WELDED TO THE ANGLES.

ELEVATION BEAM IS 320 MM DIA, TUBE, WITH WELDED PLATES FOR ATTACHING TO THE TRUSS.

A PAIR OF SUPPORT ARMS SECURES THE REFLECTIVE ASSEMBLY TO THE DRIVE MECHANISM.

THE PEDESTAL IS A STEEL PIPE 560 MM DIA., WITH A CIRCULAR FLANGE WELDED TO THE LOWER END TO INTERFACE WITH THE FOUNDATION STUDS.

TWO DOORS AT THE NORTH SIDE PROVIDE ACCESS TO THE ELECTRONIC BOX AND THE AZIMUTH ENCODER.

DRIVE MECHANISM

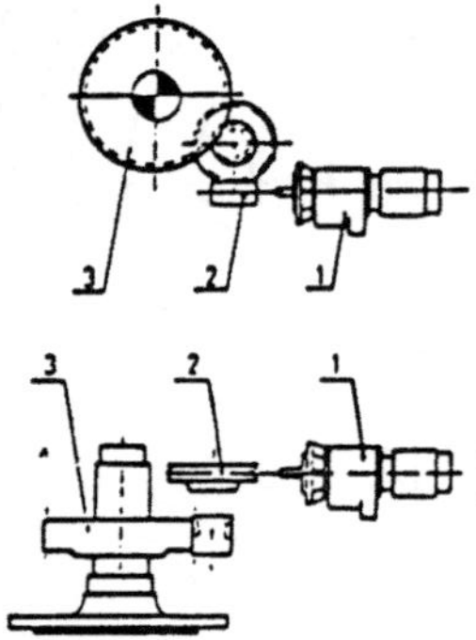

- CONVENTIONAL GEARING
- DOUBLE CASTED GEAR BOX
- 2 MOTORS: AC, 220v 1/4 HP
- BOTH AXES ARE NON-REVERSING

1 - GEAR MOTOR
2 - WORM GEAR
3 - SPUR GEAR

- OVERALL REDUCTION: 41.202
- OVERALL EFFICIENCY: 43 %

LOCAL CONTROL

- OPEN-LOOP EPHEMERIS BASED STEERING
- TWO LEVEL ARCHITECTURE (LOCAL & CENTRAL)
- ABSOLUTE OPTICAL ENCODERS OF 13 BITS
- TWO SPEED ASYNCHRONOUS OF 13 BITS
- MAXIMUM HELIOSTAT POWER CONSUMPTION 550 w
- PORTABLE DIRECT MOTOR HAND CONTROL BOX

INTECSA HELIOSTAT COMPONENT TEST UNIT

I.

REFLECT.SURFACE&OVERALL DIMENSIONS

	SINGLE GLASS	LAMIN. MIRROR
-HEIGHT	7.8 M	7.8M
-REFLECT.ASSEMB.LENGHT	7.4 M	7.2M
-REFLECT.ASSEMB.WIDTH	7.4 M	7.2M

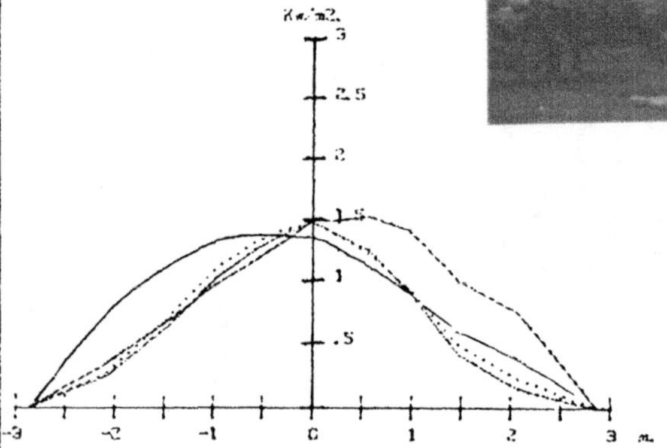

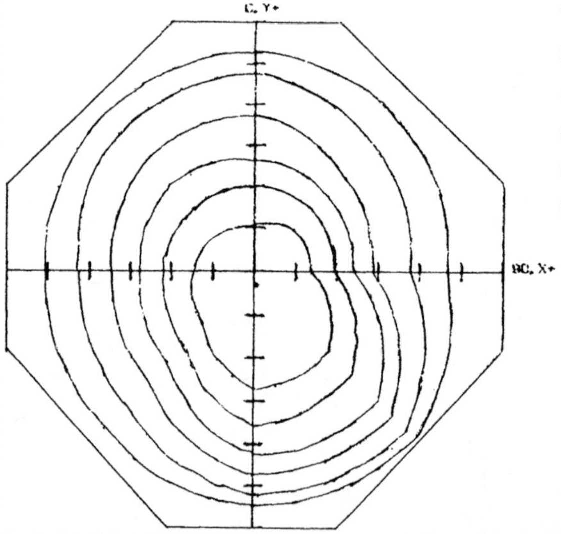

RADIACION DIRECTA	: 933 W/m^2
RADIACION DIFUSA	: 166 W/m^2
TEMP. AMBIENTE	: 0°C
VELOCIDAD VIENTO	: 0°Km/h
DISTANCIA H-B	: 302 m
ANGULO ABERRACION	: 36 grad
ANGULO INCIDENCIA	: 7 grad
CONC. MAXIMA	: 1.64
RADIO DEL 95 %	: 2.37 m

INTECSA HELIOSTAT COMPONENT TEST UNIT

II.

MIRROR MODULE

SINGLE GLASS (CYLINDRICAL)

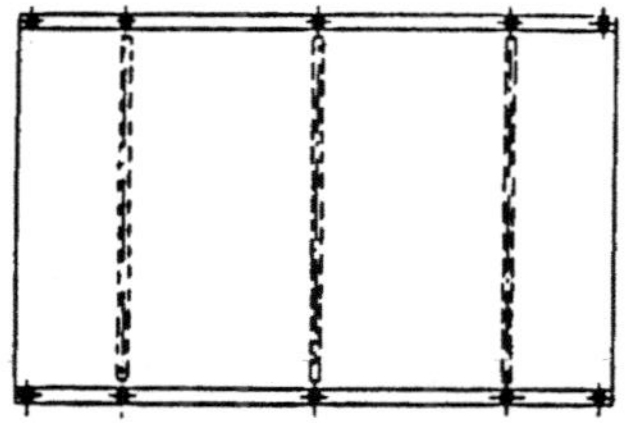

. ATTACHMENT POINTS
- MIRROR: 3 mm
- ALUMINIUM FRAME
- MIRROR CURVED BY BOLTS IN ATTACHMENT POINTS
- FACET WEIGHT: 26.25 Kg.
- REFLECTIVITY: 87%

LAMINATED MIRROR (BIPLANE)

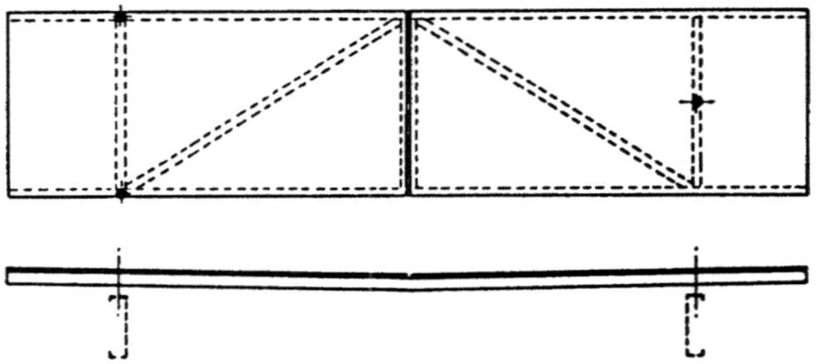

. ATTACHMENT POINTS
- MIRROR: 3+3 mm
- GALVANIZED STEEL FRAME
- FRAME ANGLE : 179.83
- FACET WEIGHT: 95 Kg
- REFLECTIVITY: 87 %

STRUCTURE

- FOUR LONGITUDINAL BEAMS (C PROFILES, GALVANIZED STEEL)
- TWELVE TRANSVERSAL STEEL TUBES WELDED TO THE BEAMS.
- ELEVATION BEAM IS A 320 MM DIA. TUBE WELDED TO LONGITUDINAL BEAMS.
- ALUMINIUM FRAME OF MIRROR MODULE IS BOLTED TO TRANSVERSAL TUBES FOR ATTACHING AND CURVING THE MIRROR.
- THE PEDESTAL IS A PRESTRESSED REINFORCED CONCRETE TUBE OF 24 INCH DIA. AND 4 INCH THICK WITH SIX STRANDS OF 0.6 INCH DIA. FOUNDED IN MORTER.
- THE PEDESTAL HAS ON UPPER PART A STEEL FLANGE TO ATTACH THE GEAR BOX.
- ONE DOOR AT THE NORTH SIDE PROVIDE ACCESS TO THE ELECTRONIC BOX.

CESA-1 HELIOSTAT FIELD
EVALUATION STATUS REPORT

Fernando Sánchez - INITEC

Summary

This paper describes the status of the evaluation activities of the -- CESA-1 Heliostat Field. The Heliostat field is composed of three hun - dred heliostats of two different types (CASA-II and SENER) distributed- in sixteen rows of a north field with a total reflective area of 11800 sq. meters.

The system has been in operation since February 1.983. The specific -- evaluation tasks began in December 1.983. A description of the on going evaluation tasks is given hereinafter.

1. HELIOSTAT FIELD EVALUATION ACTIVITIES.

The activities related with the evaluation of the CESA-1 Heliostat -- field are listed below:

- Heliostat beam quality, using two different systems, PAIS and SAIS. (Image size and tracking accuracy).
- Mirror modules corrosion. Study of variation of corroded area as well as- variation of single spot corrosion.
- Reflectivity and washing methods, variation with dust build up.
- Accuracy and resolution of the encoders.
- Tracking tests. Stability of the BIAS of the encoders and centroid variation.
- Study and analisys of the solid state relays behaviour in the heliostat - control boxes.

2. HELIOSTAT BEAM QUALITY.

Two different systems have been developped for the purpose of determining the beam quality of a single heliostat.

2.1. PAIS System, (see fig. 1).

Is based on a TV camera watching at a single heliostat image. The video signal can be sent, either to a image analizer for different display presentations or to a digitizer of a personal computer for processing the image.- In fig. 2 a typical data sheet obtained with the system is showed.

2.2. SAIS System, (see fig. 3).

Is based on radiometers mounted in a rotating bar, data logger and personal computer for optical analysis of heliostat images. See figure 4 for - a tipical data sheet of the system.

Both systems are now being used in a measuring campaing that will give a figure on heliostat field beam quality concerning image size (90% of to - tal power of a single heliostat) as well as tracking accuracy.

3. MIRROR MODULES CORROSION.

A first survey of the heliostat field shows 940 mirror modules affected with corrosion. Different degrees of corrosion in terms of percentage of --- affected area were stablished ranging from 80 mirror modules with 75% of -- affected area up to 137 mirror modules with 10% of affected area. The number of mirror modules that show small corrosion spots are 575. These data were- taken on January 84, that means 1 ½ years on site.

4. REFLECTIVITY AND DUST BUILD UP. WASHING METHODS, (see fig. 5).

Tests of mirror dust build up show an average of 1,5% reflectivity degradation per week. Different washing methods were tested. Best cleaning methods are the use of wiper or sponge, which allows to recover original clean mirror reflectivity (90%).

5. ACCURACY AND RESOLUTION OF THE ENCODERS.

Tests to determine the accuracy and resolution of the encoders are under way. The test uses a laser mounted on a heliostat that is driven via a simulator wich increments the demand of the heliostat position 1 bit con - tinuosly. The deviation of the laser beam is measured in a calibrated target and converted into gimbal axes mouvement.

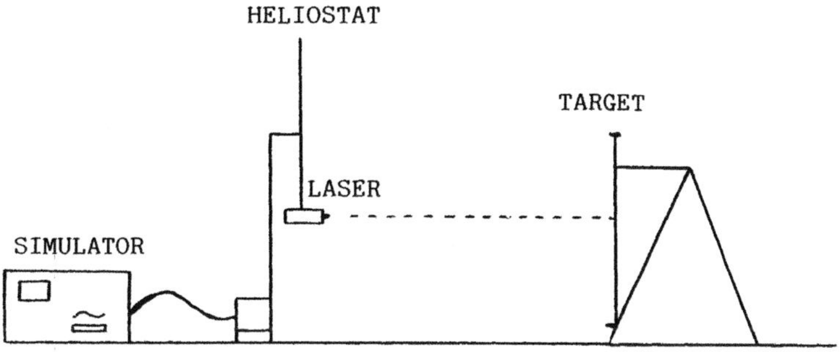

6. ANALISYS OF THE SOLID STATE RELAYS BEHAVIOUR IN THE HELIOSTAT CONTROL BOX

The average availability of heliostat is lower than expected namely due to failures in the power control circuit of the heliostat. A great number - of failures in solid state relays has been detected. The cause of these failures has been analyzed in depth. Undesired overvoltages pulses and big --- dv/dt rates induce avalanche trip of solid state relay at random, producing short circuits between phases.

The origin of the disturbances, detected via oscilloscope and Universal Disturbance analyzer, comes from power supply line inductions as well as -- conmutation and electric brake.

A modification of the circuit, adding RC filters in paralell, to derivate undesired dv/dt rates and also voltage transient suppresors (varistor or gas surge protector) is being done.

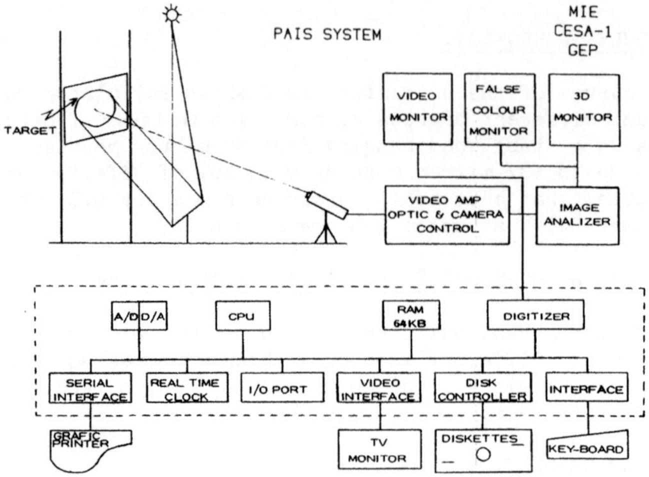

Fig. 1

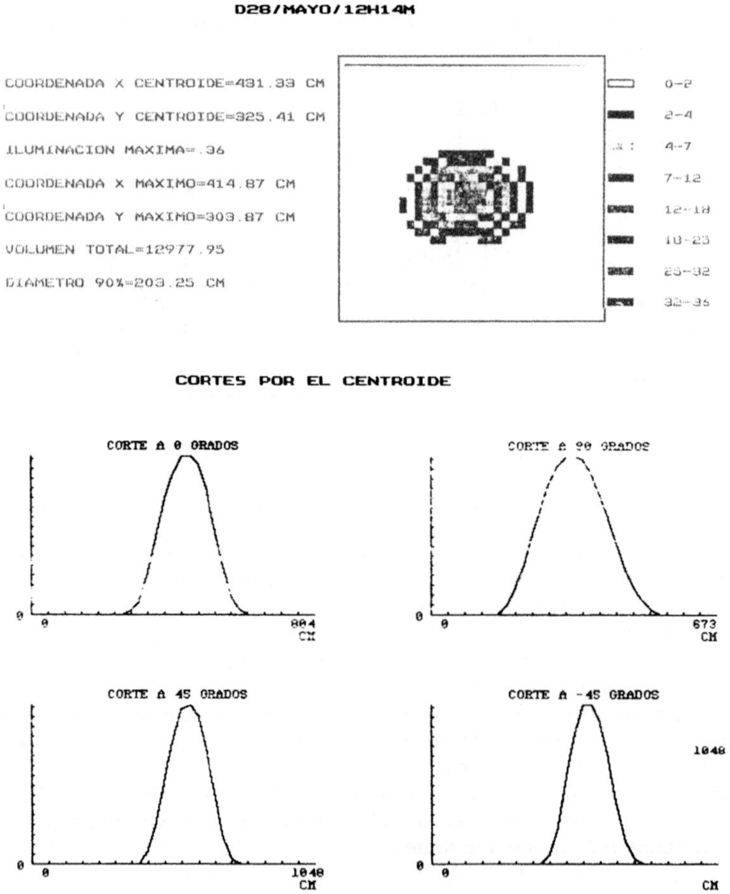

Fig. 2 PAIS DATA SHEET

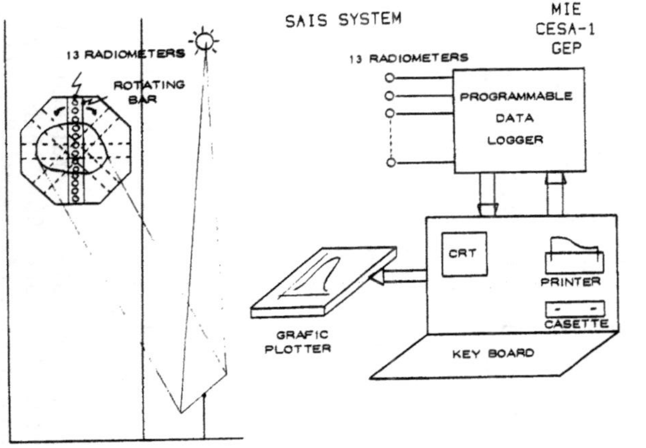

Fig. 3

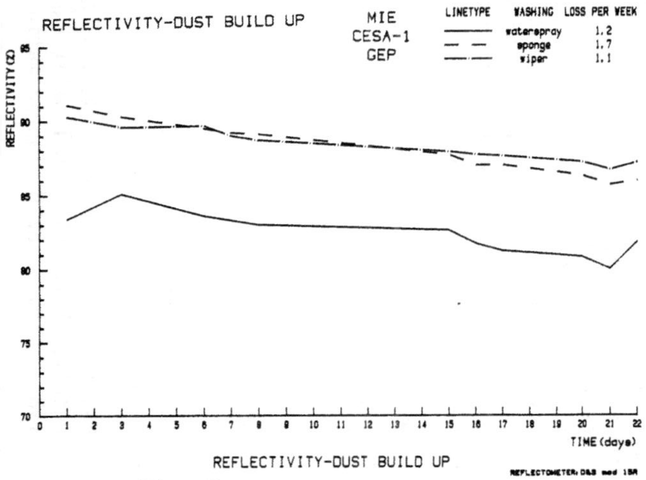

Fig. 5

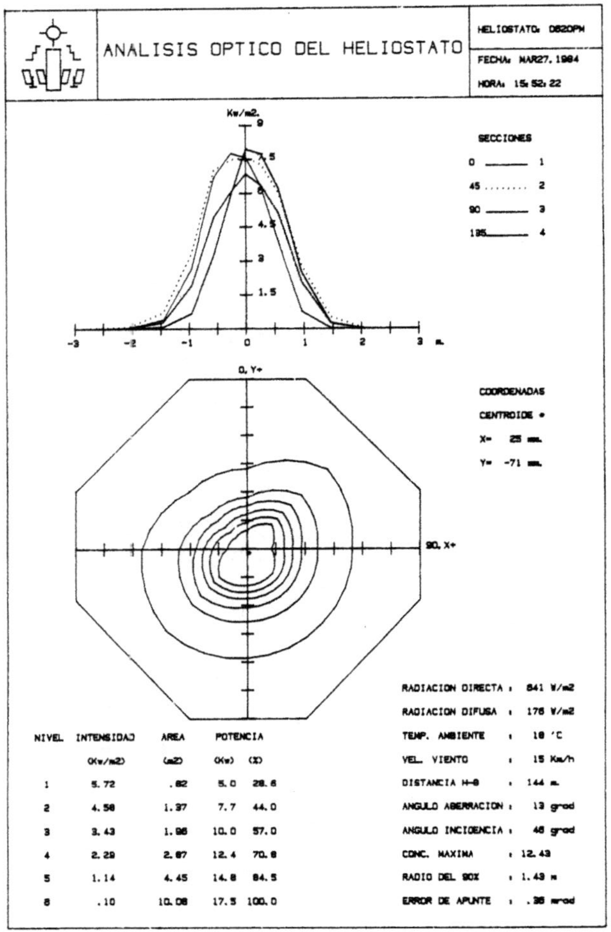

Fig. 4 SAIS DATA SHEET

10 MWe SOLAR THERMAL CENTRAL RECEIVER PILOT PLANT - HELIOSTAT EVALUATION

C.L. MAVIS
(Presented by J.B. WRIGHT)
Sandia National Laboratories
Livermore, California USA

Summary

At the 10 MWe Solar Central Receiver Pilot Plant near Barstow, California, the beam characterization system (BCS) has been upgraded and a sunshape measurement system has been added. Heliostat mirror cleanliness has been measured at 2-week intervals, and the effects of rainwashing and spray rinsing of the mirrors have been determined. Mirror module vents are being installed on almost half of the modules to dry out the water that has accumulated inside them and to halt mirror corrosion. During 1983, 94 to 99 percent of the heliostats were in operation at any one time. Maintenance hours are estimated to be 160 hours per month for 1983. The actual hours have not been determined yet. There were 817 maintenance actions during 1983 as compared with 929 in 1982.

1. INTRODUCTION

Sandia is responsible for evaluating the heliostats at the 10 MWe Solar Thermal Central Receiver Pilot Plant near Barstow, California. The three evaluation objectives are: (1) characterize heliostat performance, (2) identify areas where heliostat research and development may lead to performance improvements and (3) evaluate the need for a heliostat beam characterization system in future plants.

During the past 12 months, reports have been published detailing Barstow heliostat experiences, mirror corrosion survey results, and 1982 meteorological data. A 1983 Meteorological Report will be published in May 1984.

During the next 12 months, evaluation activities will continue; however, there will be a reduced level of effort after August 1984. An evaluation report for the 2-year plant evaluation period will be published late in 1984.

2. ACCOMPLISHMENTS

2.1 Reports

The following reports have been published during the past 12 months. A meteorological report for 1983 will be published in May 1984.

Mavis, C. L., "Monograph Series, No. 1: 10 MWe Solar Thermal Central Receiver Pilot Plant Heliostat Experiences November 1981 - February 1983," SAND 83-8220, Sandia National Laboratories Livermore, May 1983.

Noring, J. E., Mavis, C. L., Decker, E. V., and Skvarna, P. E., "10 MWe Solar Thermal Central Receiver Pilot Plant Mirror Module Corrosion Survey," SAND 84-8214, Sandia National Laboratories Livermore, March 1984.

McDonnell Douglas Astronautics Company, "Solar One Solar Thermal Central Receiver Pilot Plant, 1982 Meteorological Data Report," SAND 83-8216, Sandia National Laboratories Livermore, June 1983.

2.2 1983 Meteorological Data Summary

Meteorological data were collected at the pilot plant on 312 days during 1983. The Data Acquisition System (DAS) was inoperative on the remaining 53 days, primarily because plant controls development required use of the DAS computer.

Rainfall was over 2.5 times the normal amount (9.4 inches) during 1983. Temperatures ranged from 27°F to 108°F and the mirror module temperature was at or below the dew point on 32 of the 312 days. Insolation was available and recorded on 289 days. Of this period, 104 days were clear, 52 were light cloudy days and 132 were heavy cloudy days. Daily average insolation for 1983 was lower than in 1976 and lower than the 25-year average values for all months except July, which was a very good insolation month during 1983. The maximum wind speed during 1983 was 60 mph. The wind speed was above 30 mph on 67 of the 312 data days and above 40 mph on 21 days.

2.3 Beam Characterization System (BCS)

The BCS uses video cameras in the field to obtain images of heliostat beams on the targets located just below the receiver. These images are analyzed by means of a computer to provide heliostat tracking accuracy and beam quality data. An additional camera was added in 1983 to obtain sunshape data. The BCS was upgraded for better accuracy during 1983; it was converted to a stand-alone system and made operational in April 1984. Further improvements for beam power measurements will be incorporated during 1984. Preliminary tracking accuracy measurements with the BCS indicate that the heliostats perform as expected. Additional data will be taken in the near future to fully characterize the heliostats and evaluate the BCS.

A preliminary assessment shows that a BCS or a similar-type system will be required in future plants. A BCS is probably the best tool available for measuring tracking error and beam quality. However, BCS data cannot uniquely characterize a heliostat or indicate what is wrong, since mirror waviness, mirror canting, contour, and sunshape effects on beam quality are not separable.

2.4 Moontracking

A heliostat moontracking capability has been developed at the pilot plant to identify heliostats with large tracking errors. While moontracking, the heliostats are all aimed at one of four aimpoints just below the receiver. Visual observations are made or photographs are taken from the aimpoint to identify the heliostats most in need of a BCS measurement so that tracking errors can be corrected.

2.5 Mirror Cleanliness

The mirror soiling rate during 1983 varied between 3 and 5 percent per month. The entire field was spray rinsed once using an electrical substation insulator wash truck. The rinse increased the mirror reflectivity from about 89 percent to 95 percent of clean. Rain washes were used on seven occasions to restore the reflectivity to 95-98 percent of clean. A mechanical wash rig, which uses brushes and a water spray, was delivered to the pilot plant in 1983. Modifications and repairs are being made to make the washer usable. The cleaning effectiveness will be evaluated during the next several months.

2.6 Mirror Corrosion

Mirror corrosion was first noticed in the collector field in February 1982. Following this observation, a random sample of the mirror modules was surveyed to determine the amount of corrosion, the cause of the corrosion, and the corrosion growth rate. The cause of the corrosion is water from dew and rain which enters the mirror modules through leaks in the edge seals. During the summer of 1983, a survey of all 1818 heliostats was completed by Southern California Edison (SCE) to document the extent of the corrosion. The major findings of the survey are:

- 15 percent of the mirror modules have some amount of corrosion
- 0.015 percent of the total reflective surface is corroded
- The area corroded increases by a factor of 5 to 10 each year
- Over 11 times more corrosion was observed on mirrors that were fabricated before July 1, 1981, than on those fabricated after this date. Production process changes on this date caused a dramatic change in the quality of the mirror seals.

Mirror corrosion at the current growth rate will not have a significant impact on plant performance for several years; however, attempts are being made to dispel the water in the mirror modules through additional ventilation. Vents that were added to selected mirror modules have demonstrated that the modules can be dried out. Vents are now being added to 10,546 mirror modules. All of the vents will be installed by May 31, 1984. The vents consist of four 1/2-inch diameter aluminum tubes in each module, two on each end.

2.7 Heliostat Maintenance

Heliostat maintenance was not a major concern or problem during 1983. On any given day, between 94 and 99 percent of the heliostats have been operational. Maintenance hours are estimated to have been 160 hours per month. The actual hours have not been determined yet. The maintenance required from January 1, 1982, through March 31, 1983, was 2586 hours. The number of maintenance orders during 1983 are about the same as those for 1982, as shown in Table I.

3. FUTURE PLANS

Heliostat evaluation will continue during the next 12 months (April 1984 - April 1985); however, the level of effort will decrease after August 1984 when the plant enters a power production phase. The items to be completed and ongoing activities throughout this period are listed below:

Items to be Completed:

1. Document BCS
2. Improve BCS power measurement accuracy
3. Evaluate heliostat washing machine effectiveness
4. Analyze wind speed and wind load
5. Publish heliostat evaluation report

Ongoing Activities:

1. Measure tracking accuracy and beam quality
2. Measure mirror cleanliness
3. Monitor mirror corrosion and venting effectiveness

Table I - Heliostat Maintenance Orders

Item	Dec 81 - Dec 82	1983
Azimuth Motor	121	48
Elevation Motor	35	18
Gear Box	2	13*
Heliostat Controller	296	465
Heliostat Field Controller	34	20
Azimuth Encoder	103	138
Elevation Encoder	130	43
Mirror	21	11**
Physical	86	61
Totals	929	817

*Includes 1 Gear Box Failure
**Does not include mirrors identified by Mirror Corrosion Survey

SESSION II - PART 2

RECEIVER

Summary of the Session by the Rapporteur
M.R. GENIER, EDF, Chatou, France

Operating experience and modifications in the solar receiver of the Eurelios plant

The SSPS advanced sodium receiver : transient response

Preliminary results on the performance of the Sulzer cavity receiver and the Franco-Tosi external receiver

Relevant aspects in the design and construction of the advanced sodium receiver ASR for the IEA-SSPS central receiver system plant (Almeria - Spain)

The Themis plant : start-up of the receiver - first results

Stability of salt and corrosion resistance of circuit materials in the Themis power plant

An originally scaled test unit of a modular gas cooled receiver of 50 MW thermal power

Ceramic as a material for large receiver constructions

SUMMARY OF THE SESSION BY THE RAPPORTEUR

M.R. GENIER
EDF, Chatou, France

1. INTRODUCTION

First of all, Chairman Ortiz and myself thank the speakers of the Receiver session for the quality of their papers and we apologize for not having given more time for their presentations and for the discussions on each subject; too many papers were to be presented during the attributed time.

A lot of information was given on the receivers of EURELIOS, SSPS, SOLAR I, CESA I plants and of the GAST project; unfortunately, we had no presentations on the NIO receiver.

I shall try to gather the main points concerning the same subjects.

2. MAIN POINTS

2.1 Design Characteristics

Design characteristics depend on two important choices:

(1) heat transfer medium: water, sodium, molten salt, gas...

(2) cavity or external type.

The design of the present prototypes may involve the development of new technologies, in particular for the GAST project, where ceramic materials will be used at temperatures above 1000°C.

The use of usual materials or heat transfer fluids in non-usual conditions requires particular mechanical and chemical studies: for example, studies on the use of molten salt, HITEC, as a heat tranfer medium for the THEMIS receiver, in the range of 450°C, shows that well-known materials can be employed without great corrosion and that the chemical stability can be simply measured. Even with the common steam water technology, some developments are needed because of the particulars of the solar energy (mainly its continuous instability).

2.2 Daily Start-up Phase

The steam receivers present a longer start-up time than the sodium or molten salt ones. For the warm state, starting at sunrise, the following values have been given:

EURELIOS	90 mins.
CESA I	90 mins.
SOLAR I	60 mins.
ASR	30 mins.
THEMIS	10 mins.

The main constraints for the team receiver are due to the low fluxes admitted by the superheater or to thermal inertia.

Improvements have been considered for EURELIOS by transforming the receiver from a once-through type to a recirculation type--the time will be reduced to 60 mins. For the other ones, improvements are expected by modifying the operating rules progressively with the increase of the knowledge of their behaviors.

It would be interesting to complete the above values with those during the day and the daily average.

2.3 Transient Behavior and Cycling

The stability of the outlet temperature during solar power variations is not the only criterion of appreciation: it is necessary to analyze the stresses which can be induced in different parts of the receiver.

At the present time, this problem has been evident for EURELIOS and SOLAR ONE receivers; we don't know yet for the other plants.

The consequence for EURELIOS was breakages in the weldings of the superheater and for SOLAR I--cracks in particular areas of some tubes.

It has been solved by addition of new devices (supporting systems, shields paint, steam flow modifications, etc.) and new operating rules.

Also, computer codes have been developed in order to understand the tubes' behavior; these studies have been emphasized by the ASR speakers who pointed out the care with which they analyzed the critical parts of the receiver during transience, and how good its actual behavior is.

2.4 Efficiency Measurements

Some peak values were given more or less at the nominal conditions:

EURELIOS	70% (but not at nominal power)
ASR	93%
SULZER Receiver	85%
SOLAR ONE	76%
CESA ONE	85% (preliminary results)
THEMIS	90% (preliminary results: not at the nominal outlet temperature)

These values are very close to what was expected, which is satisfying. But the accuracies of the measurements of fluid flows, temperatures, and solar fluxes have never been given; it may be they are not well known. Nevertheless, evaluation teams should make an effort to give a margin of error in order to discuss reliable results. These values would be completed with interest by the daily average values (e.g. for SOLAR ONE, 70%).

3. CONCLUSIONS

The studies presented during this session are very encouraging. Even if problems were underestimated during the design phase, the means engaged to solve them are efficient.

Technical papers, such as those on molten salt chemistry, ASR stress analysis, and ceramic components' development are the proof that teams involved in solar energy want to go ahead in new technologies.

We should keep in mind that evaluation and operation of the different solar receivers are in more or less advanced status. So we have to be careful in comparing the present results and to avoid early conclusions.

Personally, I wish that a future workshop can point out, as soon as possible, the best receiver design--not only in itself, but also as a part of the more efficient tower power plant design.

OPERATING EXPERIENCE AND MODIFICATIONS IN THE SOLAR RECEIVER OF THE EURELIOS PLANT

F.AIELLO, B.BELLAGAMBA, G.DINELLI
ENEL - Centro Ricerca Termica e Nucleare
- Pisa -

Summary

After a brief description of the central receiver of the EURELIOS heliothermoelectric plant its main design and operating limits are considered.
In order to solve the functional and mechanical problems encountered during over 500 hours of operation of the plant in connection with the national electric grid two separate modification studies have been performed. These modifications are described and discussed.

1. DESCRIPTION OF THE RECEIVER

The receiver of the Eurelios plant in Adrano (Sicily) is of the once-through cavity type. It consists of two parallel tubes rolled up in a coil to form an open conical body in which the solar radiation is focussed. The length of each branch is 500 m.

Functionally the receiver is divided in four sections (Fig. 1):

- a feed-water preheater (ECO) with a basket structure placed in the center of the cavity;
- an evaporating section (EVP) on the conical side walls of the receiver;
- a primary superheater (SH1) on the wider angle truncated conical section, protected from solar radiation by the preheater;
- a cylindrical secondary superheater (SH2) consisting of interlaced circular and straight pipes forming an open square-cell honeycombe-like structure in which the steam is brought up to its final state (nominally 510°C at a pressure of 62 bar).

The receiver pipework is of A213 TP321 stainless steel painted with matt black Pyromark 2500 paint to increase its absorbtivity. The pipe outer diameter is 44.5 mm in the pheheater and 57 mm thereafter. The two branches are fed separately, the main feedwater pipe having a bifurcation prior to the flow meters and control valves which are at the base of the tower. The branches came together again into a common main steam pipe at the end of the secondary superheater, immediately before leaving the receiver.

Structurally the receiver pipework is supported, along four generatrices of its conical shape, by a pair of mild steel three-quarter-circular tubes.

The receiver is installed at the top of a tower at a height of 55 m, the centerline being inclined downwards 22° from horizontal towards the heliostats field. The whole receiver is mounted on a trolley which can be winched to the top of the tower on integral rails.

2. DESIGN AND OPERATION LIMITS

During the operation experience the receiver showed two types of problems: functional and mechanical.

The functional problems, which are peculiar to the once-through receiver type, consist mainly in:
- long start-up time;
- receiver temperatures instability during thermal flux variation.

The mechanical problems consist in frequent breakages in the weldings of the primary superheater tubes to the support plates.

With the aim of solving these problems two separate feasibility studies have been performed.

2.1 Receiver Functional Modification

A feasibility study has been carried out to reach the two goals:
- to reduce the start-up time of the receiver from cold conditions;
- to improve the operation stability during charge variations.

The modification which has been considered most suitable for the mentioned purposes is to transform the receiver from once-through type to assisted recirculation type with the introduction of a steam drum and a recirculation pump. The scheme of the modified receiver is shown in Fig. 2. The introduction of the steam drum has the advantage to fix in the drum the transition region beetwen evaporation and superheating even during the variation of operative conditions. In this way it is possible to reduce the control problems of the receiver which are due either to the swinging of the dryout point during the variation of the thermal charge, either to the very long transport delays which affect the receiver dynamics.

The modification of the receiver implies, as a consequence, a different start-up procedure. The start-up will be divided in two steps (Fig. 3). In the first step the feed-water is recirculated throughout all the boiler till when a temperature of 151°C (5 bar) is reached. From this moment the second step starts. It means that the flow is recirculated only in the preheating-evaporating section and a certain amount of steam is bled from the drum to be superheated. During this second stage the steam flow-rate which is bled should be low to permit a fast pressure rise, but sufficient to guarantee the cooling of the superheater.

In order to analyze the advantages of the proposed modification in terms of start-up time improvement, a computer code has been developed to simulate the start-up transient of the modified receiver. The model used in the code divides the receiver in three components: preheater-evaporator, drum, superheater. For the first two components an homogeneous model has been assumed, while a model with distributed parameters has been assumed for the superheater. The model also considers the different distribution of the thermal flux inside the receiver and its time variations. The analysis of the results shows that the modification reduces the start-up time (from tracking till the connection with the grid) of about one third (about 60 minutes instead of 90 minutes under nominal operating condition).

2.2 Receiver Mechanical Modification

The peculiar operative conditions of the receiver, the variability of the meteorological conditions and the consequent fast operation transients produce inevitable thermal cycling on the receiver tubes.

During the operations frequent breakages took place in the weldings of the primary superheater tubes to the support plates (Fig. 4 and 5).

The metallographic analysis has diagnosed these breakages as due to cyclic stresses with very high amplitude and low frequency connected with not free dilatation.

To analyse the stress distribution in the receiver pipes, a computer code has been developed using a mathematical model of the tubes and their restraint conditions.

The results of the computations have demonstrated that, in the above mentioned critical zone of the receiver, the values of stresses which take place are so high that they cannot be accepted by the material. As a consequence of this analysis, a proposal of mechanical modification has been considered. On this proposed modification a new stress analysis has been performed. The results have shown the validity of the modification to reduce the original heavy stress conditions.

The modification which already has been carried out consists in reducing the number of tubes restraints in the primary superheater and in the upper part of the evaporator, allowing the thermal expansion of the tubes. To hold-up the weight of the liberated coils, spring supports have been introduced. Besides, to guarantee a sufficient compactness and to avoid the interpenetration of the free coils during the operation, some suitable spacers have been used between one coil and the other. Some details of the modified receiver are shown in Fig. 6 and 7.

3. REMARKS

The problems which have been met in the operation of the Eurelios plant receiver have pointed out that the future design of this component should consider carefully the functional and mechanical aspects connected with the variability of the thermal flux and the daily cycling consequent from the morning start-up and evening shut-down. The peculiar characteristic-to be a component which constantly works in transient conditions - leads to study thermohydraulic schemes which fit better, from the dynamic point of view, this type of operation.

Regarding the mechanical aspects, the inevitable and heavy thermal cycling makes the structural design of this type of boiler different from that of an ordinary boiler.

For this reason it is absolutely necessary a complete and refined stress analysis that can be used as guide in the choice of geometry and tubes restraint conditions.

4. ACKNOWLEDGEMENTS

The analysis of the modifications and improvements in the solar central receiver have been carried out together with the manifacturer, Ansaldo Div. Breda; the stress analysis and the structural study have been performed in cooperation with the Department of Mechanical and Nuclear Engineering of the University of Pisa.

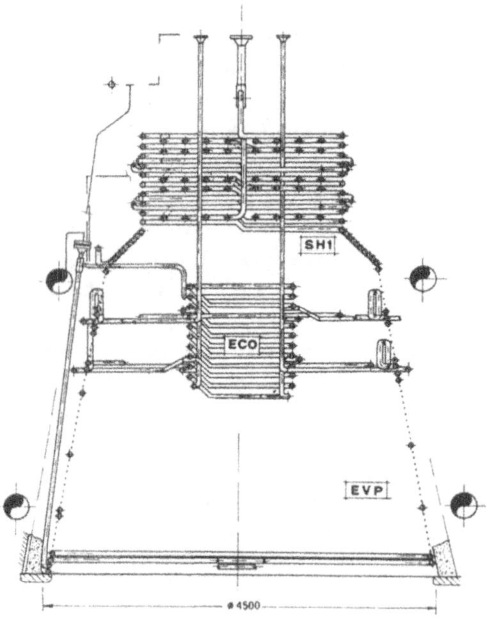

FIG. 1 - SEZIONE DELLA CALDAIA

FIGURE 1

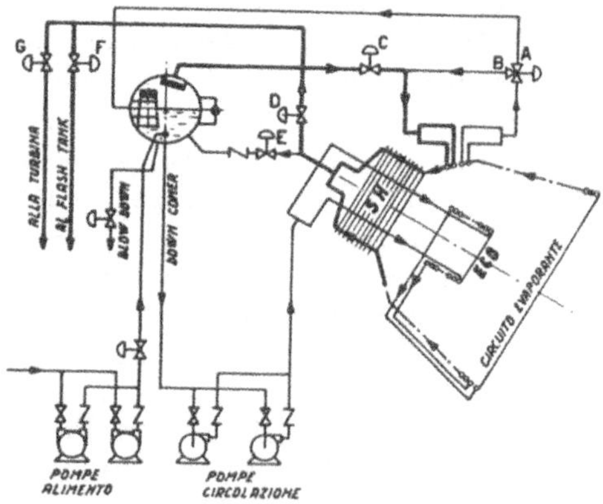

FIGURE 2

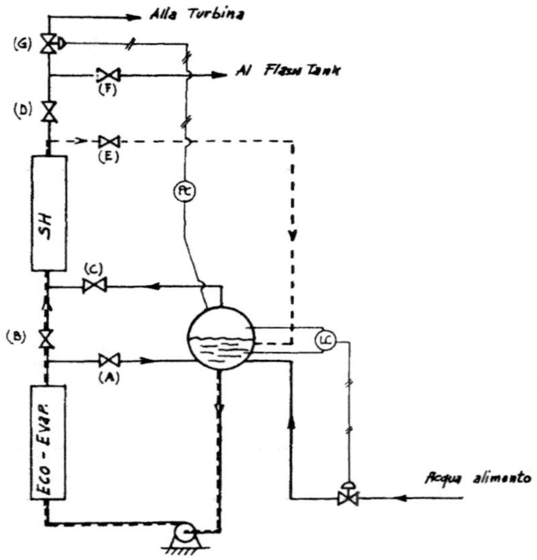

FIGURE 3

FIGURE 4

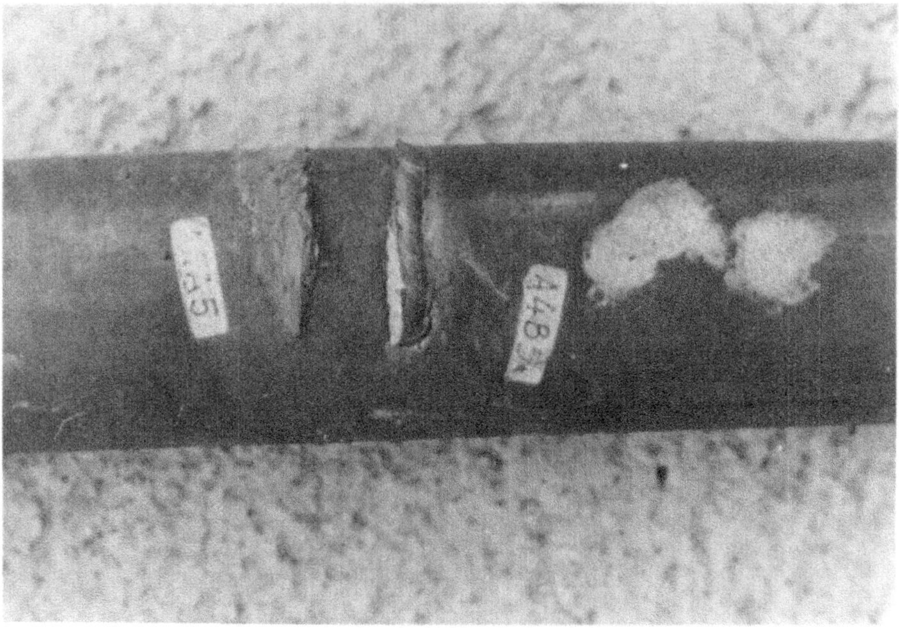

FIGURE 5

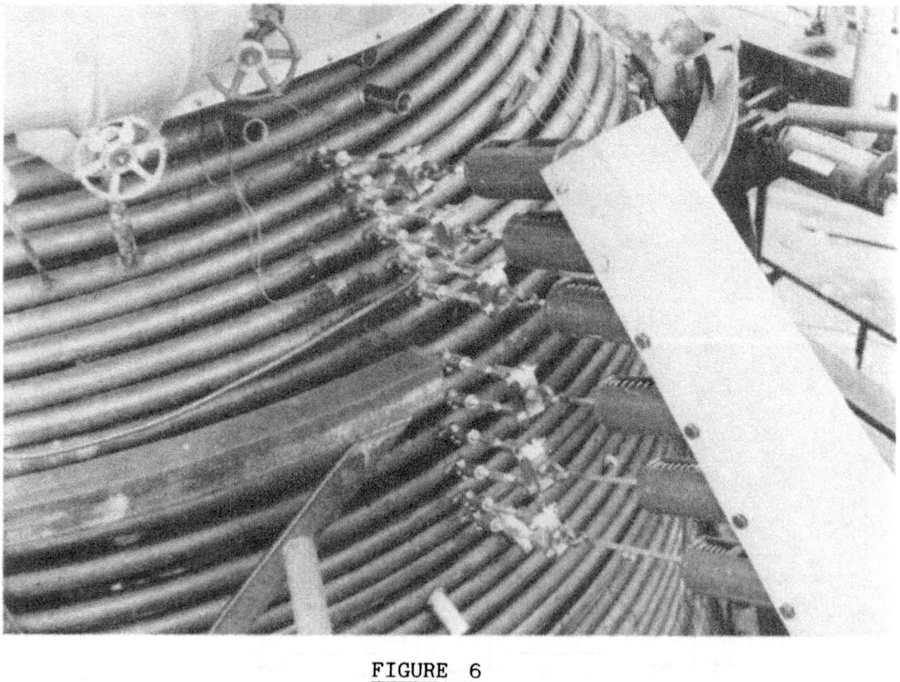

FIGURE 6

FIGURE 7

THE SSPS ADVANCED SODIUM RECEIVER: TRANSIENT RESPONSE

R. Carmona and J. G. Martín
IEA/SSPS - ITET

Summary

This paper reports on the results of work in progress at the IEA/SSPS site on the evaluation of the Advanced Sodium Receiver (ASR) transient response.

High efficiencies for solar energy collection ($\sim$90%) are possible with high flux, sodium-cooled receivers: these have low thermal inertia and short response times. Controls must keep the outlet temperature constant at changing power levels, without compromising structural integrity. The ASR and its controls are designed to accomplish this; it is of interest to determine how close performance approaches expectations.

The ASR transient response has been tested by making changes in the sodium flow rate and the heat flux reaching the receiver at different loads. From the tests, one estimates gains and time constants. These vary inversely with the flow, and are close to computer predictions.

The receiver configuration and actual response suggest a form for its transfer function. The coefficients have been identified: the temporal response evaluated from this function fits the data well. It is now possible to predict the response to arbitrary changes in the operating conditions, and to conduct systems and stability studies.

This paper also considers the control system, with reference to operation during warm-up and cloud passages. The controls do protect the receiver from overheating when the insolation rises sharply, but it is desirable to improve the accuracy of the irradiation signal.

1. INTRODUCTION

A receiver system must respond to variations in solar flux so as to maintain constant the coolant outlet temperature, T. This 'quasi-stationary' state is achieved by manipulating the coolant flow rate, W. This state must be reached quickly, and maintained, while keeping the temperature gradients in the receiver within tolerable limits.

In sodium-cooled receivers which operate at high power fluxes, sharp temperature gradients do not facilitate this goal. Those receivers, which may be very efficient (1), will have thin tubes and low thermal inertia: the ASR fabricated by Franco-Tosi and installed at the IEA/SSPS site is a good example.

Variations in either W or the heat flux reaching the receiver, q, will affect T. The response may be roughly characterized in terms of ratios between variations in T and those in W and q, and the corresponding time constants. The results of tests performed on the ASR will be reported, evaluated, and discussed here, in terms of these parameters. These results validate a dynamic computer model developed by ENEL et al. (2).

Dynamic system analysis and stability studies are facilitated if the receiver is represented in terms of a tranfer function. One is suggested here, based on the actual response and the receiver configuration. The temporal response calculated from this function represents the data accurately.

Finally, this paper will present some considerations on the time required for warm-up and observations about the performance of the receiver loop (i.e., the system composed of receiver, pump, and controls) when clouds pass over the heliostat field.

2. GENERAL CHARACTERISTICS OF THE ASR RECEIVER

The ASR is a sodium-cooled receiver, 2.75 m wide times 2.85 m high, which may be considered as a representative module for a larger receiver.

It is composed of five panels, all on the same plane (Fig.1). In each panel there are of 39 cylindrical tubes, placed vertically and in parallel. The tubes are 2.65 m long; their outer diameter is 15 mm, and the walls are 1 mm thick. The nominal outlet temperature for the receiver is 530°C.

Sodium flows into headers above each panel, and is piped along a single tube to the next. The thin pipes permit a large heat flux: the receiver is designed for a peak flux of 1.5MW/m^2 under normal operation.

To flatten the flux profile, the heliostats are aimed at three points. A typical flux plot is shown in Fig.2. For operation at rated power (2.84 MWth), nominal T, and inlet temperature of 270°C, sodium flows at the rate of about 32 m^3/hr; its residence time is about 54 seconds.

3. RECEIVER TRANSIENT RESPONSE

Transient tests have been performed on the ASR. The effect on T of a step change in W is shown in Fig.3.

The insolation at the time of this test ranged between 960 and 965 W/m^2. Before the test was made, T had remained within 1° of 490°C for more than 5 minutes, and W was 5.25 ± 0.02 Kg/sec (equivalent to 21.5 ± 0.1 m^3/hr). By operating the pump under manual control, the flow was lowered to 4.82 Kg/sec (19.7 m^3/hr) in about 2 seconds, the pump response time. T rises until it reaches 513°C.

As recorded by the Data Acquisition System (DAS), T has rapid fluctuations due to noise in transmission and discrete effects of the analog-to-digital convertor. To smooth out the noise, the data were passed through a filter of the second order: the figure shows the output data of the filter.

The response may be characterized in terms of the ratio between variation in T and variations in W. In general this gain has a steady state component, G_W, which does not vary with the period of the signal, and a dynamic gain which does. For a system where the output variable,\T,satisfies a first-order differential equation in time, t, a step change, ΔW, in the manipulated variable causes the change in the output, ΔT, to obey an exponential function. The receiver response permits a rough characterization in terms of G_W and the time constant, τ_W. A study by the manufacturer gives the theoretical justification for this (2). The data in Fig.3 may be characterized by G_W = 53°C/Kg sec and a time constant, τ_W = 57 seconds.

The next test consisted in raising the flow. The results are also shown Fig.3: the gain and time constant are the same as before.

The receiver is not a linear system: gains and times depend on the flow. Similar tests were carried out at high (8.06 Kg/sec) and low (2.3 Kg/sec) flow rates: time constants, residence and dead times, and gains are listed in Table I. The data are given in 1 second intervals: the times are approximated to the next closest digit.

The next set of tests consisted in changing the heat flux at the receiver by sending heliostats out and into track. The response to these changes may be represented in terms of a constant gain, G_q, and a time constant, τ_q. At high flow (W = 8.06 Kg/sec), the flux step down consisted in sending 6 heliostats (out of 90) out of track when the insolation was 960 W/m^2. This is a step of 165 kW (down from 2645 kW, close to design conditions). The results are plotted in Fig.4. The heliostats were then returned to track for the upward step.

Similar tests were performed at medium flows with 6 heliostats sent out of track (out of a total of 61), and at low flow with 3 heliostats sent out of track (out of 29). Fig.5 shows the data at medium flow. The gains are also listed in Table 1.

The time constants were found to be the same as those for the flow changes. In general, τ_w should be smaller than τ_q because of the delay induced by the transfer of heat across the tube walls. Here, the difference is not noticeable: heat transfer is fast.

Gains and time constants vary approximately inversely with the flow rate. They can be approximated by: τ = 250/W seconds, G_w = 218/W °/Kg sec, and G_q = 0.73/W °C/kw, with less than 6% error, when W is given in Kg/sec. The difference between G_w and G_q are due to the different units used; equal percent variations in input give rise to approximately equal percent variations T.

There are thermocouples at each header, so it is possible to compute gains and time constants for each panel for all tests.

As an example, the time constant is about 5.5 seconds for all of the panels at high flow. From Panel 1 to Panel 5 the gains are 2, 1, 7.3, 7, and 9.4 °/Kg sec, respectively.

The panels are connected in series; the temperature increase, and the gain, is proportional to the total energy absorbed. Because of the flux profile, the central panel (#5) has the highest gain.

4. COMPARISON WITH SIMULATED RESULTS

ENEL has developed a dynamic simulation model which is applicable to the ASR (2). The program has been implemented in the computer: with the model, we have simulated the tests. The model considers the effect of a constant wind on convection losses, but not gusts or effects on heliostat tracking. Fig.4 and 5 show the simulated and actual response. During the test, the wind ranged between 20 and 25 km/hr, and there were small variations in the insolation (about 5 W/m^2). The actual response curve is not as smooth as the simulation.

The actual flux shape also introduces errors. We used the standard design profile: work is in progress to use actual flux distribution data from the Heat Flux Distribution (HFD) sensing bar. Asides from these considerations, the fit between actual and simulated responses does validate the model.

5. RECEIVER TRANSFER FUNCTION

It is of interest to represent the receiver response in terms of a Laplace transfer function. To facilitate the following discussion, the simulated results will be used as reference.

The receiver cannot be represented in terms of linear differential equations over the whole range of operation. It is necessary to make linear approximations about equilibrium points.

Considerations on the receiver geometry, and the receiver response itself, suggest a general form for the transfer function. Most of the flux reaching the receiver is absorbed by the three central panels: there is a delay between the time sodium leaves one of these panels and the time it reaches the outlet thermocouple.

Fig.6a shows the response when the flow is changed from 5.42 Kg/sec to 4.78 Kg/sec in one step (a 2 second ramp). The shape of the curve suggests that it may be approximated by three main contributions, with different time delays (Fig.6b).

The first is an 'S' shape which appears about at $t \sim 4$ sec. This 'S' shape is of the type which characterize a second-order system. The first contribution comes from the panel which is closest to the outlet (i.e., Panel 5).

The second contribution appears at $t \sim 21$ sec. This is the contribution from the next to the last panel (i.e., Panel 4): sodium leaving Panel 4 reaches the sensor point at the outlet in approximately 21 sec. For simplicity, we take it to be a contribution from a first-order system.

Finally, about 43 sec after the perturbation (the time for the sodium leaving Panel 3 to reach the outlet), there is another component, also approximated as the response from a first-order system. This component is related to the contribution from the first three panels.

Taking the three contributions together, the overall transfer function may then be represented as

$$\frac{\Delta T(s)}{\Delta W(s)} = \left(\frac{b_1 s + b_2}{s^2 + a_1 s + a_2} e^{-td_1 s} + \frac{b_3}{s + a_3} e^{-td_2 s} + \frac{b_4}{s + a_4} e^{-td_3 s} \right) \frac{0.5}{s + 0.5}$$

This function has three time delays and eight coefficients. The factor 0.5/(5+0.5) is the response of a thermocouple with a time constant of 2 sec--the one at the outlet. The lags are inversly proportional to the flow rate; they may be approximated by $td_1 = 16/W$, $td_2 = 100/W$, and $td_3 = 210/W$.

The system is nonlinear and the coefficients must be identified at the different flows. They were identified by using the Powell algorithm (3), minimizing the sum of the mean square errors with a program developed at the University of Seville. The coefficients are listed in Table 2.

From these coefficients, a general simulation program also developed in Seville has been used to obtain the temporal response in time domain for low, medium, and high flows. A response is shown by the dotted line in Fig.6. The solid line represents the results from the ENEL model.

6. CONTROLS

The control system for the receiver must manipulate the flow so as to maintain T at some reference value, now chosen to be 530°C. Fig.7 shows the block diagram for the controls.

Although the inlet temperature may vary, the cold storage limits these variations; these are slow compared with the receiver response. The present ASR control system assumes that the inlet temperature is constant. Small oscillations in T are eliminated in the hot storage tank.

The most important variations are those in the absorbed heat flux. These may be large amplitude variations, with high frequency components. For example, after a cloud passes, the heat reaching the receiver rises quickly. The controls must avoid high temperature peaks after such passages.

The control system consists of feedforward and feedback loops. The inner loop manipulates the flow. This loop must be fast enough so that it does not interact with the outer ones. The intermediate loop detects the variations in the temperature differences across all panels, to speed up the control response. The outer loop, where the respnse is slower than the intermediate one, maintains T constant.

The flow rate ranges from 8 to 40 m^3/hr: the regulator parameters must vary to maintain stability in all the operating range.

The feedforward action depends on the heat flux: it limits temperature peaks when the flux rises sharply. This action is more important at low flow rates, when response times are longer. Feedforward is imprecise, because heat flux is not estimated accurately. There are ten photovoltaic sensors, mounted on heliostats in the field. The average of the output from these sensors is the signal used: there are several problems with this arrangement. The cells are mounted horizontally; the gain is not constant, but depends on sun angle. They measure global insolation: a cloud which causes a sharp drop in the flux at the receiver, may not induce as sharp a drop in the global insolation. Under some conditions, a cloud may actually increase the horizontal insolation. Another difficulty occurs because of shading of the sensors, particularly in winter.

Feedforward depends on the insolation and its time derivative. Because of the imprecise estimation of the solar flux, the derivative action is the relevant one: its main purpose is to protect the receiver from overheating when the insolation rises.

Warm-Up

From the discussion on response time constants, it may be implied that the receiver could be heated from about 270°C to 530°C in less than 3 minutes, at design insolation, at noon.(For early morning start-up, when the heat flux and the flow rates are smaller, the time required would be about four times longer).

Because of the stresses induced on the structure by rapid temperature changes, the designers recommend that temperature variation should not exceed 8°/second, which also corresponds to a warming period of less than three minutes. In fact, the POA heats the receiver in about ten minutes.

Cloud Passages

Outlet temperature and flow rate during a cloud passage around noon are plotted in Fig.8. Two clouds of short duration (about 50 seconds each) make the insolation drop to about 400 W/m^2 (the irradiance plot represents the reading from the pyrheliometer installed atop the main SSPS building). The flow drops to about 8 m^3/hr. There is a slight decrease (of about 10°K) in T. Note that, in the figure, the flow drop ('effect') seems to anticipate the irradiation drop ('cause'). The feedforward action is initiated by sun presence sensors installed in the field rather than by the pyrheliometer, and irradiance does not fall simultaneously at both locations.

The passage of a cloud of longer duration (about 100 seconds) is illustrated in Fig.9. Insolation drops to about 200 W/m^2: this larger perturbation induces a stronger feedforward action; W drops to its minimum value (6 m^3/hr) and remains there for about 50 seconds. When the cloud passes, the feedforward is strong, making the pump operate at maximum power (maximum W is about 40 m^3/hr). T begins to rise, but the induced flow surge stops and reverses the rise: T drops to about 420°C, before it rises again. The control system is safe indeed, and it has kept the receiver from overheating in spite of the large and fast insolation rise. In fact, it has done this to excess.

A better evaluation of the heat flux reaching the receiver would make it possible to adjust the feedforward action to eliminate this undesirable excess.

7. CONCLUSIONS

The compact receiver responds quickly to variation in the heat flux. The controls successfully protect the receiver from overheating and from large temperature gradients.

The dynamic simulation model developed by the manufacturer has been validated by the actual receiver response.

A transfer function has been suggested; this function represents the receiver response accurately. It is now possible, using this function, to predict what T will be for arbitrary changes in W and in flux. It is also possible to study the stability of the receiver loop.

For the receiver design, it had been assumed that the flow could be as low as 1/10 of the nominal value (i.e., 3 m^3/hr), but the present pumping system does not permit operation below 6 m^3/hr. During the acceptance tests, the control system designer noted that at lower rates the flow oscillated. This was related to the slow response of the pump controls: there was no problem with the first installed receiver, a high thermal inertia cavity type, but the pump controls had to be modified to speed the response for the ASR. The limitation on the minimum flow rises the minimum irradiance at which the receiver operates.

A more accurate estimate of the heat flux at the receiver is desirable in order to improve the response.

8. ACKNOWLEDGMENT

The simulaton program was implemented on the SSPS computer by J. Magnani. To identify the coefficients for the transfer function, the authors relied on support from Eduardo Fernández Camacho and Francisco Rodriguez Rubio of the Engineering School of the University of Seville. The authors are grateful for this help.

9. REFERENCES

1. JACOBS, H. (1984). Performance of the Sulzer Cavity Receiver and the Franco-Tosi External Receiver....Paper presented at this workshop.
2. AGIP NUCLEARE and FRANCO-TOSI INDUSTRIALE, (1983). The Advanced Sodium Receiver (ASR). SSPS Technical Report Nº 3/83. Topic Report Nº 3.
3. POWELL, H. B. (1964). An Efficient Method for Finding the Minimum of a Function of Several Variables...... Computer Journal, 7. p.155.

FLOW W, (Kg/s)	TIME CONSTANT τ (sec)	RESIDENCE TIME (sec)	DEAD TIME td (sec)	GAIN G_w, (°C/Kg/s)	GAIN G_q, °C/kW
HIGH (8.1)	31	50	2	29	0.10
MEDIUM (5.1)	48	80	3	40	0.14
LOW (2.3)	115	172	7	90	0.28
FLOW DEPENDENCE	250/W	400/W	16/W	218/W	0.73/W

Table I : Response for Different Flow Rates

	a1	a2	a3	a5	b1	b2	b3	b4
HIGH	1.48	1.33	0.12	0.037	-0.735	-11.57	-1.11	-0.42
MEDIUM	0.82	0.30	0.062	0.026	-0.744	- 3.98	-0.98	-0.38
LOW	0.47	0.07	0.032	0.013	-0.733	- 2.13	-1.04	-0.32

Table II : Coefficient Identification

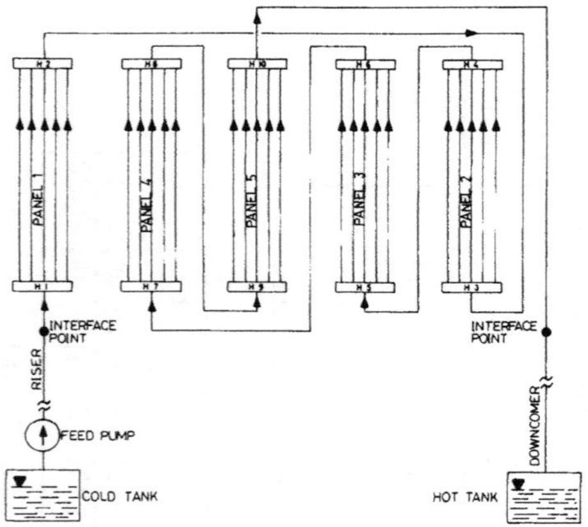

Fig. 1 : ASR Flow Diagram (2)

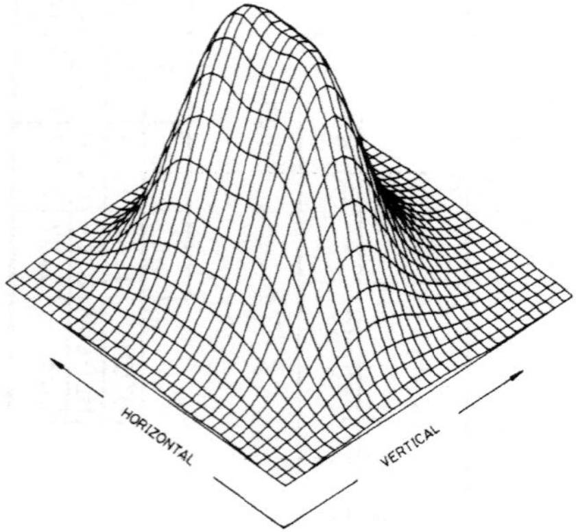

Fig. 2 : Typical Flux Plot at the Receiver Plane (2)

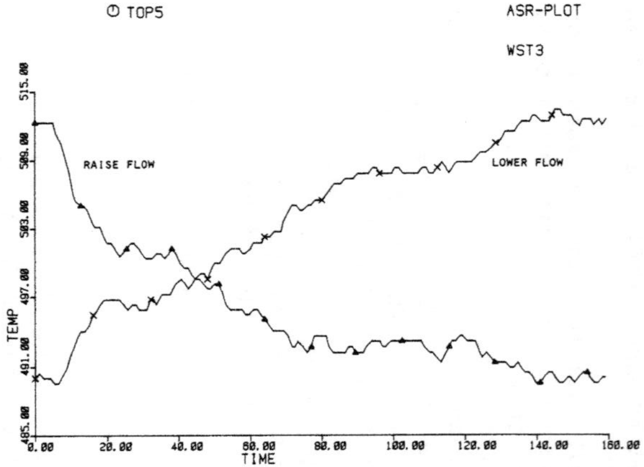

Fig. 3 : Response to Step Changes in the Flow Rate (5.25 to 4.82 Kg/sec)

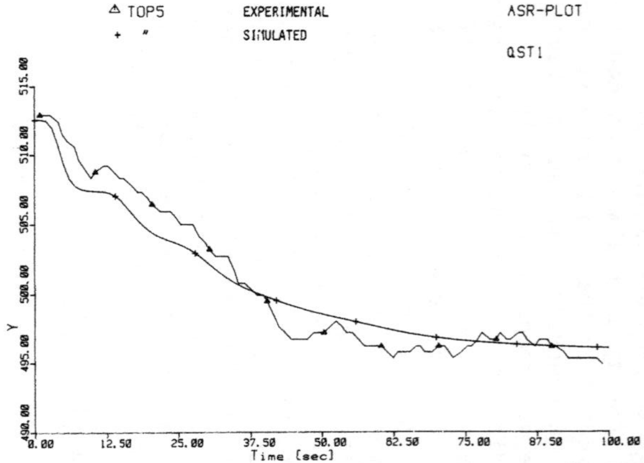

Fig. 4 : Response to a Down Step of 165 kW in the Flux, at High Flow (8.06 Kg/sec)

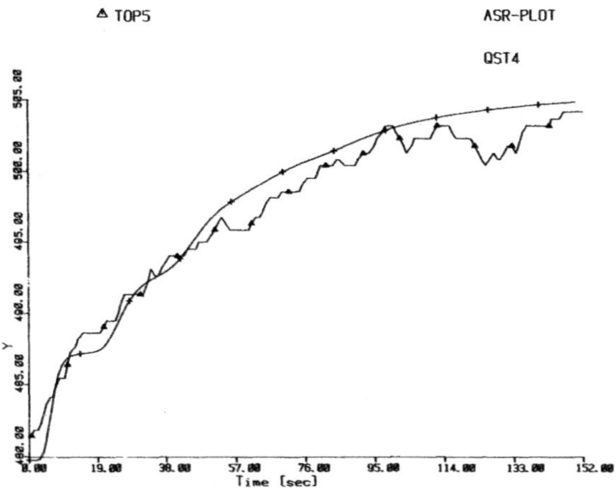

Fig. 5 : Response to an Upward Step of 165 kW in the Flux, at Medium Flow

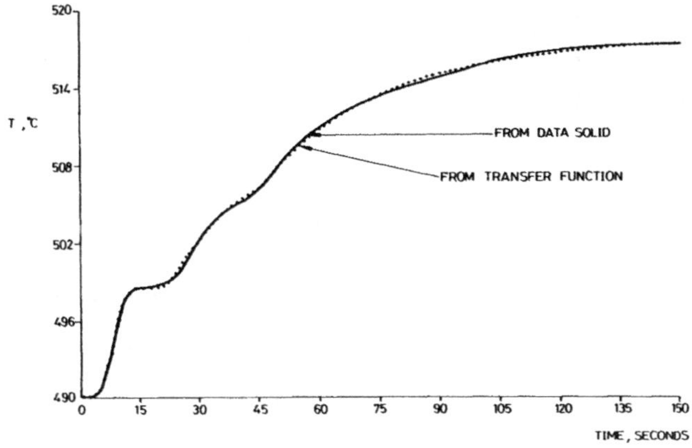

Fig. 6A : Response

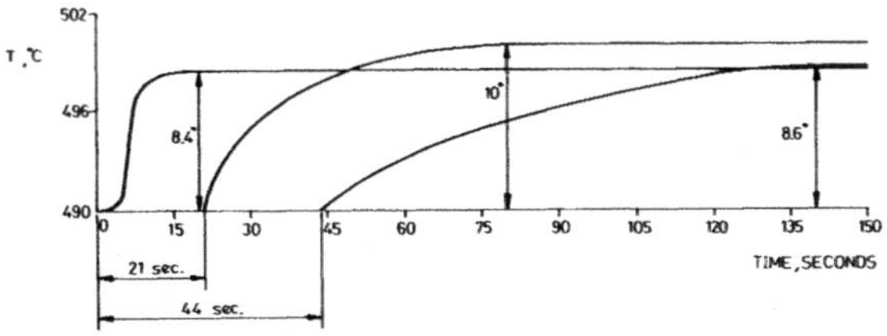

Fig. 6B : Contributions to Response

Fig. 6 : Effect of a Change from 5.42 to 4.78 Kg/sec

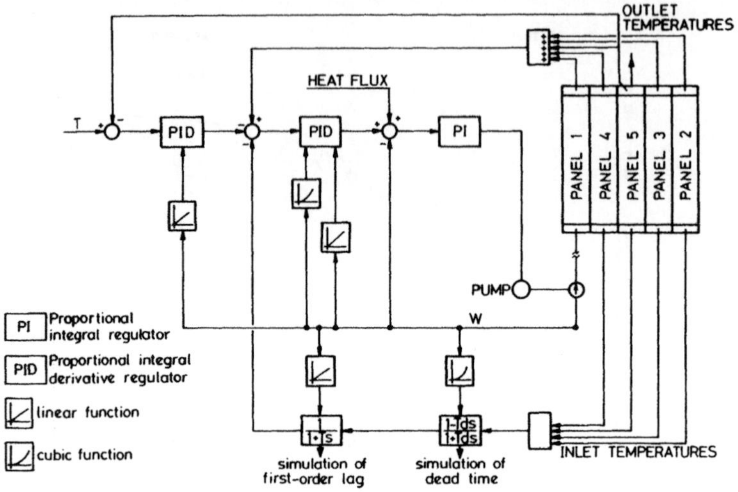

Fig. 7 : Control system (2)

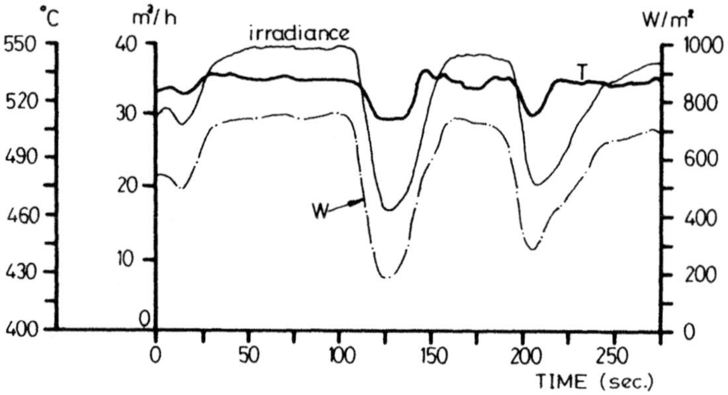

Fig. 8 : Effect of Passage of Clouds of Short Duration

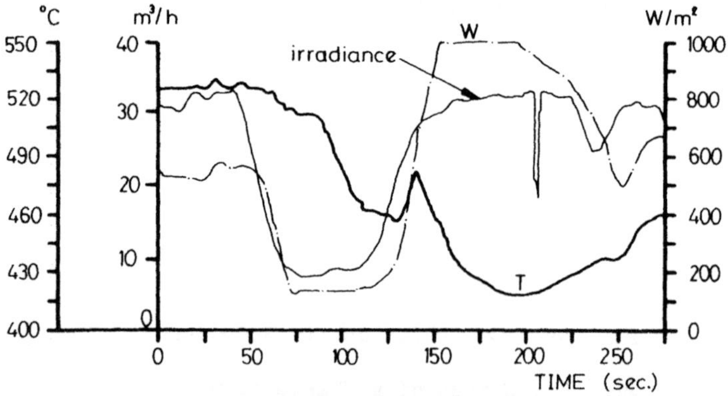

Fig. 9 : Effect of Passage of Clouds of Long Duration

PRELIMINARY RESULTS ON THE PERFORMANCE OF THE SULZER CAVITY RECEIVER AND THE FRANCO-TOSI EXTERNAL RECEIVER

C. S. Selvage and H. Jacobs
IEA/SSPS ITET

Summary

This paper presents the preliminary results of evaluation work performed on the cavity receiver (Sulzer) and the external receiver (ASR) installed at the IEA/SSPS Project.

Operational data are combined with data obtained in loss tests which were performed on both receivers.

The methods used arrive at the same results: the cavity receiver has lower efficiency than the external receiver, and the losses of the cavity receiver are twice as high as those of the external receiver.

The figures presented have a marginal error of 5 - 15%.

For more confidence on the results, and to achieve a better accuracy, additional tests and evaluation work are planned.

1. INTRODUCTION

Two different receiver concepts have been chosen for the 500 kW Central Receiver System at the Small Solar Power Systems Project in Almería, Spain (Tables I and II (6)).

One consists of a cavity receiver and the other is an external receiver. Both receivers use sodium as a heat transfer fluid.

The basic layout and thermal analysis of the cavity receiver was carried out by INTERATOM. Detailed design with stress analysis, and the manufacture and erection of the receiver on site were performed by Sulzer (acceptance testing in Spring, 1981). The contract companies responsible for the design, manufacture and erection of the external receiver were AGIP and Franco-Tosi Industriale (acceptance testing took place in Autumn, 1983).

The work concerning receivers performed on site in Almería, considers two topics: 1. evaluation of operational data, and 2. performance of loss tests.

First, results of these tests are presented in this paper. It is necessary to continue performing convection loss tests to verify these preliminary estimations, and improve the accuracy. Additional work on the influence of wind on efficiency and an evaluation of the energy needed for start-up will be performed in the near future.

2. CONVECTION LOSS TEST

The temperature drop of sodium pumped through the receiver while no solar power is delivered by the heliostat field is an indication of the convection and radiation losses (1). The receiver conditions during the test are slightly different from the normal ones. The surface temperature is lower and the temperature profile does not follow the characteristic curve. Two different tests were performed; a normal flow and a reverse flow test. The advantage of the normal flow test is that it does not require any special operational status. It was performed several times just after normal operation, the only difference being that the heliostats were out of track.

The reverse flow test is more complicated. For this reason it was performed only four times.

The procedure of the reverse flow test for the cavity receiver is

- the receiver door is closed
- the valves are switched to reverse flow
- wait for steady-state (at least 15 minutes).

For the external receiver (ASR), this procedure is

- preheat the receiver back wall with some heliostats, to avoid a long wait for steady-state
- switch valve to reverse flow (flow is a result of a difference in gas pressure in both tanks)
- steady-state after 20 minutes.

Results

Figure 1 shows the losses as a function of the temperature difference between the receiver and ambient temperatures. The curves for the two receivers are extrapolated to the nominal temperature difference in which the losses become 465 kW for the cavity receiver and 205 kW for the external receiver.

To interpret these results, it is necessary to look more carefully into the difference between test status and operation status. There are several arguments which indicate that operational losses are different than test losses.

First, the operational surface temperature of the absorber tubes is higher during the test. This indicates higher operational losses.

Second, the temperature profile during operation is different from the one shown for the loss test. During the test, the profile is more or less linear. To correct the profile influence, the receiver is divided in two parts: Panels 1 and 2, with 250°K temperature difference between the panel and ambient temperatures, and Panels 3, 4, and 5 with 370°K temperature difference.

The loss figures for both groups are taken from Figure 1, and corrected by the area:

Panels 1 and 2	97 kW . 2/5 (of the total area) =	39 kW
Panels 3, 4, and 5	205 kW . 3/5 (of the total area) =	123 kW
TOTAL		162 kW

Based on the result for the total loss, the efficiency is plotted in Figure 2.

3. RECEIVER EFFICIENCY DURING OPERATION

The other method to calculate receiver losses is to compare incoming power from the field with the absorbed power. The incoming power is calculated using the HELIOS computer code. The calculated figures are used to make a look-up table (which gives power as a function of day and solar time under design conditions) (2). Additional measurements were performed using HERMES (3) and FAS (4), and the results support this table.

The figures from the matrix need to be corrected by actual data of insolation, reflectivity, and the number of heliostats in track. The Insolation and the number of heliostats in track are recorded automaticaly by the Data Acquisition System (DAS). The reflectivity is measured on site periodically. A theoretical analysis of the data which allows a limited forecast is given in Ref.5.

The absorbed power is calculated by enthalpy and mass flow. The estimated uncertainites of HELIOS calculation (5%), operating with an undesignated number of heliostats (5%), facets' soiling (3%), and calculation of absorbed power (4%), produced an overall margin of error between 5 and 15% (6).

Results

In the following figures the results presented are always daily energy averages for a day which had at least 2 hours of uninterrupted receiver operation.

The processed data (14 days for the external receiver and 9 days for the cavity receiver) are read directly from the DAS data tapes, containing five-minute average values of all measurement points.

Figure 2 shows the efficiency as a function of absorbed power. The points are the daily averages and the lines represent the results from convection losses tests. This obviously fits the results very well.

The efficiency of the external receiver is approximately 14% higher than the efficiency of the cavity receiver.

Figure 3 shows the efficiency as a function of the energy absorbed in one day. The times are the upper limits. Due to load changing and other efficiency reducing enfluences (wind, clouds, etc.) there are points under the limit curves.

Also, if less energy is absorbed, the thermal inertia reduces the daily efficiency.

The same values used for plotting Figure 2 are used in Figure 4--Losses as a function of absorbed energy. There is no dependence of the losses on the energy. The result is convincing, because losses depend mainly on temperature differences. Table III shows the loss figures during operation and during the loss test.

The average of the loss numbers fits well to those observed during convection losses tests.

Figure 5 shows the efficiency as a function of the average insolation during operation. If the average insolation is low, the efficiency will be lower than normal because of the relatively constant losses.

Clouds, wind, or a change in the number of heliostats in track can cause a reduction in the efficiency during a day. Therefore, it also seems reasonable to establish a maximum limit line for both receivers.

4. ENERGY DISTRIBUTION IN THE FIVE PANELS OF THE EXTERNAL RECEIVER

Based on 35 days of operation, the average energy distribution for the external receiver panels appears as described in Table IV.

Panels 1 and 2 only receive a total of 6% (average) of the total amount of energy received during one day. Figure 6 shows the percentage of received energy which corresponds to each panel.

As already mentioned, the low amount of energy absorbed by the edge panels influences the loss distribution and increase the total efficiency.

5. LOSSES DURING THE NIGHT

The losses during the night for the cavity receiver are 6 - 9 kW, and 16 - 18 kW for the external receiver. The reasons for the low standby losses of the cavity receiver are the fact the the doors fit the cavity better, and the fact that the high inertia (i.e., ceramic wall) keeps the temperature drop in the sodium on a low level. However, this last points makes it necessary to heat up the wall in the beginning of the next operational period.

6. CONCLUSIONS

The external receiver (ASR), with a high heat flux, has higher efficiency than the cavity receiver (Sulzer). This is proven by comparing the results of the loss tests: the losses of the cavity receiver are more than twice those of the external receiver. One reason for this is that the smaller heat transfer area of the cavity receiver produces smaller losses. The actual area with high temperatures of the external receiver is reduced by the two edge panels (1,2). This results in only 3 of the 5 panels having really high losses.

7. REFERENCES

1. KRAABEL, JOHN S. (1983). Convection Testing of Central Receiver System - Fall 1982, and Receiver Efficiency, Fall Measurement Campaign, 1982. SSPS - CRS Midterm Workshop Proceedings. SSPS Technical Report Nº 4/83.
2. BECKER, M., ELLGERING, H., and STAHL, D. (1982). CRS-Construction Report. SSPS SR2.
3. SCHIEL, W. (1983). HERMES Measurements. Proceedings op. cit.
4. VON TOBEL, G., SCHELDEN, CH., and REAL, M. (1982). Concentrated Solar Flux Measurement at the IEA/SSPS Plant. SSPS Technical Report Nº 2/82.
5. SANCHEZ, M. (1984). A Mathematical Model for Estimating the Average Heliostat Field Reflectivity. SSPS Internal Report.
6. PESCATORE, M. (1984). Comparison Between the Sulzer Cavity Receiver and the Franco-Tosi External Receiver. SSPS Internal Report.

DESIGN DATA	CAVITY RECEIVER (first built sodium)	EXTERNAL RECEIVER (advanced sodium)
HEAT EXCHANGER	1 vertical 120° cylinder tube bundle with 6 parallel sodium-carrying tubes directed in a serpentine from bottom to top with 14 passes	5 vertical tube panels with 39 tubes, each connected serially
DIMENSIONS OF HEAT EXCHANGER	Radius: 2.25 m Height: 3.61 m Arc Length: 4.71 (active)	Height: 2.85 m (= active tube length) Width: 2.75 m
TOTAL WEIGHT	approx. 25 T	approx. 20 T
NORTH ORIENTED APERTURE	9.0 m^2, octagonal shape	9.0 m^2, rectangular shape
ABSORBING SURFACE	17 m^2	7.8 m^2
APPROX. TUBE LENGTH	87 m (66 m active)	50 m (14.3 m active)
TUBE GEOMETRY	Ø 38 mm X 1.5 mm	Ø 14 mm X 1 mm
TUBE PAINT (black)	Pyromark 2500	Pyromark 2500

Table I : Comparison of Design Data for Both Receivers

OPERATING CONDITIONS	CAVITY RECEIVER	EXTERNAL RECEIVER
THERMAL POWER - Input	2,840 kW	2,840 kW
- Output	2,508 kW	2,525 kW
SODIUM - Mass Flow	7.34 Kg/s	7.34 Kg/s
- Inlet/Outlet	270°C/530°C	270°C/530°C
- Pressure Drop	0.5 bar	1.5 bar
- Flow Velocity	1.5 m/s	1.85 m/s
HEAT FLUX - Peak	0.6 MW/m^2	1.38 MW/m^2
- Average	0.16 MW/m^2	0.36 MW/m^2
- Aiming	single point	three points

Table II : Comparison of Operating Conditions for Both Receivers

RECEIVER	LOSSES (MIN/MAX)	AVERAGE	CONVECTION TEST
CAVITY	390 kW / 535 kW	460 kW	465 kW
EXTERNAL	100 kW / 200 kW	150 kW	162 kW

Table III : Loss Values During Operation and During Loss Test

PANEL	MINIMUM (%)	MAXIMUM (%)
1	3.0	7.4
2	5.0	8.8
3	25.9	31.6
4	24.6	30.4
5	29.9	33.8

Table IV : Individual Distribution of Absorbed Energy at ASR Panels

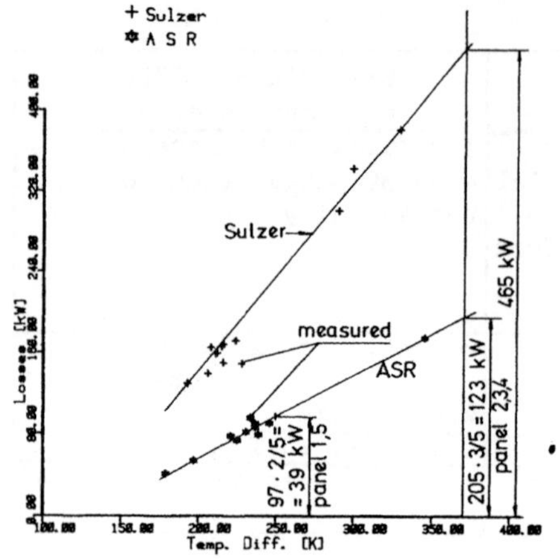

FIGURE 1 Total Losses during Loss Test

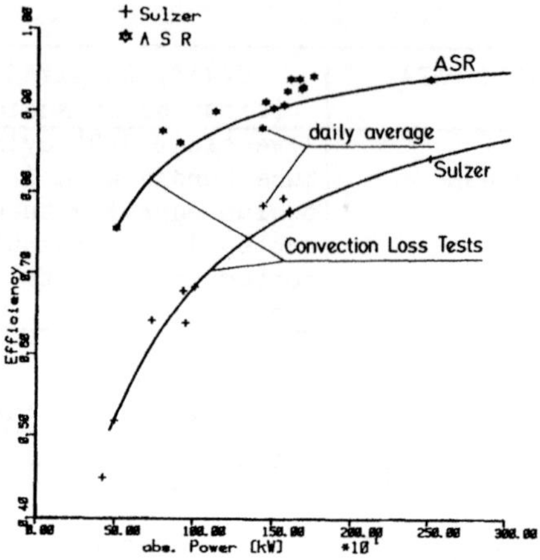

FIGURE 2 Efficiency versus absorbed Power

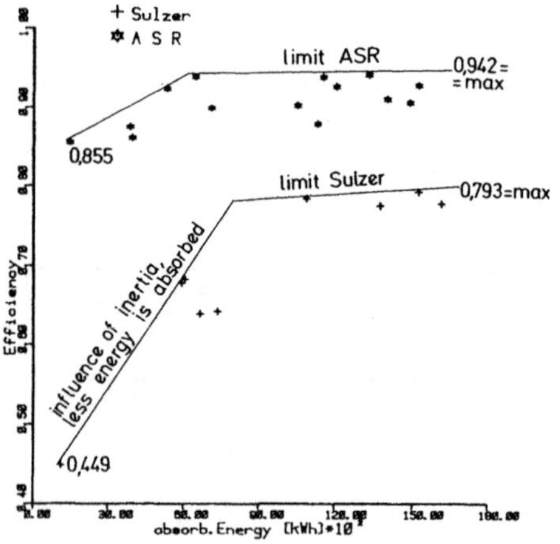

FIGURE 3 Calculated daily Efficiency from DAS record versus absorbed Energy

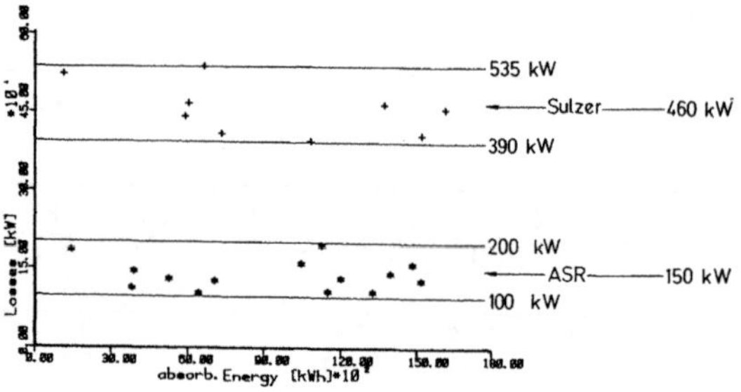

FIGURE 4 Losses during Operation versus absorbed Energy

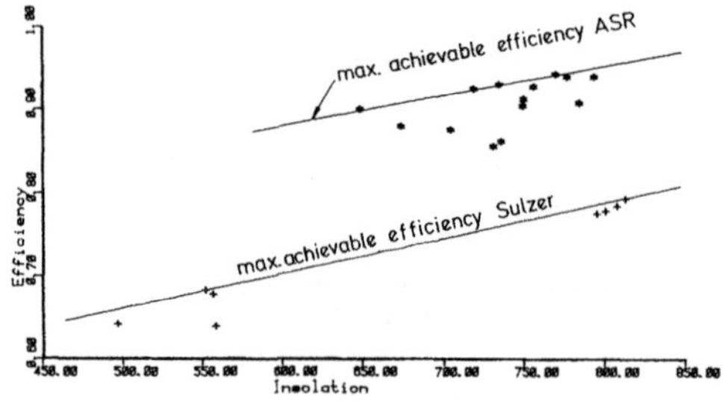

FIGURE 5 Calculated Efficiency from DAS record versus average Insolation

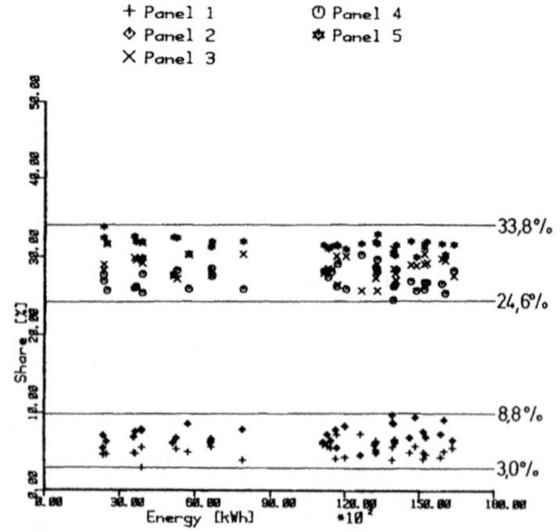

FIGURE 6 Percentage of Total Energy absorbed by each Panel

RELEVANT ASPECTS IN THE DESIGN AND CONSTRUCTION OF THE ADVANCED SODIUM RECEIVER ASR FOR THE IEA-SSPS CENTRAL RECEIVER SYSTEM PLANT (ALMERIA - SPAIN)

P. CAVALLERI, V. BEDOGNI and A. DI MEGLIO
Research and Development Division
Franco Tosi Industriale S.p.A.-Legnano (Italy)

Summary

This paper describes the main steps of analyses, design and manufacturing of the ASR (Advanced Sodium Receiver) for the IEA-SSPS CRS Plant at Almeria (Spain). The high incident heat flux density on the absorbing surface of the receiver with a peak value of about 1.4 MW/m^2 called for the solution of relevant new problems in the design and construction phases of ASR receiver. The paper summarizes all the main aspects related to the thermal and mechanical analysis carried out in order to resolve any uncertainty pertaining to the overall receiver design, creep-fatigue and cycle life. The most important results concerning the basic technological aspects of the manufacturing are described as well.

PREFACE

The receiver was designed and constructed by Franco Tosi Industriale in collaboration with Agip S.p.A. and with the determinant contributions of ENEL (Italian National Board of Electricity) for the dynamic analysis and control system design and CNR (National Research Center) as Italian official representative.

The receiver installation was completed in August 1983 and ASR has been successfully working up to now.

INTRODUCTION

ASR is an "external type" receiver of 2.7 MWt and consists of 5 panels arranged to form a rectangular shape absorber of 2.85 m height and 2.78 m width. Liquid sodium from the cold storage tank at 270 °C is pumped by the receiver feed pump through the receiver where the sodium is heated up to 530 °C.

BASIC DESIGN OBJECTIVES

The primary objectives of the ASR design are:
- incident heat flux peak density in the range 100÷150 W/cm^2;
- average incident heat flux in the range 30÷50 W/cm^2;

- upscaling aspect: ASR should include basic technological aspects of large future receivers.

RECEIVER CONCEPT

The selection of a multipanel external type receiver, together with a 3 point aiming strategy of the heliostat field, allows for the attainment of:
- peak heat flux density 138 W/cm^2 (d.p.)
- average heat flux density 35 W/cm^2 (d.p.)

BASIC DESIGN DATA

Design point: equinox noon
Maximum power input: 2.7 thermal MW at design point
Receiver sodium inlet temperature: 270 °C
Receiver sodium outlet temperature: 530 °C
Design pressure: 6 bar
Maximum pressure drop: 1.2 bar
Receiver life time: 10 years
Number of operating transients: 6343 per year

RECEIVER PANEL GEOMETRY AND MATERIAL

According to the results of 10 years lifetime preliminary analysis, absorber tube size and material were selected:
- absorber material: AISI 316 L
- tube OD/tube thickness: 14 mm/1 mm

With these dimensions and with a sodium velocity of 1.9 m/sec the maximum wall temperature of 596 °C is reached in the most irradiated section of the absorber tubes which is cooled by sodium at 495 °C.

To assure the required cooling(sodium velocity) to the tubes, these are connected in parallel in a number of 39 to form a single panel. All tubes are tangent to one another forming a 0.555 m wide flat panel.

RECEIVER ASSEMBLY AND SODIUM CIRCULATION

The receiver absorber consists of five panels laid side by side so to form a rectangular shape target (2.78 m x 2.85 m). Sodium circulation is shown in fig. 1 and allows for the attainment of two main objectives:
- receiver efficiency optimization
- the central panel will work in the typical operating conditions of a large power plant external panel as regards sodium temperature and heat flux density peak and distribution (3rd basic design objective).

RECEIVER PANEL THERMAL ANALYSIS

Detailed thermodynamic analysis has been carried out in order to calculate:

- absorber tubing temperature profile in steady state and transient condition
- receiver performance considering reflected insolation, infrared radiation and convective losses

at different partial loads by employing ad hoc developed computer codes based on finite difference method. Main results and conclusions are reported in (1).

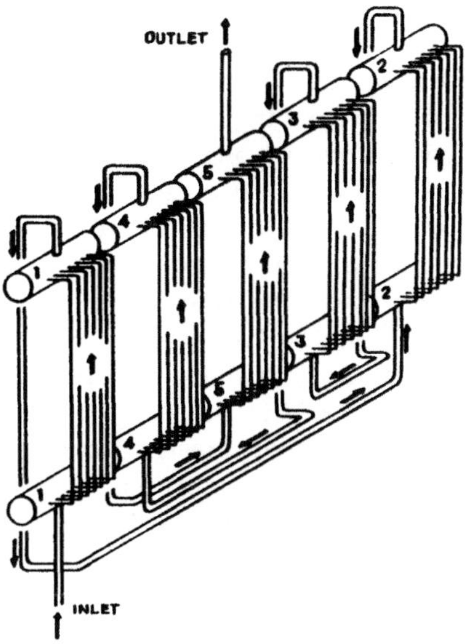

FIG. 1 - Sodium circulation in the panels

RECEIVER DESIGN

Three main parts of the construction are worth describing:
- sodium wetted parts
- main frame structure
- backwall insulation system

The sodium wetted parts of a panel are the tube bundle, the bottom and top headers and the downcomer (see fig. 2).

The fixed points of a panel are the flange of the bottom inlet header and the restraint at the downcomer sodium outlet. The top header is vertical sliding only. With this configuration no problem arises from vertical thermal growth of the panel. The irradiated tubes are three by three assembled together with a supporting plate at four levels (see detail 1 of fig. 2) so to form a "triplet". The triplets, in corrispondence of the four plates, are connected to the panel framework (see detail 2 of fig. 2) by means of pins. This kind of tube restraint, so called "stirrup system", allows the tubes to grow axially with respect to the frame and also to rotate owing to the clearance between the pin diameter and the hole in the triplet supporting plate.

In order to allow lateral thermal expansion of tubes and panels, gaps are provided between tubes in addition to manufacture tolerances. With this concept of supporting system for the irradiated tubes, each triplet is free to expand indipendently of the other ones so no problem arises from the heat flux gradient in the horizontal direction.

In the back of the tube bundle a double shield of high refractory Alumina based material, protects the back structure from the concentrated solar flux passing through the tube gaps (see detail 3 of fig. 2).

Behind the second shield a thickness of 175 mm of ceramic fiber insulates the hot parts of the receiver from the cold back structure.

The triplets, in groups of two or three, are anchored with beams (detail 2 of fig. 2) to the "single panel frame", which is verified to withstand, besides its dead weight, wind and earthquake loads, forces and moments due to panel tubes and connecting pipes.

The structure analysis is addressed to avoid displacements and distortion so to make sure a regular operation of the panel. For the same reason particular attention has been paid to limit conduction heat transfer from hot sodium wetted parts to cold panel frame through supporting elements: stirrup beams, headers supports and piping restraints mainly. Each panel frame is then assembled in the mainframe structure of the receiver.

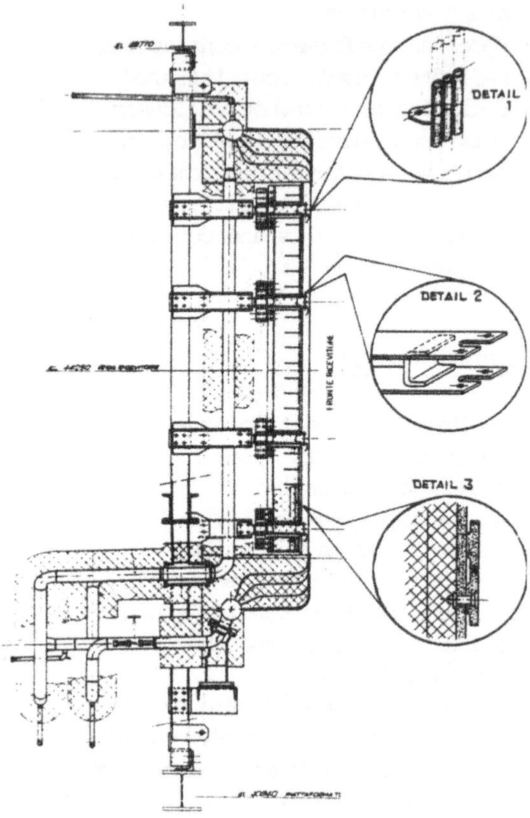

FIG. 2 - Absorber back-structure details

RECEIVER STRUCTURAL ANALYSIS

The major role of the thermal stresses and the heavy operating conditions in a sodium solar receiver requires a suitable stress analysis to verify the component structural stability against failure modes as:

. bursting
. gross distortions
. creep-fatigue damage

This analysis has been confined to the following parts which are the most subjected to solar flux transients during operation:

- panel tubing (most irradiated tube section)
- tube-supporting plate connection (detail 1, fig. 2)
- tube-header connection

For all these parts a detailed thermal and structural analysis has been performed with an extensive use of finite-element technique. The evaluation of the results has been performed utilizing criteria in accordance to ASME Pressure Vessel Code Section VIII Div. 1 and 2.

For component service lifetime evaluation, that is to evaluate creep-fatigue interaction effect, Code Case N47-17 has been assumed. Particular emphasis in this analysis was given to the most irradiated tube section, the analysis of which has been performed at first on elastic basis and then, because of the high thermal stresses beyond the yield limit, on inelastic basis (F.E. MARC Code) to check the validity of the previous analysis and in order to evaluate its safety margin. The results showed the large conservatism of the Code Case elastic creep-fatigue analysis and underlined, once more, the necessity for the development of special design standards for solar applications. (2)

Tube-supporting plate connection and tube-header connection have been carefully analyzed by FEM method in transient condition. Particularly the evaluation of the shrink effect of the supporting plate on the triplet tube has required a detailed 3D analysis. In fig. 3 the 3D mesh plot is shown with 20 nodes, brick type, finite-elements: the mesh was used both for thermal transient calculation and for mechanical evaluation.

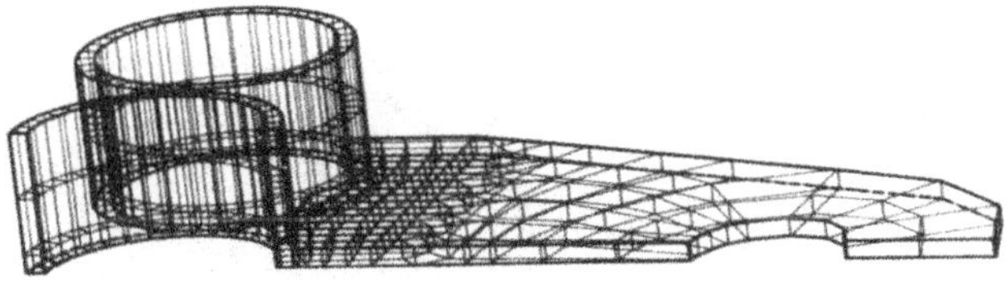

FIG. 3 - Tube-supporting plate 3D mesh plot

SUPPORTING TECHNOLOGICAL TESTS

Before starting the receiver construction an extensive development program was completed to ensure the feasibility of the proposed design. This work included mainly:

- Header prototype with the scope
  . to investigate the manufacturing problems and to set up improved technological procedures with the objective to attain thickness uniformity as required by fast sodium thermal transients.
- "Triplet" prototype with the scope
  . to investigate the influence of the supporting plate thickness and of the welding bead length on the straightness of the triplet;
  . to investigate the gaps between tubes;
  . to investigate the tearing strenght of the tube-plate connection.
- Oxidation test with the scope
  . to investigate the oxidation possibility in the pin material in order to avoid blocking in the supporting system of the triplets (stirrup system).

MANUFACTURING

The major steps of the manufacturing process are:

. tube inspection, cutting and bending
. set up "triplets" in fixture
. inlet/outlet headers fabrication
. panel support structure fabrication
. inlet/outlet headers anchorage to panel structure
. welding of triplets to headers

. gap width adjustment (by regulating the position of the stirrup beams)
. paiting, curing and vetrification
. insulation installation
. assembling of panels with interconnection piping
. preparation for shipment

Several manufacturing problems were encountered with this new project. The most interesting and the most important was tube-tube supporting plate welding. Owing to the very thin thicknesses to join, plasma jet manual welding procedure was adopted. It was found that the weld profile was very sensitive to parameter variations and as this is an important feature for the acceptance of the weld, considerable work was required to stabilise variables to obtain the desired profile on production.

A full program of destructive testing on welding specimens was completed. Similar precautions were taken for the welding of the triplet tubes to the headers nozzles. In this case an automatic orbital inert gas welding procedure was adopted. (fig. 4)

FIG. 4 - Automatic orbital welding of absorber tubes to bottom header

Considerable inspection effort has also been applied by an extensive quality control program exercised on the quality of the materials from the point of view of mechanical and chemical properties, on the manufacturing, on the welding and generally on all the most relevant activities related to the attainment of a certain confidence in the standard of the unit.

To ensure that the required quality is achieved, an extensive documentation of the performed examinations was provided too.

The method of inspection adopted for each weld started with a visual examination followed by a dye penetrant and radiographic inspection extended to 100% of the welds. For the tube-tube supporting plate welding, endoscopic visualization from the tube inner side was employed.

On the completion of each panel tube bundle, pressure tests and helium leak test were carried out.

After assembling the five panels with the connecting pipes a final pressure test and helium leak test were performed of the whole unit before conservation and packing for shipment.

CONCLUSIONS

The general approach for the development of ASR is directed towards providing a component of the highest quality through mainly:

- accurate structural analysis with particular attention to transient absorber performance (creep-fatigue interaction) which is a determining factor in selection of the best solutions;
- well-proven technological processes based on the experience gained by Franco Tosi in sodium-cooled nuclear reactor program partecipation;
- extensive quality control program.

The design basis used for the ASR continues to be successfully proven through receiver operation now in progress.

Further investigations concerning mainly the stationary and dynamic performances are reported in (1).

REFERENCES

(1) De Benedetti, A. (1984). The Advanced Sodium Receiver for IEA-SSPS Central Tower Plant: operative conditions, control system design and performances. Agip S.p.A.-Milano

(2) Bedogni, V. and De Benedetti, A. (1983). ASR Receiver Stress Analysis Consideration. Franco Tosi Industriale-Agip S.p.A. Presented at ISES Congress, Perth, Australia, August 1983

THE THEMIS PLANT START UP OF THE RECEIVER-FIRST RESULTS

B . RIVOIRE
Groupe d'Evaluation Scientifique THEMIS (GEST)

Start up of the receiver began in June 1982 by connecting the electrical tracing system. From November 1982, the next step was initiation

of the hydraulic functioning. In April 1983 focusing tests began. They have been performed at high salt trough put until now.

1. PREHEATING OF RECEIVER

Test and modifications during starting of the electrical tracing system lasted four months (summer 1982).
The weak points encoutered at this step are mainly due to imperfections in the insulation of the cavity: the insulation was completely modified (after the tracing power was increased without any result). The status of fonctioning attained is not perfect, since we had to accept unbalanced sections from a power point of view, and consequently, thermal losses and over heatings. However, we encountered the real problems in April 1983, after a successful working period.
As a matter of fact, a succession of troubles occured, making the primary loop unvailable. The problems were due to a make-shift construction which, in the case of a low temperature tracing system, would not have caused any trouble. Yet, we resolved these problems after 10 months of repairs, maintenance, analysis and modifications. The most serious and significant disorder is the consequence of two leaks due to corrosion which appeared on draining pipes, leaks which could be imputed to repeated over-heatings of the electrical tracing.
We could explain these over-heatings by the existence of local additional loops created at the installation in order to absorb excess length of the tracing cable. As soon as this default could be identified, a whole part of the equipment had to be re-examined to suppress irregular loops.
We consider that the tracing equipment has been reliable and operational since January 1984.

2. HYDRAULIC WORKING

The hydraulic working of the receiver did not pose a special problem. Apart from the above mentioned leaks and some others due to bad set joints, some obstructions of small pipes (draining and rising pipes) made us anxious for a time. But the obstructions no longer occur and we think that they were due to an agglomeration of dust , which was swept out when using these pipes.
The working of gates and pumps has no been problematic, except for the valves of the bottom column which had to be changed (their role is to prevent the discharge of the circuit while the pumps are stopped.)

3. WORKING UNDER FLUX: ADJUSTMENT OF TEMPERATURE AND SAFETY REGULATION

3.1 Constraints due to the design of the receiver

The THEMIS receiver has been designed in order to work under maximal theoretical conditions - precisely described in the specifications. In particular, the flux-map received by the panels in the most extreme case has been specified. This data has been used with confidence by the manufacturer who tried, for example, to decrease pressure drops as much as possible. Whis this purpose in mind, the speed of the salt in the different parts of the receiver has been exactly adapted; this speed is lower where the lower flux is expected. In pratice, the vertical panels of the receiver are described by two parallel circuits (HIGH circuit - LOW circuit), each of these circuits formed with six bands traversed in series. Each of these bands is made of several close tubes traversed in parallel. The number of tubes is smallest at the central band (12 tubes, high speed for the salt, high turbulence, high pressure drops but good exchange coefficient permitting the absorption of a high flux without making the temperature gradient in the tube thickness higher than the allowed value). The number of tubes is highest at the lateral bands (27 tubes, reduced pressure drops, limited capacity to absorb flux) (cf Figure 1).

RECEIVER FLOW DIAGRAM

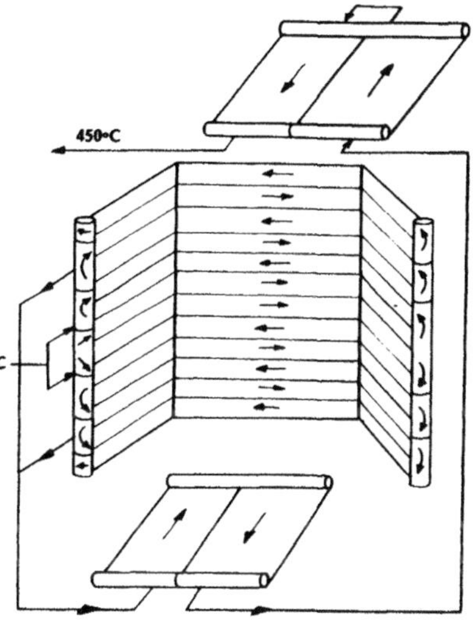

Figure 1

This design has two consequences:

1) The average temperatures of the salt in the two parts - HIGH and LOW of the receiver must be near each other. If they were not, differences of viscosity would appear, preventing a good distribution of flows.

2) The maximum flux received by each band must be lower than a threshold value dependent on the salt flow and on the maximum salt tempera-

ture in this band. The receiver has been delivered with a flow regulator which takes into account these constraints (the regulator maintains a high salt flow when a part of the heliostat field is masked.But it has been delivered especially with a protection controller which surveys the flux received by each band, the salt temperature at the exit of each band, and the general flow in order to verify, that, at any point,the received flux does not exceed the allowable flux. When this controller detects an excess flux, it immediately gives the order to unfocus.

3.2 Tests

The focusing operation began without the flow regulation (at the higher flow) and without the protection controller. Nevertherless, this protection controller had been replaced by a code which acquires and treats eighteen flux measures taken at the bottom panel (the most charged), thus allowing the simulation of its analysis. But we could quickly establish that the flux distribution was different from the expected one, preventing the establishment of the regulation. The test periods which occured in 1983 (from April to October) and in 1984 (February - March) were devoted to an analysis of the actual working conditions: We tried to modify the conditions by changing the focusing point of the nine heliostat groups and to determine new temperature working conditions and their associeted protection values.

3.3 Results

1) Focusing behavior: we obtained a flux distribution nearer to the expected one by disfocusing a little the whole heliostat field: the central group focusses at the focal point but the eight others focus at a point situated one meter in front of the focal plane (cf figures 2 and 3).

MAX FLUX MEASURED
COMPARED TO
REQUESTED VALUE
Test on april, 13, 1984
11H45

Thermal_load : 97 %

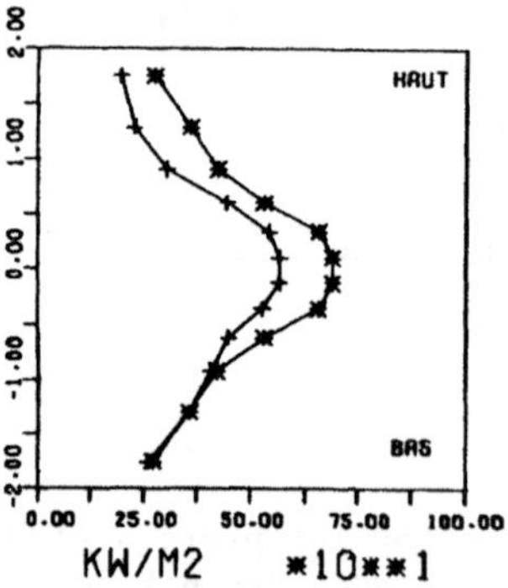

HELIOSTATS'S GROUPS
POINTING IM FRONT
OF THE FOCAL POINT

Figure 3

MAX FLUX MEASURED
COMPARED TO
REQUESTED VALUE
Test on january,17., 1984
11H32

Thermal load: 72%

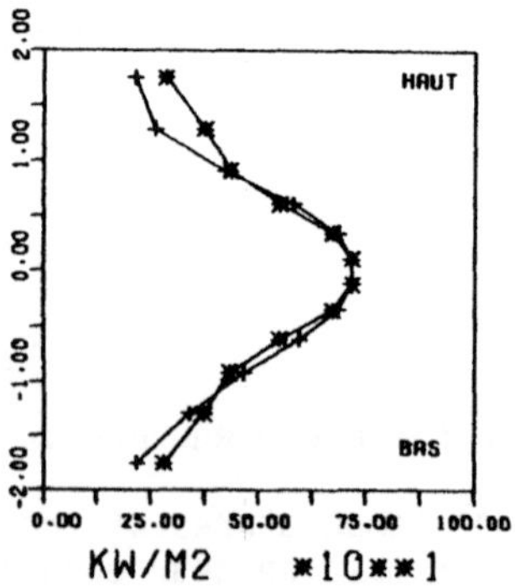

ALL HELIOSTATS GROUPS
POINTING AT FOCAL POINT

Figure 2

2) Nominal working temperature:

We have had to decrease to 430°c the nominal temperature of the hot salt in order to increase the flow ant thus increase the capacity of absorption of the different bands. This makes the real flux and the admissible flux compatible at each point, even for the highest powers specified.

3) Associeted protection thresholds: taking into account the new data (flux and temperature) we have re-computed the normal salt temperature at the exit of the bands in terms of the entrance temperature and, conse quently, we determined the threshold they must not exceed The receiver builder has, on this basis, re-calculated the permissible flux (higher than the first calculated).(Fig 4 gives the graphs of the variation of th se flux as a function of the flow and the output band temperautre.

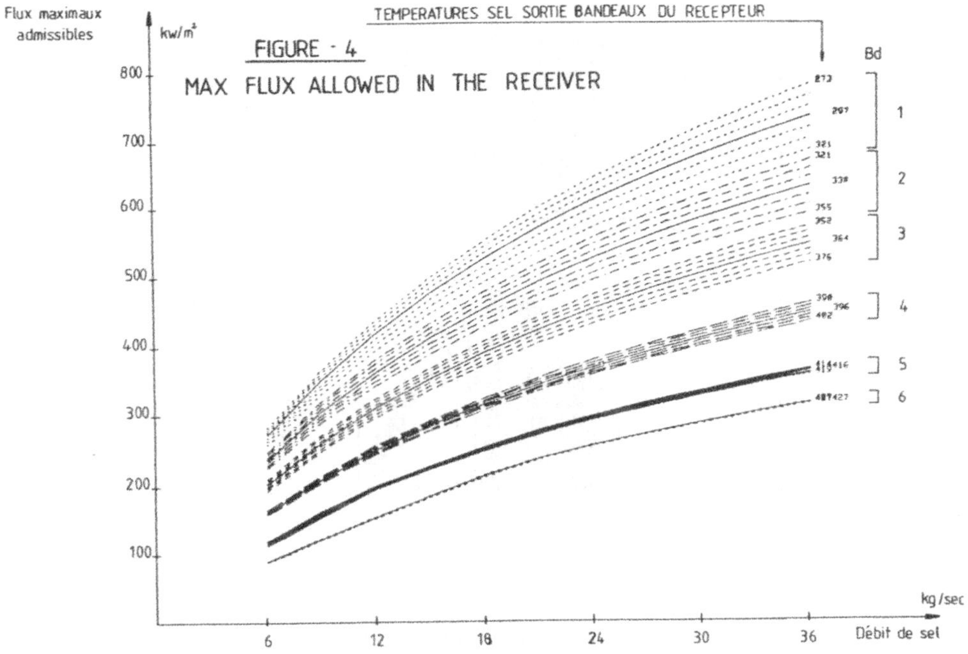

On this basis, we are modifying the protection controller and we hope to put it in service at the same time as the flow regulation (temperature fixed at 430°c) at the begining of August.

4) Installation of the focusing controller:

On the other hand,we put a controller adjusting the focusing of the groups. This controller transmits vertical motions of small amplitude to the focus point, as a function of the difference of temperature between the high and low outputs of the vertical panels of the receiver. Its action is effective and the thermal balance HIGH - LOW is perfectly realized.

5) Other results:

- Though we mainly paid attention to the points described above, we could determine the excellent working conditions of the primary loop at every power level. As a matter of fact, we could focus in April with an insolation of 1040 W/m² and with 199 héliostats.
- Ability of the receiver to take advantage of irregular solar flux (cf figure 5).

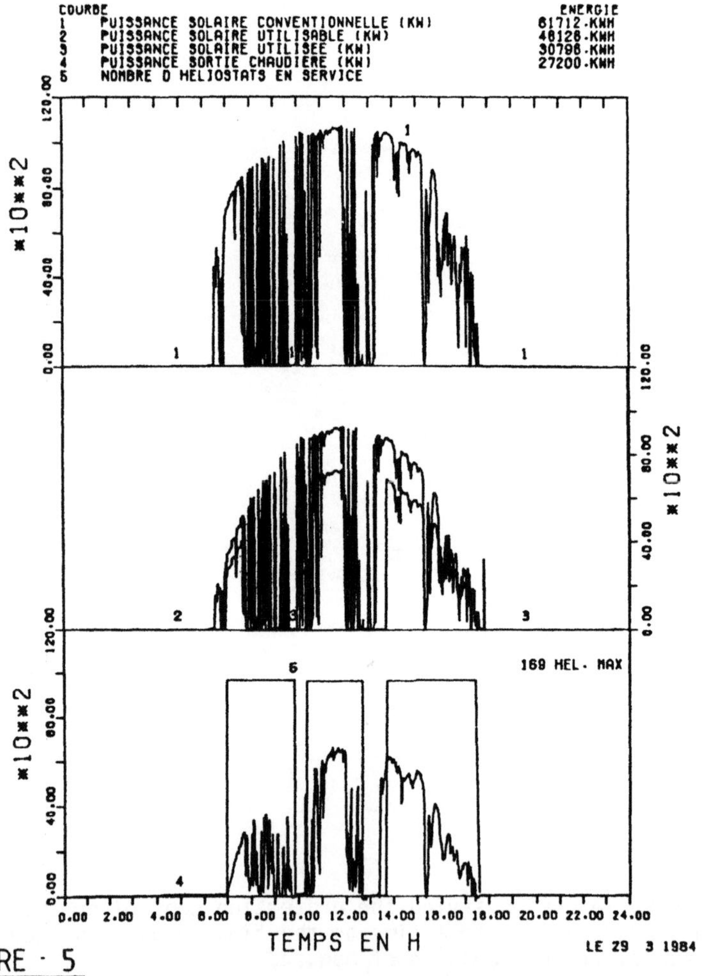

FIGURE · 5

ENERGY BALANCE IN THE RECEIVER

• the lack of steam pressure in the loop (and thus the lack of thermic leacks) permits a good temperature maintenance of the loop. The thermal inertia is very low, and we do not need more than a few minutes to reach the working temperature (cf figure 6).

SALT TEMPERATURE IN THE RECEIVER ON START UP

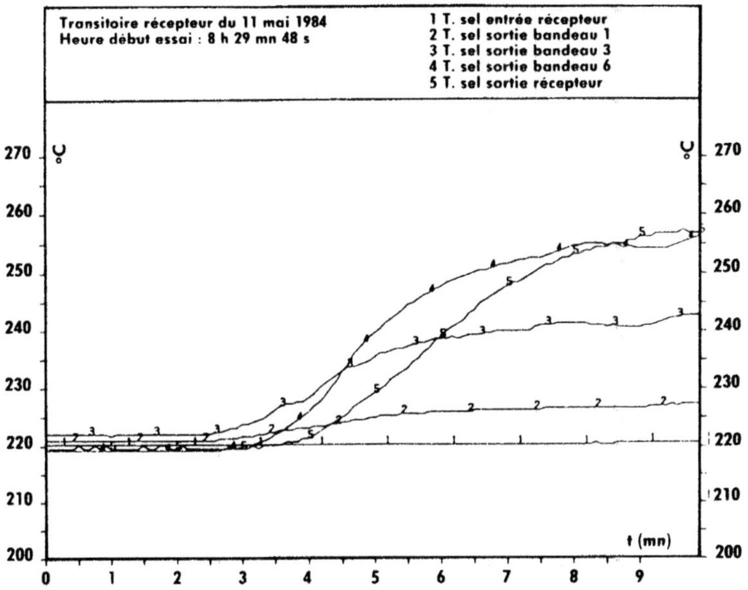

Figure 6

° the collecting efficiency, measured as on figure 5 is more than 90%, a number close to the expected one.

4. BEGINNING OF EXPERIMENTATION

The precise analysis of receiver evaluation and of its thermal working conditions has not been taken up. But, putting into service the temperature regulation, we will have the opportunity-during the operation which will now be steady-to study these points.
At present, we are implementing an infra-red camera which will give us a detailed map of the radiation temperatures of the bottom of the receiver. It will, for the first time, give us the possibility of comparing the calculated and the measured values. This apparatus will be operational in July. In an other connecting, we think that the flux measurement bar in the focal plane will soon be implanted. It will give us valuable indications concerning the evaluation of the receiver: the power at the

entrance of the receiver, infra-red and visible radiation losses.
Finally, we are interested in the precise calibrating of the measuring equipments that we have been using for a year: particulary the salt flowmeter and pyrheliometer. The thermogages of the bottom of the receiver whose measure is difficult to interpret will be replaced by photonic flux gages, developed at E.D.F and which will allow more precise measures.

With the precise instrumentation, we hope that, before 2 years, we will be able to develop a correct model for the receiver with respect to its evaluation and to its thermic contraints.

STABILITY OF SALT AND CORROSION RESISTANCE OF CIRCUIT MATERIALS IN THE THEMIS POWER PLANT

P. SPITERI
Département Etude des Matériaux, EDF Les Renardières,
77250 - Moret-sur-Loing, France

Summary

Studies on salt chemistry and materials behaviour have been carried out since 1978 at EDF in the framework of the THEMIS programme. The study of nitrate-nitrite mixtures chemistry has allowed the main thermodynamic equilibria to be identified and a method has been defined to stabilize the salt by fixing the rate of oxoacidity. Corrosion tests on the steels used for the circuit in a eutectic mixture have confirmed the good corrosion resistance of 316 L type stainless steel at its nominal operating temperature (500°C). By contrast, low alloy ferritic steels at their operating temperature (450°C) show different behaviours, depending on the oxoacidity of the salt : linear corrosion for steel without chromium in oxoacid salt, parabolic corrosion in all other cases. A deleterious effect of increasing the chloride content of the salt has also been observed on these steels.

1. INTRODUCTION

A sodium-potassium nitrate-nitrite eutectic mixture has been selected as coolant and storage fluid for the primary circuit of the French solar-thermal conversion plant THEMIS. The "Département Etude des Matériaux" of Electricité de France and the "Ecole Nationale Supérieure de Chimie de Paris" (Prof. TREMILLON's laboratory) cooperated in the study of the stability of the eutectic mixture and of the corrosion resistance of the different steels used for the salt circuit.

2. THERMODYNAMIC STABILITY OF THE MOLTEN SALT MIXTURE

The 550 tons of salt were supplied in accordance with the specification shown in table I, together with the average analysis of the salt provided by SERVIMETAL. A preliminary study (1, 2) of the chemical factors acting on the thermal stability of the nitrate-nitrite mixture has been undertaken to identify the reasons for the progressive decomposition of nitrite above 400°C (the nominal operating temperature is 450 and 500°C at the ID of some receiver tubes).

2.1 Oxoacidity

Different oxidation-reduction (redox) processes of nitrogen and of the O^{2-} ion take place in the molten salt. A potential versus oxoacidity plot allows the very complex possible equilibria to be examined ; the expression used for oxoacidity is the value pO^{2-} ($pO^{2-} = -\log[O^{2-}]$), which is defined by analogy with the pH in aqueous media.

$$\text{oxobase} \rightleftharpoons \text{oxoacid} + O^{2-}$$

The potential-oxoacidity diagram of figure 1 reveals the existence of a stability area of the nitrate-nitrite mixture whose limits are fixed :

- in the oxoacid area, by the decomposition of nitrites into nitrates and nitrogen oxide,
- in the oxobasic area, by the decomposition of nitrates into nitrites and peroxides.

The boundaries of this thermodynamic stability area depend on temperature and of the ratio of nitrate versus nitrite concentration (different between HITEC and DRAW SALT).

2.2 Study of the water/OH^- couple

Given the equilibrium 1, the oxoacidity of the molten salt mixture can be fixed by acting on water and OH^- activity.

$$\text{eq. 1} \qquad 2\ OH^- \rightleftharpoons H_2O + O^{2-}$$

The method was successfully applied at the laboratory by adding a specific quantity of sodium hydroxide and by pre-setting the partial pressure of water in the cover gas (2). This technique may be employed industrially to fix the oxoacidity of the mixture in the stability area of the salt (3).

The values of this equilibrium constant were determined between 420 and 520°C for a mixture of sodium salts and sodium-potassium salts (figure 2).

2.3 Composition of the cover gas

To prevent oxidation of the nitrites of the eutectic mixture, nitrogen is used as cover gas for operating temperatures above 400°C.

In addition, equilibria 1 and 2 fix the partial pressures of water and nitrogen oxide in this cover gas.

$$\text{eq. 2} \qquad 3\ NO_2^- \rightleftharpoons NO_3^- + 2\ NO + O^{2-}$$

Figure 3, established for a eutectic mixture of sodium salts, shows that the more oxoacid the mixture and the higher the temperature, the greater the partial pressure of nitrogen oxide. Stabilisation of the mixture by adding sodium hydroxide is here again a valuable solution to limit the highly toxic nitrogen oxide content of the cover gas.

3. CORROSION OF THE MATERIALS USED FOR THE SALT CIRCUIT

Table III lists the different steels used for the components of the THEMIS salt circuit. Low alloy ferritic steels are used for all parts whose nominal temperature remains below 450°C. Type 316 L stainless steel with controlled nitrogen content has been selected for the tubes of the solar receiver because of its adequate high temperature mechanical properties and its corrosion resistance in the salt better than that of low alloy steels.

3.1 Corrosion of austenitic stainless steels

The specimens have been weighed before immersion into the salt and after descaling of the oxide layers in order to evaluate the weight loss. Type 304 L, 321 and 316 L with controlled nitrogen content stainless steels lead to equivalent metal losses (4).

The general corrosion of these steels is in most cases parabolic, the corrosion rate decreasing as a function of time through the formation of a protective oxide layer (figure 4).

A constant corrosion rate of the steels has however been observed in a salt containing 1 % $BaCl_2$, the metal losses increasing very rapidly in that case.

Local corrosion was also observed : pitting whenever metal loss exceeds 10 µm and surface nitridation particularly noticeable on type 321 stainless steel when affected by advanced general corrosion (figure 5).

3.2 Corrosion of low alloy ferritic steels

Tests were performed on low alloy Mo and Cr-Mo steels at 450 and 500°C : AISI 4023 (with 0.3 % Mo), 1.25 Cr-0.5 Mo and 2.25 Cr-1 Mo.

Tests in an industrial salt revealed a very rapid corrosion of the 0.3 Mo steel at both temperatures and a loss of metal following linear kinetics (figure 6). Sodium hydroxide addition into the salt resulted in parabolic corrosion kinetics and metal losses equivalent for the three steels. Alteration of the medium due to the oxidation of the specimens could be responsible for the change in corrosion rate observed on 0.3 Mo and 1.25 Cr-0.5 Mo steels at the end of tests. Tests performed by SANDIA (5) at 550°C on a carbon steel also showed a decrease in the corrosion rate which could be related to a high degree of decomposition of the salt, due to reaction 2 making the salt extremely oxobasic.

Tests at 500°C in salts containing different impurities (sodium hydroxide, sodium chloride, sodium sulphate) confirm the beneficial effect of sodium hydroxide addition and reveal the deleterious effect of an increased chloride content of the salt on the general corrosion of 0.3 Mo and 1.25 Cr-0.5 Mo steels (figure 7).

Nitridation accompanying general corrosion of the steels is the deeper the higher the oxidation and can lead to embrittlement of the specimens.

CONCLUSION

The different studies carried out on the nitrate-nitrite mixture have made it possible to optimize the operating conditions of the THEMIS salt circuit in order to prevent both the deterioration of the mixture and too fast corrosion of steels.

A monitoring programme has been defined with the operating team of the plant in order to compare the laboratory results with those recorded on a large industrial system. The first measurements performed at THEMIS confirm the results which have been presented in this paper.

REFERENCES

1. FLAMENT, T. (1982). Oxoacidité et stabilité des mélanges nitrate-nitrite. Thèse de Docteur-Ingénieur, ENSCP, Paris.
2. PICARD, G., FLAMENT, T. and TREMILLON, B. (1983). Effect of oxoacidity on the thermochemical stability of molten sodium nitrate-nitrite mixtures. Proc. of the 4th International Symposium on Molten Salts, San Francisco, in press.

3. Brevet France 82 16 201 (1982).
4. SPITERI, P. (1980). Corrosion of different steels in Na-K nitrate-nitrite mixture. Proceedings of the Molten Nitrate Salt Technology Workshop, Sandia, Dublin Ca.
5. KRAMER, C., SMYRL, W. and ESTILL (1979). Corrosion of Fe alloys in HITEC at 823 K, SAND 78-8256.

Table 1 - Analysis of Themis salt (weight %)

	NO_3^-	NO_2^-	K^+	Na^+	OH^-*	Na_2O*	Cl^-	S	C	Others
Specification	36,7 38,0	26,5 27,5	20,5 21,0	14,9 15,3	$\leqslant$0,01	–	$\leqslant$0,05	$\leqslant$0,015	$\leqslant$0,010	$\leqslant$0,3
Results	37,5	26,8	21,0	15,0	–	0,015	$\leqslant$0,002	0,008	<u>0,013</u>	$\leqslant$0,003

* The chemical analysis takes both the OH^- and HCO_3^- ions into account; these values are used to determine alcalinity in the form of Na_2O content.

Table 2 - Steels used in the molten salt circuit

Component	Steel
Storage tanks	Carbon steel and AISI 4023 (0.3 Mo)
Molten salt circuit	0.5Cr-0.5Mo
Steam generator	1¼Cr-0.5Mo
Solar receiver	316 L with controlled nitrogen content

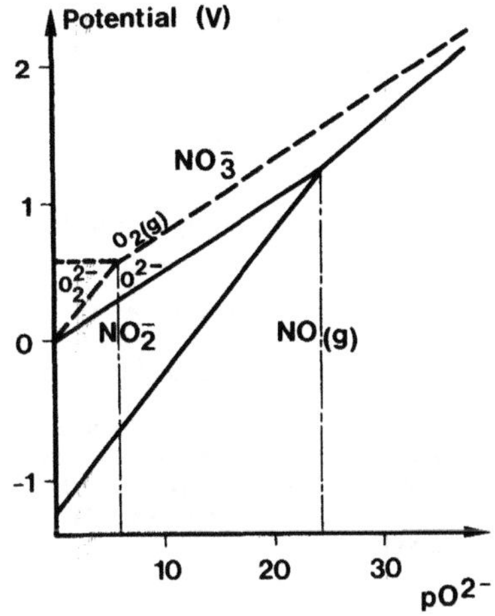

Figure 1 - Potential - oxoacidity diagram of nitrate - nitrite mixture at 227°C.

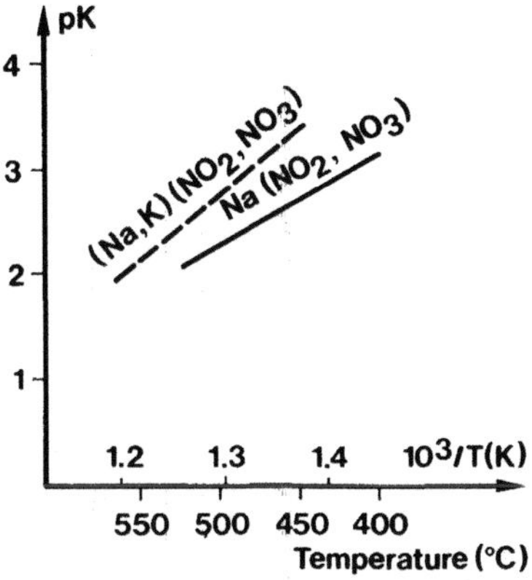

Figure 2 - Evolution of the stability constant of water as a function of temperature (sodium and sodium - potassium mixtures).

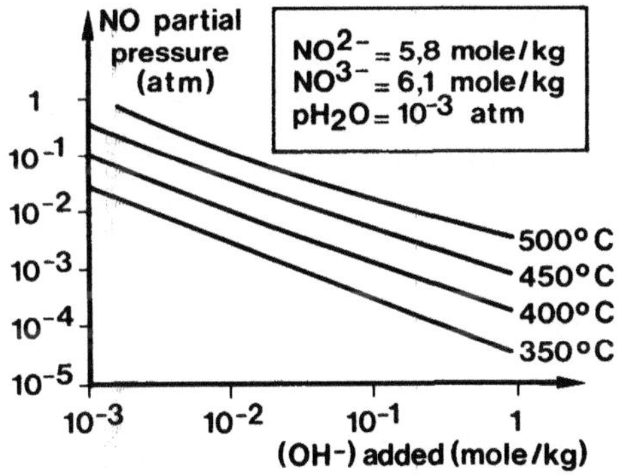

Figure 3 - Influence of sodium hydroxide addition upon NO partial pressure at various temperatures.

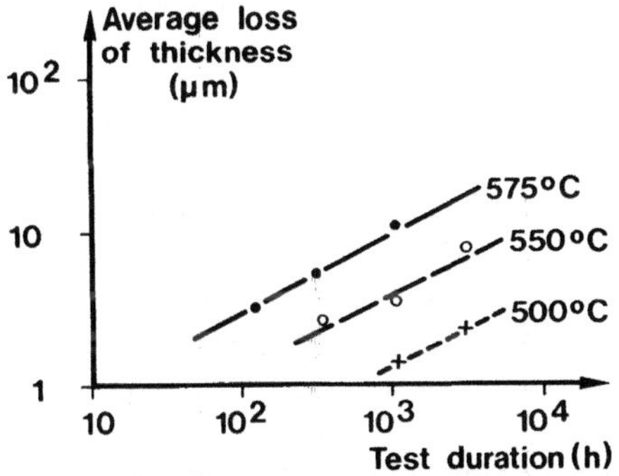

Figure 4 - Evolution of the average loss of thickness of 316L SS between 500 and 575°C versus time at various temperatures.

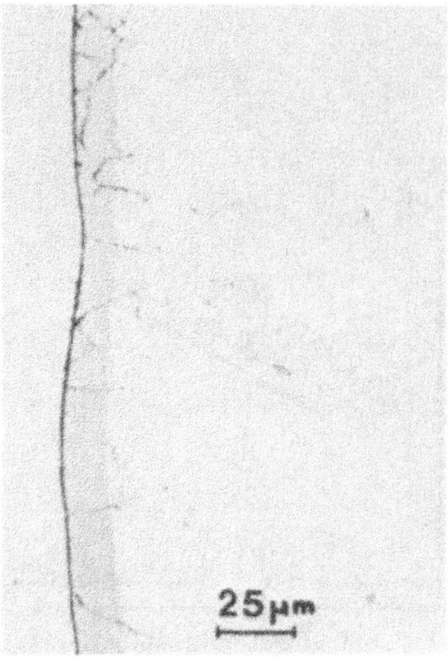

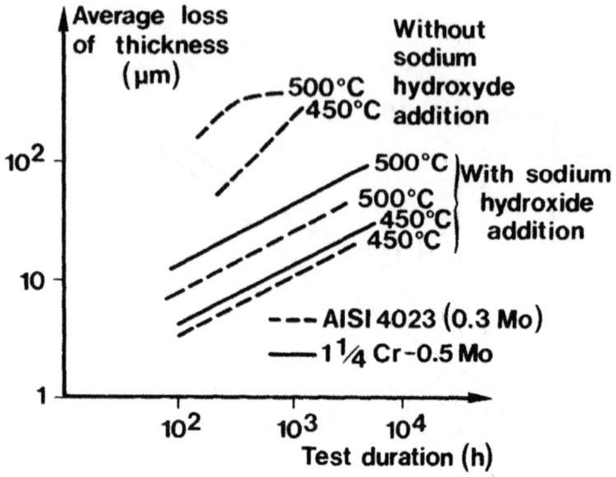

Figure 6 - Evolution of the average loss of thickness of ferritic steels at 450 and 500°C in salt mixture with and without sodium hydroxide addition.

Figure 5 - Nitridation at the surface of a 321 H SS specimen after test in salt containing 1 % $BaCl_2$ (775 h, 575°C).

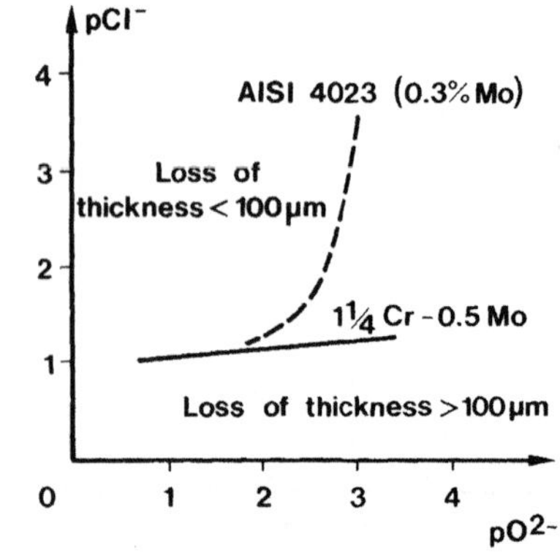

Figure 7 - Effect of oxoacidity and chloride content on corrosion of two ferritic steels (1345h at 500°C).

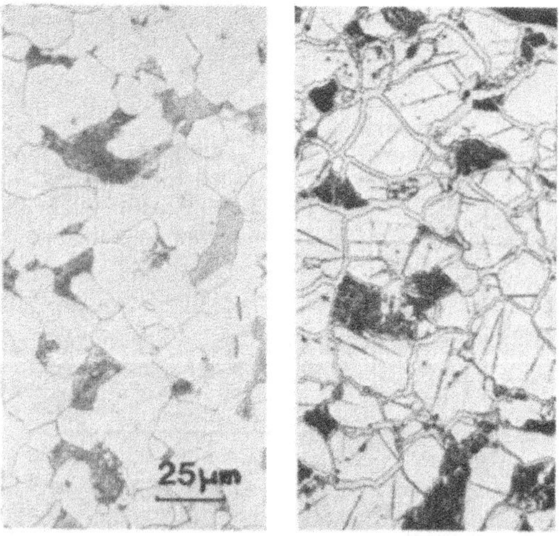

as-received after test

Figure 8 - Nitridation of the AISI 4023 (0.3 % Mo) steel specimen during test in salt without sodium hydroxide addition (100 h, 500°C).

AN ORIGINALLY SCALED TEST UNIT OF A MODULAR GAS COOLED RECEIVER OF 50 MW THERMAL POWER

P. AGATONOVIC, H. FUHRMANN, D. ROOS
M.A.N. Neue Technologie

Summary

This is a report on the receiver tests planned under the German-Spanish GAST technology programme at the Plataforma Solar in Almeria. From the beginning of 1985 it is planned to test over a period of about one year a heat exchanger unit for a gas-cooled solar receiver with an air outlet temperature of 800 °C and ceramic piping with air outlet temperatures of 1000 °C at a pressure of 10 bar. The design and construction of the test units and the planned measurements are described.

1. INTRODUCTION

German industry has been working on the GAST project (gas-cooled central receiver solar power plant) since 1978 with the support of the Federal Ministry for Research and Technology. The reference concept for this plant envisages the generation of hot air at a temperature of 800 °C at a pressure of approx. 10 bar in a solar receiver. The total output of 56 MW will be split between two indentical receiver modules, the apertures of which are about 200 m above the ground and arranged symmetrically facing the north.

The piping of the receiver is divided into a low-temperature (LT) and a high-temperature (HT) section. One LT panel and one HT panel form a train. Each receiver module contains 18 trains. The principal data of the reference receiver train, which contains the HT panel exposed to the greatest stress, and of the heat exchanger units that are to be tested, are compiled in Table 1. In addition to this receiver development the technology for ceramic heat exchanger pipes made of reaction-bounded silicon-infiltrated silicon carbide (RBSC) within the GAST technology programme will be developed further so that gas outlet temperatures of 1000 °C can be achieved.

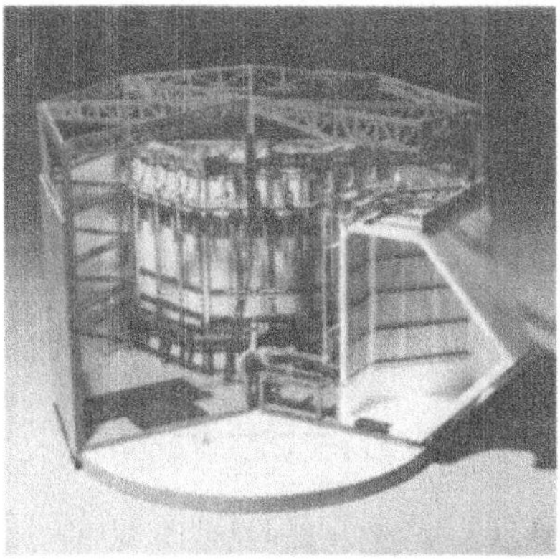

Fig. 1 Photograph of a receiver model

	Reference Receiver - trace of max. load -	Metallic Test Panel	Ceramic Test Panel
Part of low temperature:			
Inlet temperature	340 °C	-	340 °C
Inlet pressure	9,5 bar	-	9,5 bar
Outlet temperature	612 °C	-	-
Rated output	965 kW	-	-
Part of high temperature:			
Inlet temperature	612 °C	625 °C	-
Inlet pressure	9,3 bar	9,3 bar	-
Outlet temperature	800 °C	800 °C	1000 °C
Mass flow	3,28 kg/s	2,45 kg/s	0,48 kg/s
Rated output	700 kW	489 kW	360 kW
Irradiated length	8 m	8 m	6 m
Outer diameter of tubes	44 mm	42 mm	42 mm
Wall thickness of tubes	2,2 mm	2,1 mm	5 mm
Material of tubes	X10 NiCrAlTi3220H	X10 NiCrAlTi3220H	RBSC (SiC)

Main Panel Data of the Reference Receiver and of the Test Panels
Hauptdaten der Panel des Referenz-Receivers und der Test Panels — Tab. 1

2. DESCRIPTION OF THE TEST PROJECT AT THE "PLATAFORMA SOLAR DE ALMERIA"

At present two solar tower test plants, the Central Receiver System (CRS) of the Small Solar Power Stations (SSPS) from the International Energy Agency and the Spanish CESA-1 plant, are installed at the Plataforma Solar. One panel of the GAST receiver with original dimensions is to be tested on the CESA-1 plant. After this the testing of a panel fitted with ceramic pipes is planned.

A gas supply system (see Fig. 2) is being installed in the CESA-1 plant so that the tests can be carried out. The test panel with insulating back wall and structural protection will be mounted so that it hangs above the receiver of the CESA-1 plant about 75 m in front of the tower. The panel will - as is planned in the receiver - be suspended in the area of the hot gas line with internal insulation by means of a flange on this line. The hot gas line is in turn supported by a frame which is fastened to the tower. The lattice structure which holds the insulating wall behind the panel is also connected to this frame. This insulating wall has two functions:

- it is involved in energy exchange with the panel and simulates the inner wall of the receiver
- it protects the tower from radiation

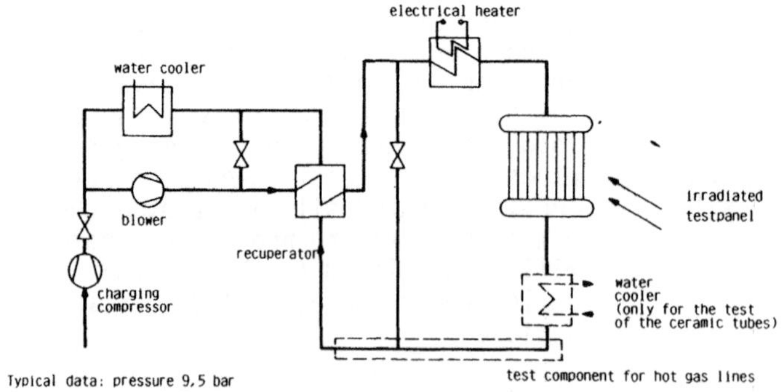

Gas Supply System for the GAST-Project Panel Tests
Gasversorgungssystem für die Paneltests im GAST-Projekt — Fig. 2

The insulating wall consists of vacuum-moulded, pre-fired box-like modules filled with fibre products. The maximum permissible operating temperature is 1200 °C.

The insolation output necessary for the metal panel amounts to approx. 1.9 MW and that for the ceramic piping to approx. 2.7 MW.

Messrs. Interatom estimate that the output from the heliostat field of the CESA-1 plant necessary for the metal panel will be reached with approx. 210 heliostats if the radiation density is 760 W/m². The CESA-1 heliostat field contains a total of 300 heliostats.

The following are to be gained with the proposed tests:

- proof that the panel functions under various operating conditions and loads relevant to the operation of the receiver
- experience in design, construction and operational behaviour of the panel

3. DESIGN AND CONSTRUCTION OF THE TEST PANEL

Fig. 3 shows a general drawing of the metal panel. Its main parts are the lateral compensator at the gas inlet, which balances out the overall expansion of the panel and a twist in the inlet collector; the two collectors with the elbows; the heat exchanger pipes on which the rays fall and an angular compensator at the panel outlet.

The planned tests are aimed at achieving component and gas temperatures in the radiated area that are like those expected in the receiver.

Fig. 5 compares the temperature curves along the pipe for the most highly stressed panel in the receiver and the test panel. The degree of similiarity achieved is good. The insolation distribution necessary for this is realised by means of a suitable destination point strategy.

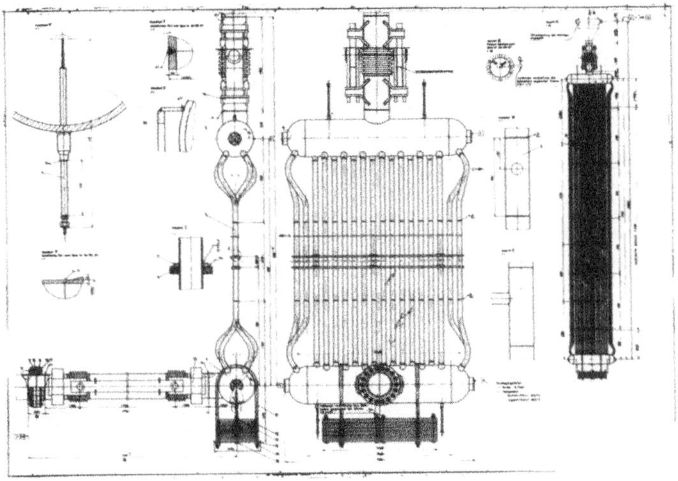

Fig. 3 General drawing of metal panel

The TEPAS programme is used to calculate the temperature fields at the periphery, across the wall thickness and along the pipe. At given irradiation this programme calculates the exchange of radiation between the surface of the pipe and the surface of the back wall as well as the heat transfer in the material and the working medium.

The energy flux is calculated, which is absorbed, emitted and reflected by each surface node in the visible and infrared ranges of the spectrum bearing in mind the view factors in each case. Beyond it the programme considers the convective heat transfer to ambient air in a simplified manner.

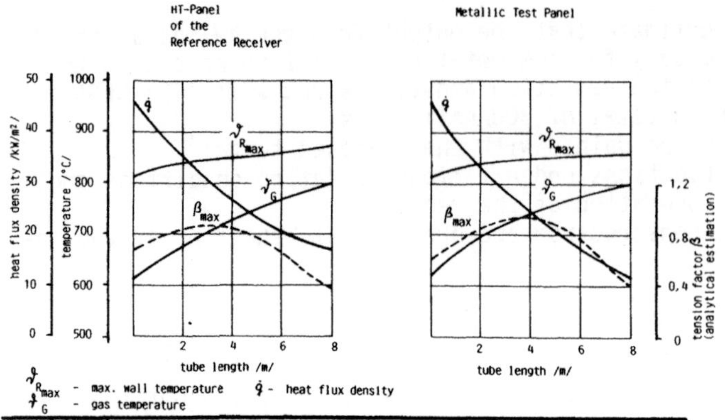

FIG. 4 | Comparison of Steady State Conditions for Max. Loaded HT-Receiverpanel and the Metallic Test Panel
Vergleich der stationären Zustände für das max. belastete HT-Receiverpanel und das metallische Testpanel | Fig. 5

The programme also comprises the conduction of heat in the pipe in radial and tangential directions, in the back wall and, in addition to the convection of heat to the coolant, the exchange of radiation in the pipe between zones of different temperature on the periphery of the pipe.

The temperature fields dependent on time because of the fluctuation in insolation from time to time are determined by the RODYN programme, which basically takes account of the same interactions as TEPAS. To determine the stress which occur in the components the test panel is subjected to a detailed stress analysis which covers the loads from heat stress, internal pressure, own weight and prestressing. The permissible stresses have been determined on the basis of material tests taking into account the damage to the material as a result of creep, fatigue and oxidation.

The stresses thus calculated must not exceed the permissible stresses for the required service life of 10 years.

The state of affairs is represented by a stress factor β, which is defined as the relationship of operational stress to the permissible stresses.

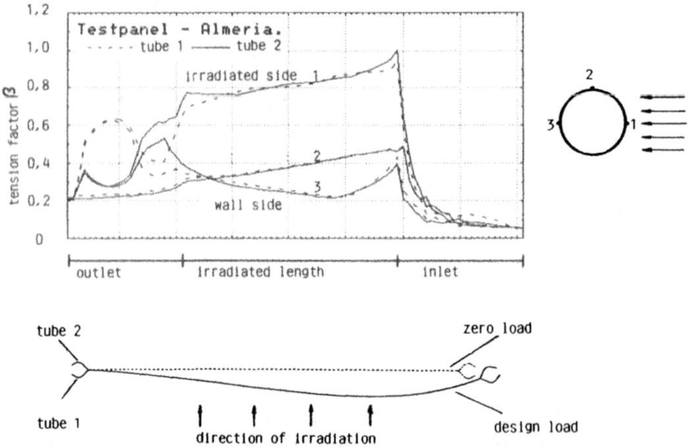

Tensionfactor and Deformation of Metallic Tubes
Spannungsfaktor und Deformation der metallischen Rohre | Fig. 6

Fig. 6 shows the progress of deformation of two heat exchanger pipes held together by pipe racks and the progress of the stress factor along pipe 1 and pipe 2, the bends of which point in opposite directions. Supplementary calculations were carried out for the pipe bends to determine the stress concentration factors; in the result of these the stress factor has also not exceeded value 1.

Another critical component is the outlet collector, which is subject in particular to a quick alternation of load. A calculation for the collector to design codes (ASME, TRD) shows that it satisfies the design requirements for stationary loads. To determine the non-stationary loads tests were performed on a collector test rig. The results of measurements show that the non-stationary thermal loads exceed the purely mechanical loads and therefore they define the strength of the collector.

Construction of the metal panel is complete. Fig. 7 shows the production status achieved at the end of April 1984.

Fig. 7 Construction status of metal panel, April 1984

4. PLANNED EXPERIMENTS AND MEASURING INSTALLATIONS

After the commissioning tests, which will be used to check the functionability of the testing and measuring installations, the real test series under low load will commence. The inlet temperature will be set at 340 °C and the irradiation increased from an initial 50 % of nominal load to 100 % nominal load. This increase in the load will then be repeated with the panel inlet temperatures increased in stages until the nominal outlet temperature of 800 °C is reached.

The objective is to arrive at stationary conditions under the differing conditions of load. At these stationary conditions in particular the comment temperatures are to be compared with the theoretically determined values.

The possibilities of running up certain non-stationary test conditions when the insolation is stationary with the help of the field control are limited. To investigate the non-stationary behaviour, therefore, meteorologically influenced fluctuations in the insolation are used.

To measure the most important test parameters there are various measuring installations. The VISIR measuring system is envisaged for measurement of the surface temperatures and the insolation; this system records from the ground via the mirror telescope the radiation (temperature) emitted in the infrared range of the spectrum and the radiation (insolation) reflected in the visible range. Since the insolation is a key quantity in the theoretical calculations a redundant OSIRIS system from the Spanish company ASINEL which works in a similar way to the flux analyzing system from the Swiss Federal Institute for Reactor Research (EIR) and uses an additional 15 heat flux gauges which can be slewed in at three levels above the height of the panel.

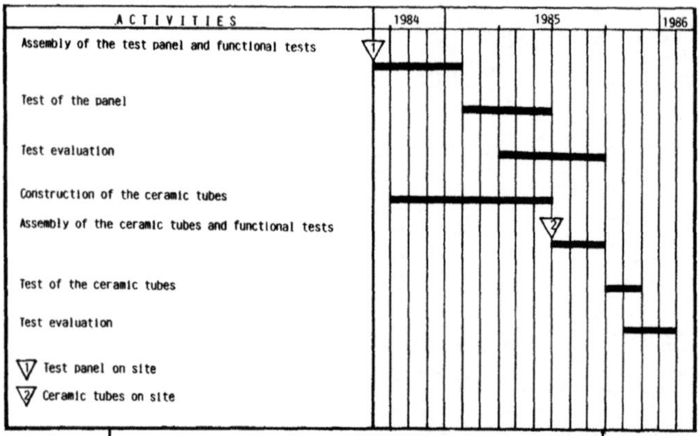

Time Schedule for the GAST-Projekt Paneltest
Zeitplan für die Paneltests im GAST-Projekt

Fig. 8

Special thermocouple probes installed in three pipes in the panel make it possible to measure the pipe wall temperatures at the periphery of the pipe and the gas temperature.

A displacement pick up system is linked to the inlet collector to determine its thermal expansion and the primary creep and twist (bending out of the pipes). In addition the attempt is to be made to record the deformation of the pipe with a video camera set up at the side of the panel and then to evaluate this via the image analysing system of the VISIR measuring system.

5. TIMETABLE

The timetable for performance of the tests is shown in Fig. 8.

6. ACKNOWLEDGEMENT

This work was carried out on behalf of Messrs. Interatom with the support of the BMFT.

The authors would like to thank the Board of Management of Messrs. M.A.N., who made available company funds for this work, and all colleagues who were involved in the project.

CERAMIC AS A MATERIAL FOR LARGE RECEIVER CONSTRUCTIONS

H. GEHRKE
Dornier System GmbH, Postfach, D-799 Friedrichshafen

SUMMARY

Within the Spanish-German "GAST" project - a cooperation sponsored by the German BMFT - ceramic piping for the high temperature part of the GAST receiver is being developed and build and is planned to be tested summer 1985 at the plata forma solar in Spain. The panel embodies all elements typical for a ceramic construction: long ceramic pipes, ceramic-ceramic joints, ceramic to metal connections, expansion compensation and is thus able to demonstrate the practicability of this new technology. The additional questions to be anwered for a proper functioning are discussed as well: material selection, stress analysis by FEM-methods, and life time prediction.

1. INTRODUCTION

The GAST-technology program aims to develop and test key components of a solar tower power plant. This means among others the use of ceramic materials for the development of a high temperature receiver panel with wall temperatures exceeding 1100 ^{o}C.

Compared with similar goals of gas cooled receivers in the US (ceramic dome receiver of MIT/PNL, ceramic matrix receiver of Sanders, volumetric receiver of PNL, and ceramic pipe receiver of B. + V.) the GAST receiver has to match with a gas pressure of appr. 10 bars. The chosen design for the test panel is shown in fig. 1. The panel figures as a representative cutout of the 30 MW_{th} GAST-receiver. The irradiated tube length is 6 m, width is 1.5 m, solar flux density is 200 KW/m^2 peak, the total heat input appr. 350 KW. Aim is to demonstrate the practicability of a ceramic panel technology. Answers have to be found for such important questions like proper material selection, sufficient functioning of ceramic-metal connections, strength and reliability of ceramic-ceramic joints, expansion compensation. These and further design and construction elements are outlined below.

2. Material selection

A ceramic material for high temperature receiver application has to fulfill the following requirements

- sufficient strength up to 1300 ^{o}C
- sufficient thermal shock resistance
- high thermal conductivity
- resistance against oxydation
- availability of semifinished components
- gastightness
- cheapness

Among the various ceramics only SiSiC (silicon infiltred silicon carbide) matches these requirements, nevertheless a big effort had still to be made to reach the goal (improvement of the process, reproducibility of dimensions, grain structure, general characteristics, dimensions), these are described elsewhere [1, 2]. Tab. I shows a summary of the most important characteristic data of the chosen material. The cost are thought to lie in the region of 50 - 100 DM per meter pipe at higher production rates.

3. Connection and joining technique

To realize a heat-exchanger with straight pipes the following techniques have to be mastered: connection of the ceramic heat-exchanger pipes to the metallic periphery on the hot as well as on the cold side, and joining of the ceramic tubes with each other (these are located within the irradiated section, a continious pipe of 6 m length is not possible to manufacture [3].

The transition ceramic-metall is possible by direct brazing in principle, but here with the required dimensions and changing temperatures not workable. For that reason a detachable flange connection was designed. Fig. 2 shows test elements for the cold and hot ends of the pipe. Finding this design one has to be aware of avoiding concentrated loads, realize uniform wall thickness of the flange pieces, admit different thermal expansions characteristics, establish uniform sealing power, limit sealing temperature to below 400 °C. The chosen design realizes these criteria, for this purpose we had to pass to an internal insulated pipe at the hot end of the panel (Fig. 3). The conical transition pieces reduce temperature of 1000 °C inside to values compatible to sealing, guy rope, and spring. Both flange connections have been tested under cyclic load up to 1000 °C and 10 bar in the meantime and thereby show the expected characteristic. We are quite aware that for a commercial plant other more elegant solutions should be found.

The siliconiced pipes are joint together by means of a special brazing technology. The two saw cutted ends of the pipes are provided with a brazing paste and pressed together. In vacuum they are heated just below the melting point of silicon. The result is a component of high strength. They are proof tested with 100 bar, corrosponding to a strength of 36 N/mm^2.

The following problems arose during the development or are under investigation

- high requirements to the quality and condition of the surface
- high sensivity against changes in grain structure especially concerning the free silicon
- during brazing stress is induced which has to be limited by corresponding process control
- the reproducibility of the strength values is unsufficient and needs further development efforts
- the time characteristic of creep strength is still not known

The behaviour under cyclic thermal shock loads is under investigation.

4. Compensation of thermal expansion

In spite of the relatively low coefficient of thermal expansion ($\beta = 4.5 \cdot 10^{-6} K^{-1}$) the tubes while in operation will expand appr. two centimeters. This movement is matched for each pipe individually by flexible hose compensators in two directions. A compensation of the total panel is not possible because the tubes get a different heat flux and thus connected expansion differences. The resulting peak stress loads cannot be reduced by spontaneous flow and creep processes (as in metalls).

5. Stress analysis

Ceramic materials are very brittle by nature and therefore a careful stress analysis is absolutely necessary.

The dominant role is played by the temperature induced stresses. It is therefore necessary to know the exact temperature distribution of the workpiece. This can be calculated for steady-state conditions with sufficient accuracy, for transient phenomenons these calculations have to be verified and supported by adequate tests. The performed FEM-calculations show in the region of the brazing joint maximal axial stresses at the inner tube surface of appr. 20 N/mm^2, this correspondents to a proof pressure of 54 bar. With a chosen proof pressure of 100 bar there are enough reserves for peak stresses and unaccuracies in the stress analysis.

6. Life time prediction

Important for the design of the panel is an estimation of the probable lifetime. For ceramic materials an important factor for this analysis is the so called subcritical crack growth [4, 5]: Very small microcracks within the grain structure may grow under load in the course of the time until they are big enough to cause the abrupt failure of the component. This time dependent behavior was investigated at small samples undergoing standardized bending tests. The results show nearly no subcritical crack growth for the SiSiC-ceramic. This means the material fails at once when loaded or not at all, a time dependent creep effect must not be taken into account.

This fact is used by proof testing by which the tubes are pressurized with a certain load above operational condition. With help of a strength-probability-time (SPT-)diagram Fig. 4 it is now possible to determine the probability of failure at given stress and lifetime. For the GAST-components the probability of failure is planned to be smaller than one promille.

[1] Willmann, G. and Heider, W.: Herstellung von Komponenten aus Siliciumcarbid für das Sonnenturm-Kraftwerk GAST. Sprechsaal 114 (1981) p. 758

[2] Willmann, G.: Keramische Werkstoffe für Solarerhitzer cfi/Ber. DKG 58 (1981) p. 153

[3] Röttenbacher, R. and Willmann, G.: Joining of SiSiC-Components. Science of Ceramics 12 (1983), p. 155

[4] Richter, H., Willmann, G., and Heider, W.: Ermüdungsverhalten von SisiC. Z. Werkstofftechn. 13 (1982) p. 355

[5] Kleer, G. et al.: Bruchmechanische Charakterisierung des Ermüdungsverhalten von SiSiC. Paper presented at the German-French Kolloquim on Special Ceramic Materials; Lyon March 1983.

Table I: Material properties of SiSiC

property		temperature °C	unit	material characteristic SK 5011	SK 308
general properties	density	-	g/cm^3	3,05	3,05
	porosity, open	-	Vol. %	0	0
	gastightness	-	cm^2/sec	10^{-8}	10^{-8}
	Si-content (free)	-	weight	12-15	8-12
mechanical properties	bending strength (4 Point)	RT	N/mm^2	320	300
	"	1200	N/mm^2	350	340
	Weibull-Modul m	-	-	10-12	8-10
	creep rate (130 N/mm^2)	1350	h^{-1}	$3,9x10^{-6}$	-
	critical stress intensity factor K_{IC}	RT	$N/mm^{3/2}$	130	100
	"	1200	$N/mm^{3/2}$	160	160
	crack velocity parameter n	RT	-	220	120
	"	1200	-	115	125
	dynamic E-moduli	RT	kN/mm^2	360	340
	"	1000	kN/mm^2	340	320
	static E-moduli	RT	kN/mm^2	330	310
thermal and electrical properties	thermal expansion coeff.	RT-1000	1/K	$4,5x10^{-6}$	$4,5x10^{-6}$
	therm. conductivity	RT	W/mK	150	180
	"	1200	W/mK	40	45
	R_1-value ($\sigma_2/E\cdot\alpha$)	200	K	150	120
	R_2-value (λR_1)	200	W/m	13500	12000
	Spec. electr. resistance	RT	µmΩ	>2000	>2000
	"	1000	µmΩ	400-600	400-600
	oxydation (100 h)	1300	mg/cm^2	0,2	0,18

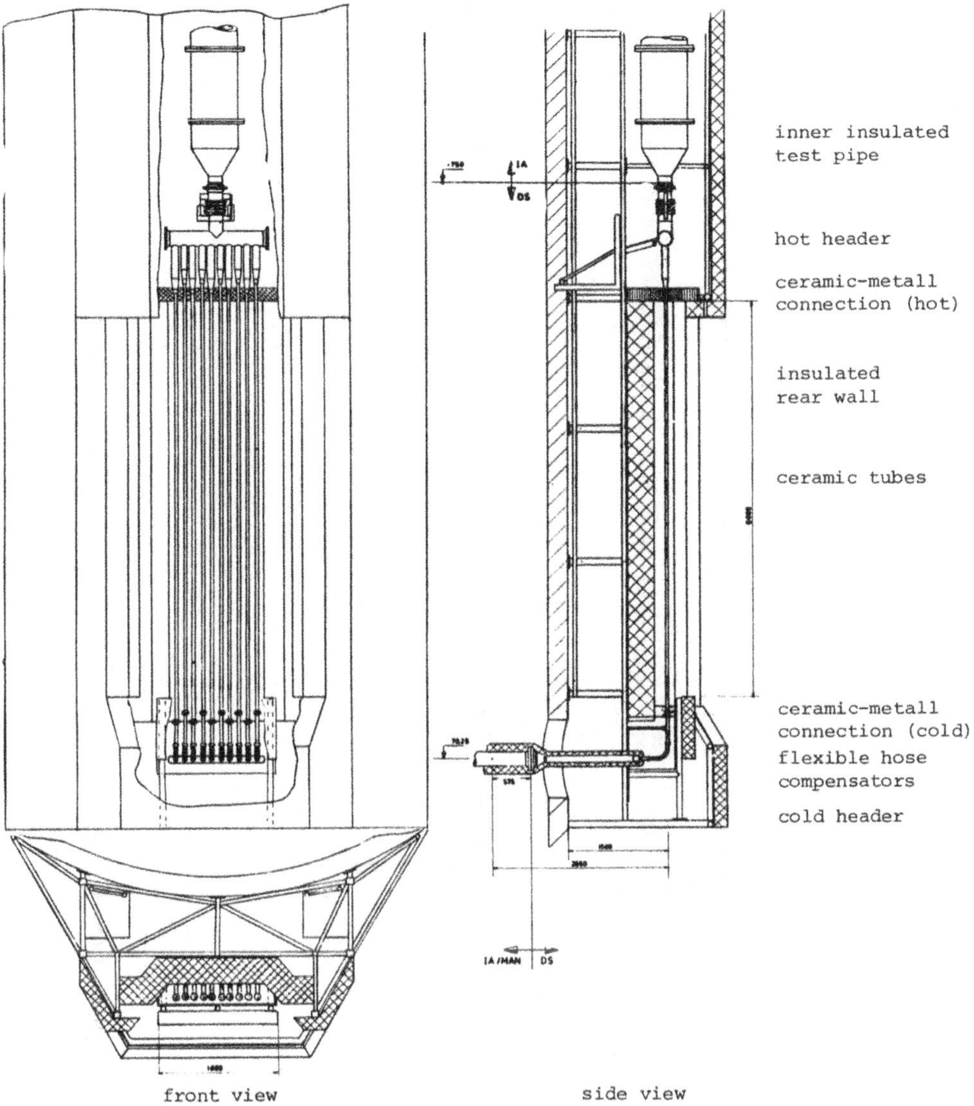

Fig. 1: GAST-ceramic test panel with 10 tubes 42 x 5 mm, each 6.7 m long.

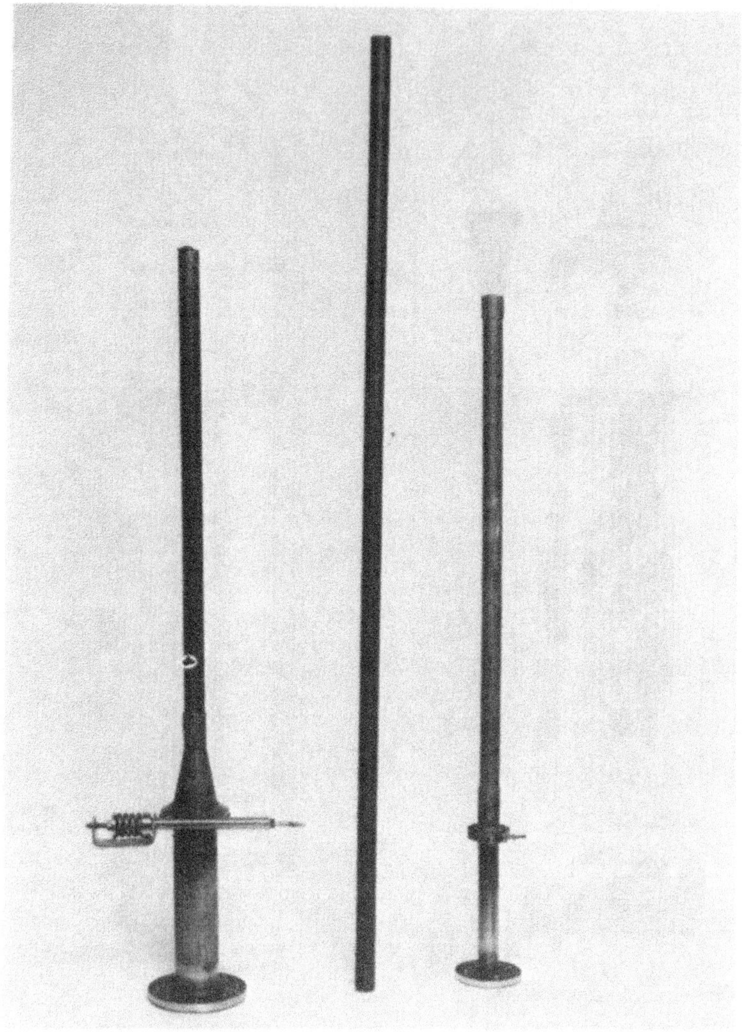

Fig. 2: Panel test elements with hot and cold ceramic-metall connections.

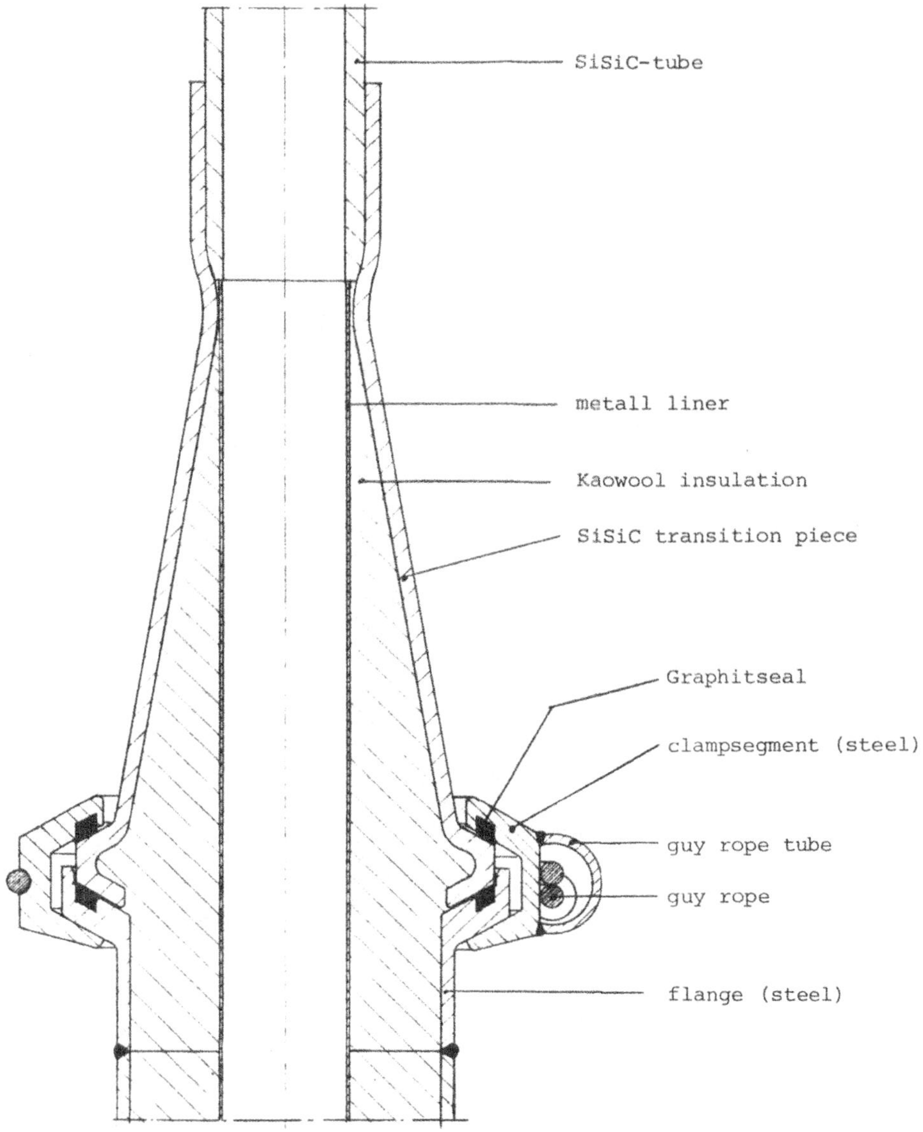

Fig. 3: Internal insulated hot connection between ceramic and metall tube

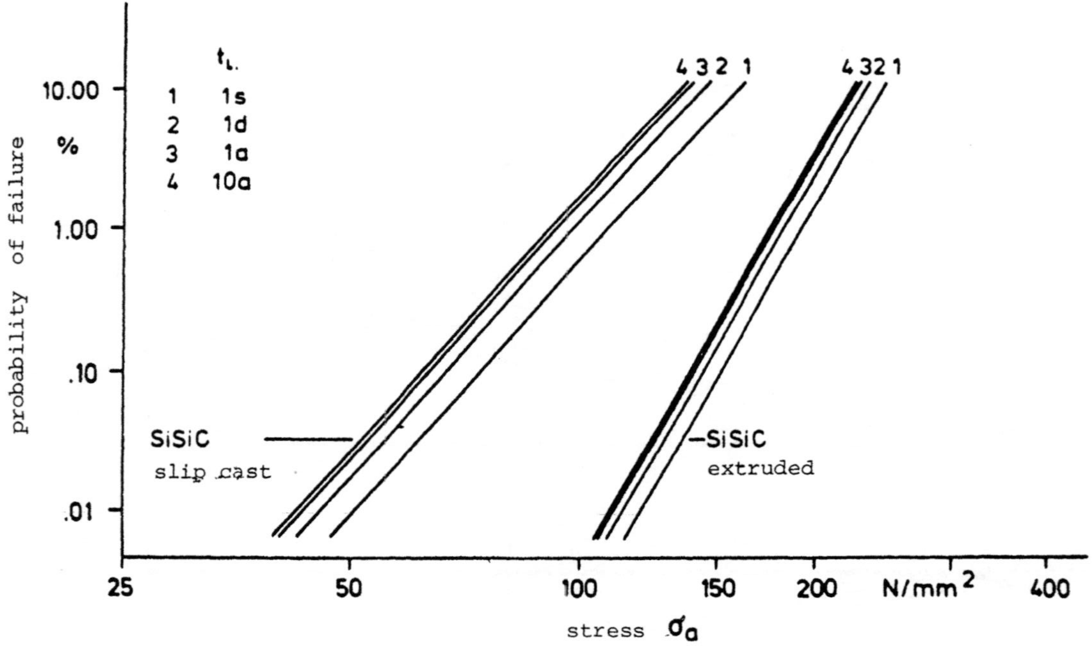

Fig. 4: Strenght-Probability-Time (SPT)-diagramm for SiSiC (room temperature)

SESSION II - PART 3

CYCLE, SYSTEM, CONTROL

Summary of the Session by the Rapporteur
R.L. GERVAIS, McDonnell Douglas, Huntington Beach, USA

Evaluation of sodium and steam system design for the solar power plant with central receiver tower

The advanced sodium receiver (ASR) for the IEA/SSPS central tower plant : operative conditions, control system design and performances

The front end processor of Eurelios supervision systems

The concept of the integrated design of process, control and operation in solar central receiver plants

CESA-1 - Control, system and cycle - Operation and status report

SUMMARY OF THE SESSION BY THE RAPPORTEUR

R.L. GERVAIS
McDonnell Douglas, Huntington Beach, USA

1. INTRODUCTION

The Cycle, System, and Control Session was dominated by control, electronic and data papers. The sophistication of the control systems presented were indication of the complexity of the overall plant systems for which control is required.

In all discussions, the control and data acquisition systems are separate. Most control systems use digital as opposed to analog approaches. Anticipatory receiver control is required by all systems. Water/steam receiver control appears to be more difficult than sodium because of the evaporation zone fluctuations, although this problem was not encountered at Solar One.

The use of integrated design strategy appears to be fundamental. This strategy could lead to the use of a supervisory, automated, control system which is enhanced by the use of a system integration laboratory checkout approach and semi-automated receiver collector field feedback control.

The requirements for in-situ receiver flowmeters were identified as well as the requirements for instrumentation accuracy.

2. PAPERS

The following papers were presented at the session:

1. "Evaluation of Sodium and Steam System Design for the Solar Power Plant with Central Receiver Tower"; H. Yoshikawa and N. Ikeda.
2. "Advance Sodium Receiver for IEA/SSPS: Operating Conditions Control System Design and Performance"; A. de Benedetti and C. Sala.
3. "The Concept of the Integrated Design of Process, Control, and Operation in Solar Central Receiver Plants"; C. Maffezzoni.
4. "The Front End Processor of Eurelios Supervisor Systems"; M. Maini, G. Pandini, and U. Zoni.
5. "CESA-1 Control, System and Cycle"; R. Balanza and J. Ramos.
6. "THEMIS, Cycle and Control"; P. Cachet.

2.1 Yoshikawa and Ikeda

The NIO Plant operating mode was revised from the standard load pattern, rated output of 2 hours in the evening and reduced output in the daytime, to rated output during the day and reduced output (low pressure operation) in the evening for reasons of electrical production optimization.

The control system was revised to meet this new mode of operation. The most important control parameter is now water level control limit affecting receiver signals to the variable feedwater receiver pump regarding flow and pressure control and the condenser. Constant pressure exists between the accumulator and the impulse turbine. The new control system is now automated defined as manual input with manual override.

A feasibility study of a sodium system versus the reference saturated steam, at the NIO plant, was conducted. The sodium system showed an increase in thermal efficiency, a saving in heat loss by compact equipment arrangement at the top of the tower, and a cost savings through weight reduction of equipments and piping. The output power of the sodium system will increase about 80% (1876 kWe) in comparison with the saturated steam NIO plant using a 550°C maximum operating temperature and a dual sodium-water loop system.

2.2 De Benedetti and Sala

Both cavity and external receivers were considered to determine how an average incident flux of 300-500 kWe/m^2 and a peak flux of 1000-1500 kWe/m^2, together with acceptable spillage and distribution, could be obtained on the receiver absorbing surface. Analyses conducted with the Helios Code indicated that these fluxes could not be accommodated with a cavity receiver. A flat vertical target was then considered with a single aim point. Analysis indicated the possibility of matching the reference peak flux but the resulting average flux is too low and the flux profile too peaked. Target tilts toward the field produce only limited improvements. In order to obtain uniform flux distributions, limited ΔT among adjacent tubes, a three point aiming strategy was adopted; back row of heliostats aim to target center, first row heliostats aim to two lateral points whose position depends on heliostat field error. After the decision to go ahead with a field of 93 heliostats and changing the total beam quality and tracking error interval of 2.5 to 3.6 mrad versus the reference 2.6 mrad, a peak flux of 1380 kWe/m^2 and an average of 320 kWe/m^2 were obtained with acceptable distributions. The ASR active surface was optimized at 2.75 m wide and 2.85 m high and was carried out with five tube panels connected serially, each panel consisting of 39 side panel parallel tubes welded to inlet and outlet headers.

Outlet receiver temperature (530°C) is regulated by pump motor voltage (speed) to obtain required flow rate. The intermediate loop, sensitive to the change of the ΔT between panels inlet and outlet, delivers the main control action. A feed forward signal, based on the measure of global radiation incident on the heliostat field is included to speed up the flow rate increase in case of heat flux jumps to maximum starting from low fluid flow rate conditions.

2.3 Maffezzoni

Addressed on the receiver controls for the IEA/Almeria Project and the Eurelios Project. The requirements of the control were discussed including the need for automatic operation, start up and cloud transients and the constraints on receiver temperature leading to the identification of the dynamic properties and the degree of controlability. Using these operational requirements, dynamic analysis is recommended to clarify design parameters and dynamic properties as well as to identify achievable performances and automated control designs using the basic tools of process, modeling and dynamic simulation codes.

The results of this design analysis resulted in a new control concept for the Ameria ASR and partly for the Eurelios receiver. It requires the use of a digital control system based on feedback of intermediate temperature measurements filtered by a suitable predictor of the receiver tube temperature response. The resulting control system is called Predictive Feedback Temperature Controller (PFTC). The concept was successfully applied to the IEA project. In water cooler receivers (Eurelios), the application of the PFTC is not straightforward because of the complex behavior of the evaporation zone.

Application of the PFTC permits operation of solar receivers during cloud passages without need for direct radiation power measurements; it is also used for controlled morning start up. Basic instrumentation required for its implementation include:

* temperature sensors with sufficiently prompt response (seconds)
* fluxmeters for radiation measurements
* digital systems for non-trivial control algorithms
* possible use of enthalpy meters in the evaporation zone of water cooled receivers.

Furthermore, it has been practically proven in both projects that automatic operation is a prerequisite for acceptable receiver performance, particularly during transient operation.

2.4 Maini, Pandini, and Zoni

The design of the Data Acquisition and Supervision System (DASS) for Eurelios includes a front end processor for plant data acquisition and formatting, and a central processing system. The front end processor was developed by ENEL using a LS 11/23 computer (16 bit computer) while the central processing system was developed by the Belgium consortium "TEI-EN" using a SEL 32/27 computer (32 bit computer).

The Front End Processor (PIU) scans, acquires, converts, formats, and displays in real time analog and digital plant values within a basic 1 second cycle time. The input signals coming from the plant are high and/or low level analog voltages and/or currents and free contact closures and/or TTL level digital signals. The input scanner can accommodate 400 analog and 1000 digital signals. The entire PIU software is table driven. The physical link between the PIU and DASS is 16 bits parallel, high speed, optical decoupled and based on DMA channels available on both computers. The logical links between the PIU and the DASS provide a bi-directinal dialogue.

The DASS unit which only accepts digital signals provides:

* a plant operator console (2 displays, 1 keyboard)
* an experimentors console (2 displays, 1 keyboard)
* a programmers console
* a real time data base
* a set of data processing functions, for example, alarm management or menus for plant supervision displays.

This solution for this kind of non-conventional plant appears suitable because the PIU effects all plant changes and allows the DASS to run on a more general software package.

2.5 Balanza and Ramos

The subsystem and associated controls status and operation are presented. The power conversion system has been synchronized for 65 hours, experienced 964 hours of operation and produced 30400 kWe-hr gross. Problems have been encountered in the turbine electro-hydraulic control system; the thrust bearings, the turbine casing and the turbine exhaust have been modified.

Regarding operating modes, the following modes work satisfactorily: receiver to turbine, receiver to thermal storage charging, and receiver to turbine and thermal storage charging. Thermal storage discharging to turbine is in the startup phase and the remainder of operating modes have yet to be initiated.

The thermal storage system, while operative, has experienced some problems; that is, thermal inertia has made it difficult to reach hot tank nominal temperature, salt freezing and pipe blockage due to drainage problems and heat tracing failures and fluid leakages in heat exchanger headers and salt crack pipes.

The control system for the cycle can be characterized as a conventional analog electronic system with man in the loop and a programmable interlocking system that gives flexibility for design modifications. Furthermore, a large percentage of instrumentation is dedicated to DAS. Regarding the receiver control system, the feedwater control has not experienced any relevant problems. The temperature control has experienced difficulty in the automatic operating mode due to superheater sensitivity to heat flux variations and metal temperatures close to trip limits. The start up control loop has been limited to date to manual operation in an effort to reduce start up time. Experience with the thermal storage control system has been limited to date. Some problems have been encountered with salt freezing in control valves, instrumentation installation and heat tracing of process tubing.

3.6 Cachet

A description of the GILOTHERM cycle was presented. GILOTHERM (oil) is the working fluid of the parabolic dishes. The energy of the GILOTHERM is then transferred to water/steam (250°C) where it is stored in an 80 m^3 tank, using the thermocline principle, and then used for turbine feedwater heating, plant trace heating and initial salt heating.

The THEMIS control system was described and consists of a redundant host computer based, digital/microprocessor based, three bus systems. Remote stations situated throughout the plant are utilized where the digital commands are manipulated and converted to analog signals. The system also uses the concept of supervisory control wherein overall plant control is supervised by a central host computer.

EVALUATION OF SODIUM AND STEAM SYSTEM DESIGN FOR THE SOLAR POWER PLANT WITH CENTRAL RECEIVER TOWER

H.YOSHIKAWA and N.IKEDA*
Tokai University and New Energy Development Organization*

Summary

An evaluation between the sodium and steam system for solar power plant with central receiver tower is reported in order to increase a thermal power efficiency and to decrease a construction cost. These system designs involve three kinds of design output temperature of sodium loop in receiver including the sodium safety devices and the steam-water design for one MWe solar power plants. Layout designs of main components which are based on above mentioned design system are performed. For the evaluation of four layout designs with sodium loops one MWe plants, it is considered worthy to reduce the weight of hardwares and the construction cost of plant of which the components arrenge on the top area of the receiver tower and/or the ground level. The technical evaluations are performed by following items: (a) Thermal loss from receiver and sodium loop, (b) Method of construction and installation, (c) Compact layout of the plant, (d) Maintenance and operation, (e) Total technical evaluation. From the result of the evaluation, solar power system with sodium loop is prospective for many technical advantages to result in higher output power efficiency and lower relative costs.

1. INTRODUCTION

Solar thermal power plants with central receiver tower are evaluated at prsent as a most excellent type. A prototype and/or pilot plant of solar power plants are constructed and operated in several countries. However, it will be necessary for a commercial solar thermal power plant to reduce an electricity cost as low as practicable to that of a fossil power plant and a nuclear power plant. The advance technologies shuold be developed in the field of construction cost and the thermal efficiency.

The conceptual design of solar thermal power system expected the following technical advantages are presented in this paper;

(a) increment of thermal efficiency by high temperature sodium system,
(b) saving the heat loss from equipments and thier pipeing surfaces, by compact arrangement of them at the top area of the receiver tower,
(c) cost saving through the weight reduction of equipments and piping according to application of low pressure system.

It is expected from the results of this conceptual design that the output power will increase about 80 % in comparison with the reference plant at Nio. The design procedures of this conceptual design are shown in Fig.1.

2. ADVANTAGES OF SODIUM LOOP

2.1 Features of sodium loop

The system is able to operated higher temperature without any pressurization, because boiling point of sodium is 881°C at atomosphere pressure as described in Tab.2.1. For the volumetric expansion of sodium at high temperature, argon and nitrogen gas are filled in the buffer volume which is located at the top portion of receiver in order to keep a margin of boiling temperature by slightly pressurization. The sodium technology in this conceptual design are based on the domestic technology which are developed by Japanese Fast Breeder Reactor of " Monjyu" project.

2.2 Start up and shutdown

Sodium has the paticular characteristics which are a solid at below 98°C and are very active chemically under the contact between water steam and sodium. Therefore, the loop has to keep vacuum,dry and to keep above 200°C by heat tracing and/or insulation,when sodium is feeded to primary loop. In case of cooling down below about 100°C, sodium in the loop has to be dumped to the sodium tank installed a heater to avoid to be a solid immediatly by operating electromagnetic valve. Then,the sodium loop has to be filled with argon gases as soon as possible. When turbine system would be the shutdown and/or load reduction,the heat accumlators have to maintain the flexibility of operation.

2.3 Sodium system safety Feature

Fig.2.1 presents the emergency system in an accident of reaction between sodium and the steam generator and/or superheater. A hydrogen gas generated by its reaction is pressurized before the ruptured disk and introduced to the hydrogen combustion chamber after the ruptured disk breakes out by the accumulated hydrogen pressure. At the hydrogen combustion chamber, the hydrogen gas are burnt out prior to release the atmosphere. The equipment and its piping which are involved the liquid sodium,are arrenged inside, the reaction between air and system are caused to fire.

3. SYSTEM DESIGN

1000 kWe central receiver solar thermal power plant at Nio is selected as a reference plant for this conceptual design study. Special consideration of this design are as follows:(Tab.3.1)

(a) the output heat from receiver is 2×10^6 kcal/h,
(b) the electric power output is calculated from the heat balance diagram of turbine plant,changing the average temperature of sodium loop as parameters,
(c) Maximum operating temperature of this sodium system is selected the experimental information of the FBR project "Monjyu".

A few system flow sheets under the design study are shown in Fig.3.1 -3.4.

4. LAYOUT DESIGN

4.1 Layout design criteria

(a) the equipments and piping involved sodium are arrenged in the compartment which are filled with nitrogen.
(b) the sodium loop piping lengthes are minimized in order to decrease the heat loss from their surfaces,
(c) the electromagnetic punp for sodium loop is only replaceable,
(d) the turbine overhaul is applicable on the top portion of the tower by using a movable crane,
(e) the life of the static equipments are designed for the same life of the plant such as crane, superheater, steam generator and tanks,
(f) the sodium feed operation has to be done on the ground,
(g) the emergency system and components are located at the top of the tower.

4.2 Equipment design basis

The system equipments have been designed by the cosideration with the reduction of allowable stresses and the creep of materials at the high design temperature. The typical sample of the steam generator are shown in Fig. 4.1. The material for sodium loop equipment has also selected low carbon austentic stainless steel (304L). The weight and size of the equipment become decreased according to good heat transfer on the surface at high temperature.

4.3 Layout design basis

The equipment and main piping of the primary and secondary loop are arranged just below the receiver floor compactly which are located at the top portion of the tower about 70 m hight. Fig. 4.2 illustrates an typical arrangement of equipments of system.

4.4 Evaluation of the layout design

The technical evaluations according to the following items are shown in Tab.4.1;

(a) thermal loss from the receiver and sodium loop,
(b) method of construction and installation of the equipments,
(c) compact layout,
(d) maintenance and operation.

The thermal losses have been evaluated by the calculated total surfaces of the equipments and piping. With the aspect of construction, designs are evaluated from the calculated total weight of the equipments and piping including insulation and heat tracing.

The list of the main components and piping weight which are related directly sodium loop, are presented in Tab.4.1. It is found that case III is the most light weight design in these conceptual systems. The compact layout has been evaluated by the point of the necessary space for installation and maintenance space area. The maintenance has been evaluated by replace only in the event failure of dynamic components such as sodium pump.

The results of the technical evaluation by 5 points about each items are presented in Tab.4.2. It is found from these evaluations that case D may be excellent layout shown in Fig.4.3.

5. ACKNOWLEDGEMENTS

The authors would like to express their gratitude for this helpful cooperation and also for the contribution of Mr.Y.Ohishi and Mr.H.Katano who are the students of TOHKAI University.

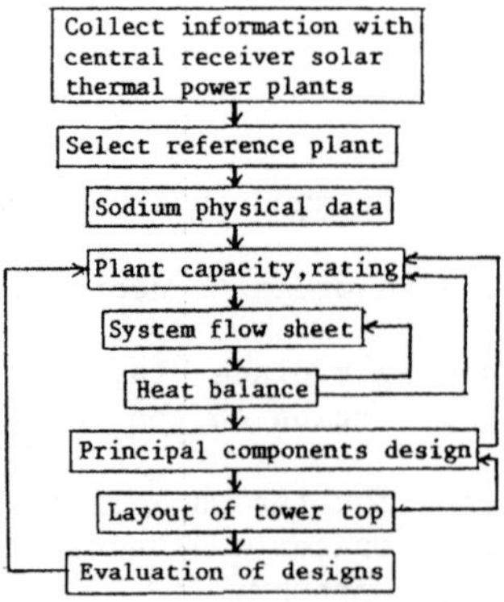

Fig.1.1 Procedure of the conceptual designs

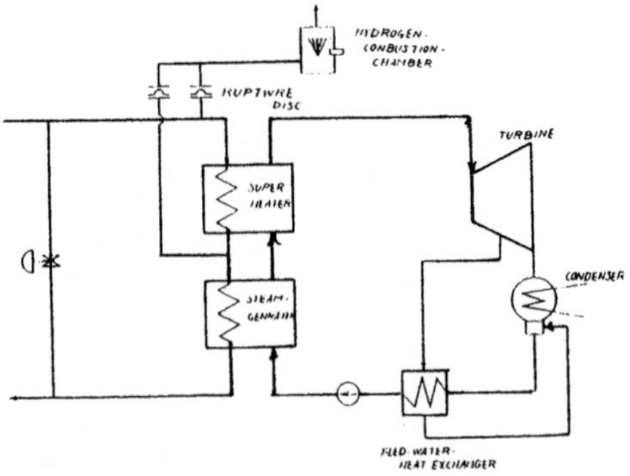

Fig.2.1 Safety System Diagram of Na System Reaction

Table 2·1 Principal parameter of Na and Water

		Na (300°C) 1 kgf/cm²	Water (300°C) 87·6 kgf/cm²
Boiling point	°C	881	300
Density (at 300°C)	kgf/m³	800	713
Specific heat	kcal/kg°C	0·316	1·36
Viscosity (at 300°C)	kg/m·s	$3·43 \times 10^{-4}$	$0·92 \times 10^{-4}$
Dynamic viscosity (at 300°C)	m²/s	$0·392 \times 10^{-4}$	$0·129 \times 10^{-4}$
Heat conductivity	kcal/mh°C	66·0	0·462

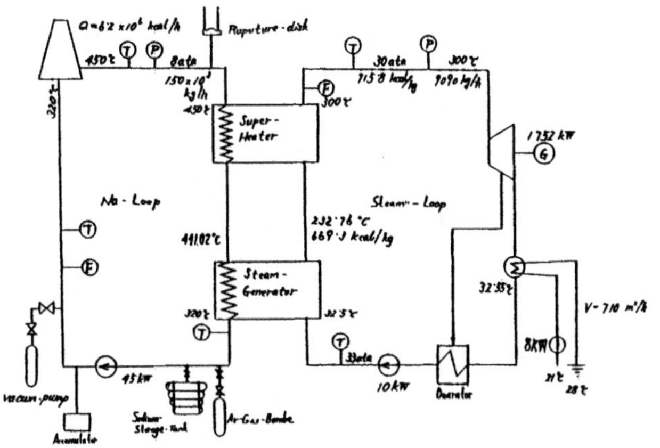

Fig.3.1 Design system : Case I.

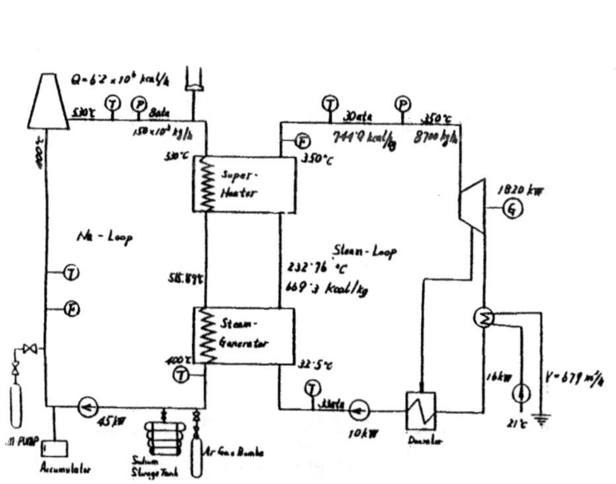

Fig.3.2 Design system : Case II.

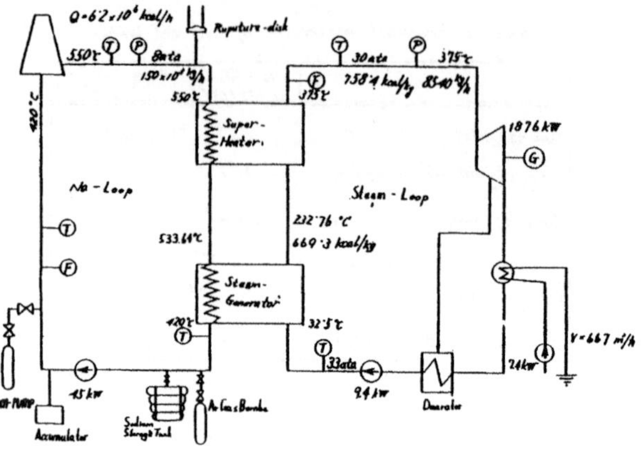

Fig.3.3 Design system : Case III.

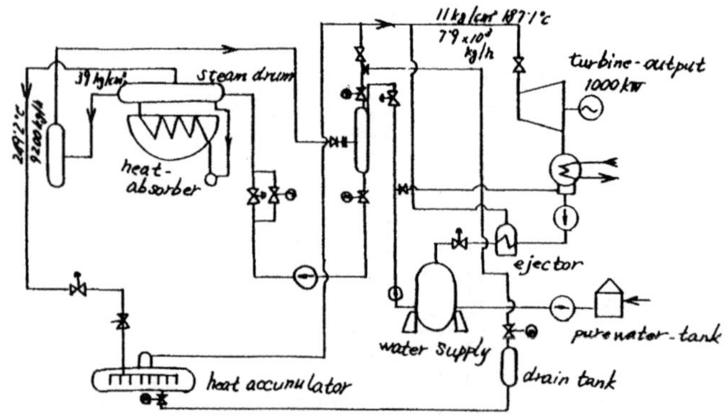

Fig.3.4 Reference system (Nio plant)

Tab.3.1 System parameters

		Examined systems CASE I	CASE II	CASA III	Nio plant (Japan)	HITEC LOOP (TEMIS - France)
Input Collector Capacity Kcal/h		6.2 × 10⁶				
Gather Heat Apparatus		Tower Swing Frame Outside Accept Light Forced Circulation System			Tower Swing Frame Internal Accept Light Forced Circulation System	[illegible]
Working Fluid	Primary	Na			Pure Water	HITEC (KNO₃ 53% NaNO₂ 7% NaNO₃ 40%)
	Secondary	Steam - Water				
Temperature (in, out) ℃	Heat Collector	in 320 out 450	in 400 out 530	in 420 out 550	in 187.1 out 249.2	in 250 out 450
	Steam Generator	in 32.5 out 232.76	in 32.5 out 232.76	in 32.5 out 232.76		out 430
	Super Heater	in 232.76 out 300	in 232.76 out 350	in 232.76 out 375		
Steam Generating System		Tower, Steam Generator, Super Heater			Tower	
Max. Continuous Evaporation kg/h		9090	8700	8540	9200	
Na - Flow Rate kg/h		150 × 10³				
Heat Collector Exit Pressure kg/cm²		8			39	
Steam Turbine	Type	Impulse One Culinder Single Flow Condencing Turbine				
	Rating Output kW	1752	1820	1876	1000	2200
	Rating Steam Pressure kg/cm²	30			11	41.32
	Rating Steam Temperature ℃	300	350	350		
	Rating Steam Flow Rate kg/h	8180	7830	7686		
	Exhaust Pressure kg/cm²	0.05			0.082	0.1
Solar Receiver	Heliostats Number	807				200
	Plane Mirror Dimension	4m × 4m (1m × 1m × 16)				
	Mirror Total Area	12912				10740
	Plane Mirror Architecture	White Board Glass t = 5 mm				Flat Glass [illegible]
	Plane Mirror Reflection Factor	0.88				0.9
	Solar Tracking System	Calculating Machine Control Tracking				

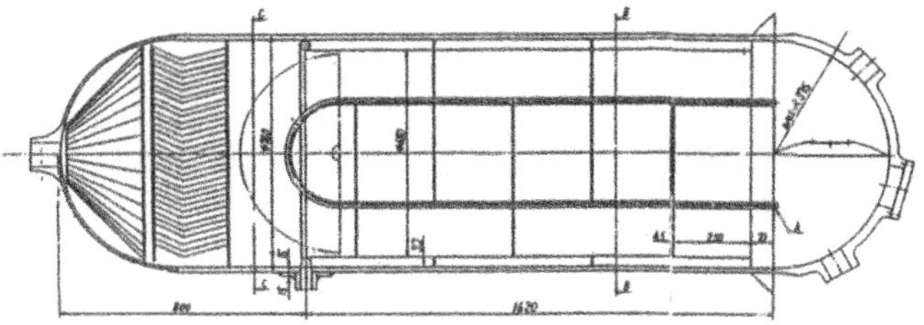

Fig.4.1 Steam generator

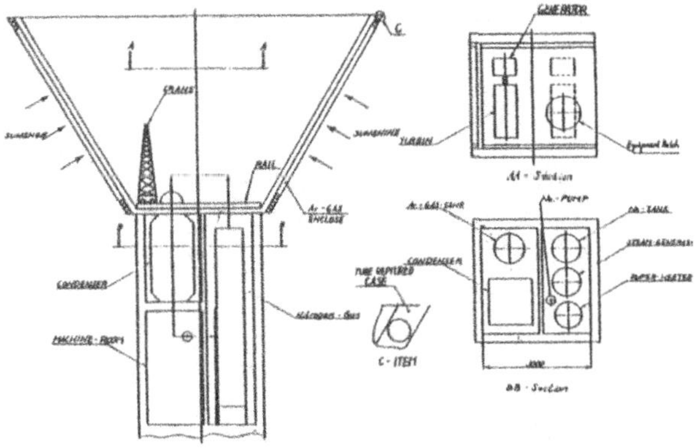

Fig.4.2 An typical arrengement at the top area

Tab.4.1 Weights of various systems

(Kg)

	S/G	Heater	Piping	Na-Tank	Total
CASE I	1960	1650	300	80	3990
CASE II	1515	2000	330	80	3925
CASE III	1260	2211	360	80	3911
Reference Plant	11000	—	2300	—	13300

Tab.4.2 Evaluation of system layouts

	Thermal Loss	Na Effectiveness	Compact	Maintenance	Handling	Building	Total
CASE A	1	2	1	4	3	4	15
CASE B	5	3	4	3	5	3	23
CASE C	5	3	5	3	5	3	24
CASE D	4	3	5	3	4	3	22

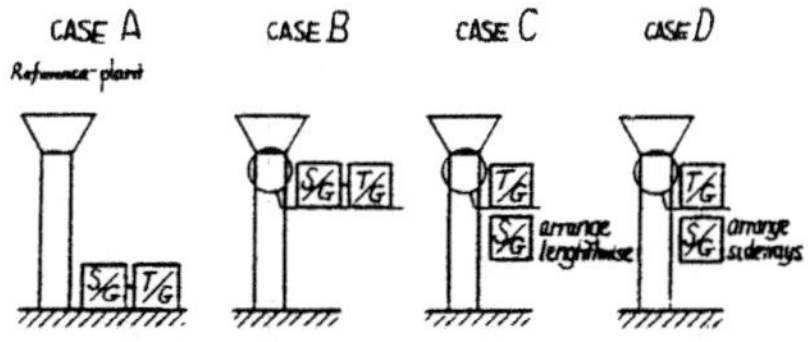

Fig.4.3 Various layouts of systems

THE ADVANCED SODIUM RECEIVER (ASR) FOR THE IEA/SSPS CENTRAL TOWER PLANT: OPERATIVE CONDITIONS, CONTROL SYSTEM DESIGN AND PERFORMANCES

A. DE BENEDETTI and C. SALA
AGIP S.p.A. MILANO

Summary

The high incident heat flux intensity characterizing the ASR design, joint with time dependency and fast transients, claims for a detailed and refined approach of the thermal flux distribution on the receiver. The basic input both to the mechanical and the control system design derives from optical analysis and thermo-fluido dynamics computations. Relevant and peculiar aspects of the design steps are described and presented. Computed results are then compared with preliminar experimental values collected during the functional tests in start-up phase of ASR, which is now operating in the SSPS-CRS plant, confirming in daily routine both the design basic assumptions and the expected perfomances.

1. INTRODUCTION

The paper hereafter presented deals with the Second Sodium Receiver, contractually named the Advanced Sodium Receiver (ASR), which has been routine operated since December 83 at Almeria (SPAIN), where two different solar plants (500 kWe each), outcome of an International effort managed by the German DFVLR as Operating Agent, are runned by "Compañia Sevilliana de Electricitad" and evaluated by an International team.

The testing and experimentation of a high performance and refined design receiver is contributed in kind by AGIP-S.p.A. and F. TOSI Ind.; ENEL (Italian Electricity Board) contributed the receiver dynamical model and the control system synthesis; CNR (National Scientific Council) is a partner in the Italian Contracting Parties.

Referring to the project technical objectives and goals, ASR main design task is to operate at high incident flux in order to optimize the collection of energy and its conversion, getting benefit by the good thermal characteristics of sodium as low operating pressure cooling medium. Larger size central receiver plants can draw advantage from these design aspects that, making possible the use of light and compact heat transfer surfaces, not only semplify the tower structure but also improve performances and daily operation. It stands to reason that the severe impinging flux makes the receiver highly stressed and claims for accurate computations and analysis in design, for ingenuity in constructive provisions and for a careful and precise work in manufacturing.

2. INCIDENT FLUX, TARGET POSITION AND HELIOSTATS AIMING

Both cavity and external receivers were considered in order to investigate how an average flux in the range 300/500 kW/sqm and a peak flux of 1000/1500 kW/sqm, together with acceptable spillage and distribution, could be obtained on the absorbing surfaces. A parametric stress analysis previously carried out in fact demonstrates that these flux levels are acceptable on receiver tubes, provided that a suitable selection of the design is made. Computations performed with Helios code (1) point out the impossibility to meet the above mentioned figures with a cavity shaped receiver; for instance on a cylindrical vertical axis surface, 1.5 m radius, 186 (290) kW/sqm average, 740 (1140) kW/sqm peak are obtained (in parenthesis are indicated the results obtained with the 160 heliostats field, result of Project Phase N° 1). To fulfil the abovementioned flux figures, a flat vertical target was then considered with a single aiming technique. Computations at different distances from the aiming plane point out the possibility to match the reference peak flux (Fig. 1 shows the flux peak behaviour vs distance D in case of 160 heliostats), but the resulting average value is low and the flux profile too peaked. Target tilts toward the field produce only limited improvements.

In order to obtain the flux distributions as uniform as possible, to limit temperature delta among adiacent tubes, a multiple points aiming strategy has to be considered. Three aiming points are a satisfactory compromise between the above mentioned requirement and semplicity. Back row heliostats aim to the target center, first row heliostats aim to two lateral points (Fig. 2), whose position depends on heliostat field error. After the decision to go ahead with a field reduced to 93 heliostats and the change of the reference total error (beam quality and tracking) from 2.6 mrad (design value) to 3.1 mrad (in field value), the optical analysis has been finalized and updated, in order to proceed with the ASR detailed design. As a consequence of the reduced heliostat precision, the two lateral aiming points are drawn near to the target center (10 cm. both in X and Z direction) obtaining 1380 kW/sqm peak and 320 kW/sqm average. An additionally performed sensivity analysis checks the heliostats error impact on the computed flux distributions and shows that aiming points can be arranged in the interval 2.5 - 3.6 mrad, always obtaining acceptable distributions. Compromising spillage with losses and considering receiver edges thermal conditions, ASR active surface was optimized (2.75 m. wide, 2.85 m. high) and carried out with five tube panels series connected, each panel consisting of 39 side by side parallel tubes welded to inlet and outled headers (2, 3, 4).

Despite the good uniformity obtained on the target, the absorber tubes are subject to high flux gradients which have been computed in detail sufficient for stress analysis.Figures from n. 3 to n. 8 present the incident peak flux, the incident power and the spillage behaviour vs time (in 3 typical days) and day and settle the basic input data both for the receiver design and for the performances estimation. The simmetry of the incident flux distribution at different instants and in case of disturbances (clouds) is also achieved with this aiming solution.

Thermal fluxes on the 45° angled refractory border, on the receiver frontal casing and on the door frame (0.3 and 0.6 m. before the tubes plane respectively) result (3.5 mrad figures in parenthesis)

- 21/6, noon : 71, 45, 15 kW/sqm
- 21/6, 6.15 : 46 (78), 47, 20 (21) kW/sqm

As a consequence of the tubes arrangement (side by side without welding), a detailed evaluation of the thermal flux passing through the gaps between the tubes and impinging on the backwall was performed too.

3. INFLUENCE OF FLUX DISTRIBUTIONS ON CONTROL DESIGN

As ASR operation and its control present same peculiar aspects related to:

- uncontrollability of the thermal power entering the coolant, which is subjects to sudden and large variations
- uncontrollability of the thermal distribution impinging on the receiver
- automatic operation in an unusually wide load range (10% - 100%).
- high operating temperature, low thermal capacity of the active parts and high impinging flux
- relevant time lags, due to the long connections between the panels.
- unavailability of a direct estimation of the flux impimping on the receiver

the process modelling was very accurate and inclusive of all the relevant dynamic phenomena. (5, 6, 7). To enable such a modelling, situations affecting the design (fast transients, asymmetries, emergencies,...) have been carefully identified and analyzed.

Fig. 9 shows the block diagram of the three cascaded loops temperature control system selected among the different possible control configurations analyzed. (8)

Sodium temperature at the receiver outlet (controlled variable with a set point value of 530 °C) is regulated by the action on the pump motor voltage (speed) to obtain the required sodium flow rate. The intermediate loop, sensitive to the change of the temperature differences between panels inlet and outlet, delivers the main control action. In addition to the fast response, this loop is not affected by flux map shape variations, as the sum of all the temperature delta is considered. A feedforward signal, based on the measure of the global radiation incident on the heliostat field, is included as necessary to speed up the flow rate increase in case of heat flux fast jumps to maximum starting from low fluid flow rate conditions. In this case in fact, in spite of the metal temperatures sudden increase, the process is very slow and the temperature variation at the outlet of the panels much delayed.

4. START-UP PRELIMINARY RESULTS AND DESIGN DATA COMPARISON

During start-up and functional tests period, receiver operation has been checked and monitored in order to detect malfunctionning or critical conditions.

Some figures, strictly connected with the receiver safety and integrity (control behaviour during transients, power sharing among the panels, panel temperature distribution across the tubes at the outlet, edges temperatures) have been observed with particular care.

Other relevant characteristics (temperature axial profiles in the tubes, efficiency, thermal inertia) have been preliminary compared with design data.

At the beginning power distribution among the different panels (Fig. 10) showed changes in time larger than expected however remaining within the design limits (the five panels are independent, but a maximum power on the same panel of about 800 kW - 32% of the total corrisponding to 70 °C of temperature delta at the design flow rate, - is accepted). Also temperature difference among the tubes pointed out this sensible and unexpected time dependency, reaching (early in the morning and late in the evening) the 40 °C (design value) in the lateral panels, while it keeped lower than 10 °C in the central panel. Some time later the Plant Operating Authority noticed an error in the site latitude managed by the field controller and modified the figure. Since the right latitude has been introduced, the heliostats image results to be stable and only expected shape changes verified. Temperatures at the receiver borders also result more balanced reaching a maximum of about 275°C.

With regard to the ASR dynamical behaviour and control system, during functional tests clouds have been simulated defocusing and focusing some rows of the field (from 25% to 100%), as the clouds passage on the heliostats field is expected to be the most critical and frequent disturbance in receiver operation.The responce of the control (without the action of the feedforward signal and with the motor velocity controller slow and not yet tuned) is summarized in Table I in case of 25% and 50% disturbances.

Preliminary estimations based on nearly steady-state conditions show a good agreement both with the computed temperature profiles in the tubes and with the expected efficiency (88% at design point conditions).

5. CONCLUSION

The objective to manufacture a solar receiver able to operate at elevated temperature with high thermal flux impinging on and to follow sun radiation fluctuations can be achieved only with a really coordinated and specially designed plan. The receiver operation in fact takes advantage of the low thermal inertia in the active parts (obtained with a refined mechanical design) only in case of a perfectly tuned and correctly synthetized control system. And in its turn a detailed optical/thermal analysis can be justified only if the final construction (receiver and controller) is able to take care and include the reasonable information coming from.The results from SSPS/CRS Plant - Plataforma Solar - Almeria, seem to show that this design integration has been achieved.

6. ACKNOWLEDGMENTS

The authors would like to express their gratitude to Mr. W. Grasse (DFVLR - SSPS Operating Agent), Prof. F. Reale (CNR) and Mr. C. Micheli, who always supported the ASR activity, and to the Plant Operator (Cia Sevillana de Electricitad) for the valuable support during ASR installation and start-up.

REFERENCES

1. F. BIGGS, C. VITTITOE (1979). The Helios model for the optical behaviour of reflecting solar concentrators. SAND76-0347.
2. A. DE BENEDETTI, V. BEDOGNI et alii (1983)
   The Almeria Advanced Sodium Receiver (ASR): Basic Design and Operating Condition. ISES Solar World Congress - Perth, 1983
3. V. BEDOGNI, A. DE BENEDETTI (1983)
   Almeria Advanced Sodium Receiver: Stress Analysis Considerations. ISES Solar World Congress - Perth 1983
4. V. BEDOGNI, P. CAVALLERI, A. DI MEGLIO (1984)
   Relevant aspects in the design and construction of the Advanced Sodium Receiver (ASR) for the IEA/SSPS central receiver system plant. Second International Workshop on the Design, Construction and Operation of Solar Central Receiver Projects - Varese (Italy)
5. C. MAFFEZZONI, G.A. MAGNANI, S. QUATELA (1982)
   Integrated Process and Control Design of high temperature solar receivers. 21st IEE-CDC Conference - Dec. 1982 - Orlando (Florida)
6. C. MAFFEZZONI, G.A. MAGNANI, S. QUATELA (1982)
   Process and Control Design of high temperature solar receiver: integrated approach. IEEE Transaction on Automatic Control (in course of pubblication)
7. C. MAFFEZZONI (1984)
   The concept of the integrated design of Process Control and Operation in Solar Central Receiver Plants. Second International Workshop on the Design, Construction and Operation of Solar Central Receiver Projects - Varese (Italy)
8. G.A. MAGNANI, A. QUATELA, A. DE BENEDETTI (1983) Dynamic Analysis and Control of the Almeria Advanced Sodium Receiver (ASR). ISES Solar World Congress - Perth 1983

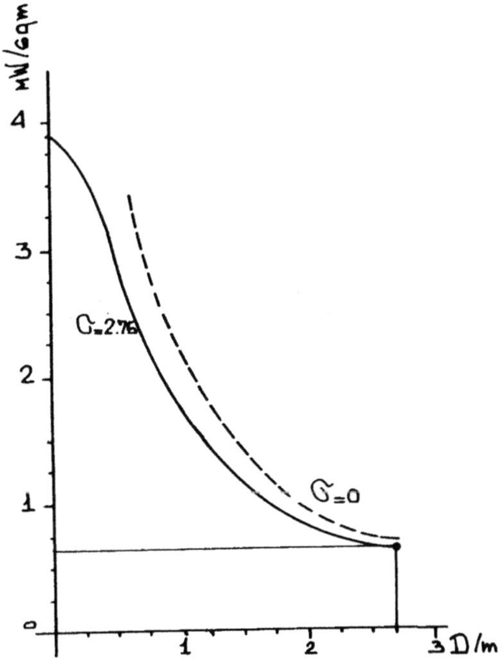

FIG.1 INCIDENT PEAK FLUX

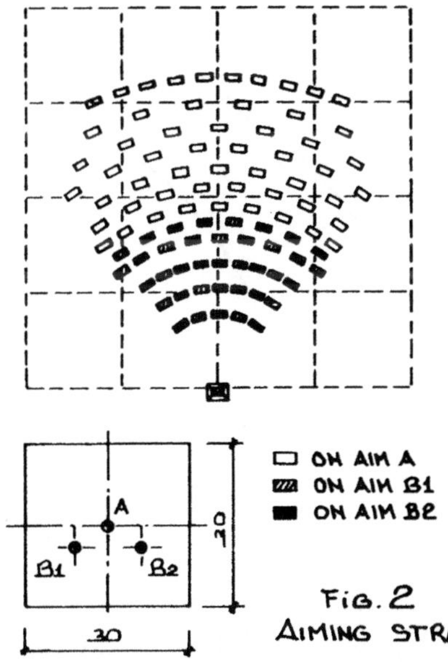

FIG. 2
AIMING STRATEGY

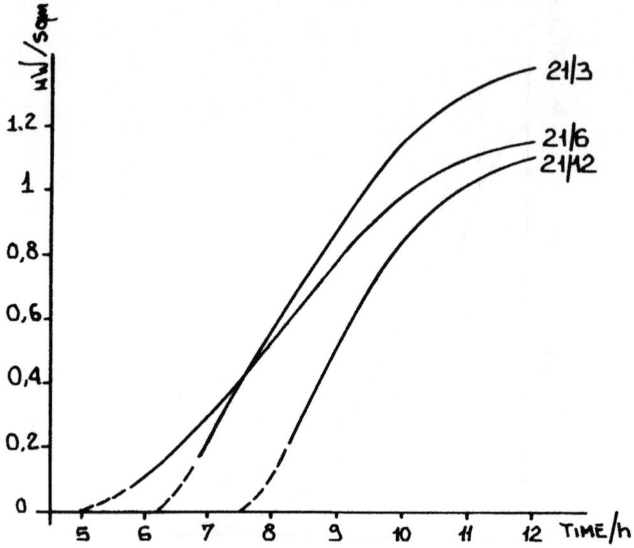

FiG. 3 INCIDENT PEAK FLUX

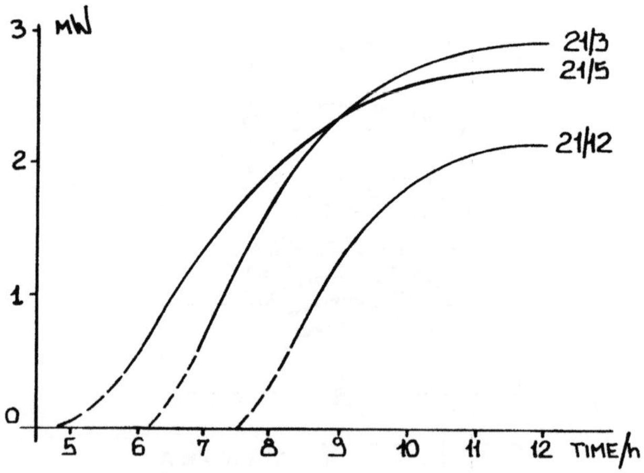

FIG. 4 INCIDENT POWER

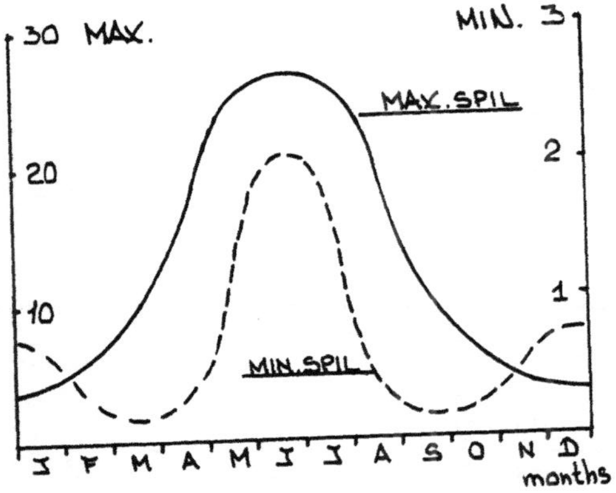

FIG. 5 SPILLAGE %

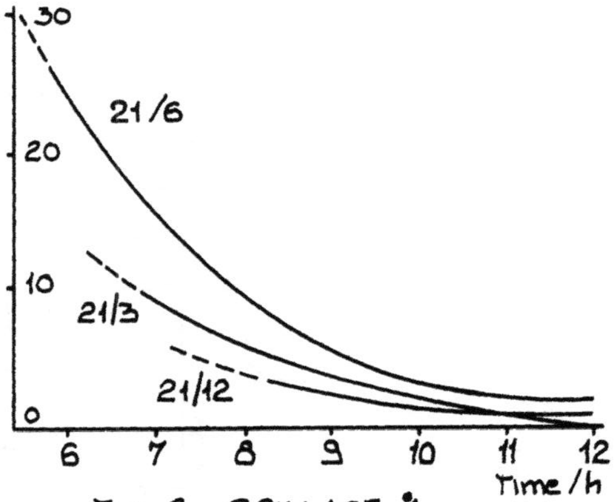

FIG. 6 _ SPILLAGE %.

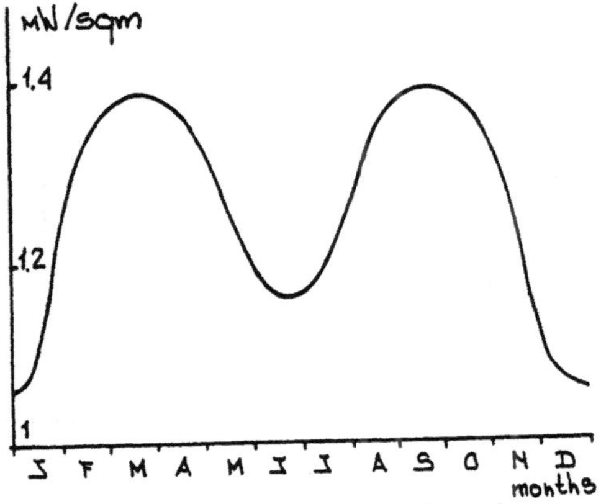

FIG.7 INCIDENT FLUX (MAX)

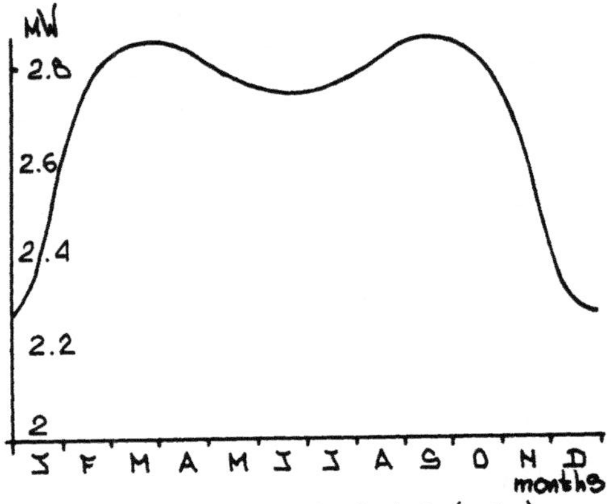

FIG.8 INCIDENT POWER (MAX)

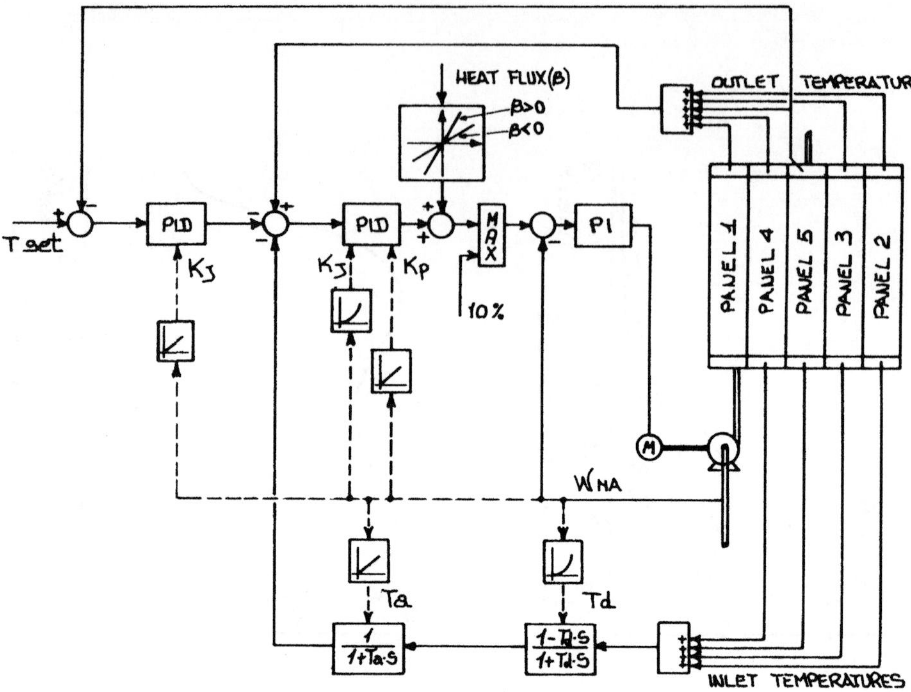

FIG.9 CONTROL CIRCUIT DIAGRAM

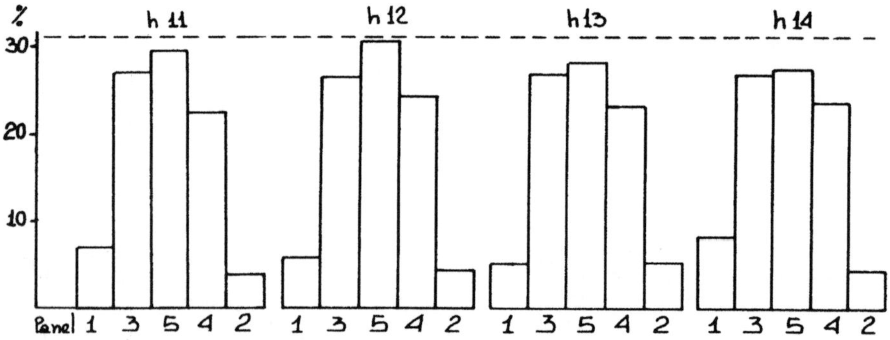

FIG.10 _ POWER DISTRIBUTION AMONG THE PANELS

Load variation	Temp.overshoot	°C	Delay sec.
25 % decrease	receiver outl.	18	11
	tubes	16	8
increase	receiver outl.	14	11
	tubes	14	8
50 % decrease	receiver outl.	22	13
	tubes	19	12
increase	receiver outl.	20	15
	tubes	21	11

TABLE I

THE FRONT END PROCESSOR OF EURELIOS SUPERVISION SYSTEMS

M. MAINI, G. PANDINI, U. ZONI
ENEL - Italian Electricity Board
Research and Development Dept. - Automation and Computing Research Center

Summary

The design of the "Data Acquisition and Supervision System" (DASS) for EURELIOS solar plant includes a Front End processor for plant data acquisition and formatting. The paper presents the most relevant hardware/software features of this system.

0. INTRODUCTION

The design of the whole DASS System started during the first operational period of the EURELIOS solar plant (early 1981). Therefore it has been positively influenced by real experience gained "on field", mainly as far as concerns data collection, presentation and processing.

The basic decision of splitting the whole system in two sub-systems:
- the Process Interface Unit (PIU) to be developed by italian partner "ENEL" around a LS 11/23 computer;
- the central processing system (DASS) to be developed around a SEL 32/27 computer by the Belgian consortium "TEI-BN", selected through a bid;

was taken very early.

For that reason the working team (who has been in charge of this job since the beginning) was able to balance the specifications of the DASS and the design of the PIU, matching them with the actual needs of the plant.

Scope of this paper is to present the PIU as a front end subsystem for the digital supervisor of EURELIOS solar plant. The design of this subsystem mainly takes into account the non conventional and experimental nature of solar plant.

The result of this special design is an unit more flexble than some conventional products but also cheaper, which could be applied in other similar experimental plants.

1. BASIC DESIGN CRITERIA

The design of the dual computer PIU-DASS system has been carried out following some basic criteria and some quantitative specifications reported in the next list.

1.1 The Front End Processor (PIU) scans, acquires, converts, formats and displays in real time analog and digital plant values whithin a basic 1 sec. cycle time. The input multiplexer is a solid state, high speed, type.

1.2 The inputs signal coming from plant are:
- high and/or low level analog voltages and/or currents
- free contact closures and/or TTL level digital signals.

The input scanner can be easily configured for very different sets of plant signals up to 400 analog and 1000 digital.

1.3 The PIU provides also the output of a limited set of analog and digital signals for driving lamps, recorders, alarms and so on.

1.4 The whole PIU software is completely "table driven": this means that all software activities are configured off line and driven on line by a set of tables which are available to the user, in order to guarantee a complete configurability on field, following the plant needs.

1.5 The physical link between PIU and DASS (Fig. 1) is 16 bits parallel, high speed, optical decoupled and based on the DMA channels available on PIU and DASS computers.

1.6 The logical link between PIU and DASS provides a bi-directional dialogue, with: fixed format messages at 1 sec. rate; timeouts for avoiding hang-up situations; leadership in opening/closing the dialogue given to PIU, so that it is able to continue operating also if DASS unit goes wrong or it is put out of service; error detection and management.

1.7 The DASS unit (Fig. 1) provides:
- a plant operator's console (two displays and one keyboard) mainly for supervision of the plant;
- a experimenter's console (two displays and one keyboard) mainly for managing and documenting experimental tests on solar plant;
- a programmer's console for programs and data base housekeeping and developing;
- a real time data base which contains: plant data coming from PIU; data and commands exchanged with the aforementioned consoles; time data from Plant Time Reference; configuration data like plant and system parameters, plant structure information, etc.;
- a set of data processing functions (like alarm management or chaining of plant supervision displays) for giving the operator a synthetic view of the plant status and evolution.

2. INTERPROCESSOR COMMUNICATION

2.1 In order to guarantee a reliable, high speed, easy to manage interprocessor link we chosen the 16 bits parallel solution, based on the two standard interfaces: DRV11-B from LS 11/23 side and MPCI from SEL 32/27 side. In fact those interfaces generate and recognize a 16 bit bi-directional bus plus some handshake lines, while (Fig. 2) the related software drivers manage the handshake protocol in the same way.

Therefore we introduced only minor modifications like: optical decoupling on MPCI card; an adapter card for re-shaping some handshake signals and for cables matching.

2.2 Because the PIU must play the "master" role, all inter-processor transactions can be initiated only from DRV 11 side, by sending an interrupt to MPCI, which in turn raises the attention of SEL 32/27. The protocol of this dialogue is based on two fixed format records (1024-16 bits words per record): the first one going always from PIU to DASS and the second from DASS to PIU (Fig. 2).

Each record contains:
- an header with identification label and with time and record count info;
- a 960 words buffer of analog and digital data, where each datum is followed by a validity flag;
- a set of command words for requesting the other computer to execute some

special activities;
- two final longitudinal CRCC words for errors detection.

The transmission from PIU to DASS is <u>always</u> followed by the answer from DASS to PIU, which contains either digital and analog data to be output from PIU and also error status flags, concerning the previous PIU-DASS transmission. In case of errors the PIU repeats the transmission without losing the basic 1 sec. cycle rate.

The whole communication cycle is surveilled from both PIU and DASS against hang-up situations by means of timeout services. This allows to automatically retray the transmission mantaining the link also in case of random noises or temporary malfunctions.

Finally, the communication cycle PIU-DASS and DASS-PIU takes about 20 ms in normal condition. In case of CRCC errors it takes 40 ms.

3. PLANT INTERFACE

3.1 The whole set of plant measurements (analog) and status (digital) are available at plant interface in some standard format (0÷10 V, 4÷20 mA, on-off free contacts). The input subsystem of PIU is splitted in three cabinets, where the aforementioned signals are ordinately plugged.

Each cabinet has its own power distribution and input subassembly (i.e. interface cards and control logic), so that each cabinet and related plant points could be separately tested.

The main cabinet, which contains the multiplexer and the A/D converter, provides also a set of test resources like: voltage reference generator; digital voltmeter; digital input simulator panel; which allow to check the plant signals and to supply simulated inputs for calibration and test purposes (Fig. 3).

3.2 The interface cards provide high impedence input, filtering, amplification, and common mode noise rejection.

3.3 The solid state multiplexer and the A/D converter provide the analog digital conversion (13 bit plus sign) and transfer into the computer memory. The logic data are packed and transferred to computer memory as a string of 16 bits words.

4. SOFTWARE FEATURES

4.1 The whole cycle of activity of PIU software is managed by a "scheduler module" (written in Fortran) which in fact acts as a specialized executive program. It activates the internal steps of the cycle (acquisition, formatting, trasmission to DASS, etc.) following the content of a table available to the user via the operator's console. Finally it allocates part of free time of the cycle (actually $\simeq$ 50%) to auxiliary activities like: operator's console, video outputs, printouts, on line diagnostics and so on.

4.2 Data processing is completely carried out on the basis of a set of tables (available to the user) which define:
- the subset of plant points to be acquired;
- the validity of each point;
- the formatting parameters;
- the contents of the records exchanged with DASS computer (Fig. 4).

This kind of on-line data base is created off line on disk via a configurator module; then it can be accessed and modified on line allowing the user to follow all plant needs, mainly during experimental tests.

4.3 The on-line operator's console allows the user to access the data base, the actual plant values, the scheduler parameters and so on (fig. 4). Besides, it allows to monitorize a pre-defined set of points updated at each cycle. Therefore it is possible to follow some vital plant variables also in case of out of service of DASS computer.

4.4 The whole software is written in Fortran; actually it makes use of 64 KB of RAM memory.

The execution times of various activities are: acquisition (400 analog points, 600 digital point) : 50 ms; formatting: 400 ms; dialogue with DASS: 20 ms; services: 30 ms; free time: 500 ms.

The floppy disk is activated only during program and data base loading; the on line software is completely memory resident.

5. ON FIELD ACTIVATION AND TEST

5.1 The PIU has been accurately step by step tested in laboratory environment during its developement phase. Because the nucleus of PIU (i.e.: CPU, multiplexer, A/D converter, video console) is physically very compact, it was sent (summer '82) to Bruxelles at TEI-BN facilities where the hardware/software link (see Chap. 2) with SEL 32/27 (DASS) has been tested and tuned accurately. In the same time the I/O cabinets (see Chap. 3) and input subsystem were tested point by point at ENEL laboratory facilities in Milano, using an identical CPU as a driver.

5.2 As a consequence of this extensive laboratory test program, the on field activation c/o Adrano plant (spring-summer '83) has been carried out quickly and without maior changes. In fact we had to solve on field only those kind of problems related to industrial environment i.e.: cabling, noise on analog signals, over temperature situations and so on.

Finally, because the total number of plant signals available during the DASS test were very limited, we put into the PIU a set of routines for simulating in real time static and dynamic plant behaviour, in order to test the whole set of functions of DASS unit, like live synoptics, X-T diagrams, alarm pages and so on.

6. CONCLUSIONS

The above solution for this kind of non conventional plant appears suitable, mainly because it bounds to PIU the effects of all plant changes and modifications, and allows to run on DASS a more general software package.

The actual availability of spare memory space and CPU time allows the PIU to support more complex duties in solar or other experimental plant.

7. AKNOWLEDGEMENTS

The authors want to specially mention the valuable cooperation of Mr. A. Brambilla and Mr. P. Groppelli in building, testing and activating the PIU system.

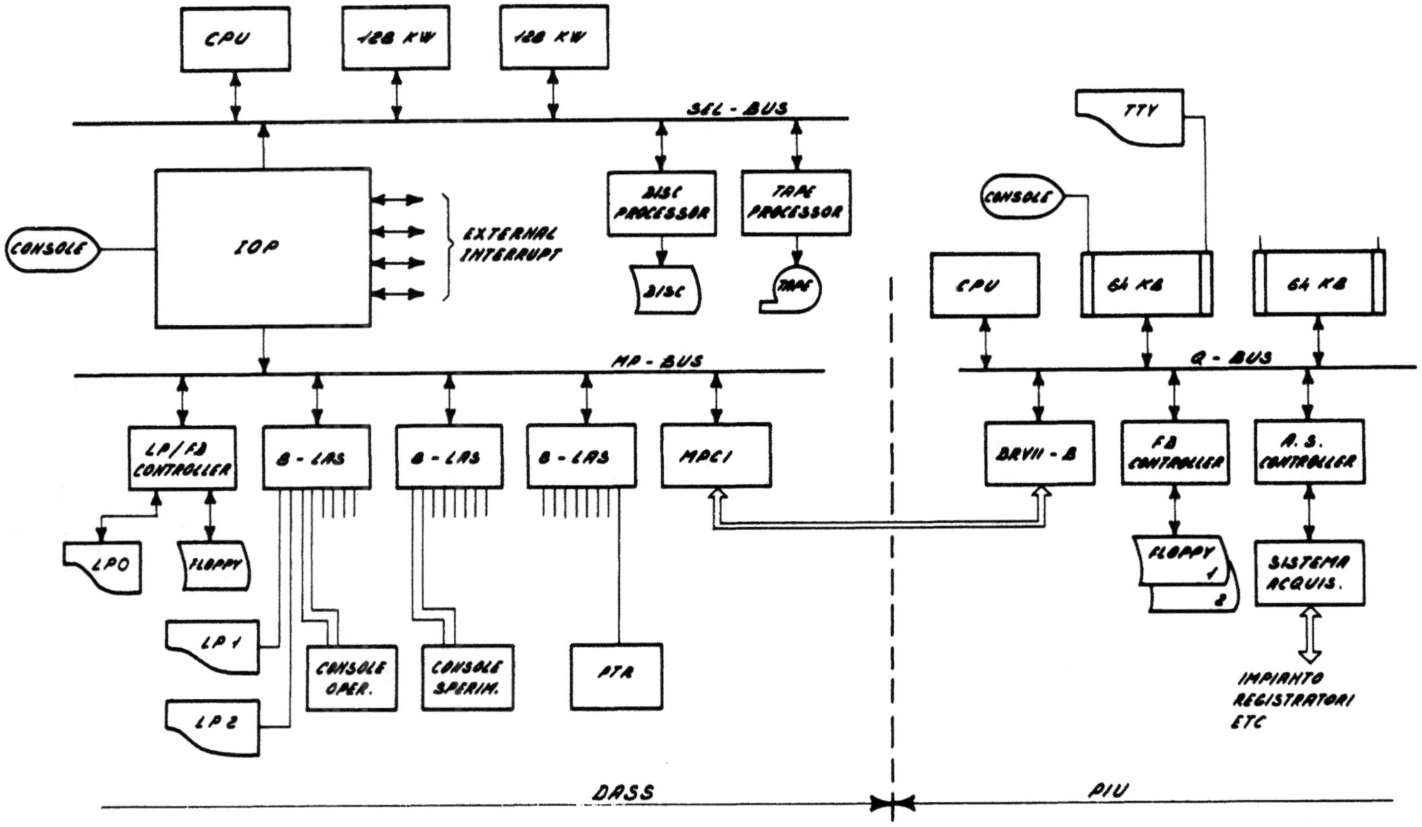

Fig. 1 - Configuration of DASS system

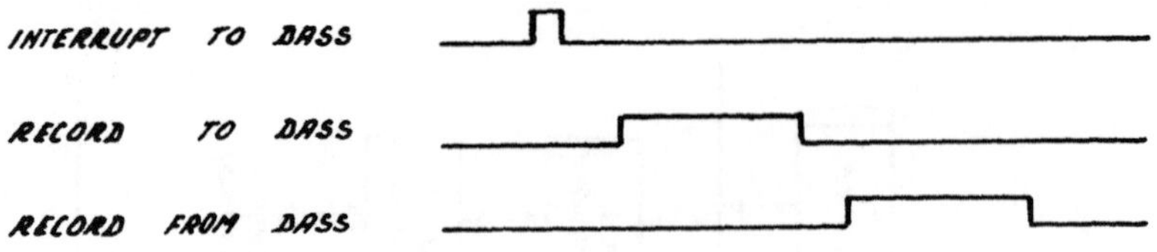

Fig. 2 – PIU–DASS handshake protocol signals

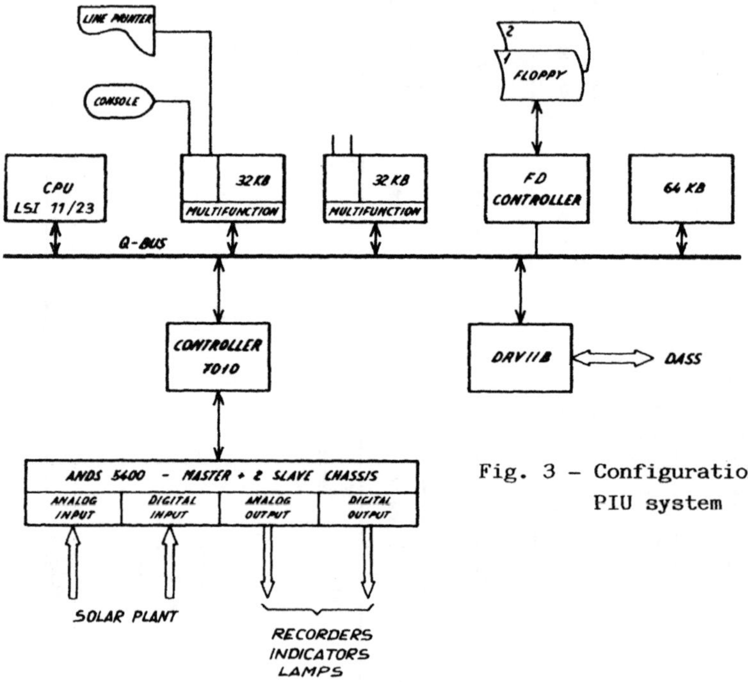

Fig. 3 – Configuration of PIU system

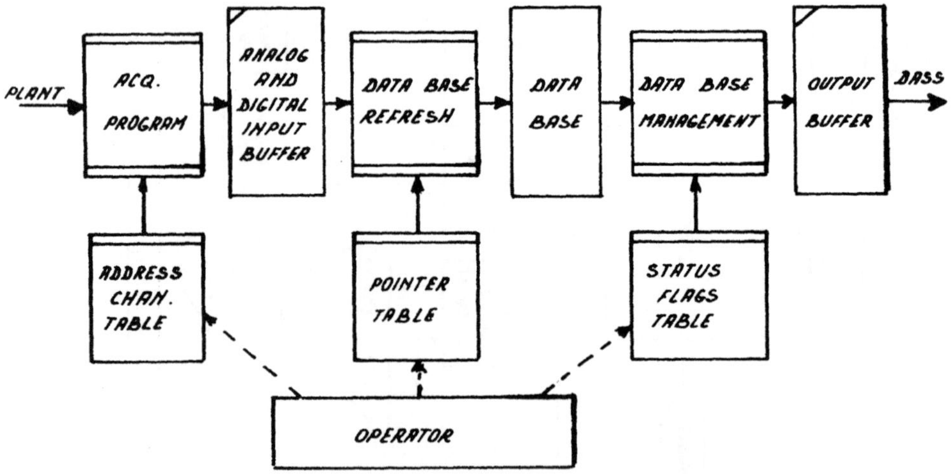

Fig. 4 – Plant to DASS data flow driven by tables

THE CONCEPT OF THE INTEGRATED DESIGN OF PROCESS, CONTROL AND OPERATION IN SOLAR CENTRAL RECEIVER PLANTS

C. MAFFEZZONI
ENEL - Italian Electricity Board
Research and Development Dept. - Automation and Computing Research Center

Summary

The experience gained from the participation in the Eurelios and in the IEA/Almeria Projects showed how process dynamics concepts and automatic operation requirements play an important role in designing Central Receiver Plants. In particular, this presentation aims at clarifying basic interactions between design parameters of the solar receiver and its continuity of operation (in automatic control) in sundays disturbed by cloud passages.

1. INTRODUCTION

Operation requirements of Solar Central Receiver Plants (SCRP) are mostly determined by the peculiarity of the energy source which imposes both periodic daily variations and irregular large disturbances due to cloud passages. In spite of this, solar receivers must be operated at constant temperature as long as possible during the sun hours. The experience gained with the Eurelios and Almeria ASR Projects clearly showed that automatic control is necessary in all the operation phases, in particular when cloud passages occur and for the morning start-up.

In view of the severe constraints on the receiver temperature variations, the possibility of achieving satisfactory automatic operation not only depends on the application of an effective control system but also on a suitable design of the receiver process which must be based on accurate dynamic analysis. This is the reason why process, control and operation of SCRPs should be designed in an integrated way, looking at the performance of the plant in the real environmental conditions rather than optimizing the process efficiency at the stationary design point. Basic tool for the Integrated Design (ID) is the dynamic modelling and simulation of the process, whose importance was discussed both for the Eurelios SCRP (1), (2) and for the Almeria ASR (3), (4).

This paper is a concise discussion of guidelines for ID as suggested by the principal results obtained in the framework of the two mentioned Projects and is organized as follows: Section 2 describes the basic dynamic properties of solar receivers, Section 3 is devoted to the identification of design parameters mostly affecting dynamics, Section 4 illustrates the fundamental control concepts, Section 5 aims at establishing guidelines for ID based on the analysis presented in the preceding sections.

2. ROLE OF DYNAMIC ANALYSIS

The role of dynamic analysis for the design of solar receivers is much more important than for other heat exchanger processes because solar receivers are required to normally operate in dynamic (transient) conditions (star-up, cloud passages). Moreover, static design criteria and manufacturing simplicity can often suggest a receiver design not suitable for practical continuous operation.

To clearly define the relevant phenomena affecting receivers dynamics, it is worth recalling that tubular heat exchangers generally exibit transport delays and/or storage lags, both resulting in a retardation of the process response to a perturbation of any exogenous variable. Furthermore, if the receiver coolant is subject to phase transition (this is the case of water-cooled receivers), quite peculiar phenomena are due to the interaction of hydrodynamics and energy balance within the evaporation zone.

The dynamics of receivers cooled by a liquid metal (like sodium) is generally dominated by transport delays, which are essentially determined by the time τ required by the fluid to cross the receiver (τ = lenght of the receiver/fluid velocity). Considering an ideal receiver made of a single panel with uniform radiation flux, the response of the fluid outlet temperature T_o to a small step variation ΔW of the fluid flow rate is approximately that of Fig. 1, where $\tau' \cong \tau(1+C_m/C_f)$ is the "apparent" crossing time, with C_m and C_f the total thermal capacities of the metal and of the coolant respectively.

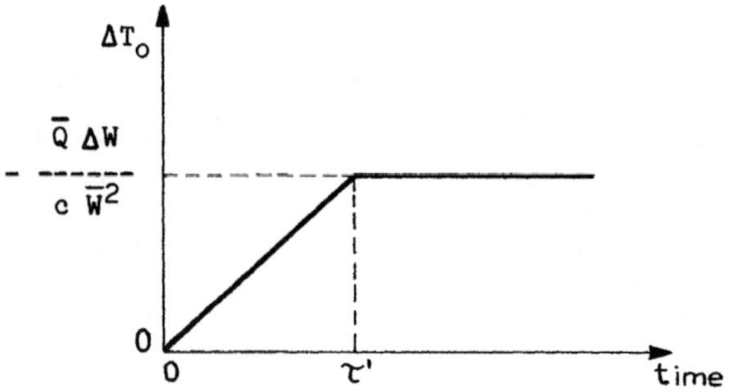

Fig. 1 - Response of a single-panel receiver with uniform flux.

The behaviour of Fig. 1 is due to the following properties:

(1) liquid metals allow very high heat exchange coefficient with the tube wall;

(2) receivers using nonboiling fluid are operated at low pressure so that tube wall is very thin and the metal thermal resistance is small.

In fact, due to the above properties, the tube metal is dynamically "coherent" with the adjacent fluid and, as a consequence, the relevant transport delay τ' is significantly greater than the mere crossing time τ (for instance, for the ASR of Almeria $\tau' \cong 2.7\tau$).

When a receiver consists of different panels in series, the dynamic responses of the different panels must be combined to obtaine the outlet temperature response. For instance, a receiver consisting of two panels in series with the same geometrical data as that of Fig. 1 will approximately exibit the response of Fig. 2, where it appears that the "prompt" temperature increase (from t=0 to t=τ') is proportional to the power $\bar{Q}_2$ absorbed by the outlet panel and the "delayed" temperature increase (from t=τ' to t=2τ') is proportional to the power $\bar{Q}_1$ absorbed by the inlet panel.

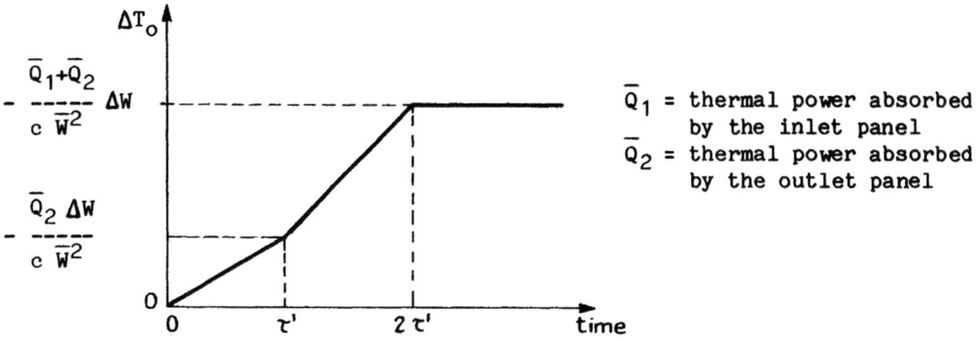

Fig. 2 - Response of a two-panels receiver

To be more precise, the effect of the metal wall is not simply that of lengthening the transport delay τ but also that of smoothing the temperature response at the edges of the two-times response.

It can be easily shown that the automatic (or manual) operation of the receiver with constant outlet temperature (as is usually required) becomes much more critical when:

1. the delay τ' is large,
2. the "delayed" components of the response are larger than the prompt one.

In fact when property 2 is verified the input-output transfer function is generally with non-minimum phase and, consequently, the temperature control by conventional concept (feedback of the outlet temperature) can not achieve a very high performance like that required in occasion of cloud passages. Once the ratio $\bar{Q}_1/\bar{Q}_2$ is given (see Fig. 2), the settling time achievable with the mentioned traditional control is roughly proportional to τ'.

The above reasoning is extendable to multi - or single - panel receivers with more complex radiation flux distribution: "delayed" components of outlet temperature response are more important (and such are nonminimum phase effects) whenever a significant part of the absorbed thermal power is concentrated close to the receiver inlet (3), (4).

More complex is <u>the dynamics of receivers cooled by water</u>, where evaporation occurs within the receiver. This is essentially due to the presence of an evaporation zone (EZ) between the subcooled water zone (SWZ) (where the behaviour is similar to that of liquid metals) and the superheated steam zone (SSZ) (where the behaviour is similar to that of gas). The main point here is that quite irregular movements can affect EZ when the receiver is not carefully designed to ensure a very high stability of the initial point of EZ, that is to minimize the movements of that point. It can be easily realized (1), (2) that stabilization of EZ is mostly achieved by:

(1) operating the receiver at constant (well controlled pressure);
(2) designing the flux map on the receiver so as to have a smooth thermal power distribution in SWZ and EZ with sufficiently high flux at the beginning of EZ.

Indeed, <u>pressure variations</u> causes a sort of "flash" effect in EZ, consisting in a forward or backward motion of the beginning of EZ and, consequently, a motion (much larger) of the end of EZ in the opposite sense. On the other hand, an irregularly-shaped flux map (especially in

SWZ) causes irregular temperature "waves" which propagate with the fluid velocity; those temperature "waves" are responsible of "delayed" movements of EZ resulting in nonminimum-phase temperature responses.

The experience of the Eurelios Project showed that, even with well controlled receiver pressure, large and slow oscillations of EZ arising in occasion of solar disturbances were essentially due to the sharp concentration of thermal power at the beginning of SWZ and to the excessive lenght of the receiver tubes (in particular of EZ and SWZ) which, together with the very low speed of the fluid, were responsible of slowly-moving temperature waves.

3. DESIGN PARAMETERS MOSTLY AFFECTING DYNAMICS

From the analysis of Section 2 it results that design parameters mostly affecting dynamics are:

(1) the thermal power distribution along the fluid stream;

(2) the lenght of the receiver or, more precisely, the receiver crossing time;

(3) the thermal capacity of the receiver metals.

With respect to dynamics, optimizing the thermal power distribution means avoiding concentration of power close to the receiver inlet or, in general, providing uniformly increasing power densities along the fluid stream (except for receiver zones with low heat exchange coefficients like the post dry-out zone in water cooled receivers).

The crossing time τ of the receiver is reduced, ceteris paribus, by using receivers with many parallel tubes, to the extent allowed by the flux map and by manifacturing constraints.

Using more parallel tubes also reduces the total mass M_T of the receiver metal (it can be shown (3), (4) that both τ and M_T are inversely proportional to $\sqrt{n}$, being n the number of parallel tubes).

4. CONTROL SYSTEM CONCEPT

Though receiver dynamics can be optimized in the design phase, practical constraints and manufacturing considerations often lead to a receiver design where significant non-minimum phase effects (delayed response components) still affect the dynamic response. This was the case of the Eurelios receiver and also, yet in a reduced way, of the Almeria ASR. For that reason, receiver temperature control by traditional concepts demonstrated to be not acceptable.

A new control concept was developed (3), (4) and applied to the Almeria ASR and, partly, designed also for the Eurelios receiver. Its realization requires the use of a digital control system essentially based on the feedback of intermediate temperature measurements "filtered" by a suitable predictor of the receiver tube temperature response. The resulting control system was called Predictive Feedback Temperature Controller (PFTC), because of the special use of temperature measurements. Its concept is schematically illustrated in Fig. 3 referred to the practical realization of the Almeria ASR, where it was successfully applied.

The application of PFTC to water cooled receivers is not straightforward because of the complex behaviour of the evaporation zone, where, moreover, temperature measurements are not significant.

Application of PFTC allows the operation of solar receivers in occasion of cloud passages without the need of a very precise direct radiation power measurement, which otherwise is the only practical means to compensate such solar disturbances. The use of a reliable estimate of the absorbed thermal power is still necessary, but quite realistic accuracy is sufficient when PFTC is used. PFTC was also used for controlled morning start-up (with simple operator duty).

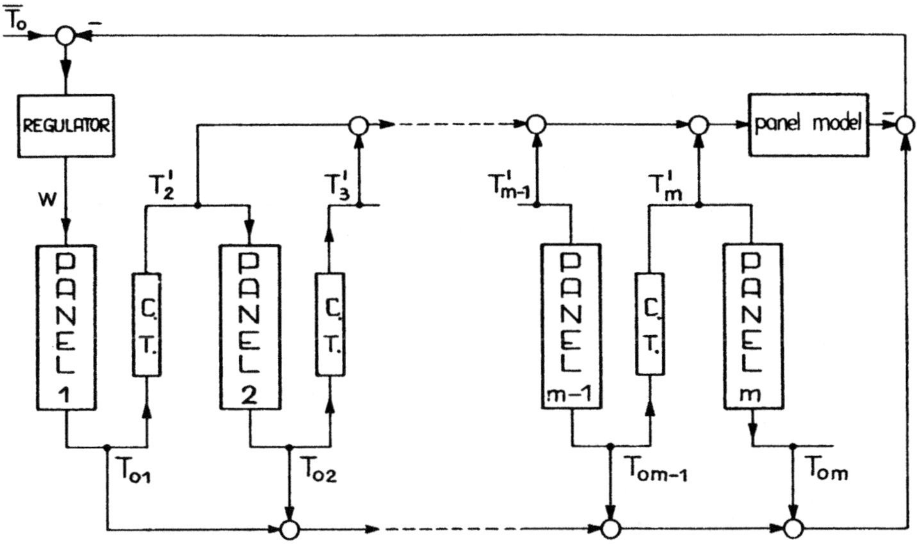

C.T. = connection tube & header

T_{oi} = outlet temperature of panel i

T'_i = inlet temperature of panel i

$\bar{T}_o$ = receiver outlet temperature set-point

W = coolant flow-rate

Fig. 3 - Practical realization of PFTC

Basic instrumentation for control system implementation are:
- temperature sensors with sufficiently promt response (a few seconds);
- fluxmeters or alternative means (pyroheliometers etc.) for radiation measurement;
- digital systems for the realization of non trivial control algorithms (PFTC).

Significant improvements in the control of water cooled solar receivers would be allowed by the use of enthalpy-meters or quality-meters applied in the evaporation zone (to substitute useless temperature measurements).

It is worth noting that the worse is the receiver dynamics the higher must be the control performance and, therefore, the more expensive the control system technology. Moreover, it was practically proved in the two mentioned Projects that automatic operation is a prerequisite for acceptable receiver performance, particularly during cloud passages and plant start-up.

5. CONCLUDING REMARKS

Dynamic analysis of solar receivers shows that receiver dynamics can significantly be affected by design parameters and that a bad dynamics can prevent the practical operation of the plant. On the other hand, automatic control is necessary for solar receivers to withstand the unavoidable disturbances of the energy source and must be designed to fit the receiver dynamics in order to obtain acceptable performances. The Integrated Design (ID) of the receiver process (dynamics) and its control system (based on the concept of PFTC) appears to be the only reliable solution to verify that the resulting (controlled) receiver can practically be operated without requiring non-realistic control performances and/or instrumentation costs. In particular, the receiver design must provide the possibility of installing prompt temperature sensors at intermediate receiver points, expecially at points immediately after receiver sections with high radiation densities. Of course process modelling and simulation are basic tools for ID.

REFERENCES

1. C. MAFFEZZONI and F. PARIGI (1982). Dynamic analysis and control of a solar power plant, part I: dynamic analysis and operation criteria. Solar Energy, Vol. 28, No. 2, 105-116.

2. C. MAFFEZZONI and F. PARIGI (1982). Dynamic analysis and control of a solar power plant, part II: control system design and simulation. Solar Energy, Vol. 28, No. 2, 117-128.

3. C. MAFFEZZONI, G. MAGNANI and S. QUATELA (1982). Integrated process and controller design of high temperature solar receivers. Proceedings 21st IEEE Conf. on Decision and Control, Vol. 2.

4. C. MAFFEZZONI, G. MAGNANI and S. QUATELA. Process and control design of high temperature solar receivers: an integrated approach. To appear in IEEE Trans. on Automatic Control.

CESA-1. CONTROL, SYSTEM & CYCLE OPERATION & STATUS REPORT

R. Balanza & J. Román

Summary

The purpose of this paper is to explain the main characteristics, the present status and the operation experiences related to the CESA-1 control system and cycle.
Concerning the power conversion system, its operative parameters and the most significant problems evident up to date will be exposed. The different operating modes and their actual status will be commented. Main incidents apparent in the thermal storage system during its functioning will be presented. Finally, the control system features will be explained. The status and operation of the receiver and storage control systems as well as the programmable interlocking and the data acquisition systems will be exposed.

1. POWER CONVERSION SYSTEM.

The power conversion system is a regenerative water/steam Rankine cycle. It has a dual admission turbine with two extraction points, an air cooled condenser and a boiler (receiver), with appropiate pumps, valves, controls, safety systems,...

The gross electric output, with high pressure steam, is 1200 Kwe. The nominal steam conditions are 5962 Kg/h, 520ºC and 110 Kg/cm^2. The calculated gross cycle efficiency is 27.7%.

The gross electric output, with low pressure steam, es 840 Kwe. The nominal steam conditions are 6901 Kg/h, 330ºC and 15.5 Kg/cm^2. In this case, the expected gross cycle efficiency is 21%.

Moreover, is possible the operation with high and low pressure steam mixture.

Since the first synchronization in October 1983, the number of synchronization hours has been 65.

Several efforts were made in order to get the nominal conditions of temperature and pressure.

The most relevant problems apparent in this system have been the following:

- Thrust bearing failure, due to a loss of lubrication oil.
- Turbine exhaust pipe modification. An expansion joint had to be installed to compensate the vacuum forces.
- Turbine casing distortion. Causes of those deformations are still under investigation.
- Electrohydraulic control system failures. Several adjustements has been needed.

These modification works caused several interruptions, which reduced the periods of possible power production.

2. OPERATING MODES.

In CESA-1 system there are seven different modes of operation (figure 1):

- Mode I: direct operation. The steam flow produced in the receiver is expanded in turbine HP. It works correctly. The number of operation hours, up to May 31st. 1984, has been 65. The reasons for this short number have been different problems in turbine, already mentioned, and adverse weather conditions.
- Mode II: charging storage. The steam generated in receiver is used in thermal storage charging. It works satisfactorily, but nominal conditions in hot tank still have not been reached. Number of operation hours has been 72.
- Mode III: discharging storage. Secondary steam is produced by thermal storage discharging. This mode is still in the start-up phase.
- Mode IV: direct operation + charging storage. Generated steam is distributed between turbine and thermal storage charging. To be tested.
- Mode V: direct operation + discharging storage. Primary and secondary steam is simultaneously sent to turbine. Not yet experimented.
- Mode VI: buffered operation. Steam generated in receiver is used in storage charging and storage discharging is utilized to produce secondary steam for turbine LP. This mode has not been tested yet.
- Mode VII: combination of all above mentioned modes, has not been tested yet.

Summarizing, some modes are still in start-up phase. On the other hand, transitions among different modes are still under experimentation.

3. THERMAL STORAGE SYSTEM.

This system can store 159900 Kwh_{th}, corresponding to an electric energy output of 3360 Kwh (4 hours at 840 Kwe level). It consists of two tanks, 200 m^3 each, that contain 300 tons of molten salt (Hitec: 53% NO_3K, 40% NO_2Na, 7% NO_3Na) as a storage medium. The tank is filled with salt headed at 340ºC and the cold tank temperature is 220ºC. This storage system is able to produce low pressure steam (330ºC, 15.5 bar).

Charging loop is already operative, though there are many problems in order to get nominal conditions of flow and temperature of the hot salt. Besides this, thermal inertia makes difficult to reach hot tank nominal temperature.

Discharging loop is still in start-up phase.

The most relevant problems in this system have been the following:

- Salt freezing and pipe blockage due to problems in draining lines and failures in heat tracing.
- Fluid leakages in heat exchangers headers and salt pipe crack due to defective pipe material.
- Two rupture disk failures in heat exchanger.

4. CONTROL SYSTEM.

For heliostat field, there is a computerized control system with a lot of man-machine interface utilities.

Receiver, cycle and storage process controls are conventional analogic electronic systems. All of them are "man in the loop" controls. A great deal of instrumentation is dedicated to data acquisition system.

Moreover, a programmable interlocking system is used. It gives great flexibility for design modifications. Alarm system is conventional in both, control desk and data acquisition system.

4.1.Receiver control system.

Receiver control is made by three loops:

- Feed water control.
- Superheater steam temperature control.
- Start-up control.

4.1.1.Feed water control.

Feed water to drum steam is controlled from the fixed set point and -- actual level in drum steam. This is the one-element control mode. As well, control can be made from steam flow produced, instantaneous feed water flow and the difference between actual level and set point fixed. This mode is - named three-elements control.

Both modes work satisfactorily.

4.1.2.Superheater steam temperature control.

This control loop regulates temperatures in superheater tubes introducing water attemperation.

Its automatic operation is difficult due to the high sensitivity of -- superheater to heat flux variations. Besides this, the metal temperature is close to trip limits.

4.1.3.Start-up control.

This loop fits generated steam flow from a pressure-time curve beforehand fixed.

Big efforts to reduce start-up time have determinated manual operation of this loop, though automatic start-up is ultimate objetive.

4.2.Storage control system.

Control loops seem to work as estimated, in automatic and manual modes although there is little experience up to now.

Some problems of salt freezing have appeared in control valves, as well as the instrumentation process tubing, which had to be heat traced.

4.3.Programmable interlocking system.

The main functions of this system are the following:

- Switching on/off all the motor driven equipments in the plant (pumps motorized valves, fans,...), except the heliostat field.
- Information about the status of all above mentioned equipments.
- Performing all the interlocking among equipments and switches (level, pressure, flows, ...).
- Autodiagnosis of the system.

This system is more versatile than a conventional one. It allows to make modifications in logical system without changes in hardware.

For the time being, it works without problems. During start-up phase,- some changes in the interlocking system design were implemented.

Safety interlocking between receiver and heliostat field works correctly.

4.4. Data acquisition system.

It is a computerized system having the next general objetives:

- Supervision of the plant subsystem and component.
- On-line calculation needed to evaluate the plant performance.
- Graphic and numeric display of the status of different subsystems.
- File functions for further evaluation off-line.

The system architecture is showed in figure 2.

In general, its behaviour is good. During start-up period, little modifications in design were implemented.

Variables groups are a great help for plant operators and their importance is bigger than expected.

Also, some modifications in performance calculations were implemented.

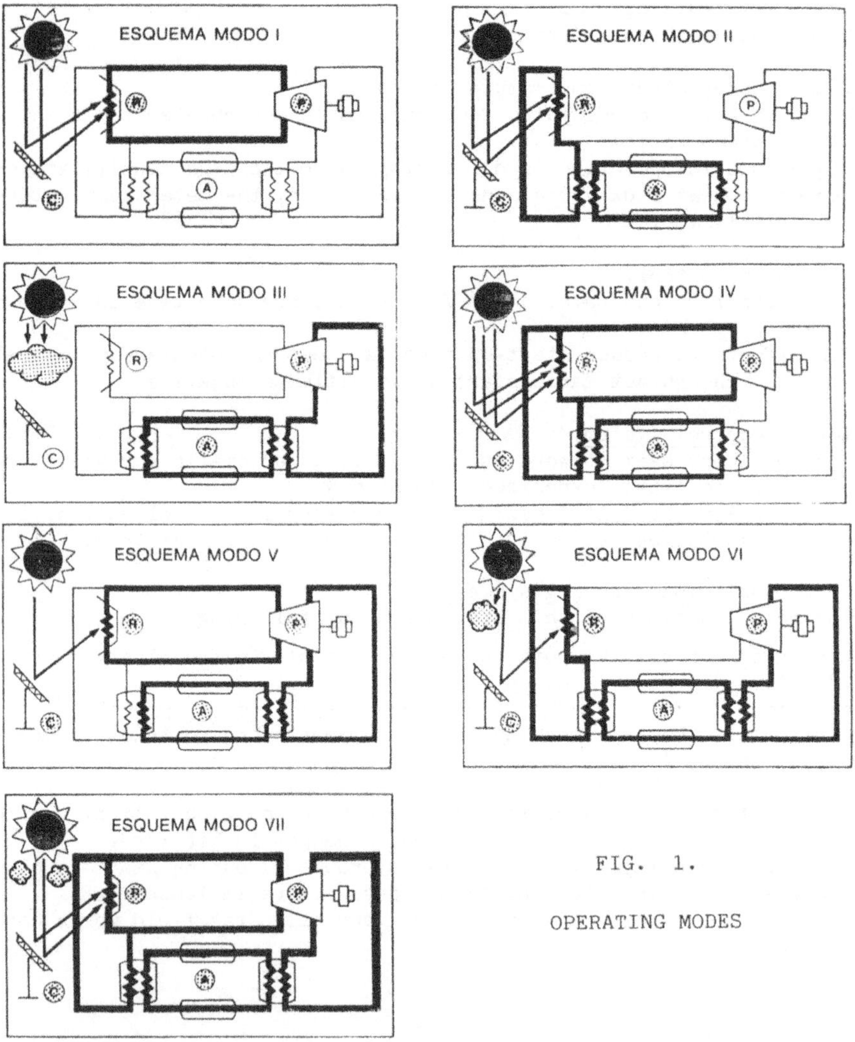

FIG. 1.

OPERATING MODES

D A S

SYSTEM ARCHITECTURE

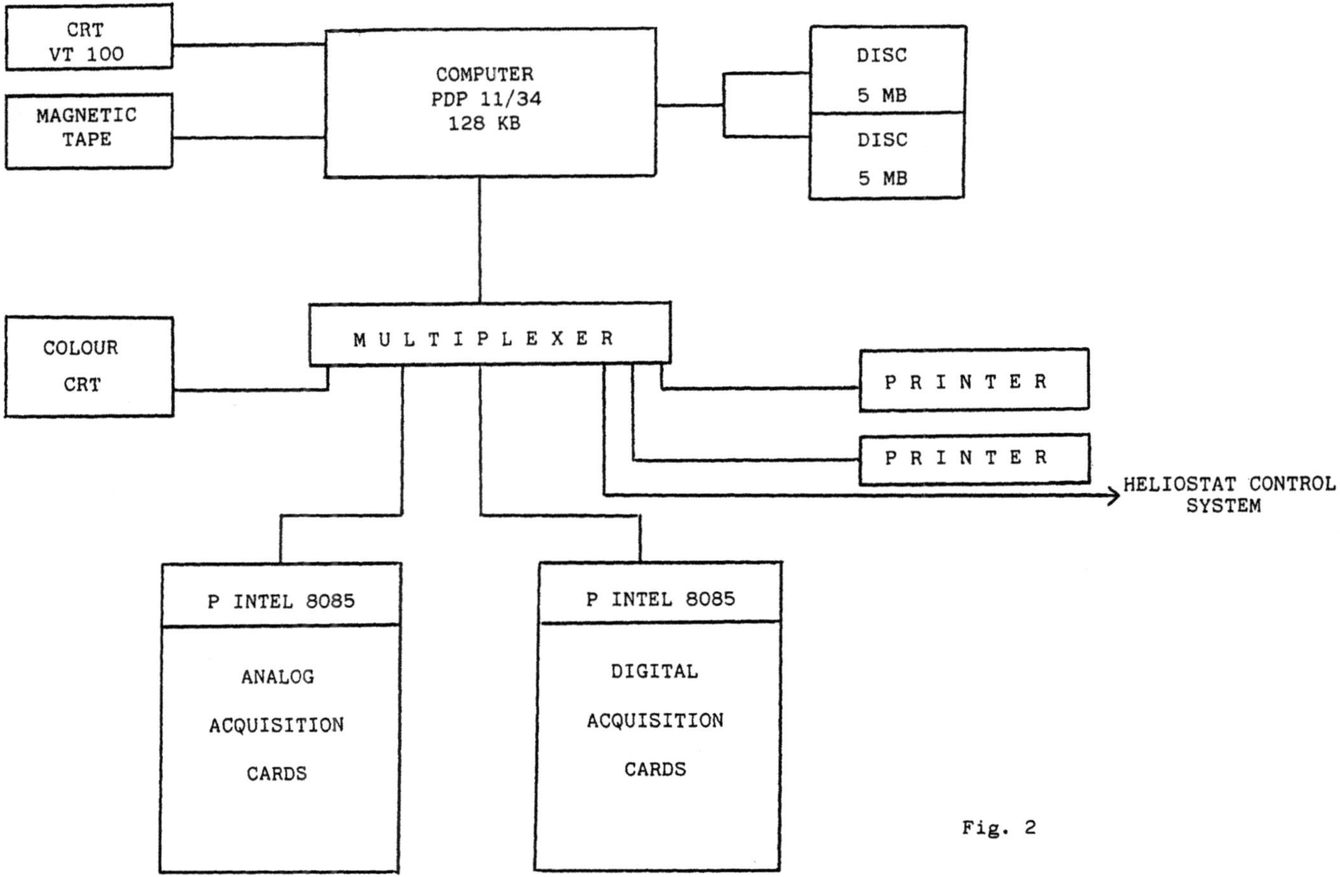

Fig. 2

SESSION II - PART 4

OPERATION, MAINTENANCE, STAFFING

Summary of the Session by the Rapporteur
L. CIVITANO, Rapporteur, ENEL-CRTN, Pisa, Italy

Operation and performance status of the Eurelios power plant

CESA-1 staffing, operation and maintenance status report

SUMMARY OF THE SESSION BY THE RAPPORTEUR

L. CIVITANO
ENEL-CRTN, Pisa, Italy

1. INTRODUCTION

The operation, maintenance, and staffing needs of the following central receiver power plants are compared: CESA-I, SOLAR ONE, THEMIS, EURELIOS, NIO, and IEA/SSPS.

2. STAFFING

Table I compares the staffing needs of the above listed central receiver power plants.

3. OPERATION

Table II summarizes the operation figures of the respective central receiver power plants.

4. MAINTENANCE

In SSPS/CRS, the following unusual maintenance problems occurred:

* cold sodium tank leakage (solved)
* steam motor (solved reducing efficiency from 27.2% to 23%)
  lubrication
  water drainage
  design
* lightning protection
* trace heating failures
* some problem of the Data Acquisition

There was also a long start up time due to thermal inertia of the system.

In EURELIOS:

* mirror corrosion (solved)
* heliostat control failure
* replacement of electrical cable of CETHAL heliostats
* receiver tube leakage (3 times)

Bad weather conditions also caused problems.

In THEMIS:

* heliostat central control system
* support salt piping in the tower
* electric trace heating
* main transformer and generator

TABLE I. STAFFING

	CESA-1	SOLAR-1	THEMIS	EURELIOS		NIO	IEA / SSPS	
				1981-83	1984		1982-83	1984
OPERATION + MAINTENANCE MANAGER	1	1	1	1	1	3	1	1
OPERATIONAL CREW	21	18	21	10	5	10	11.5	5.5
MAINTENANCE CREW	7	11	13	6	6	5	4.5	4
ADMINISTRATIVE STAFF	4	3	4	1	1	7	2.5	2.5
OTHER	-	11	-	-		-	-	2.5
	33	34	39	18	13	25	19.5	15.5
OPERATIONAL SHIFT	2	2	2	2	1	2	2	2
OPERATION DAYS/WEEK	5	7	7	6	5	7	7	5
WATCHING SHIFT (NIGHT)							1	1

TABLE II. OPERATION

	CESA-1	SOLAR ONE	THEMIS		EURELIOS		NIO	IEA/SSPS (CRS)
PERIOD	July '83-Apr. '84	1983	May '83-Apr. '84	Feb. '84 Apr. '84	May '81 Sep. '83	Apr. '83 Sep. '83	Sep. '81 May '84	1982 - Apr. '84
INSOLATION HOURS	2500	-	2400	536	5400	-	-	5660
GRID CONNECTION HOURS	42	1047	186	734	525	280	1344	154
TOTAL ENERGY PRODUCTION	17 MWh	6550 MWh	213 MWh	168 MWh	128 MWh	58 MWh	873 MWh	60 MWh
TOTAL ENERGY ABSORBED BY AUXILIARY	500 MWh	6330 MWh	2100 MWh	724 MWh	570 MWh	130 MWh	400 MWh	1600 MWh

In SOLAR ONE:
* mirror facets falling
* mirror corrosion
* mirror washing
* heliostat beam alignment
* limit switch failure
* leaks on flange and tube
* pyromark painting.

In CESA-I:
* heliostats AC motor (commutation solid state circuit)
* mirror corrosion and contour distortion
* vent tube cracking
* condensate within the superheater (steam traps)
* overnight heat and pressure loss
* turbine problem
* storage big thermal inertia
* salt freezing
* electrical trace heating

CESA-I also had bad weather problems.

5. GENERAL CONCLUSIONS

It's necessary to define better the characteristic of direct insolation in the aspect regarding the time distribution intensity and its continuity.

A deep energetic analysis on storage systems must account for the energy spent to maintain these systems versus the improvement of the percentage of solar radiation utilized.

The reliability of solar plant components must be increased to reduce the outage hours and the cost in terms of people and money.

Heliostat washing and environmental impact are felt in different intensities in the different plants and countries.

The general tendency is to reduce the staffing and auxiliary absorption.

Analysis of the generated and absorbed energy values doesn't take into account commisioning, experimental activity, technological infant mortality, and modification in order to achieve original design characteristics.

OPERATION AND PERFORMANCE STATUS OF THE EURELIOS POWER PLANT

A. CHILLE'*, L. CIVITANO*, G. DINELLI*, P. PERRONE**

*ENEL - Centro Ricerca Termica e Nucleare
- Pisa -

**ENEL - Compartimento di Palermo
Centrale Eurelios
- Adrano -

ABSTRACT

An overwiew on staffing, operation and maintenance aspects at Eurelios power plant is presented.
The principal operational data are compared with the expected values and the main sources of failures for the components are reported.

1. STAFFING

Until September 1983 the operation of the Eurelios plant has been conducted by nine people, covering two shifts, and by two maintenance groups made up of three people each(mechanical crew and electronic/pneumatic crew). A plant superintendent and a clerk completed the plant staffing. The plant has been operated 6 days a week.

The operation experience gained to date has suggested to reduce the working days to 5 days per week and the operation staff, which is in charge also of the electrical maintenance, to five people; the mechanical crew and electronic/pneumatic control crew remaining of three people each as before. The plant superintendent and a clerk complete the plant staffing.

During the night, except for the electronic control circuits, the plant is swiched off and does not need any technician shift; a number of selected alarms are sent to the gard site, in case an alarm is activated, the fard-man calls immediately the plant superintendent or his representative.

2. PLANT OPERATION

2.1 Actual Plant Performance

In the period from May 1981 to September 1983 the Eurelios power plant has 525 hours of connection with the grid, generating about 128,000 kWh; in the same period the auxiliary absorbtion has been about 570,000 kWh (Tab. 1). Thus the ratio between the gross generated energy and the absorbed energy has been about 0.22. The maximum generated power was about 500÷600 kW close enought to the actual nominale plant power.

In the period from April 1983 to September 1983 the plant had 280 hours of grid connection, the gross electric production was about 58,000 kWh and the auxiliary absorbtion about 130,000 kWh, the ratio between these values increased to 0.45. Fig. 1 shows the energy generated and absorbed by ausiliaries from September 1981 to September 1983.

Since June 1983 several actions have been performed in order to reduce the light and auxiliaries power absorbtion.

The low average power generated in the period April 1983÷September 1983 is caused by several reasons, the principal of wich are:

- grid connection held even with very low generated power in order to verify the behavior of the subsystems at derating values;
- 10÷20% in heliostats failure (Fig. 2);
- low solar direct beam intensity, due to the presence of haze, during the months of June and July 1983.

The solar beam intensity during these two months was much less than the values obtained from measurement in 1979 and used as reference values (Fig. 3).

Mirrors reflectivity was within the expected limits: the minimum reflectivity values was 0.70 versus a value of 0.80 for clean mirrors, despite the absence of manual washing. The rain washing effect is shown in Fig. 4.

Under conditions equivalent to a power production of 350 kW the main power losses are caused by (Tab. 2):

- heliostats unavailability	10%;
- mirror reflectivity	25%;
- receiver losses	30%;
- thermoelectric cycle losses	83%.

The high losses of receiver and thermoelectric cycle are caused by the reduced plant nominal power, about 500 kW instead of 1 MW, because the heliostat field area had been reduced from 8900 m^2 of preliminary project to 6200 m^2 and being the plant construct for the main purpose of testing the unconventional component, the thermoelectric cycle was not optimized.

2.2 Expected Plant Performance

The expected yearly energy balance is shown in Tab. 3; it should be noticed that the expected value of the heliostat unavailability, 5%, is considered a target value. The difference between the direct normal energy and the useful direct normal energy is a consequence of the start up time of the plant and because in the plant the storage subsystem is not used.

Concerning the storage utilization and assuming a storage subsystem with 1 hour capacity the percentage of useful energy would rise from the actual 59% to 65% of the global energy input to the mirror field. Taking into account the energy spent to maintain in operation the storage subsystem, the energy budget would be negative.

On the other hand the percentage of the useful direct energy would rise to 73% if the plant start-up would be reduced to 60 minutes.(Fig.5).

The expected operation data at Eurelios site are (Tab. 4):

- useful solar radiation: 1170 hours/year;
- grid connection: 900 hours/year;
- grid connection at nominal power (500 kW): 600 equivalent hours/year.

3. UNCONVENTIONAL MAINTENANCE

Different sources of failure were experienced for the Cethel and MBB fields. From a general point of wiew the main problems arising in the Cethel field came from the groups controller and, at heliostat level,from the proximity detectors. All the cables connecting the heliostat electronic box to the heliostat motor and limit switch had to be replaced; at present the main cause of unavailability is due to connectors corrosion.

For the MBB heliostat field the main failures concern: encoder lamps, power supply set and mirror corrosion. The last two problems were solved by substitution trasformers in the power supply set and the mirror modules with components of higher reliability.

The receiver is the other unconventional component that was source of unavailability of the plant as conseguence of leakages from the tubes at the end of evaporating zone. The receiver has been repaired three times and during last intervention the connection of the receiver tubes to the supporting frame has been modified to allow a better expansion of the tubes at different temperatures.

4. CONCLUSION

The expected plant performance values correspond to the actual design parameters of the plant. In order to reduce the start-up time which penalize heavily the power production a study has been done to convert the actual once-through receiver to a recirculation drum-type receiver.

On the other hand in order to obtain a yearly production of electric energy of 180,000 kWh which is the value of the expected yearly auxiliaries absorbtion, the mirror field surface should be increased of about 30% which means a total surface area of about 8,000 m^2, still smaller than the preliminary project value. Such intervention has not been considered attractive by the Commission of European Communities and by Enel,which are the co-proprietors of the Eurelios plant,because of the high associated cost.

TAB. 1 - EURELIOS POWER PLANT - MAIN PRODUCTION AND ABSORTION VALUES

PERIOD	GRID CONNECTION HOURS (H)	GROSS ELECTRIC PRODUCTION (kWh)	AUXILIARIES ABSORBTION (kWh)	ENERGY RATIO	AVERAGE POWER (kW)
MAY 1981-SEPT 1983	525	128000	570000	0,22	244
APR 1983-SEPT 1983	280	58000	130000	0,45	207

TAB. 2 - MAIN POWER LOSSES

	HELIOSTATS FAILURE	MIRRORS REFLECTIVITY	RECEIVER LOSS	THERMOELECTRIC CYCLE LOSS
RELATIVE LOSS TO SOLAR ENERGY TO THE MIRROR FIELD	10%	25%	30%	83%
RELATIVE POWER LEVEL	90%	67%	47%	8%

TAB. 3 - EURELIOS POWER PLANT - YEARLY ENERGY BALANCE

YEARLY DIRECT NORMAL SOLAR ENERGY (ON 6216MQ) (THRESHOLD VALUE 450 W/MQ)	$8.7 \cdot 10^6$ kWh	100%
YEARLY DIRECT USEFUL NORMAL SOLAR ENERGY (CONTINUOUS DIRECT INSOLATION MORE THAN TWO HOURS)	$5.1 \cdot 10^6$ kWh	59%
YEARLY ELECTRIC PRODUCIBILITY (7DAYS/WEEK) (CLEAN MIRRORS; NO HELIOSTATS FAILURE)	$0.32 \cdot 10^6$ kWh	3,7%
YEARLY ELECTRIC PRODUCIBILITY (7DAYS/WEEK) (MIRRORS REFLECTIVITY 0,73; HELIOSTATS FAILURE 5%)	$0.23 \cdot 10^6$ kWh	2,6%
YEARLY EXPECTED PRODUCTION (5DAYS/WEEK) (PLANT AVAILABILITY 0,70)	$0.115 \cdot 10^6$ kWh	1,3%

TAB. 4 - EURELIOS POWER PLANT - EXPECTED VALUES

SUNSHINE HOURS/YEAR	2250
USEFUL SUNSHINE HOURS/YEAR (CONTINUOUS DIRECT RADIATION MORE THAN 2 HOURS)	1170
GRID CONNECTION HOURS/YEAR	900
NOMINAL POWER EQUIVALENT HOURS/YEAR	600

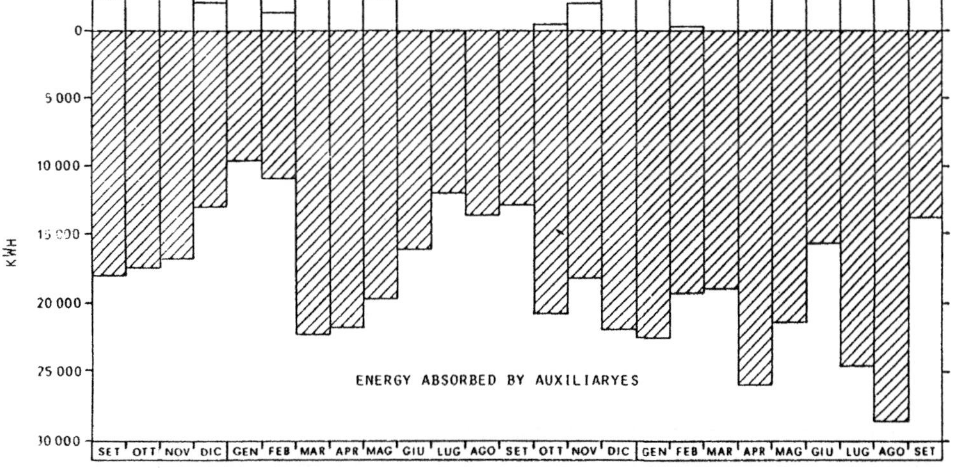

FIG. 1 - C.LE EURELIOS: GENERATED AND ABSORBED ENERGY - MONTLY VALUES.

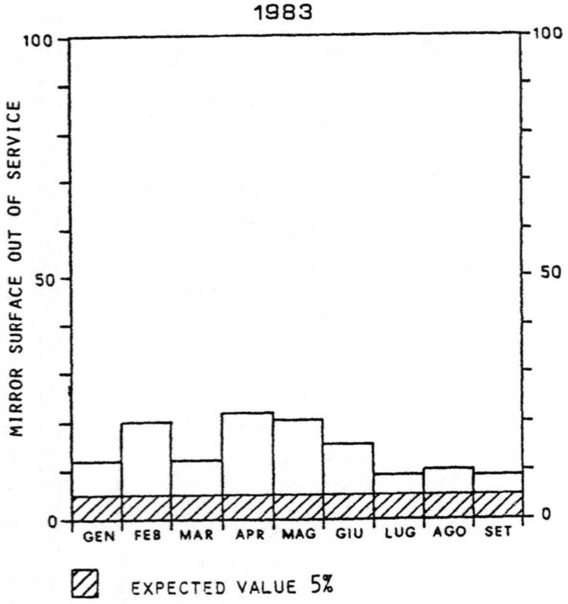

FIG. 2 - MIRRORS SURFACE OUT OF SERVICE
AVERAGE MONTLY VALUE

FIG. 3 - MEASURED SOLAR INTENSITY VERSUS TIME →

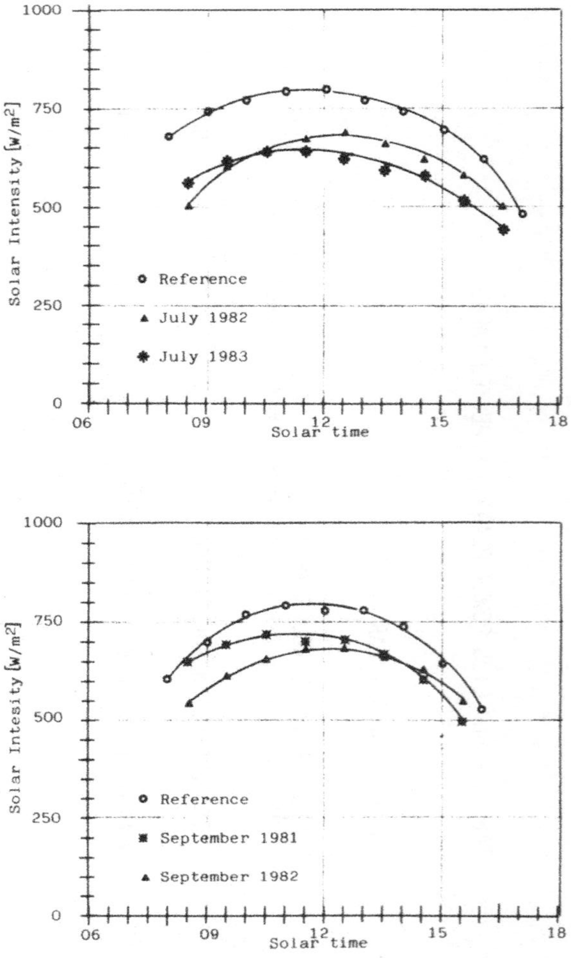

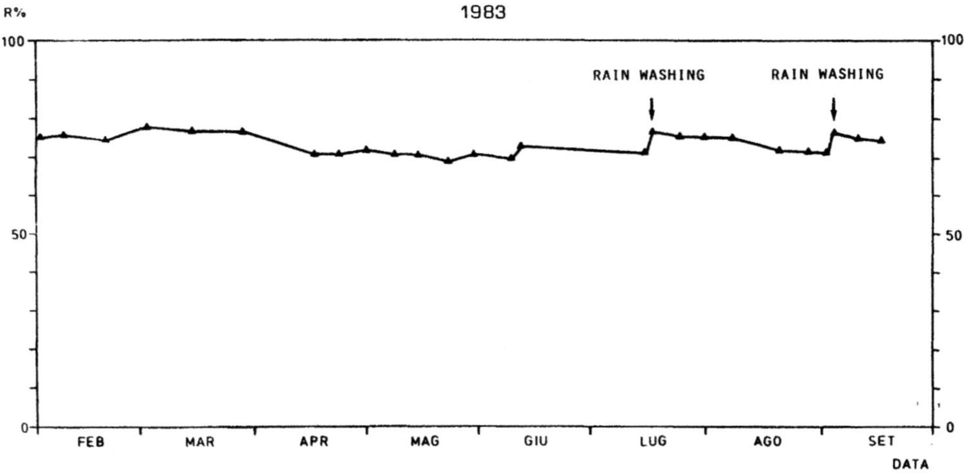

FIG. 4 - MIRROR FIELD AVERAGE REFLECTIVITY VERSUS TIME

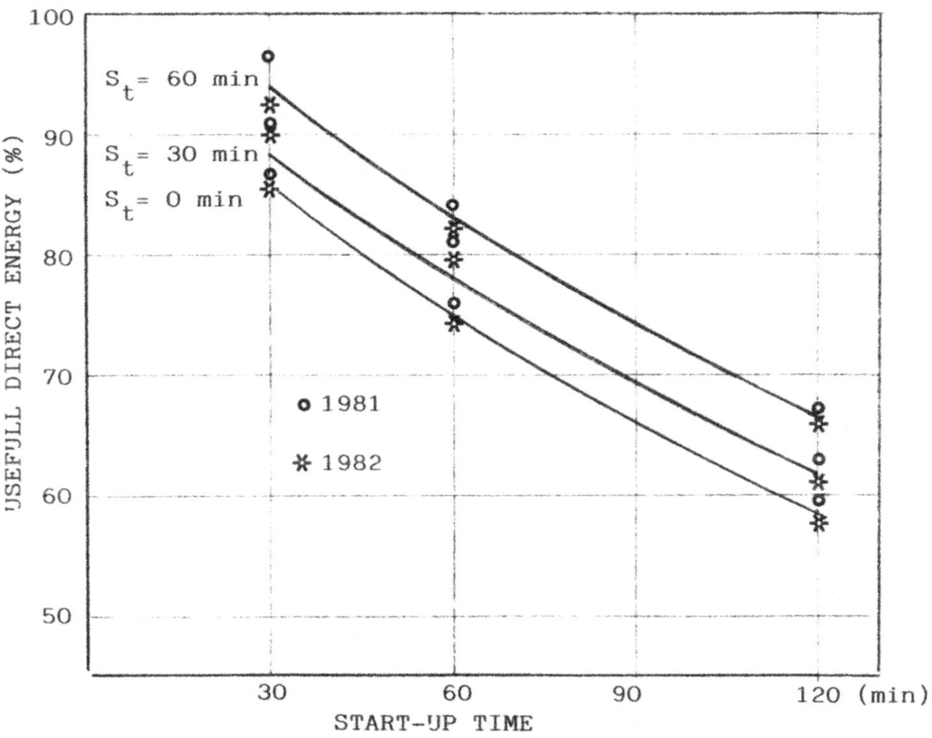

FIG. 5 - Percentage of usefull direct energy versus start-up time as influenced by the storage capacity (S_t).

CESA-1 STAFFING, OPERATION AND MAINTENANCE STATUS REPORT

Francisco Martínez Córcoles
CIA. SEVILLANA DE ELECTRICIDAD, S.A.

Summary

This paper describes the operation and maintenance status of the CESA-1 plant as well as the staffing organization and experiences.
Running strategies, statistics, highlights and costs are given.
The CESA-1 plant was taken over for operation after the official inauguration (June 22th, 1983) and the first grid connection took place on October 25th, 1983.

1. STAFFING ORGANIZATION.

The plant running staff is composed of 33 persons organized as fig. 1 shows.

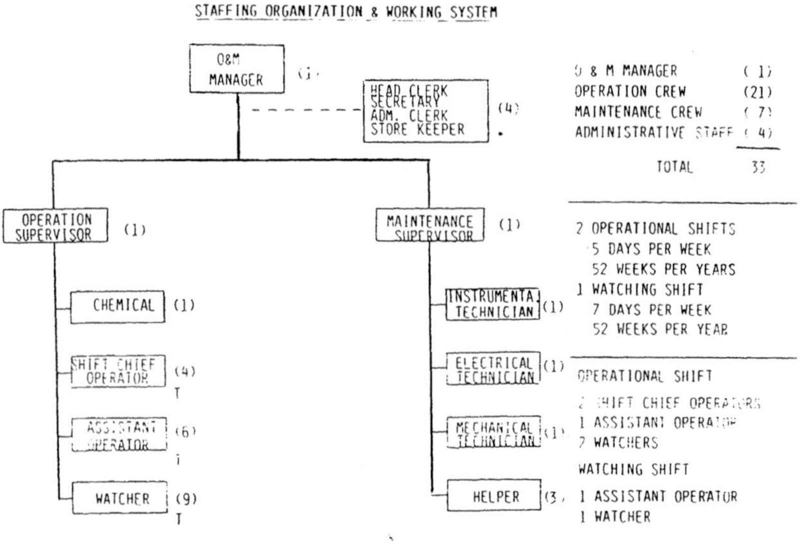

* ADMINISTRATIVE STAFF SHARED WITH EVALUATION GROUP
T SHIFT WORKING STAFF

Fig. 1

The plant is being operated 5 days per week, 2 operational shifts per day. An overnight as well as weekend and holidays shift is supervising the facilities during no operational periods.

The O & M Manager, Maintenance team, Chemical, Operation Supervisor - and Administrative staft work split journey.

The reason for the 5 days instead of daily operation has been the need of full time supervision of the receiver during operation (1 man) which has forced such reduction.

Three major lessons with respect to the staffing have been learned:

- Experimentation and technological innovations require higher qualified and flexible people than conventional systems.

- Prototypes and new components under test ask for bigger maintenance crew than commercial plants.
- Operation team size difficult to optimize due to peak manpower re-- quirements during daily start up, sunlight seasonal variations and- labour regulations constrains.

2. RUNNING STRATEGY.

As a general objective the system was operated as an electric thermal plant integrated into the electrical network.

Nevertheless a complete program of evaluation tests is being performed in order to better understand the behaviour of the different plant subsys-- tems and to optimize the general performance.

The available operational modes are as follow:

Mode 1: direct operation (collector-receiver-turbine HP).
Mode 2: charging storage (collector-receiver-salt charging loop).
Mode 3: discharging storage (salt discharging loop-turbine LP).
Mode 4: direct operation + charging storage.
Mode 5: direct operation + discharging storage.
Mode 6: buffered operation (charging storage + discharging storage).
Mode 7: full plant operation.

The adopted operational strategy was to use preferently modes I and VI, due to the fact that turbine and heat storage commisioning has been the major goals up to date.

As much steam as possible was got from the receiver by means of an expensive operation of both the heliostat field and the solar receiver.

Reduced load level of the collector-receiver in comparison with expected values induces a loss of performance, mainly due to the heliostats availability. Therefore, maximum turbine load achieved has been 70% of the de-- sign value for the time being.

3. OPERATION AND MAINTENANCE STATISTICS.

The operational data summary for the period July, 1983-April, 1984 is as follow:

	DAYS	HOURS
Calculated sun hours	-	3527
Real sunshine (measured)	296	2528
Collector field tracking	250	2134
Receiver operating	139	865
Turbine operating	32	96
Synchronization	22	42
Heat storage charging	26	67
Heat storage discharging	15	54
Gross energy production	17016 Kwh	
Gross peak power	850 Kw	
Available heliostat (daily average)	83 %	

The above values are not representatives of the real plant performances due to the fact that the system is still under functional tests.

Detailed information about monthly operational hours and production is shown in fig. 2, 3, 4, 5 and 6.

4. OPERATION AND MAINTENANCE HIGHLIGHTS.

4.1. General.

- Bad weather conditions:

- Lower insolation than expected.
- Clouds.
- High wind speed.

4.2. Collector field.

- AC motors commutation circuit: fuses and solid state relays destroyed (RC+Varystors)
- Drive mechanism Sener heliostat: azimuth transmissions get free (creasing of the shaft).
- Out-door encoders failure due to humidity (solved with a seal improved version).
- AC motors failures due to humidity.
- Split torque tube Casa heliostat: reflected image certain deformation due to mirror and structure weight.
- Mirror modules silver corrosion.
- Mirror modules contour distorsion and glass breakages due to thermal-stresses.
- Computer, peripherals and console interfase failures and blocking.
- September, 8th 1983 a triple power failure took place (grid, ups, diesel); the field remained fixed into the receiver for several minutes. Not receiver circulation pump nor venting valve provided cooling. Later inspection showed no particular damage. Venting valve control circuit was changed to fail open with remaining air pressure.

4.3. Receiver.

- Recirculation pumps mechanical seal failure.
- Vent tubes crackings (3).
- Big thermal inertia of the superheater headers during start up and clouds transients caused floading of such a headers with condensate sealing the tubes. Loss of cooling fluid induced increasing metal temperature on a very short time. Superheater tubes permanent enlargement due to thermal expansion was observed during funtional test and commissioning. Steam traps were installed to drain automatically condensate. A detailed study of the trip temperatures was conducted by the designer and a higher alarm limit was allowed. The implementation of a so called "group monitor" in the data acquisition system (real time presentation each 2 second of up to 40 variables) has be found essencial for the day to the day operation of the receiver. Manual operational strategies have to be applied to control superheater metal temperatures (high sensitivity to heat flux distribution).
- Heat, pressure and fluid losses overnight higher than expected, avoiding hot start up:
  - Considerable efforces devoted to solve the problem.
  - Modifications performed allows now a warm start up from 10 bar aprox.:
    - Valve leaks repairing.
    - Installation of a general drain valve.
    - Thermal insulation of the receiver door improved.
    - Installation of a valve to isolate the steam drum from the superheater preventing the steam to go trough the tubes getting there condensated.
- Several seals breakages.
- Steam temperature control loops (attemperators) too slow to meet the quick solar fuel variations. Normally operated manually.

4.4. Turbine.

- Thrust bearing failure caused by loss of lubricating oil (there were- no inlet holes due to manufacturing and quality assurance error).
- Electrohydraulic control circuit required several adjustements.
- Gland steam consumption higher than expected.
- Turbine casing distorsion.
- Exhaust pipe modification to compensate the vacuum force.

4.5. Heat storage.

- Big thermal inertia of the salt system in adittion to the relatively big start up time of the receiver, charging and discharging loops as well as the limited operational time and main steam flow lower than - expected, have prevented the charge and mostly the discharge subsys - tems to reach nominal temperatures.
- Salt freezing with blocking of the heating and cooling subsystem pipes mainly in the valves.
- Salt leakages (flanges and pipe crack).
- Troubles with the electrical trace heaters (breakages, low heating -- capacity on several zones, ...).
- Rupture disk breakages. Reasons not yet determinated. An existing --- steam leak in a heat exchanger is suspected and currently investigated.
- Instrumentation problems.

5. PLANT PERFORMANCE OPERATION EXPERIENCES.

5.1. General.

- As a general experience more people than foreseen is necessary to ope rate the plant. Good knowledge of the installation and high training level of the operational crew is helping considerably.

5.2. Collector field.

- Availability of the heliostats is roughly 80% to 85% mainly due to -- electronic problems. RC cells as well as varystor seem to be necessa- ry to protect the solid state relays againt over voltages.
- The man machine interfase is enough flexible in terms of adressability Command capabilities and feedback information to allow very good --- field control.
- There is a considerable amount of mirror modules showing corrosion - (small sptos, zones or most of the surface affected). Investigations within this matter is currently carried out by the experimentation - team.

5.3. Receiver.

- After several modification the receiver is doing well is general terms
- However three points have to be noticed:
  - Overnight heat and pressure losses force to a morning poor hot start up with very low reamining pressure ($>$ 15 bar) which takes 2½ hours to reach nominal pressure and temperature conditions.
  - High sensitivity to heat flux distribution of the superheater tubes requires one operator full time attention.
  - Manual operation strategy (manual start up and manual temperature - and pressure (flow) control) has to be implemented in close coope - ration with the collector field operation on a "individual helios- tat" basis.

5.4. Turbine.

- No problems related to the routine operational procedure.
- Performances of the turbine lower than designed due to several conventional problems already mentioned (gland steam, hidraulic circuit, casing distortion, ...).

5.5. Heat storage.

- Great thermal inertia problem as described before, seems to be the most critical topic.
- The use of fluids that get frozen at ambient temperature induces a lot of maintenance action to "liquefy" the salt again after the blockages.
- In adittion, electrical trace heaters give also problems due to design and heating capacity.
- Problems with the instrumentation (level transmiters, ...).

6. RUNNING COSTS.

Fig. 7 , 8 shows the running costs of the plant within the period -- July, 1983 to April, 1984.

The total running cost has been roughly 80 Mpts, equivalent to half a million of U.S.$.

The most of that money has been devoted to operation labor cost (67%)

On the other hand the materials cost has been almost no significant - (4%).

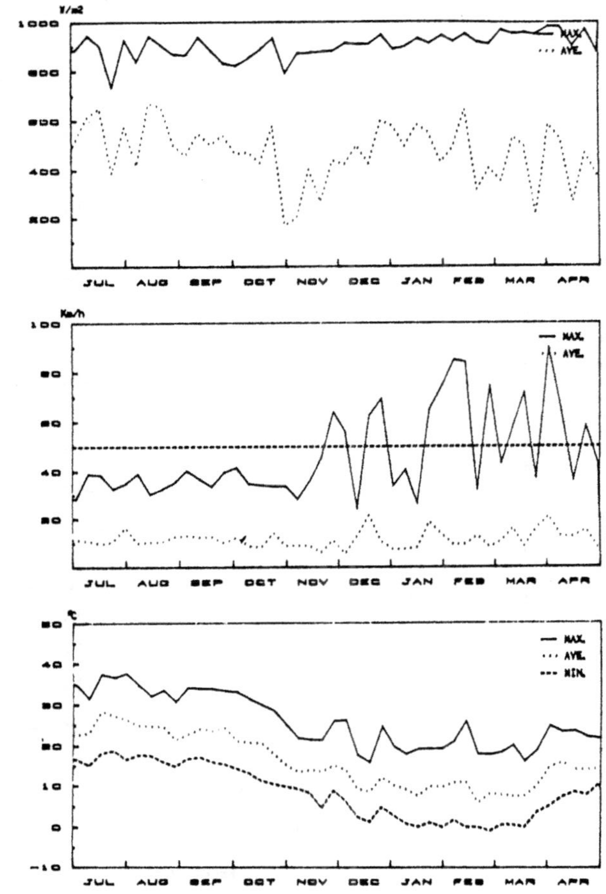

Fig. 2 METEOROLOGICAL CONDITIONS

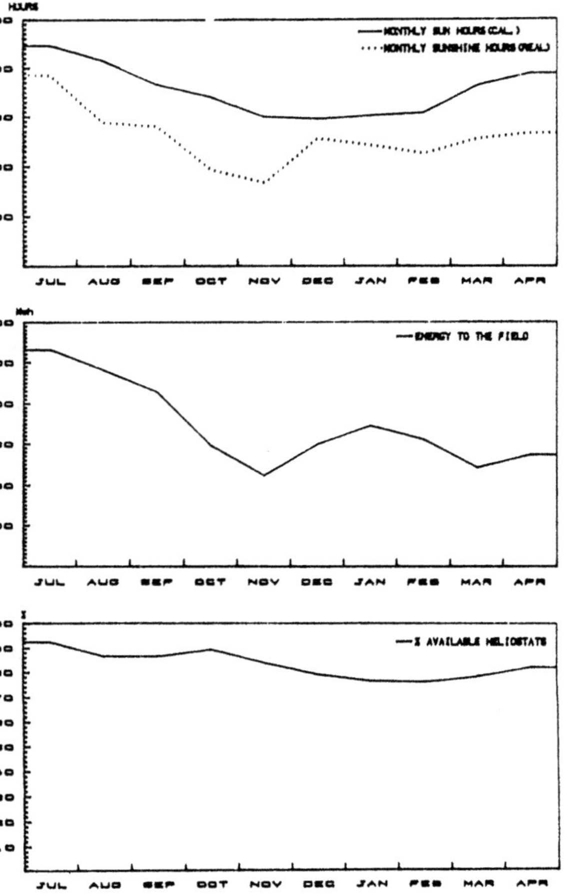

Fig. 3

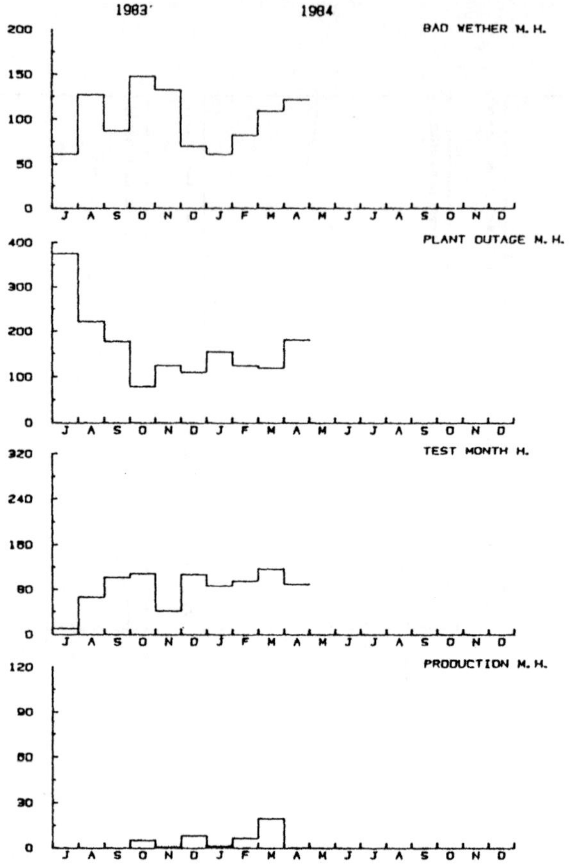

Fig. 4 SERVICE HOURS BREAKDOWN

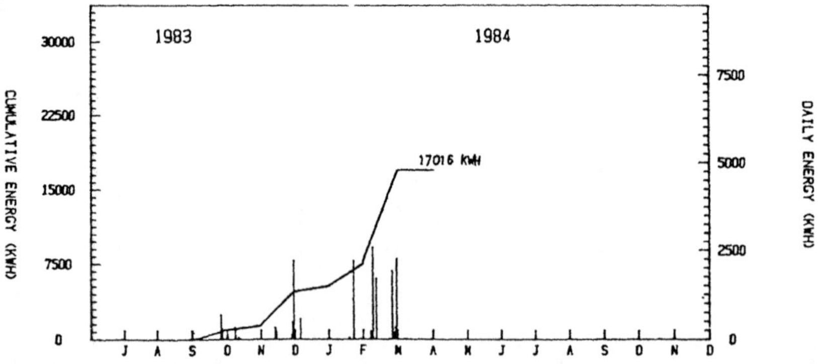

Fig. 5 ANNUAL ENERGY PRODUCTION

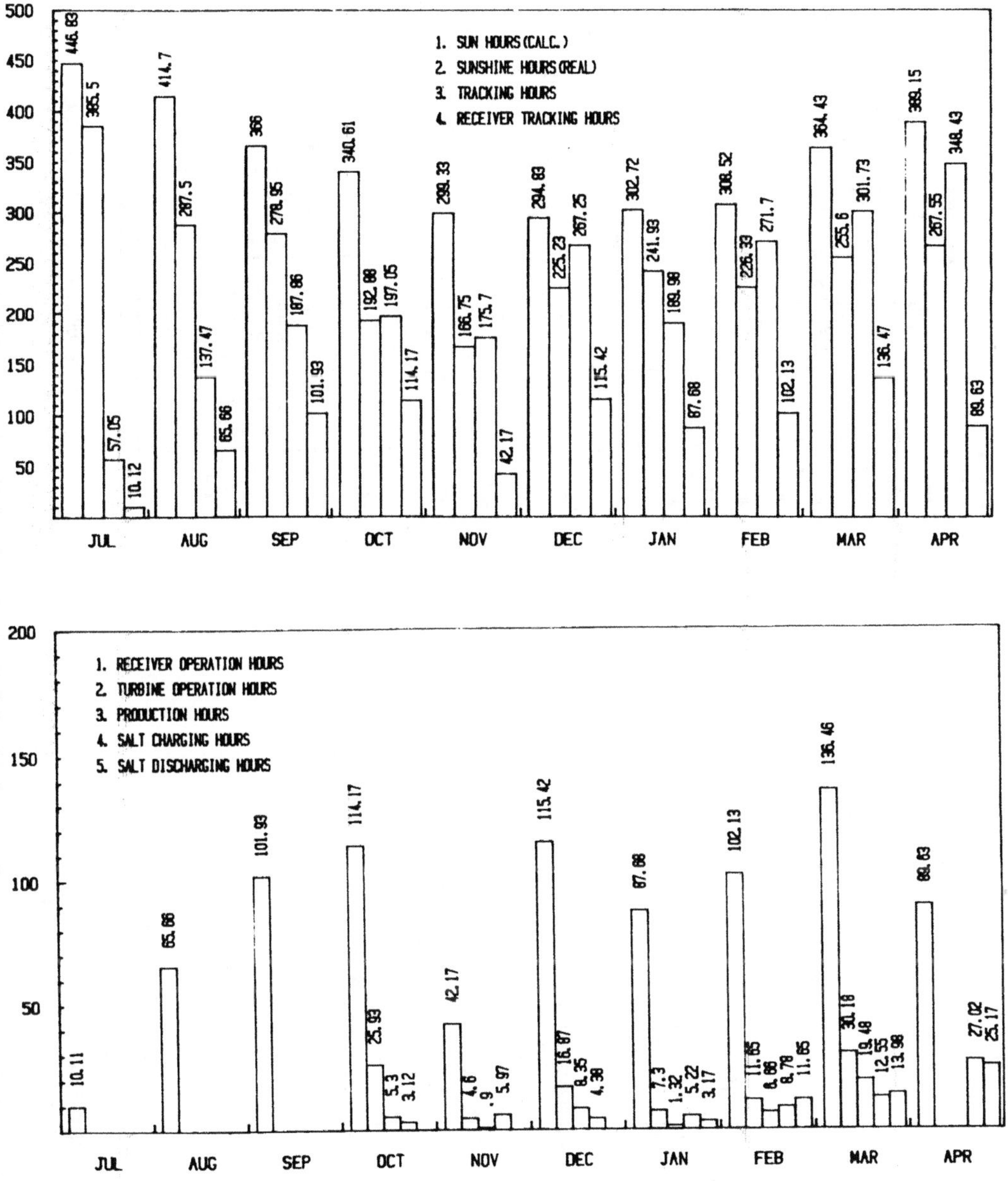

Fig. 6 MONTHLY OPERATIONAL HOURS PLANT SYSTEMS

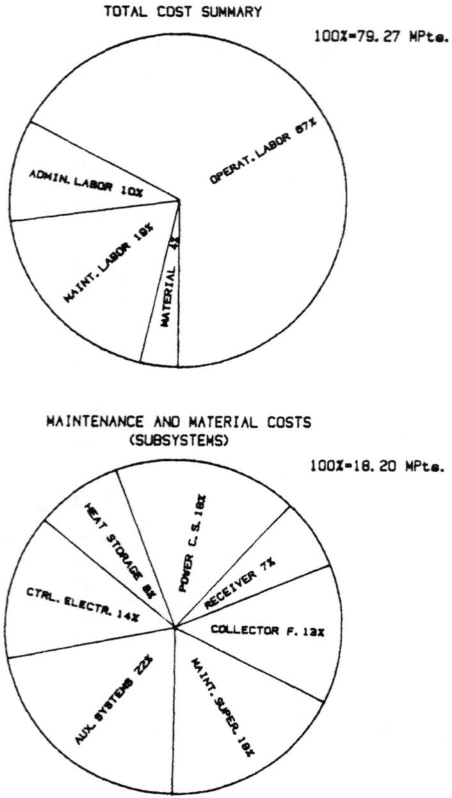

Material costs includes consumables used in plant operation

1 MPte. equivalent to 6500 US $ (April, 1984)

Fig. 7 COST SUMMARY
(JULY.1983–APRIL.1984)

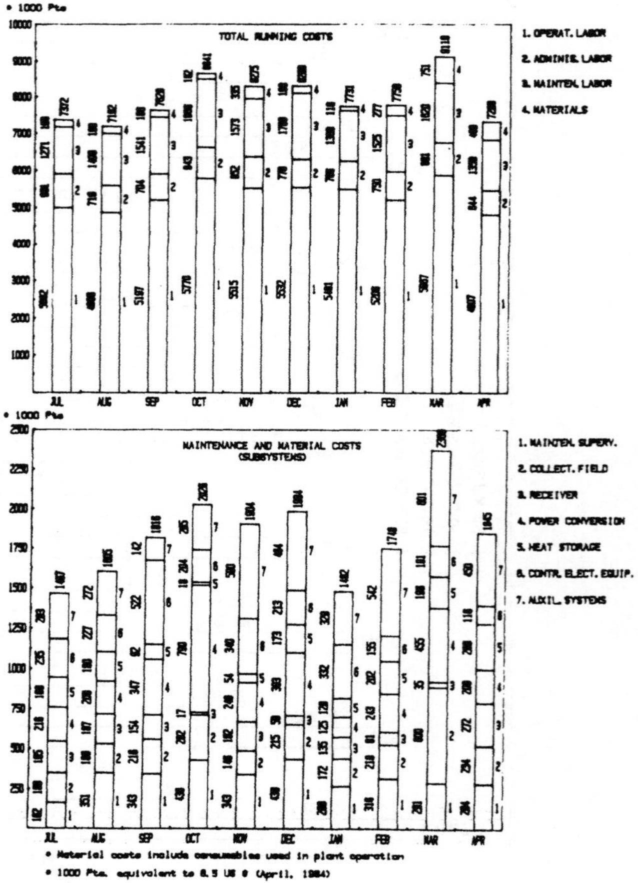

Fig. 8 MONTHLY COST DIAGRAMS

SESSION II - PART 5

ECONOMICS

Summary of the Session by the Rapporteur
G. FANINGER, ASSA, Vienna, Austria

Solar radiation measurements in the Alps

Economic modelling of molten salt solar central receiver plants

Comparison of solar thermal and photovoltaic electricity generation using experimental data from the IEA SSPS project

Development of an economic assessment model

10 MWe solar thermal central receiver pilot plant total capital cost

Analysis of Themis experimental plant costs

Central receiver costs for electric power generation

SUMMARY OF THE SESSION BY THE RAPPORTEUR

G. FANINGER
ASSA, Vienna, Austria

1. INTRODUCTION

At the beginning of the "Small Solar Power Systems--Project Definition Phase," the conditions for solar thermal power generation were favorable:

- Further increase of energy consumption and the foreseen exhaustion of mineral oil, presently the main energy source, made people conscious of the limits of energy supply and initiated the demand for the exploration and utilization of new and inexhaustible energy sources.
- Due to price escalations, particularly on the crude oil market, high investments in power production by solar systems appeared competitive in the foreseeable future.
- Supplying developing countries with energy for industrialization and for improving living conditions seemed only to be assured from novel energy sources, because the increasing living expenses for importing fossil were quickly exhausting the capabilities of Third World economies.
- Furthermore, aspects of conservation spurred the efforts to investigate solar energy as a desirable alternative.

Among the various possibilities for solar thermal power generation, particular attention was paid to the Central Receiver System (CRS) and the Distributed Collector System (DCS). From the economic point of view, the CRS facility was considered to be more effective in higher power units and the DCS facility in the lower power range.

Now six CRS facilities have been planned, constructed, and tested. The major objectives of national and international research projects were to demonstrate the technical feasibilities of solar thermal power generation and also to gather experience for further development of solar-specific components and improving future planning concepts for commercial solar power plants.

According to the state-of-the-art at that time, as much as possible use was made of already tested solar-specific components, such as collectors and heliostats, and of conventional power plant technology for standard components such as turbines and generators.

The funds made available for the research projects excluded in the past a "solar multiple"; so a maximized/optimized operation period of power plant exploitation could not be anticipated. Thus, investigations in the testing phase concentrated on technical testing of overall systems rather than on economic optimization.

The results of more than two years' operation of solar thermal power plants allow a first assessment of the projects' objectives:

* The reliability of the "solar" components come close to the requirements anticipated.
* The expected efficiencies of solar thermal power generation, as well as energy outputs, could not be realized.
* The priority target of the projects, i.e. demonstration of solar thermal power generation, could not be reached during the past test phase.
* Definite conclusions on the economic efficiency of solar thermal power plants cannot be drawn. The reasons for this are that technical troubles prevented routine operation; facility components are not optimally adapted to each other and, due to the research native of the projects, they were not optimized to reduce maintenance and service efforts; and finally, the "solar multiple" indispensible for optimum operation, was lacking.

For the above reasons, it is impossible to make well-supported economic efficiency calculations, especially with a view to future commercial use. A compiled forecast from the figures available from the present CRS facilities currently implemented would be subject to certain restrictions. However, the operational experience gained does provide an opportunity to study economic efficiency considerations in comparison with calculations made prior to the testing of solar thermal power plants and to draw conclusions concerning further research and development work.

2. ECONOMIC ASSESSMENTS

During the session, there were four economic assessments presented:

1. H. Klaiss et al.: "Development of an Economic Assessment Model."
2. T. Durand et al.: "Economic Modelling of Molten Salt Solar Central Receiver Plants."
3. P. Toggweiler and R. Minder: "Comparison of Solar Thermal and Photovoltaic Electricity Using Experimental Data From the IEA SSPS Project."
4. W. Durisch et al.: "Solar Radiation Measurements in the Alps."

2.1 Development of an Economic Assessment Model

The cost analysis of power plants is one of the most important instruments for decision finding. For the cost analysis three main interests exist: those of the government, those of the utilities, and those of the consumers. Different calculation methods are used. Therefore, it was necessary to carry out a study on the appropriateness of the different calculation methods. This has been done in a working group initiated by the IEA-Program SSPS ("Small Solar Power Systems"). The dynamic annuity method and the cash value method have proven to be the last approaches for comparing different concepts.

These methods are generally used for conventional power plants. They may also serve as a basis for sensitivity analysis providing valuable indications for further development and planning concepts.

The evaluation of electricity production costs is leased on selected data sets for the foreseen US 100 MWe Plant. Other data for a reliable economic efficiency calculation are not yet available.

Regarding the results of the cost-analysis, the following statements can be made:

* Different calculation methods lead to different costs: 33 to 52 Pfg/kwh ($.15 to .25/kwh).
* The detailed cost breakdown shows the prevalent part of the investmentment costs.
* The largest share of the investment costs is assigned to the heliostat costs (27.7%). This share is lower than those given in the previous studies (50%). One of the reasons is the lower price of heliostats (400DM/m^2).
* The sensitivity analysis shows the great influence of the discount factors, the total investment costs, and the life expectancy. The influence of the discount factor is particularly high for power plants with high investment costs.
* The sensitivities for personal costs, maintenance, and repair costs are not so significant.
* For operating and maintenance there is a minimum staff of about 30 persons necessary. For a 100 MW plant you need 70 persons, not more than for comparable conventional plants.

An example for a 100 MW CRS-Plant shows that the solar thermal power plants are still too expensive; the factor is amounting to between 3 and 5 compared with "conventional" power plants. It seems reasonable to have an additional period of about 5 to 10 years for research and development to reach a competitive status. These facts and the advantages of the solar energy regarding its overall advantages provide an absolute privilege for solar energy to be promoted by the governments for the next year.

2.2 Economic Modelling of Solar Central Receiver Plants

The main objective of the study has been to provide a simple tool for optimizing the Return on Investment cost (ROI), which is defined as:

$$ROI = (\text{Net Profit})/(\text{Total Construction Cost}).$$

The net profit is mainly determined by the local demand structure. For example, the peak load in California lies at noon, whereas in Europe the peak load occurs mostly in the evening hours.

The total construction cost is broken down into heliostat-field costs, storage costs and power conversion system costs. Not included are operating and maintenance costs in order to keep the model simple.

The following conclusions were drawn:

* No optimal size could be found for the Central Receiver System,
* There was a sharp increase of return on investment in the 1-100 MW range,
* Beyond 150 MW, the return on investment assumed a flat shape,
* Due to the rough assumptions, the exact postioning of the optimum size could not be identified within the range from 50-150 MW.

2.3 Comparison of Solar Thermal and Photovoltaic Electricity Generation

From February 1982 to April 1983, photovoltaic panels had been installed at the site of the IEA-SSPS project in Almeria. The measurements performed at these panels have been used to calculate--by means of a numerical model--the behaviors of a hypostatic 500 kw photovoltaic power plant (PVS) at this site and to compare it with the results of the CRS--not DCS--facilities. The results can be summarized as follows:

* On sunny days (direct radiation sum of approximately 9 kwh/m^2) the outputs of CRS and PVS are roughly equal. At the conditions where CRS net electricity output is just short ($\sim$ 3 kwh/m^2), the PVS-power production is still about 2/3 of maximum production.
* In addition to the comparison of single steps, a comparative study for one full year has been carried out. The main result of this comparison between the "idealized" CRS facility and the photovoltaic plant (PVS) is prescribed by the so-called "electricity production figures." For the time period under investigation (October 1982 to September 1983), it was obtained a figure of

$$E_{net}(CRS)/E_{net}(PVS) = 531 \text{ MWh}/888 \text{ MWh} \approx 0.6$$

* A comparison for a Swiss alpine site (Devos) shows that the electricity production figure is also about 0.6:

$$E_{net}(CRS)/E_{net}(PVS) = 384 \text{ MWh}/647.5 \text{ MWh} \approx 0.6$$

The annual electricity production of the "idealized" CRS is at Almeria as well as at Devos, approximately 60% of the production at a PVS with comparable design point characteristics. The reason for this figure is that Alpine areas are characterized with respect to solar radiation, by an increased annual sunshine duration as well as an increased average direct radiation intensity compared to the Central European lowlands.

The solar thermal power plants are much more sensitive to radiation conditions than photovoltaic systems. In particular, the direct radiation intensity during sunshine conditions and the frequency of cloud transients are main key factors.

2.4 Solar Radiation Measurements in the Swiss Alps

For a better comparison between CRS and PVS and in order to obtain design data for solar power stations in the Alps under actual local conditions, a specific meteorological station has been erected at a possible site for a solar power plant at 2100 m in the Swiss Alps. The autonomously working meteorological station delivers data on global and diffuse solar radiation in a horizontal plane, global radiation in an inclined plane facing south, sunshine duration, wind speed, wind direction, and temperature. The first measurements were made on November 1st of last year, and it is planned to continue with them until the summer of 1985.

SOLAR RADIATION MEASUREMENTS IN THE ALPS

W. DURISCH, P. KESSELRING, D. BARANDAO, M. BUEHLMANN
Swiss Federal Institute for Reactor Research, EIR
CH-5303 Würenlingen, Switzerland

Summary

In order to obtain design data for CRS power stations in the Alps under actual local conditions, a specific meteorological station has been erected at a possible site for a CRS plant at 2100 m a.s.l. The autonomously working meteostation delivers data on global and diffuse solar radiation in a horizontal plane, global radiation in an inclined plane facing south, sunshine duration, wind speed, wind direction and temperature. All quantities are continuously measured and five minute mean values (except for sunshine) are recorded. The data is stored on tape and collected every month. Evaluation is then performed on the in-house computer. The first measurements were made on November 1st last year, and it is planned to continue with them until the summer of 1985. The measured data to date show an interesting behaviour. As was expected, the maximum normal incident power is higher than at lower locations. Comparisons concerning the yearly radiation yields are however not yet available.

1. INTRODUCTION

Meteorological data are required not only for the choice of proper sites for solar power stations but also for the design of such plants. These data are also needed to assess the power station's performance and their economic efficiency. Furthermore they are necessary for the estimation of the potential of solar power stations in regions such as the Swiss Alps.

In Switzerland about 60 automatic meteorological stations exist /1, 2, 3/. They deliver highly valuable data, e.g. on the sunshine duration, the global radiation (horizontal plane) and in some cases on the diffuse radiation (horizontal plane) /4, 5, 6/. However, concerning the normal incident radiation very little measurements and related information is available /7, 8, 9, 10/.

In addition, the available normal incidence measurements come from places which are not interesting for solar thermal power stations and interpolation and extrapolation of radiation data is difficult. Therefore, it is very important to perform on-site measurements at potential locations for solar thermal power stations.

In the course of the SOTEL*-project /11, 12, 13/, a power station-specific meteorological station was erected and began operation in the fall of 1983 /14, 15/. The station is located at a potential site for a solar thermal power pilot plant in the southern part of the Swiss Alps at a height of about 2100 m a.s.l. In this station emphasis is on the amount and time distribution of solar radiation data, sunshine duration and wind conditions. Because the station is located at a remote area, on-line power and daily

* Solar Thermal Electricity

inspection in winter is not possible. Therefore, many special modifications of the sensing elements were required in order to have an unattended continuous operation for one month. The power supply of the station was successfully managed by a photovoltaic power system /16/.

2. MEASURING METHODS/INSTRUMENTATION

2.1. Global Radiation

The global radiation (on horizontal plane) is measured using a Kipp and Zonen Pyranometer CM 10 (temperature-compensated) which was calibrated at the World Radiation Center (WRC) at Davos. The pyranometer produces a voltage signal proportional to the incident global radiation. The signal is continuously transmitted to the data acquisition system where five-minute mean values are calculated according to:

$$\bar{U} = \frac{1}{T} \int_{t-T}^{t} U(t')\, dt' \qquad (1)$$

These mean values are stored on a magnetic tape. A five-minute sampling represents a fairly good resolution. In order to avoid icing of the pyranometer dome as well as the deposition of rain, dew, snow and rime the dome is constantly (24 hours) supplied with slightly warmed air. This is managed by a specially developed ventilation box made by METEOLABOR. As expected, no pollution of the dome has yet been observed. According to specifications, the maximum relative error of CM 10-instruments is approximately $\pm$ 2 % /8, 17/.

2.2. Diffuse Radiation

The diffuse radiation (on a horizontal plane) is being measured using a CM 10-pyranometer and a shadow ring, both from Kipp and Zonen. In order to obtain maintenance-free operation of the diffuse pyranometer for at least one month, the shadow ring had to be modified. The corresponding ring correction factors were calculated assuming an isotropic distribution of the diffuse sky radiation /18, 19/. To keep the pyranometer dome free of rain drops, dew, snow and rime, a special ventilation device was developed and applied. This device constantly supplies the dome with slightly warmed air. The sampling interval and the calculation of the corresponding mean values are the same for the global pyranometer. Extensive investigation of the uncertainty of the shadow ring correction factors have been performed by Ineichen /8/. He concludes that the ring correction factor calculated according to the isotropic distribution model has an uncertainty in the order of $\pm$ 5 % /20/. Together with the maximum error of the pyranometer signal, this results in a total maximum relative error of $\pm$ 7 % for the diffuse radiation.

2.3. Normal Incident Radiation

For solar thermal applications the most important component of solar radiation is the normal incident radiation coming directly from the sun. It can be measured continuously with normal incidence pyrheliometers (NIP) mounted on solar trackers. However, the trackers available on the market are not maintenance-free but must be checked and aligned regularly, nor are they robust enough for alpine applications. The available pyrheliometers are not weather-resistant enough for alpine applications without maintenance because of the possible accumulation of rain, dew, rime and snow on the cavity's window.

For these reasons it was decided to apply the indirect method to measure the normal incident radiation. This method is based on the measurement of the global (G_H) and diffuse (D_H) radiation, both on a horizontal plane. The normal incident radiation I_N is then calculated according to

$$I_N = (G_H - D_H)/\sin h \qquad (2)$$

where h is the height of the sun above the horizon at the time of measurement. The sun height h can be calculated exactly /21/. The indirect method for I_N has the advantage that the measuring instruments required can easily be modified so that they need no regular inspection for at least one month, even under rough weather conditions. This is very important for sites in which access is difficult in winter. The indirect method has however the disadvantage that at small values of the difference $G_H - D_H$, e.g. at sunrise and sunset, the relative error of I_N becomes large. However, on these occasions the normal incident radiation I_N is small as well and therefore not important for solar thermal applications. The indirect method suffers furthermore from the "cloud effect" for which the difference $G_H - D_H$, and therefore I_N, becomes too large. The effect is caused by the shadowring, which blocks not only the sun's direct radiation and a part of the clear sky radiation (for which corrections are made) but also a certain amount of cloud-reflected radiation which is absorbed by the global meter. While non-blocked cloud-reflected radiation is absorbed by both pyranometers it is compensated by the difference $G_H - D_H$, but the blocked radiation is not compensated. It is practically impossible to take this effect into account because a corresponding correction is dependent on the continually changing cloud configuration. However, this effect usually occurs only for weather conditions which are not relevant for solar thermal power stations. An effect similar to the "cloud-effect" is caused by the reflection of snow-covered hills and mountains higher than the horizontal pyranometer plane. Because this effect occurs on sunny days when a solar thermal power station would be in operation, this effect (albedo) has to be taken into account.

The indirect method to measure the normal incident radiation has successfully been applied, for instance, by Ineichen et al /8/.

From the maximum relative errors of the global (G_H) and diffuse (D_H) radiation measurements, the maximum relative error of the indirectly measured normal incidence radiation (I_N) can be deduced from equation (2). It amounts to $\pm 9\%$. One has to have in mind that this is the "worst case" and that most I_N measurements are much more accurate. Nevertheless, whenever maintenance and weather conditions permit, the much more precise pyrheliometer method should be used for the measurement of I_N.

2.4. Sunshine Duration

Sunshine duration is measured by means of a HAENNI Solar 110 sunshine detector. This instrument has the advantage that its output can be recorded digitally and that it can operate for at least one month unattended. The sensitivity of the device has been set according to the recommendation of the World Meteorological Organisation (WMO). At sunshine the meter delivers one impulse per second. Sunshine duration data as well as frequency and duration of sunshine periods are some of the most important information required, for solar energy systems.

3. DATA ACQUISITION AND EVALUATION

For data acquisition, an ELMES Combilog system has been chosen. The most important reasons for this choice was for its low power consumption (approx. 3 W) and the ability to supply it with direct current (24 V). As photovoltaic powering was the only feasible way to supply the meteorological station with electricity, low power consumption was a boundary condition not only for the acquisition system but for all electricity consumers. The chosen Combilog furthermore has the advantage that conventional magnetic cassetts with high storage capacity can be used. For safety reasons a second acquisition system has been installed for the two most important data-lines G_H and D_H. For this purpose a solar integrator from Kipp and Zonen has been chosen. It records ten-minute mean-values and daily sums on a strip chart. Both acquisition systems have enough storage capacity for one month of unattended operation. A flow diagram of all data lines leading to the acquisition and logging system is shown in Fig. 1. Approximately every month (depending on weather conditions) the station is inspected and the full data cassette is replaced by an empty one. The cassette's data are then copied on half inch magnetic tape and processed on the in-house computer, Fig. 1. After preliminary evaluation of the data each month's information is stored on separate discs (data base) for statistical evaluation at the end of the measurement campaign.

4. FIRST RESULTS

Every month the daily sums of the irradiated energy are evaluated and plotted /22/. As an example, Fig. 2 and 3 show the daily sums of the global (G_H), diffuse (D_H) and normal incident (I_N) radiation for the month of March 1984. As can be seen in this Fig., there were two sunny periods: one from the 4th to the 7th and one from the 9th to the 14th. The corresponding monthly integrals of G_H, D_H and I_N are 120, 44 and 161 kWh/m^2 resp. From the sunshine duration data a monthly diagram was prepared according to Egger /23/, Fig. 4. The two diagrams, Figs. 2 and 3, give a good monthly overview. The evaluation and results of the wind conditions and temperature will be reported later. Fig. 5, 6 and 7 show the daily variation of the global, diffuse and normal incident radiation for a "perfect", "medium nice" and overcast day. As can be seen from Fig. 5, a normal incidence power of 1000 W/m^2 even over several hours seems to be quite possible, contrary to places near sea level. The daily course of the global and diffuse irradiation show no extraordinary behaviour.

5. FUTURE WORK

A sun tracker from Li Cor and a pyrheliometer from Eppley will be modified so that both can resist the severe alpine weather conditions. The complete device will then be installed at the meteorological station described in this paper. A second meteorological station at another possible region for a thermal solar power station in the Alps will be taken into operation this coming fall. The statistical evaluation will be started as soon as one year's data set is complete. This evaluation will include not only the usually applied methods but will also include the determination of the following distribution functions:

$$t(I,\tau) = (t)_{I_N \geq I \wedge \tau_i \geq \tau} = (\Sigma\tau_i)_{I_N \geq I \wedge \tau_i \geq \tau} \tag{3}$$

$$E(I,\tau) = (E)_{I_N \geq I \wedge \tau_i \geq \tau} = (\Sigma\int I_N dt)_{I_N \geq I \wedge \tau_i \geq \tau} \tag{4}$$

$$n(I,\tau) = \frac{1}{N_o} N(I,\tau) = \frac{1}{N_o} (N)_{I_N \geq I \wedge \tau_i \geq \tau} \tag{5}$$

Equation (3) gives the cumulated number of hours (per week, month, or year) available at intensities I_N higher than the chosen level I and periods τ_i, longer than the fixed period τ.

Equation (4) gives the accumulated energy (per week, month or year) available at intensities I_N higher than the chosen level I and at periods τ_i longer than the fixed period τ.

Equation (5) finally gives the relative number of periods (per week, month or year) for which I_N is higher than I and at the same time τ_i is longer than τ.

All three distribution functions are of fundamental importance not only for the design but also for the energy yield of solar thermal power stations.

However, to get the limit functions of the distributions (3) to (5) a data base containing continuous measurements of several years is required.

6. CONCLUSIONS

The meteorological station presented in this paper has proven reliable even under the severe weather conditions of the alpine winter, and is able to operate unattended for one month. The indirect method to measure the normal incident radiation is sufficiently precise for solar thermal engineering purposes. However, when ever possible, pyrheliometer measurements are preferred. Extensive comparisons with other data are not yet possible. To obtain information on the local variation of the isolated energy at potential sites for solar thermal power stations, further meteorological stations must be erected and taken into operation. To collect a representative set of data for a detailed statistical evaluation, the local meteorological data have to be measured carefully over a long time period (at least a few years). The measured data can be used to check radiation calculation models as proposed, for instance, by Ambrosetti /24/. These data can also be used to check the transferability of radiation values from place to place and to develop transfer models. They can also be used to correlate satellite data with terrestrial measured data.

REFERENCES

1. T. Gutermann: ANETZ, ein neues Messnetz automatischer Wetterstationen in der Schweiz, STZ, Nr. 15, 26. Juli 1979, 937
2. P. Valko: Solardaten für die Schweiz, Richtwerte der Global- und Diffusstrahlung für Ingenieure und Architekten, Docu-Bulletin, Sonderdruck aus Heft 3/1982, Schweizer Baudokumentation.
3. Anon.: Meteodaten für die Sonnenenergienutzung, SIA-Dokumentation 64, Zürich 1983

4. M. Schuepp: Sonnenscheindauer, Klimatologie der Schweiz, Heft 3/I, SMA 1962.
5. F. Ambrosetti: Strahlung 1. Teil, Klimatologie der Schweiz, Heft 17/I, SMA 1978.
6. F. Ambrosetti und P. Valko: Strahlung 2. Teil, Klimatologie der Schweiz, Heft 26/I, SMA 1980.
7. Anon.: Heliostat Test Activities in the Swiss Alps, Final Report, Davos, Sept. 23, 1980.
8. P. Ineichen: Quatre Années de Mesures d'Ensoleillement à Genève, These 1983, Genève.
9. C. Fröhlich: Solar Radiation at the earth's surface, Solar World Forum, Solar Technology in the 80's, edited by D.O. Hall and June Morton, Vol. 3, Pergamon Press 1982, p. 213.
10. W. Durisch: Solarstrahlungsdaten; Solarthermisches Elektrizitätswerk SOTEL, 1981, p. 31, ISBN 3-85677-002-X.
11. Anon.: Solarthermisches Elektrizitätswerk SOTEL, Projektvorschlag, ausgearbeitet durch ein Schweizer Konsortium, Oktober 1981, ISBN 3-85677-002-X.
12. W. Durisch, P. Kesselring, W. Nägeli: Vom Heliostatversuch zum Pilot-Solarkraftwerk, EIR-Bulletin Nr. 47, Sept. 1982.
13. W. Durisch, P. Kesselring: The Swiss Solar Central Receiver System Project SOTEL, internal report, Dec. 1982.
14. S. Khandkar: Development of a Data Acquisition System for the Measurement and Analysis of Solar Radiation in the Alps, interner Bericht, TM-13-83-11, EIR, 1983.
15. B. Hofer: Entwicklung einer vollautomatischen batteriebetriebenen Wetterstation (mit Hauptgewicht auf Solarstrahlungsmessung) und Analyse von Messungen in Würenlingen (23.7.83 bis 23.9.83), interner Bericht, TM-13-83-15, EIR 1983.
16. M. Bühlmann: Energieversorgung einer Meteostation mittels einer Fotovoltaikanlage, interner Bericht, TM-55-84-1, EIR 1984.
17. K. Dehne: Messinstrumente zur Beobachtung der Strahlungsintensität, Bd. 3, Udo Pfrimer-Verlag, München 1977.
18. A. Drummond: On the measurement of sky radiation, Arch. Meteor. Geophys. Bioklim. Ser B7, 413 - 436, 1956.
19. Van Vely: private communication, Kipp and Zonen
20. P. Ineichen et al: Study of the corrective factor involved when measuring the diffuse solar radiation by use of the ring method. Solar Energy Vol. 32, No. 5, pp. 585-590, 1984.
21. U.S. Nautical Almanac 1983, 1984.
22. D. Barandao: Evaluation des mesures d'ensoleillement d'une station météorologique dans les Alpes, interner Bericht, TM-13-84-5, EIR 1984.
23. H. Egger, TAGES-ANZEIGER, Mittwoch, 3. Aug. 1983.
24. P. Ambrosetti: Beschreibung einer Methode für die Umrechnung des Sonnenstrahlungswerte von einer horizontalen Fläche auf eine beliebig orientierte und geneigte Fläche, interner Bericht, TM-33-83-6, EIR 1983.

ACKNOWLEDGEMENT

The authors would like to express their sincerest thanks to all those who have contributed to the successful operation of the meteorological station described in this paper. In particular they thank Mr. P. Toggweiler (Data acquisition system), Mr. W. Nägeli and his team (installation), Mr. B. Housley and Mr. P. Tobé (Data evaluation). Thanks also to Miss Kim Wheeler for English corrections and B. Heierling for typing this paper.

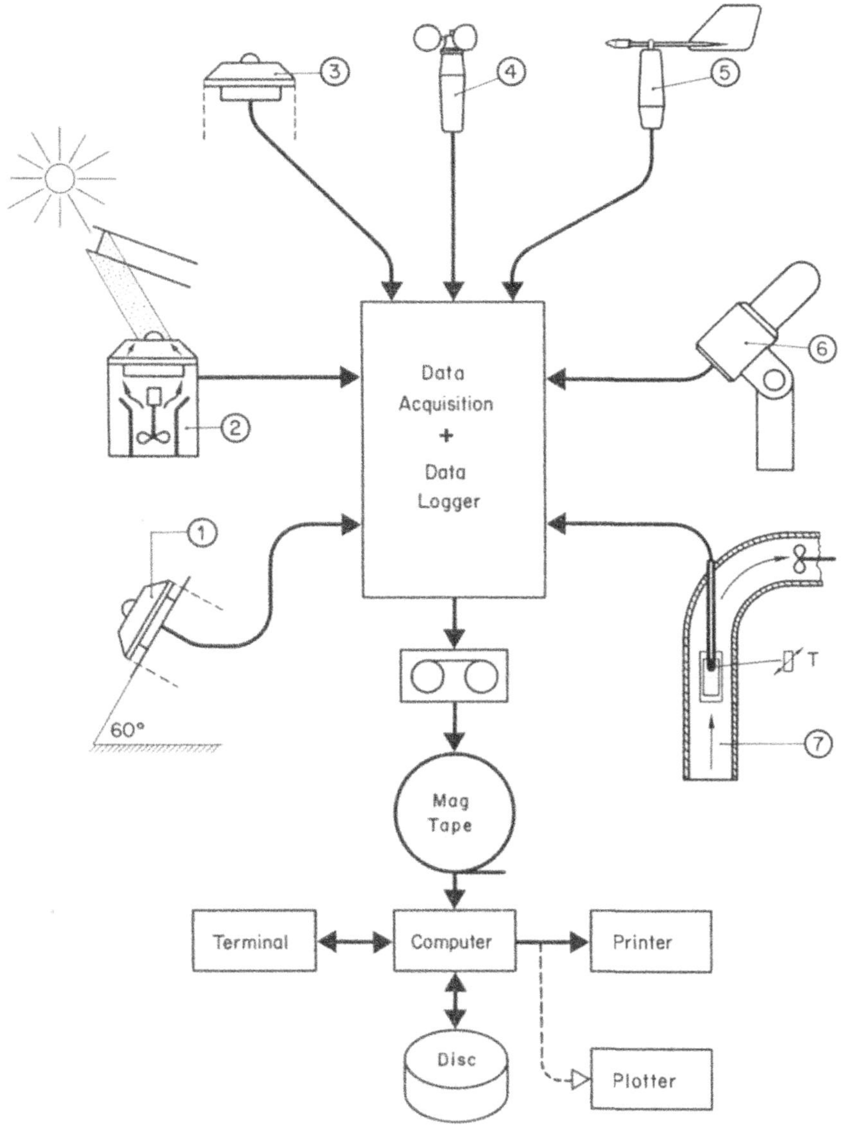

Fig. 1: Sensors, data acquisition and evaluation systems used with the CRS-specific meteorological station in the Swiss Alps. 1 Inclined global pyranometer, 2 Diffuse pyranometer, 3 Horizontal pyranometer, 4 Anemometer, 5 Wind direction vane, 6 Sunshine detector, 7 Platinum resistance thermometer with radiation shields and ventilation. All pyranometers are ventilated (dew, rime, snow). The inclined pyranometer facing south delivers data for photovoltaic power stations.

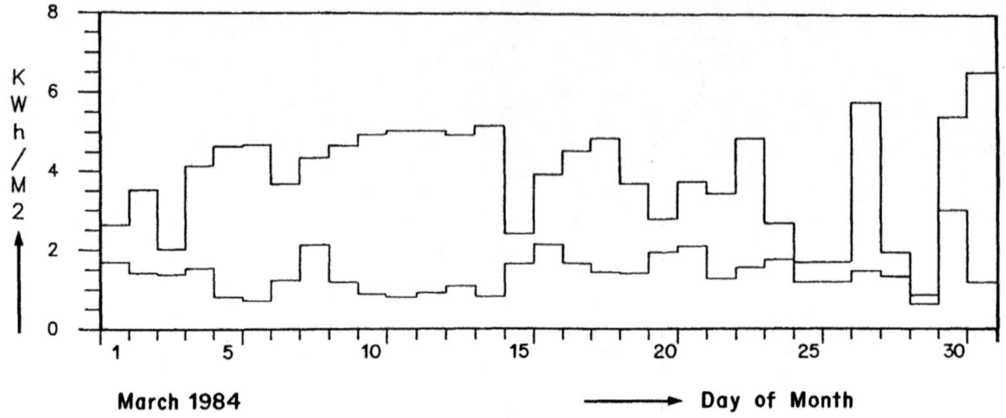

Fig. 2: Daily integrals of the global (upper line) and diffuse (lower line) irradiated energy on a horizontal plane.

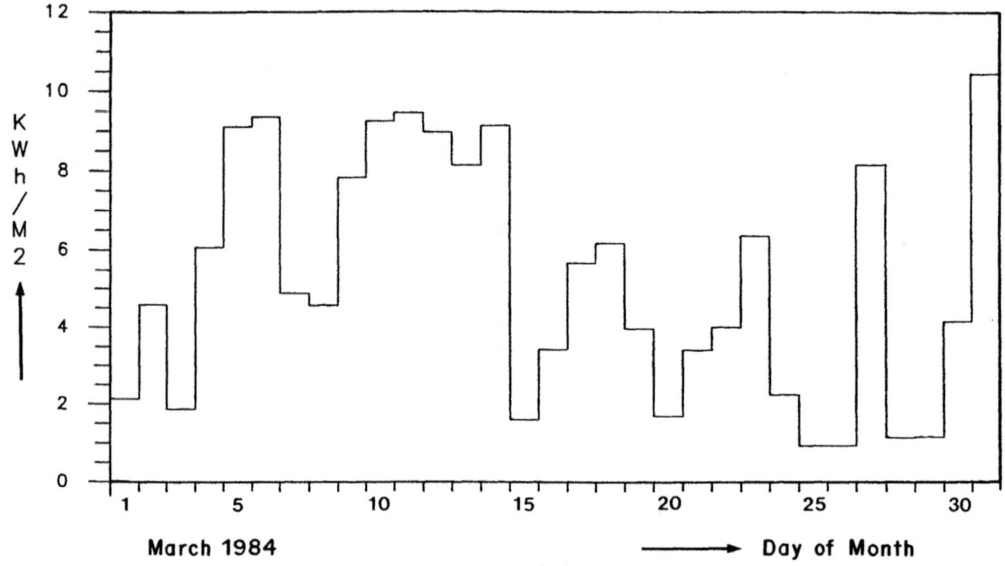

Fig. 3: Daily integrals of the normal incident radiation (direct radiation) on a plane following the sun's course.

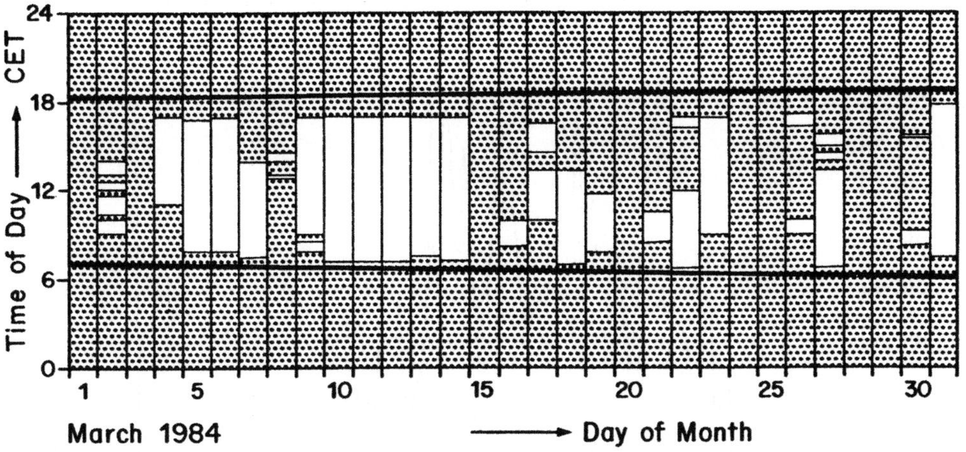

Fig. 4: Monthly sunshine duration chart. Clear spaces represent sunshine, the dotted area sunless periods. The lower solid line represents the astronomical sunrise, the upper one the astronomical sunset.

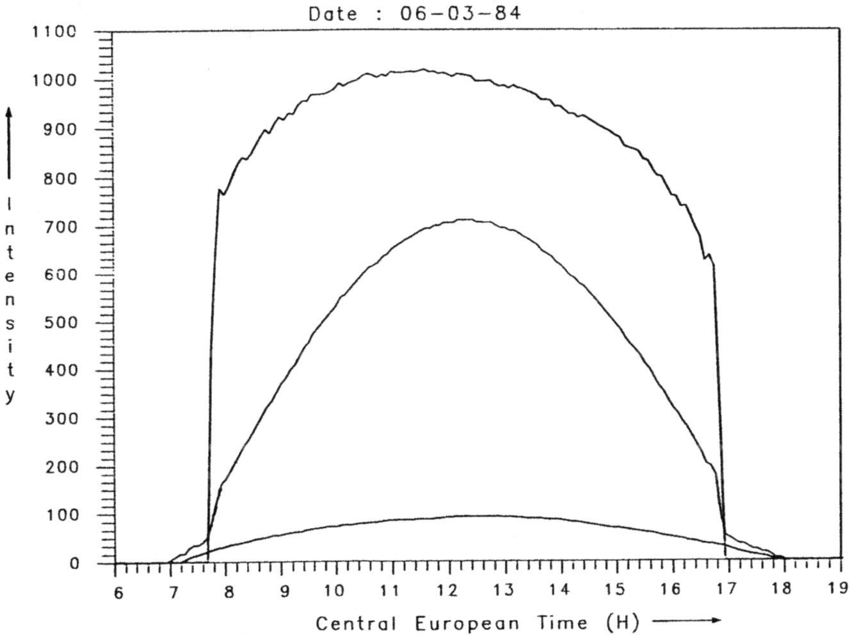

Fig. 5: Daily variation of the normal incident (upper line), global (middle line) and diffuse irradiation for a "perfect" day.

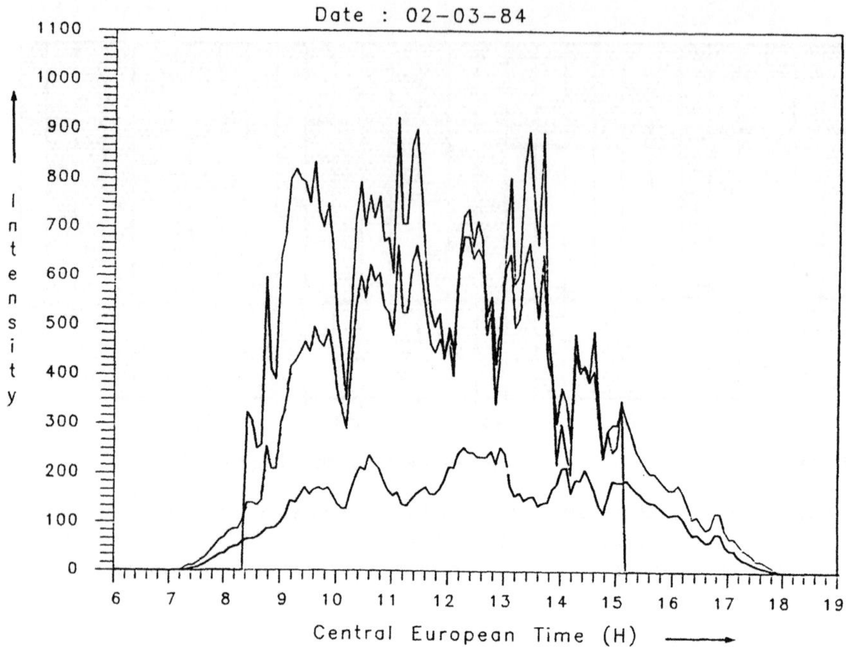

<u>Fig. 6:</u> Daily variation of the normal incident (upper line), global (middle line) and diffuse irradiation for a "medium nice" day.

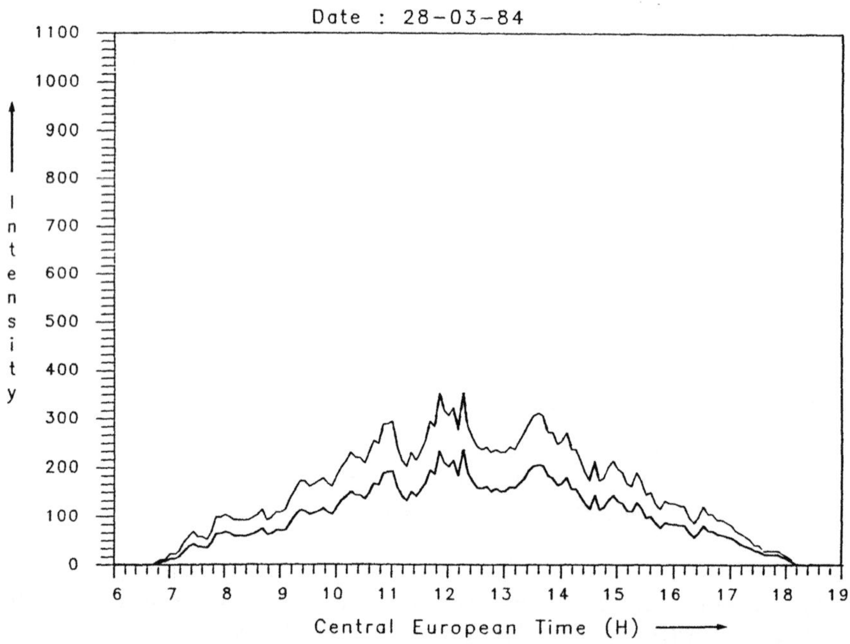

<u>Fig. 7:</u> Daily variation of the global (upper line) and diffuse (lower line) irradiation for an overcast day (no direct irradiation).

ECONOMIC MODELLING OF
MOLTEN SALT SOLAR CENTRAL RECEIVER PLANTS

Thomas DURAND, Saïd IBRAHIMI, Duc PHAM HI *
Claude ETIEVANT**

* Laboratoire L.E.I.S., Ecole Centrale de Paris, France
** Mission Maîtrise de l'Energie, Ecole Centrale de Paris, France

SUMMARY

In order to rationalize decision-making on Central Receiver Systems (C.R.S.), both financial and physical data must be integrated in the design and optimisation of a project. The model presented here has been developped for that need, at the Ecole Centrale de paris, under contract with the C.N.R.S., The French National Agency for Research.

Theoretical considerations will first explicit the interactions between budget constraints, subsystem construction costs, weather conditions and variable price of electric energy, among other parameters. Income and costs are mathematically evaluated as functions of total heliostat surface area, capacity of thermal storage, turbine power etc. Return on investment is optimized through a simplex algorithm applied to a scanning method. The cases of three countries (France, U.S.A. and Israël) are then investigated with this tool and conclusions are drawn from the graphs obtained correspondingly.

1. INTRODUCTION

There are several major decisions that have to be made by any investor willing to finance a Central receiver solar plant. Answers have to be given to questions such as : what is the optimal size of the plant ? What is the optimal amount of money to invest ? How to share optimally the available financial resources between the various subsystems of the plant (Collector, Storage, Turbine-generator) ? How should any solar facility be run optimally ?

The present paper is an attempt to model the economics of a molten salt Central receiver solar plant. The model has been developed at the ECOLE CENTRALE de PARIS (France) in order to give on the basis of a specific economic criteria, a quantitative answer to the above questions.

There is a range of economic criteria that can be proposed to help the investor in making his choices. Examples of targets of such economic criteria are as follows :

(a) Minimize the total cost of the plant
(b) Minimize the pay back time
(c) Maximize the net present value
(d) Minimize the electricity cost
(e) Maximize the return on investment (ROI)
...

Criterion (e), ROI, is selected as the objective function of the present economic model.

The return on investment is defined as follows :

(ROI) = Net profit/Total construction cost

The problem of maximizing (ROI) is rather complex because there are actually two levels of optimizations :

- The total construction cost has to be evaluated for any plan size and for any of subsystem parameters ;
- The net profit taken into account cannot be obtained from a straight forward evaluation. It is itself the result of a sub-level optimization by optimal dynamic control methods, under constraints coming from the limiting capacities of the various subsystems.

  The complexity arises from the fact that relative sizes of subsystems are simultaneously, variables of the overall optimization on plant size, and constraints of the sub-level optimization on net profit.

2. MODELLING

2.1 Modelling the construction cost

Figure I shows the conceptual CRS plant considered. The following equation is used for modelling the plan construction cost :

$$B = a_1 A + a_2 P^2 + a_3 W^{2/3} + a_4 W + a_5 P + a_6 \quad (1)$$

Where the three variables A, P, W characterize the total heliostat surface area, the thermal storage capacity and the turbine generator design power respectively. The coefficients a_1, a_2, a_3, a_4, a_5, a_6 have to be calculated by matching equation (1) with actual costs of projets of CRS plants. A more detailed model of a CRS construction cost could have been indefferently used instead of equation (1), but at the present stage of our study there is no real need for more complexity.

MODELLING

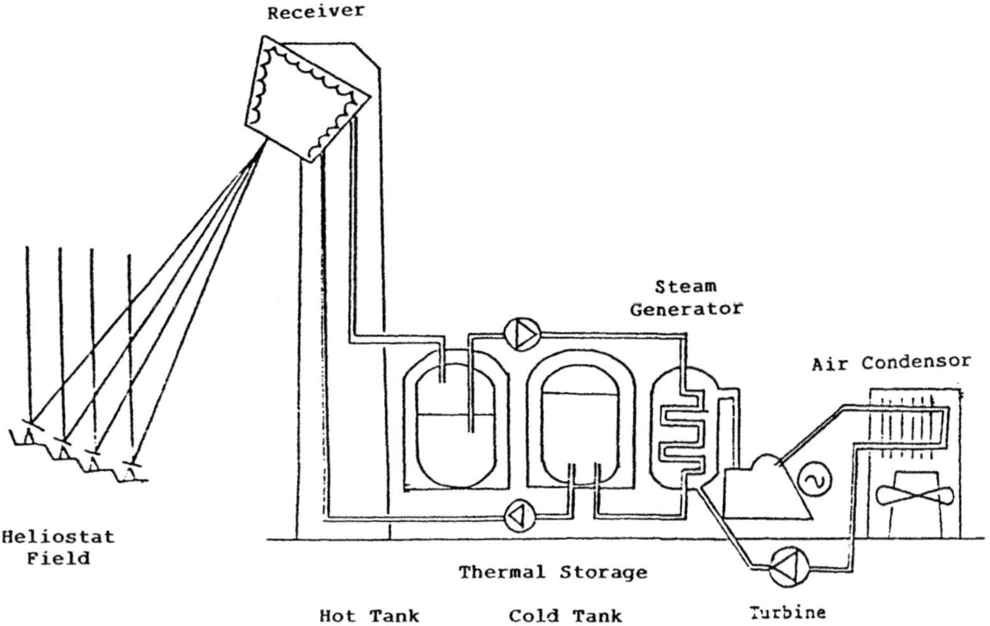

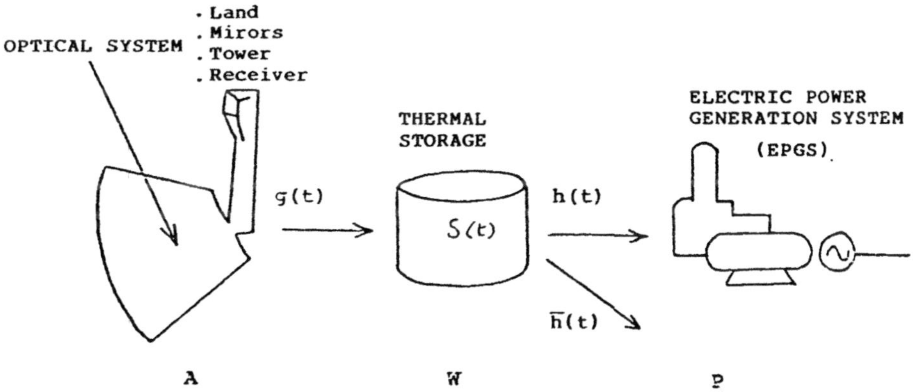

FIGURE 1

CONSTRUCTION COST

$B = a_1 A$		Heliostat Field Cost
$+ a_2 P^z$		Turbine Cost $z = 0.8$
$+ a_3 W^{2/3}$	Storage	Tanks Cost
$+ a_4 W$		Salt Cost
$+ a_5 P$	Other	Variable Cost
$+ a_6$		Fixed Cost

MAJOR VARIABLES / PARAMETERS

TOTAL RECEIVING AREA	A	(m^2)
TURBINE POWER AT DESIGN POINT	P	(M We)
STORAGE CAPACITY	W	(M W th)
DECISION VARIABLE	h(t)	

2.2 Modelling the net profit, Optimal control

Assuming a given location of the CRS plant, it is possible to evaluate the solar energy input on the collector field by using the meteorological statistics of the site. On the other hand the value of electricity sales is a function depending on both the electricity production rate h(t) and the electricity value W(t) versus time. Over a period T the sales value of electricity can be

$$J(h) = \int_0^T h(t)\,W(t)\,dt \qquad (2)$$

Where h(t) is the variable that describes the management of energy produced by the plant. h(t) is the decision variable on which the model will optimize.

The objective function of the optimal control problem is J(h). This optimization must be done taking into account the following constraints :

. The energy stored in the hot tank can not exceed the thermal storage capacity limit ;
. The energy stored can not be negative ;
. The power produced by the plant can not exceed a given power limit (i.e. Turbine-generator design point) ;
. The plant output energy flow is non negative ;
. Continuity equation on energy determines the relationship between the solar energy input, the stored energy and the electricity production.

At this point, main subsystem efficiencies are taken into account.

2.3 Mathematical Model

A mathematical model has been built, providing for any set of variables A, P, W both the total cost B of the plant and the corresponding valorization of electricity sales J(A,P,W). The model finally determines the optimal set of A, P, W values corresponding to the maximum possible ROI, see figure 3. The model requires the following inputs :

. Electricity price versus time during the day and its seasonal variation : W(t) ;
. Meteorological statistics giving the variation of the direct solar energy flux versus time over typical days of each month or season of the year : g(t) ;
. Main subsystems efficiencies (Heliostat field and receiver, thermal storage, turbine generator) ;
. Construction cost function coefficients : a_1 to a_6.

The set of equations describing the search for best net profit by optimal control is solved by using linear programming techniques. Given the two optimization problems, on size and on optimal control, the numerical routine consists in scanning the A P W triplets space, optimizing the control through a simplex algorithm for any given CRS characterized by A, P and W, and computing the ratio J/B for each triplet explored.

3. RESULTS

The model has been used to investigate the optimal specifications of CRS plants in several countries of interest : France, U.S.A. California, Israël ...

OPTIMAL CONTROL

Assuming a Plant exists with Budget/Size defined by A, P, W :

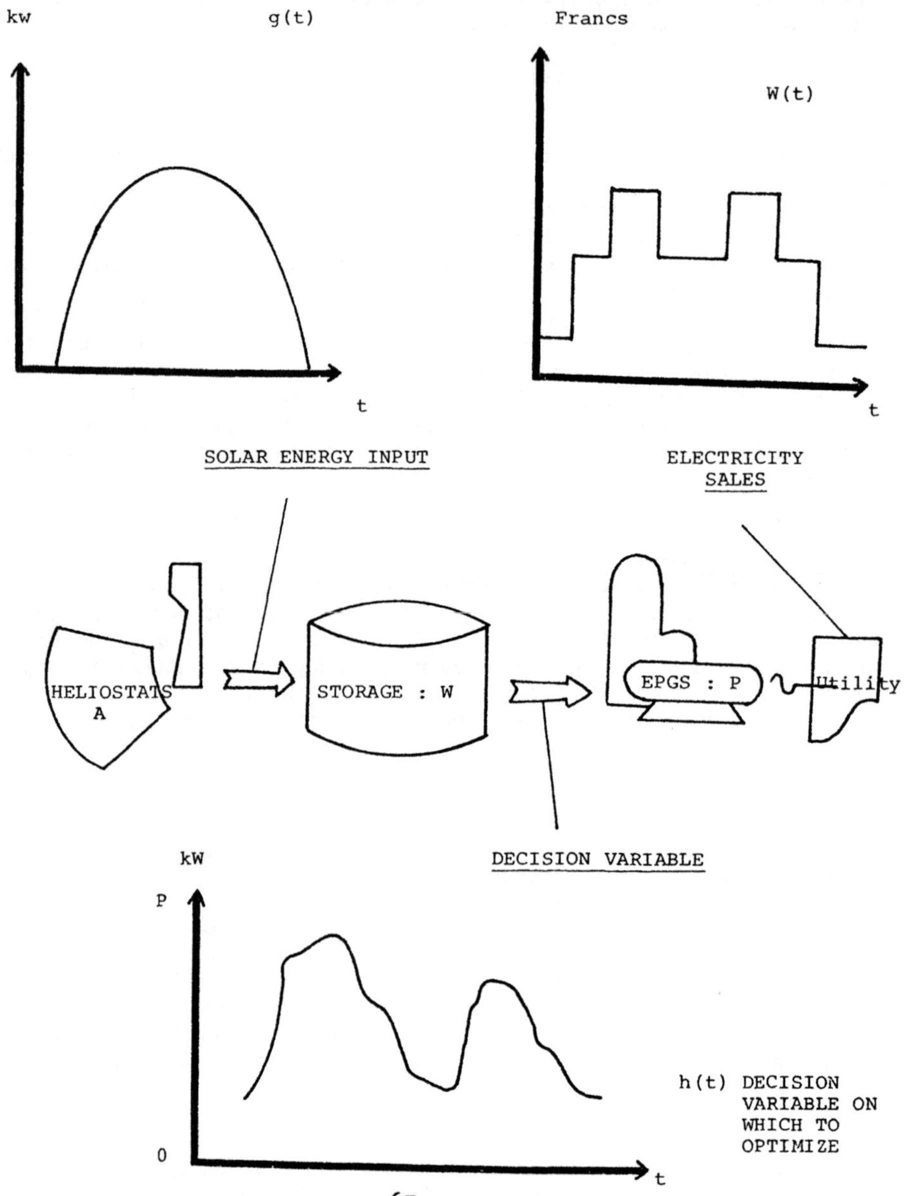

OBJECTIVE : MAXIMIZE $J(h) = \int_0^T h(t)\, W(t)\, dt$

FIGURE 2

MATHEMATICAL MODEL

SIZE OPTIMIZING :

$$\max_{A,P,W} r(B) = \frac{J(A,P,W)}{B}$$

A, P, W

where $B = a_1 A + a_2 P^z + a_3 W^{2/3} + a_4 W + a_5 P + a_6$

and where $J(A,P,W)$ is solution of :

Optimal Control

$$J = \max_h \int_o^T h(t)\, w(t)\, dt$$

$$S(t) = S(o) + \int_o^t (g-h-\bar{h})(u)\, du$$

$$o \leqslant h \leqslant P$$

$$o \leqslant S(t) \leqslant W$$

$$o \leqslant \bar{h}$$

$$S(T) = S(o)$$

J(A,P,W)

FIGURE 3

3.1 France

Inputs

The meteo data used are those of Carpentras, a typical mediterranean town in the South of France whose weather station provides fully established statistics over many years.

The component and subsystem costs are those estimated by technico economic studies described in [1]. The price of electric energy versus time corresponds to the specifications of "La Note Bleue d'E.D.F." [4].

Results

Size effect : for each optimal set of A, P, W values, (ROI) is plotted versus P in fig. 4. A drastic decrease of (ROI) is observed for plants with a turbine generator design point below 100 MW.

Influence of weather data on (ROI) : in figure 5 are represented two curves, one obtained with actual weather data from Carpentras and the other with hypothetical sunshine conditions (summer insolation all year long).

Optimal heliostat field area : for any P value, there is a set of associated optimal values for W and A. In fig. 6 P is fixed, W is kept at optimal value and A is made to vary around from the optimal value. The variation of (ROI) shows a sharp maximum at the optimal value of A.

Optimal storage capacity : similarly, for given values of P and associated optimal values of A, the thermal storage capacity W is made to vary around the optimal value showing a rather flat maximum (fig. 7).

3.2 United States California

The case of CRS plants installed in southern California has been investigated using our model.

Inputs

The meteo data are those recorded in Barstow [5]. The cost coefficients for plant construction are evaluated on the basis of informations taken from project SOLAR 100 [6]. The price of electric energy as a function of time is obtained from the Southern California Edison Tariff schedules [7].

Results

Fig. 8 displays the variation of (ROI) versus Collector fields area for a 100 MW CRS plant. Each point is represented with the assoicated optimal value of the thermal storage capacity.

Fig. 9 displays the variation of (ROI) versus thermal storage capacity for a 100 MW CRS plant with an area of collector field fixed at optimal value.

3.3 Israel

Similarly the case of a CRS plant installed in Israel has been investigated.

Inputs

Construction cost coefficients : same as case 3.2. Meteorological data see ref [8].

Variation of electricity value versus time see ref [9].

Results

Fig. 10 displays the variation of (ROI) versus collector field area for two values of the turbine generator power, 100 and 150 MW.

Fig. 11 shows the variation of the thermal storage capacity apart from optimal value for a 100 MW plant.

FRANCE

ROI VS Turbine Design Point P

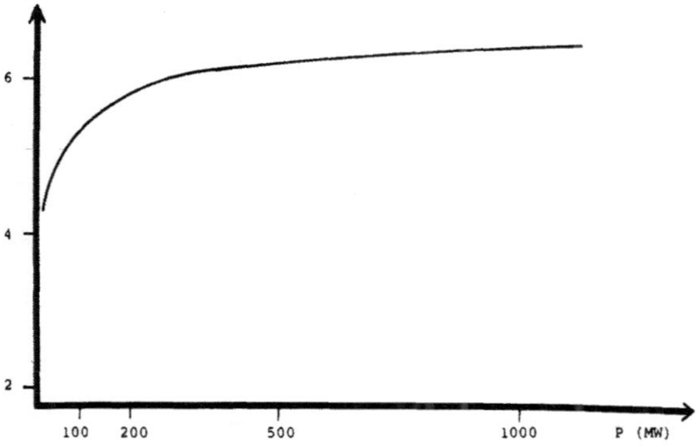

FIGURE 4

FRANCE

ROI FOR DIFFERENT METEO DATA

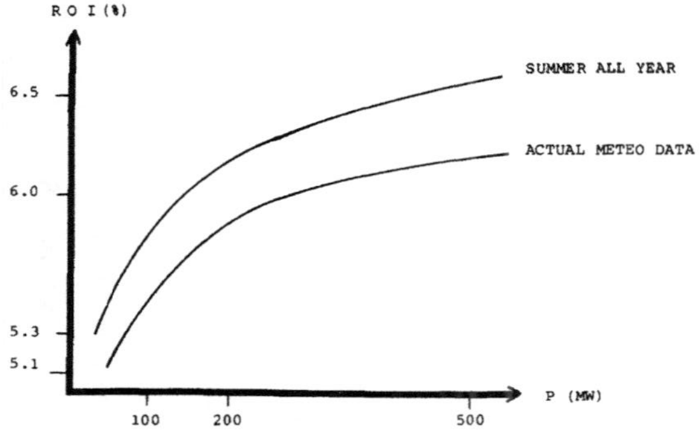

FIGURE 5

FRANCE

ROI VS HELIOSTAT FIELD AREA A

P = 40 MW

FOR OPTIMAL STORAGE SIZE

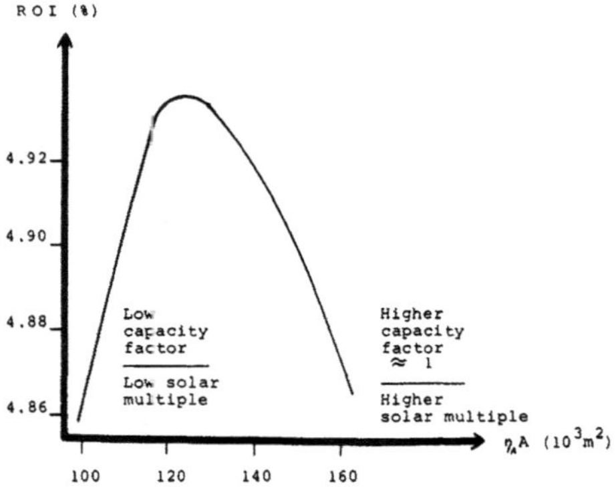

FIGURE 6

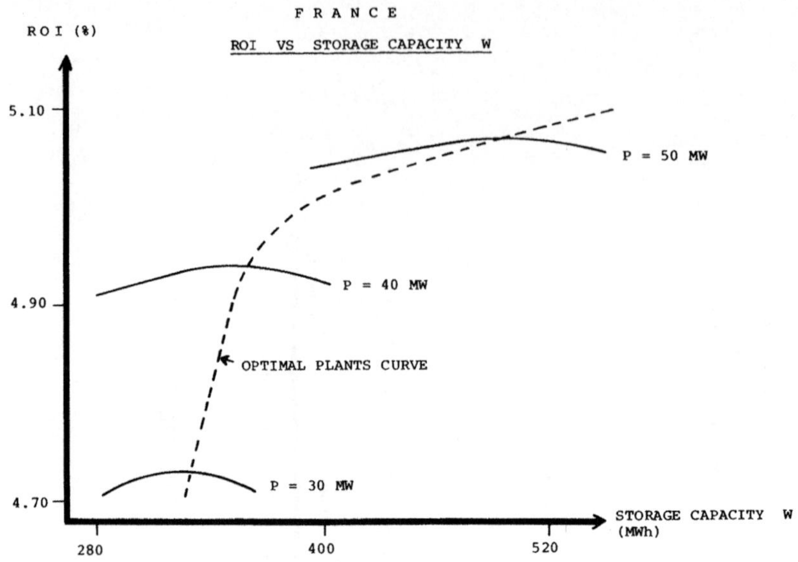

FIGURE 7

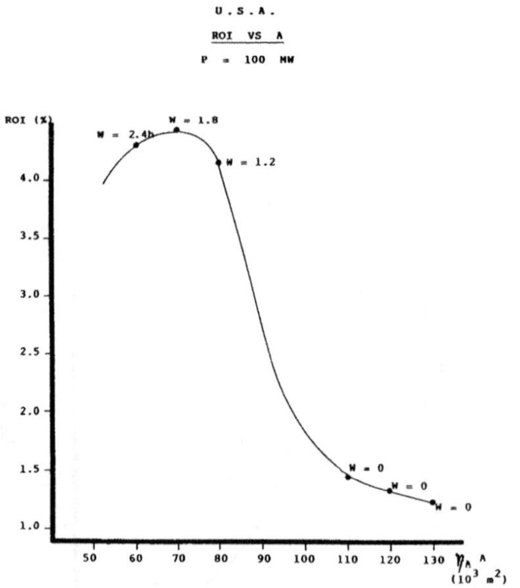

FIGURE 8

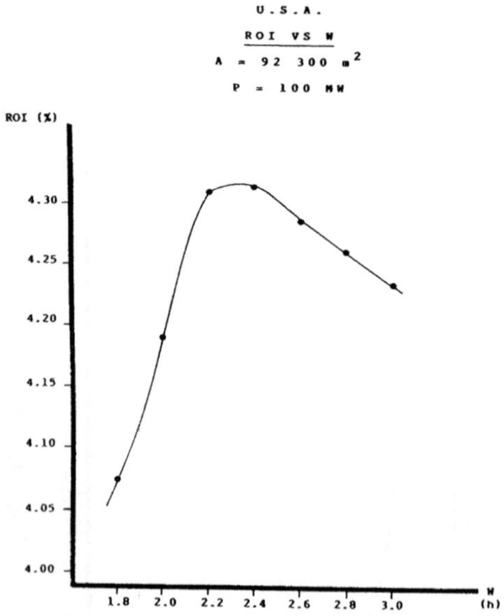

FIGURE 9

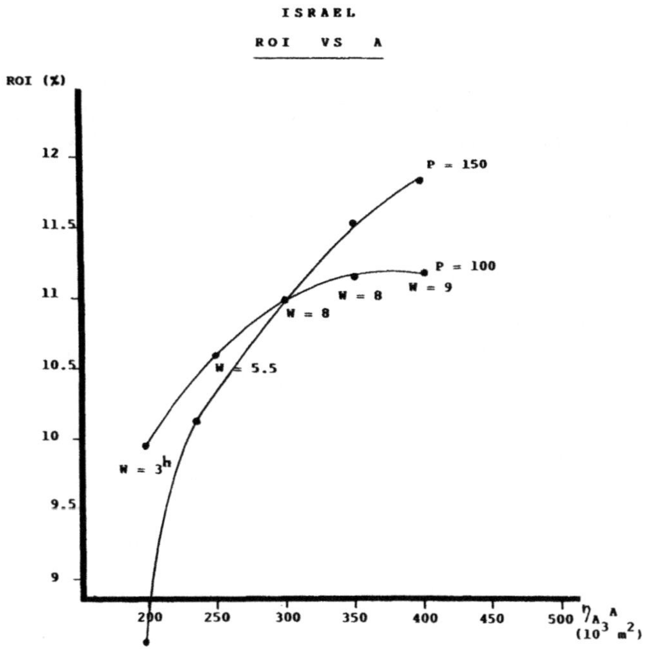

FIGURE 10

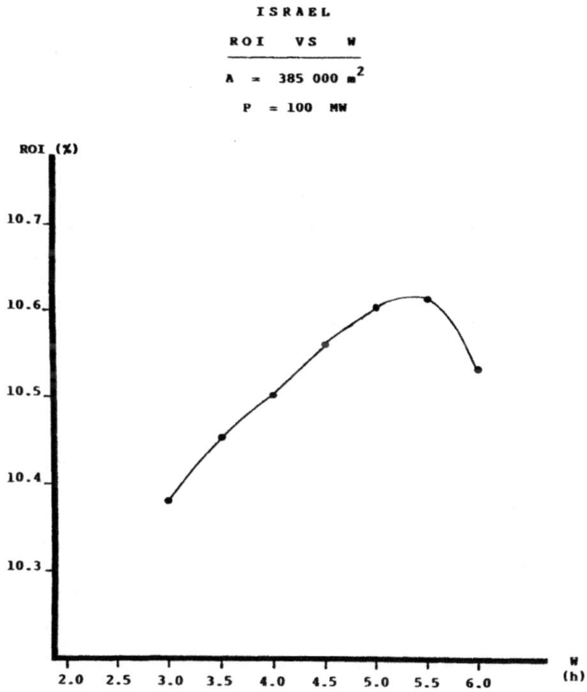

FIGURE 11

4. CONCLUSIONS

The modelling of a CRS plant at Ecole Centrale appears to be a very useful tool for optimizing the parameters of a project taking into account the meteorological and economic environments of different countries.

The inputs needed for running the model are the following :

- . Meteo data
- . Electricity value as a function of time
- . Subsystems efficiencies
- . Construction cost function.

From the first results obtained for several countries evidence is given that the optimal power for a CRS plant is rather more in the 60 - 150 MW_e range than in lower values.

The optimizing criterion we have selected, that is, finding the maximum of (ROI), leads to substantial parameter difference with other criteria such as minimum cost of produced energy.

It should be considered that the construction cost coefficients used in the above examples, correspond to the case of a prototype plant. This reason explains why the computed values of (ROI) in these examples are rather low (5 to 12 %).

This figure should improve for the following plants taking into account the reduction of costs due to the learning effect.

This optimization model may be improved ; but it is already working tool.

Acknowledgements : This work was partly funded by C.N.R.S. under contract C.N.R.S./E.C.P., 14.1.1980.

REFERENCES

1. C. ETIEVANT and M. ROCHE. Conversion Thermodynamique de l'Energie Solaire. Comparaison technique et économique de différentes filières. Entropie, Spécial issue on Central Receiver Systems, N° 103, 1982.
2. Duc PHAM HI, Le Model AMPHI, Mémoire D.E.A.-E.C.P, 1980-1981.
3. Duc PHAM HI, Saïd IBRAHIMI, Thomas DURAND, Etude Technico Economique prospective de dimensionnement de Centrales Electro Solaires à Tour, Rapport L.E.I.S.-E.C.P., 3 sept. 1983.
4. E.D.F., Tarification de l'électricité, Note Bleue.
5. C.L. MAVIS, Barstow Insolation Data, Sandia Report, Feb. 22, 1984.
6. Project SOLAR 100, executive report, Southern California Edison, Mc Donnel Douglas, 1980
7. SCE standard contract, Document N° 1505C/1506D, December 5, 1983, and SCE Schedule N° TOU 8, January 1, 1984.
8. A. MANES, A. TEITELMAN, I. FRUEHLING, Solar Radiation and Radiation Balance at Bet Dagan, Central Meteorological Institute, Series A, Meteoroligical notes, 1970.
9. A. ROY, Electricity cost in Israël (private communication).

COMPARISON OF SOLAR THERMAL AND PHOTOVOLTAIC ELECTRICITY GENERATION USING EXPERIMENTAL DATA FROM THE IEA SSPS PROJECT

P. TOGGWEILER and R. MINDER
Electrowatt Engineering Services Ltd., Zurich, Switzerland

Summary

From February 1982 to April 1983, photovoltaic panels had been installed at the site of the IEA SSPS project in Almeria, Spain. The measurements performed at these panels have been used to calculate, by means of a numerical model, the behaviour of a hypothetic 500 kW photovoltaic power plant at this site. A detailed study has then been carried out comparing measurement results of both the Distributed Collector System and the Central Receiver System with the data of the photovoltaic plant. By comparison of daily electricity production figures, the influence of radiation characteristics, in particular diffusivity for all three solar power schemes is discussed. The results obtained with the Almeria data finally are extrapolated to an alpine site with a higher direct radiation component.

1. INTRODUCTION

In the past five years, various solar thermal as well as photovoltaic power plants in the capacity range of .1 to 10 MW have been built. Qualitatively, the advantages and disadvantages of the different technologies used are well known, but quantitative comparisons are barely available. The study presented in this paper could profit of the unique situation at the IEA SSPS site in Almeria, where two 500 kW solar thermal power plants are being operated and evaluated simultaneously since Sept., 1981. By installing a few photovoltaic panels at the same site and recording their electrical characteristics, it was possible to extrapolate the behaviour of a hypothetic 500 kW photovoltaic power plant (PVS), and to compare it with the DCS and CRS results. The scope of this paper is to present the main results obtained in Almeria and to draw general conclusions for other sites and conditions.

2. METHODOLOGY

The methodology applied for the investigation is best described by the work flow chart shown in figure 1. Most of the work packages shown are self-explaining. The comments to the methodology therefore shall be limited to a few important remarks.

The work package "Computer modelling of the 500 kW PV power system" was performed using the program "PHOPO" developed by Electrowatt. The

model has a modular structure and can be used to simulate photovoltaic power systems at any given site.
The "Comparative study of DCS, CRS and PV for single days" is the most important work package of the study. The measurements made in Almeria on the CRS and DCS at selected days have been analyzed and compared with the calculated results of the 500 kW PV system.

Only short uninterrupted periods of operation data in Almeria are available (particularly in the case of the CRS). Therefore, the following work package "Comparative Study for one full year" had to be carried out by extrapolation using the available meteorological data.

"Extrapolation for Swiss alpine sites" included a calculation of the behaviour of CRS and PVS in areas which are being studied in Switzerland for installation of solar power plants. Detailed meteorological data from the World Radiation Center in Davos were used for this purpose.

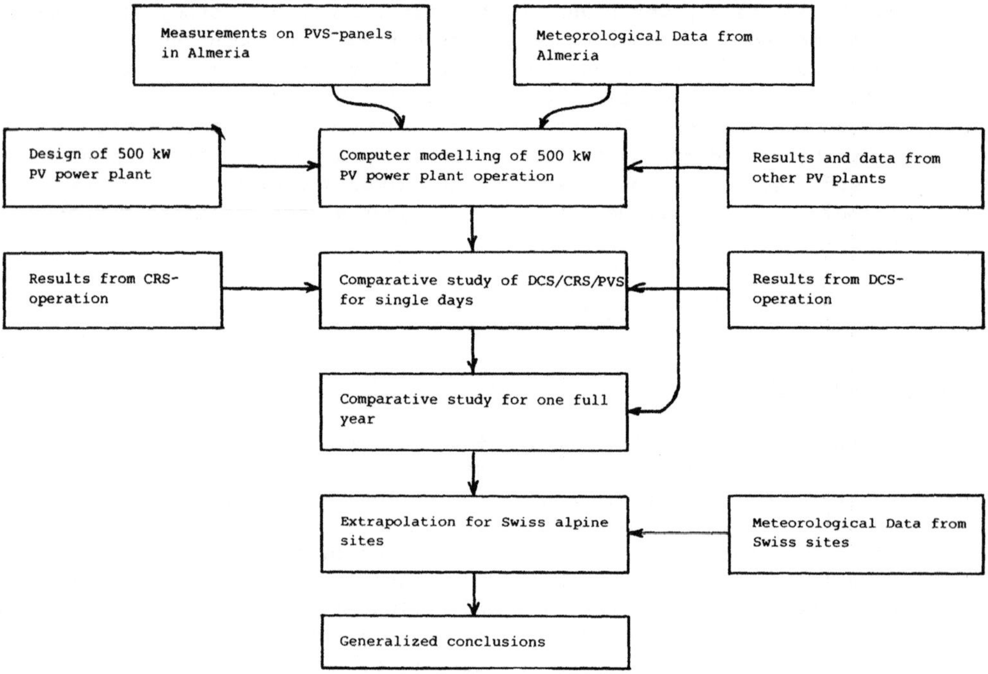

Figure 1 : Work flow chart

3. RESULTS

In this section, the main results of the study shall be presented. These include the comparison of electricity production figures for selected days, the extrapolation from these figures to one full year, and the extrapolation for Swiss alpine sites.

3.1 Comparison of daily electricity production

For comparison of the daily electricity production of CRS, DCS and PVS, the results of only a few days, when the CRS and DCS both worked satisfactorily could be used. Figure 2 shows the net daily electricity production of the three plants vs. the daily sum of direct normal radiation. In order to interprete correctly this figure, some comments are to be made:

a) In order to obtain realistic figures for the CRS, the electricity production was calculated from the available thermal energy. A power conversion efficiency of 23% corresponding to the steam quality was assumed.

b) The DCS-values, if measured again with the new additional collector field probably would come close to the CRS values shown.

c) The net electricity has been calculated as gross electric output minus only those parasitic loads absolutely necessary for plant operation..

The results presented in figure 2 can be summarized as follows: at very sunny days (direct radiation sum of approx. 9 kWh/m2) the outputs of CRS and PVS are roughly equal. At conditions, where CRS net electricity output just starts (approx. 3 kWh/m2), the PVS power production is still about 2/3 of maximum production. The explanation of this behaviour is, that normally at days with low direct radiation, total radiation which determines PVS output remains very high. This fact is illustrated by figure 3 showing the relation between direct normal and total (30° south) radiation on one hand and CRS, DCS and PVS production figures on the other hand for four selected days.

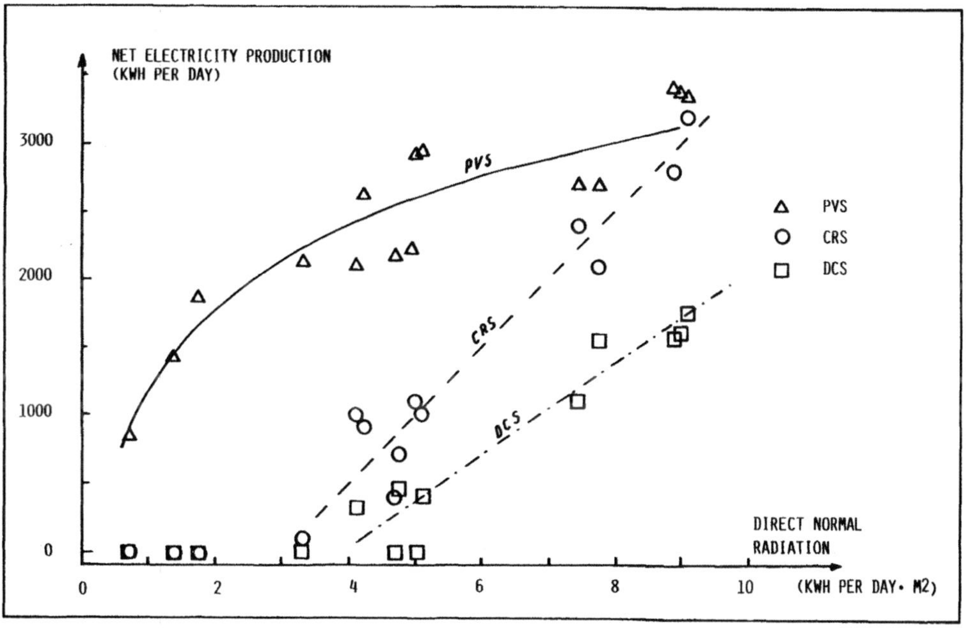

Figure 2: Net electricity production per day for CRS, DCS and PVS vs. direct normal radiation.

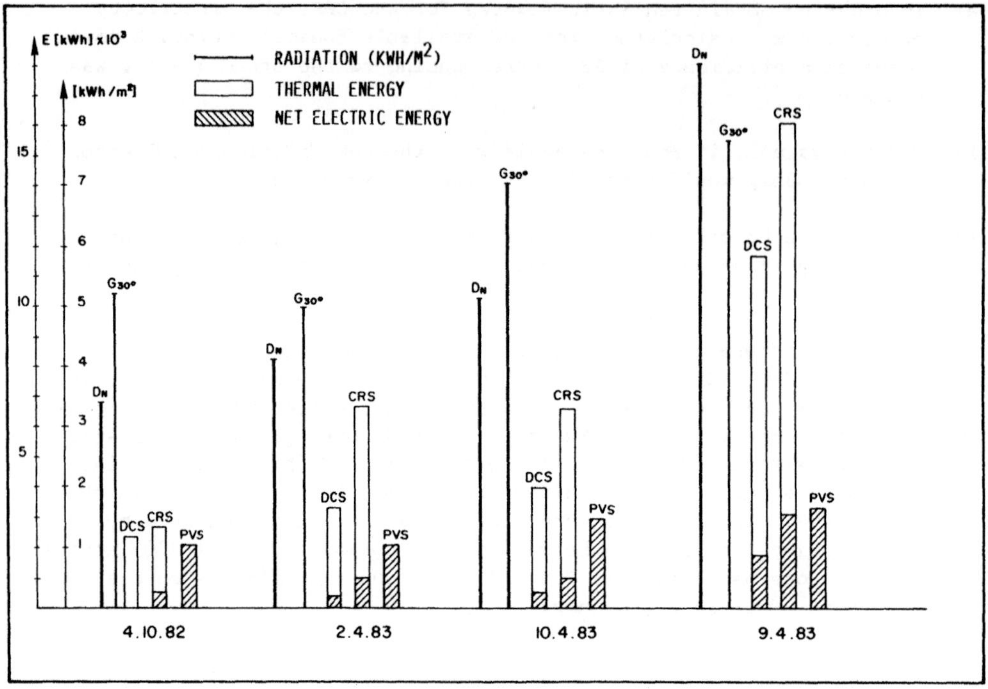

Figure 3: Comparison of direct radiation (D_N), total radiation (G_{30}) with thermal and net electrical energy values for 4 selected days.

3.2. Comparative study for one full year

In addition to the comparison for single days, a comparative study for one full year has been carried out. The problem of this study consisted in the fact, that from relatively few days when the CRS was perfectly working one full year had to be reconstructed. This reconstruction was made using the experimentally determined function "CRS output vs. direct radiation" shown in figure 2, together with the statistical distribution of daily direct radiation sums.

The main result of this comparison between the "idealized" CRS and the PVS is represented by the electricity production figures. For the time period under investigation (Oct. 1982 - Sept. 1983) we obtained E_{net} (CRS)/E_{net} (PVS) = 531 MWh/888 MWh $\approx$ 0.6.

3.3 Comparison for a Swiss alpine site

Alpine areas are characterized, with respect to solar radiation, by an increased annual sunshine duration as well as an increased average direct radiation intensity as compared to the Central European lowlands. For this reason, alpine sites for solar power plants are being studied since several years (1,2). In the frame of this study, we have made a comparison by calculating the electricity production figures of the CRS (idealized as before) and the PVS under alpine conditions. For this simulation calculation, measured radition value for the year 1979 from Davos obtained from the World Radiation Center have been used. The comparison between the electricity production figures in this case is E_{net} (CRS)/E_{net} (PVS) = 384 MWh/647.5 MWh $\approx$ 0.6.

The monthly figures are shown in table 1:

Month	Total radiation (45°, south, Wh/m2)	PVS net electricity production (kWh)	Direct normal radiation (Wh/m2)	CRS thermal energy (kWh)	CRS net electricity production (kWh)
JANUARY	73'608	30'772	59'035	140'528	16'356
FEBRUARY	92'356	39'792	65'895	150'840	20'788
MARCH	116'087	50'434	67'992	143'708	17'088
APRIL	157'065	69'210	94'876	196'229	29'683
MAY	190'004	84'716	157'272	324'913	58'765
JUNE	140'034	60'102	108'829	208'240	32'445
JULY	153'077	66'013	120'849	238'636	38'921
AUGUST	146'944	64'140	121'811	259'961	43'826
SEPTEMBER	158'445	71'252	161'324	383'660	72'792
OCTOBER	112'307	48'926	97'649	222'971	35'318
NOVEMBER	83'162	35'485	66'716	160'461	21'456
DECEMBER	64'725	26'722	52'979	123'181	12'882
TOTAL	1'487'814	647'564	1'175'227	2'553'328	383'964

4. CONCLUSIONS

The results of the study can be summarized in the following statements:

a) The annual electricity production of the "idealized" CRS is at both investigated sites (Almeria and Davos/Swiss alps) approx. 60% of the production of a PVS with comparable design point characteristics

b) In Almeria, the main reason is the low direct normal radiation intensity at sunshine conditions. The average CRS output power per sunshine hour is 180 kW.

c) In Davos, the intensity of direct radiation is substantially higher. CRS operation however is affected by frequent cloud transients which do less affect PVS production. In this case, average CRS output per sunshine hour is 248 kW.

The present study has shown quantitatively that solar thermal power plants are much more sensitive to radiation conditions than photovoltaic systems. In particular, the direct radiation intensitiy during sunshine conditions and the frequency of cloud transients are key factors. The consequences of these facts are: (i) Solar thermal power systems must be large in order to minimize relative thermal inertia losses, and (ii) only sites with excellent direct radiation characteristics are acceptable.

Acknowledgements

The study presented in this paper has been supported by the "Nationaler Energie-Forschungs-Fonds NEFF" and by the Federal Office for Energy Economy.

References

(1) Utilisation de l'énergie solaire pour la production de l'électricité dans les alpes, Série de publications de la Commission fédérale de la conception globale de l'énergie, No. 7, Battelle, Genève (1976)

(2) Solarthermisches Elektrizitätswerk SOTEL, Projektvorschlag, ausgearbeitet durch ein Schweizer Konsortium, edited by EIR, Würenlingen (Oct. 1981).

Performance data of CRS and DCS had been obtained from various SSPS publications and internal ITET reports.

DEVELOPMENT OF AN ECONOMIC ASSESSMENT MODEL

H.Klaiß, M.Geyer[1)]
Institute for Technical Physics
DFVLR - Stuttgart

Summary

The cost analysis of power plants is one of the most important instruments for decision finding. Different cost calculation methods, several data sets for solar power plants and the rating of the results will be discussed. The most significant cost calculation methods are analysed, and an economic assessment computer model is presented using data of a 100 MW-solar power plant as an example. Besides, the sensitivity of parameters is discussed and compared with data of conventional power plants.

1. INTRODUCTION

The discussion of costs and prices of electricity produced by nuclear power plants, coal fired plants, solar power plants, etc. is a favourite point all over the world. This phenomenon arises from the fact that the costs of a product belong to the most important points for the decision what kind of plant will finally be the most favourable one. Furthermore, people of today are used to think in terms of money, and so all feel to be able discussing also about costs.

Even so a discussion about cost will be held. The main items of this paper are the problems of cost analysis, the presentation of an economic analysis model and the development of sensitive parameters. The discussion of the resulting prices should be left to detailed analysis, when more reliable data will be available.

For a cost analysis three main interests exist: Those of government, the utilities and the consumers.

The government, in principle, has the task to intervene into the market, if the objectives of the government are touched by the construction or operating of a power plant. If there is any chance to improve the ways to fulfill the objectives of the country, the government may intervene, but need

---

1) This report is a result of a cooperation with G.Faninger, ASSA, Vienna and W. Bucher, DFVLR, Cologne.

not do so. It is well accepted known that especially the solar energy has the ability to make a good contribution regarding the national economoy goals, such as:

- reducing fuel imports (balancing of terms of trade)
- reducing the influence of fluctuating energy prices
- play a positive role on the labour market (decentral employment of labour; structural effects (lower wages); anticyclical effect (predating of expenditures in a time of high unemployment)
- increase flexibility of the energy demand and supply structure
- reducing or avoiding of environmental pollution problems (what will be a problem of increasing importance)

All these items give reasons enough to favour the renewable energy by means of more subsidies.

The utilities want to get a maximum of electrical output at minimum costs. The consumers want to pay a low price for the current without environmental pollution.

2. FINANCIAL ITEMS

There are various ways to give subsidies,e.g. as investment tax credits, by reducing the depreciation period, by decreasing the interest rate, etc., but it is important to know that these tools are different between different states and countries, and there is a difference too as far as the renewable energy systems are concerned. That means, there will be another tax credit in California than in Florida or Germany, but there may be another one in California for photovoltaic than for wind energy systems(for instance). Additionally, the tax rate varies through the years.

Therefore, for a comparative cost analysis, it is not opportune to publish results with respect to subsidies. Of course, this is an important financing problem, but if you take tax credits into account, there is no possibility to compare the results with other cost estimates in other countries, and it is obviously clear that there is a decrease of costs by getting credits.

The most important objective of the utilities is to maximize the profit or the cash flow.Three fields have to be analyzed: at first the original or the direct costs of the plant in the year of start of operation or all over the life expectancy have to be defined. This we call analysis of directly related cost or simply cost analysis. The next step could be to look for possibilities to get tax credits for the investment or for the operation of the plant. But in the last section we found that this represents a relevant point for financing, but it is not useful for making comparative analysis because tax credits are varying in all regions. The third point relates to the specific problem of the utilities themselves. The impact of tax credits is primarly a function of the actual tax rate of the company. If the company makes high profits per year it is better to have a great expenditure in a short time. If the company made deficits on contrary, it is better to extend the expenditure over many years. Tax credits are of little help for this company in the latter case. Because tax rates of the utilities are relatively low , the idea was born to finance the power plants by the method of third party financing. The third party financing may be a suitable way to promote solar energy, but the effect of this method differs not only from state to state, it is even different from company to company largely depending on the stock holders. Considering all these items we decided to limit the task to a verified cost analysis without regarding tax credits. Basing on this idea we have developed a cost calculation computer program for the cost methods being used most frequently.

3. ECONOMIC METHODS

For the economic analysis principally two sets of calculation methods are in use:

o "static" calculation methods, and

o "dynamic" calculation methods.

Because the static methods do not consider that capital has different values at different times (in simple words: 100 DM have another value today than in ten years), we do not consider the static methods.

Dynamic methods of economic analysis take into consideration capital supplied at different times in different amounts as well as the time dependence of the capital's status. The capital flows occuring at different times are related by means of discounting to the same reference year and are thus be made comparable.

If dynamic methods are used two main different methods are discussed. Both methods are common:

o annuity method

o present value method

Annuity method

The annuity method starts from the fact that expenditures done only in one or in several selected years (e.g. investment and decommissioning costs) have to discount to one year (mostly start of operation). This related valud K_p is spread over the whole depreciation time in this manner that through the whole time the same expenditure K_F results. The sum of all these expenditures K_{Fi} in all years i discounted to the reference year is exactly the value K_p.The factor to come from the value K_p to the yearly expenditures K_F is called the annuity factor

$$a = \frac{d \times (1+d)^n}{(1+d)^n - 1}$$

where d = discount rate

n = number of years

The inverse value of the annuity factor is called capital recovery factor. In addition to this annuity expenditure K_F different costs K_{Vi} have to be included every year, which may be fix and variable costs as well. The total costs are the sum of these both costs. To determine the specific costs the total costs are divided by the annual output of energy (Fig.1).

This calculation method was preferably employed in Germany. Its advantage is that it is relatively easy to be understood and that it shows the approximate development of the existing costs as well. As a disadvantage one can see that the costs cannot be related to a selected year. The K_V are related to nominal costs in the respective time period, but the costs K_F cannot be assigned to a selected year. A still greater disadvantage comes from the fact that when curves of alternative investments are crossing, no statement regarding the most profitable investment can be made; it only can be determined which alternative may be the best one in the different years. The reason is that by this method not the total costs occuring during the life time are taken into account, but only the costs during the considered period.

Present Value Methods

In the present value method the calculation starts in the same manner as if the annuity method were used,but in this case all costs (fixed and variable) are related to the reference year and not only investment costs and similar cost. All costs over the whole life expectancy (depreciation time period) of the plant are taken into account. The sum of all these costs is the present value (here BWE = Barwert). If the present value for different investment alternatives is calculated the alternative with the smallest present value has to be chosen because this plant has the lowest cost over the depreciation time.

The problems in the calculation of the present value turns out in the determination of a specific cost value. Once more it must be pointed out that for the estimation of alternative systems the present value is completely sufficient. A specific cost value (e.g. Pfg./kWh), however, is a more evident quantity. For the calculation of these characteristic quantities two different methods are employed. Because in both methods, costs and revenues of the whole life expectancy are included, often the expressions "life cycle method" or "life cycle costs" are used.

The two methods are the nominal levelized present value method and the real levelized present value method. The idea behind both methods is that over the life expectancy the revenues must be equal to the costs. This is a minimum requirement. In this case there do not exist any profit and losses for the utility. The finding of the costs starts with the calculation of the present value. There are first the costs K_o in the reference point. The costs are increasing throughout the years with an escalation factor r (r = 1+e, e = escalation rate) and related to the reference point with the discount factor q. Thus, the present value of the costs is the sum of all costs K_o over all years (t).

$$BWK = \sum_{t=1}^{N} \sum_{i=1}^{n} K_{oi} \cdot r_i^t \cdot q^{-t}$$

The revenues are calculated in the same manner.

$$BWE = \sum_{t=1}^{N} E(t) * k * q^{-t}$$

where E annual energy output (kWh/a) (constant over the whole life expectancy);

k price of kWh.

Now , the difference between the nominal and the real value method lies in the assumption for the price k. We get the nominal costs when we assume k being nominally constant over all years ($\bar{k}$). The solution of the equation BWK = BWE leads to

$$\bar{k} = \frac{\sum_{t=1}^{N} \sum_{i=1}^{n} K_{oi} \times r_i^t \times q^{-t}}{\sum_{t=1}^{N} E(t) \times q^{-t}}$$

We get the real costs assuming k is as increasing as the expenditures over the years with an escalation factor, mostly taken with the inflation factor p (p = 1 + i, i = inflation rate). One gets (extending in the numerator with the inflation factor p) for the year t_{NN}:

$$k_{o,NN} = \frac{\sum_{t=1}^{N} \sum_{i=1}^{n} K_{oi} \times \left(\frac{r_i}{p}\right)^t \times \left(\frac{q}{p}\right)^{-t}}{\sum_{t=1}^{N} E\ (t) \times \left(\frac{q}{p}\right)^{-t}}$$

For each optional year N to be chosen there is then

$$k(t)_N = k_{o,NN} \times p^t$$

The comparison of these three methods shows the relationship of the specific costs calculated by the annuity method as well as by the nominal real and levelized present value method (Fig.2)

- The nominal levelized present value $\bar{k}$ gives the highest specific costs, which are equal over the whole life expectancy. That means that the real value is decreasing.
- The lowest costs (in the year of start of operation) results from the real levelized present value method k(t). The higher the inflation rate the steeper the cost curve of the real levelized present value method proceeds and the smaller is its value in the year of start of operation. In this case the real expenditures are equal over the whole life expectancy.

- The annuity costs k_T lie in the middle of the range. The higher the share of the investment costs on the specific costs the smoother the annuity curve proceeds and the more the costs come close to the value of the nominal levelized present value method.
  (As an example: nominal costs (Pfg./kWh) 51,7 Pfg./kWh
  annuity costs (Pfg./kWh) 47,8 Pfg./kWh*
  real costs (Pfg./kWh) 33,3 Pfg./kWh*
  *in the first year

Derivated cost method

The net present value method (NPVM) ist familiar with the present value method. The difference is that in the NPVM the whole revenues over the life time are involved. The revenues are based on assumptions and not calculated. This investment alternative should be taken preferably which leads to the highest difference between revenues and costs. It is impossible to show specific costs by this method.

The internal interest method is a relatively complicated method and of interest for problems in the financing field and for stock holders. It is also based on the NPVM.

The pay off method, in principle, is not a calculation method of its own. The pay off period (amortization time) is the time during which the capital input of an investment has been payed back again. The investment having the shortest pay off period is the most favourable one. The pay off period, therefore, is an indicator for the investment risk and contributes substantially to the decision making process for the engagement of capital. As the future of an enterprise bears many uncertainties, the pay off periods should be as short as possible.

The break even point is defined as the cross point of the costs function with the revenue function. For the periods following the break even point, a profit can be assumed. It is possible to determine the break even point with the annuity method.

4. MODEL PERFORMANCE AS AN EXAMPLE

The cost program is a computer program implemented in a modular program system PROSES (Parametric Representation of Selected Energy Systems). It is written in PL 1 and installed into the operational system MVS on an IBM computer system 3081. It works in dialogue operation on colour or black and white terminals. Special emphasis was given to easy handling and simple use.

The best way to show the calculation model is a presentation using the data of a 100 MW CRS power plant, because this project is very similar to a commercial solar power plant. The input data have been selected from 100 MW plant proposal (study of the McDonnell Douglas Corporation) which contains values coming close to those of a commercial plant, some data have been modified, other values have been adapted to "German fiscal and financing conditions". They should be seen as one consistent example how to use the calculation method and not as a statement of "real" costs of solar electricity.

An extract of a data list is given in Table 1. The overall costs are $1{,}125 \cdot 10^9$ DM ($\sim 400 \cdot 10^6$ \$) or 11 250 DM/kW_n. The prices are based on the year 1981, the start of operation is assumed to be in 1986, the life expectancy amounts to 30 years, the discount factor to 9 %, and the price of heliostats to 400 DM/m^2.

Regarding the results (Table 2, fig. 3) the following statements can be made:

- Different methods lead to different costs (33-52 Pfg/kWh)
- The detailed cost break down shows the prevalent part of the investment costs
- The largest share of the investment costs is assigned to the heliostat costs (27,0 %). This share is visibly lower than those given in previous studies (about 50 %); one of the reason is the lower price of heliostats (400 DM/m^2).

5. SENSITIVITY ANALYSIS AND COMPARISON WITH CONVENTIONAL POWER PLANTS

- The comparable costs of a nuclear and coal fired power plant (100 % desulfurization, start of operation in 1989, price based on 1982, rated hours 5000 h, depreciation time 20 years) are in the range of 12 - 15 Pfg/kWh (first year of operation). This means that by taking a positive basis for the solar tower plant, there is a difference between solar tower plants and conventional plants in the range of a factor 3 (Tab.3)
- The sensitivity analysis shows (Fig.4) the great influence of the non-technical parameter discount factor, the total investment costs and the life expectancy (for the short life time periods).
- The effect of the discount factor variation is one of the reasons why we cannot compare cost data without having knowledge of the economic parameters. The influence of the discount factor is particularly high for power plants with high investment costs. This problem will become conspicuous when comparing the costs of a coal and a nuclear power plant. At high discount factors the ranking between plants with high and low investment costs will be changed. At a reduced discount factor (6 %) specific costs of 25,1 Pfg./kWh can be calculated for the 100 MW CRS plant (Table 4).
- The specific investment costs are too high for solar power plants in relation to other power plants (Table 5), but it is remarkable that the prices of the conventional power plants are increasing over time, and also for small plants (Table 6).
- The influence of the heliostat prices on the total costs ranges at 25 %. That means doubling the costs results in a cost increase of 25 %.
- The sensitivities for personal costs, maintenance and repair costs are not so significant. The relation of these costs of the conventional power plants is shown in Table 3.

6. CONCLUSIONS

A comparison of costs for different power plants has to reveal the employed methods and the assumptions for the relevant data as well. For comparison it is necessary to use identical economic parameters. This problem is solved by an interactive computer program for cost analysis of solar and conventional power plants. An example for a 100 MW CRS plant shows that the solar power plants are still too expensive, the factor is amounting between 3 and 5. It seems reasonable to have a additional period of about ten years for research and development to reach a competitive status, accompanied by increasing fuel prices. These facts and the advantages of the solar energy regarding its overall advantages provide an absolute privilege for the solar energy to be promoted by the governments for the next years.

ADDITIONAL REFERENCES

F. Krawiec, et al: An Investigation of Learning and Experience Curves, SERI, Golden, April 1980

P.Munjal,S.Leonard: Third-Party Arrangements for financing photovoltaic power plants, The Aerospace Corporation, El Segundo, Dec. 1982

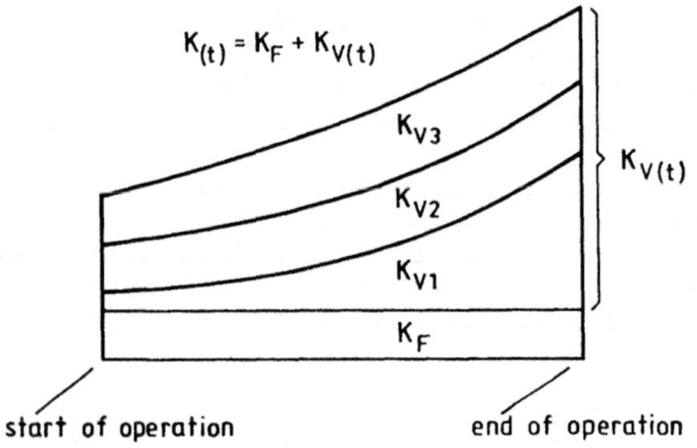

Fig.1: The annual costs (annuity method)

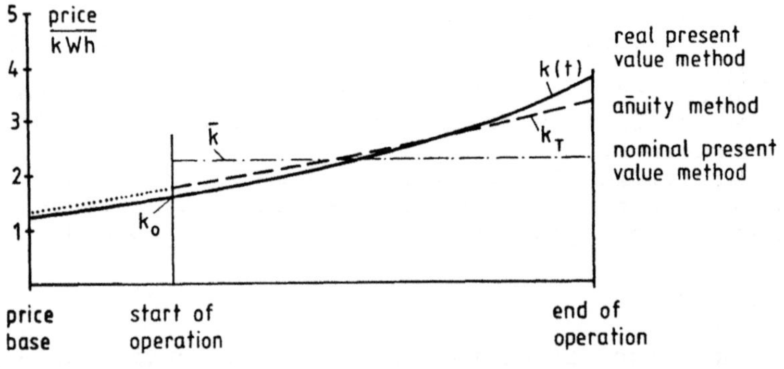

Fig.2: Comparison of different costs methods

Tab.: 1

Technical and Economic Data of a 100 MW CRS Plant

Financing Data	
Basic year	1981 a
Start-up year	1986 a
Price based on (results)	1983 a
Depreciation period	30 a
Construction time	4,5 a
Discount rate	9 %
Inflation rate	4 %

Technical Data	
Gross power	110 MW
Net power	98,3 MW
Gross electric output	419,7 GWh/a
Net electric output	373,5 GWh/a
Parasitics	11 %
Net overall efficiency	17,8 %
Annual availability	95,0 %
Capacity factor	44 %
Rated hours	3815 h
Direct Insolation	2500 kWh/a m²
Heliostats	15500
Mirror area	883500 m²

Reference:
G.R. Roland, K.M. Rass, Solar Central Receiver Technology Development and Economics-100 MW Utility Plant Conceptual Eng.Study, 1983, Eigene Berechnungen

Investment Data	DM/kW	%
Heliostat field	3540	31,5
Receiver	1050	9,3
Cycle	610	5,4
Tower	660	5,9
Storage	1400	12,4
Site	210	1,9
Engineering, home office, overhead costs	1210	10,8
Plant master control, Balance of plant switchyard/transmission	1300	11,5
Spare parts/ additional contigency	1270	11,3
TOTAL	11250 = 1125 · 10^6 DM	100

Operating and Maintenance Data	DM/ kW.a
Personnel	65
Repair and maintenance	39
Machinery costs	35
TOTAL	139 ≙ 13,9 . 10^6 DM/a

Tab.: 2

Spec. Costs Break Down (100 MW CRS-Plant)

	Pfg./kWh	% 1)		
Total costs (real present value method)	33,3	100		
Fix costs	32,1	97		
Investment costs	28,6		85	
- Heliostat field	9,0			27,0
- Receiver	2,7			8,1
- Cycle	1,6			4,8
- Tower	1,7			5,1
- Storage	3,6			10,8
- Site	0,5			1,5
- Electrical equipment	3,0			9,0
- Additional contingency 2)	6,5			19,5
Personnel costs	2,2		7	6,6
Repair and maintenance	1,3		4	3,9
Variable costs	1,2	3	4	3,6
Total costs (nominal present value)	51,7			
Total costs (annuity method)	47,8			

Remarks: 1) related to 100 % 2) + overhead costs
Reference: Eigene Berechnungen

Fig. 3: Costs Break down of an 100 MW CRS plant

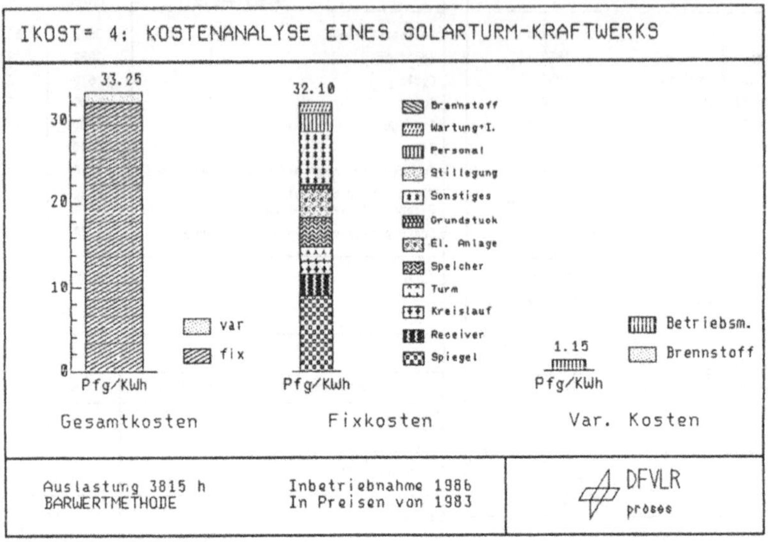

Tab.: 3

Specific Data of Nuclear-, Coal-,Heavy oil- and Solar Power Plants (1982/83)				
Items	Nuclear	Bituminous	Heavy oil	Solar Tower
Net power MW_{el}	1300	2 x 680	30	100
Parasitics %	5 - 6	7 - 8	8 - 9	11
Spec. investment costs DM/kW	3000[1)]	1700[2)]	2300[2)]	10000 - 20000
Percentage of %				
- Machinery equipment	55	65	60 - 65	70
- Electric equipment	20	10	10 - 20	10
- Construction/building	20	10 - 25	15 - 25	10
- Design/miscellaneous	5	5 - 10	5 - 10	10
Share of investment costs on total costs %	70	30	30	95 (70)[3)]
Construction time a	8	6	4 - 5	4 - 5
Site m²/kW	0,2	1 - 2	2	25 - 40
Labour P	270	240	60	70
Spec. value P/MW	0,2	0,18	2	0,7
Percentage of %				
- Operators	40	60	60	40
- Maintenance	40	30	30	40
- Supervisor/Security	20	10	10	20
Repair and maintenance 1/a (Percentage of investment costs %)	2,5	4,5	5	0,5
Machinery costs Pfg./kWh	0,1	0,2	0,2	0,9

Remarks: 1) For the export market ./. 20 - 40 %;
2) For the export market + 20 %
3) Without heliostats

References:

U. Hansen, Kernenergie und Wirtschaftlichkeit, Verl. TOV Rheinland, Köln 1983.
W. Jack, Die Wirtschaftlichkeit der Kernenergie in den OECD-Ländern, Bericht über eine OECD-Studie, atw, Mai 1984.
Badenwerk AG, Karlsruhe; Kraftwerk Union, Mülheim; Institut für Kernenergetik und Energiesysteme, Universität Stuttgart, Pers. Mitt. 1984; Der Minister für Wirtschaft, Mittelstand u.Verkehr, Stromerzeugung im Kostenvergleich von Steinkohlen- und Kernkraftwerken, Batelle-Studie, März 1984;
J.R. Roland, K.M. Ross, Solar Central Receiver Technology Development and Economics - 100 MW Utility Plant Engineering Study, McDonnell Douglas, presented to the 10th Energy Technology Conference, Washington, D.C., 28.2.-2.3.1983.
Eigene Berechnungen.

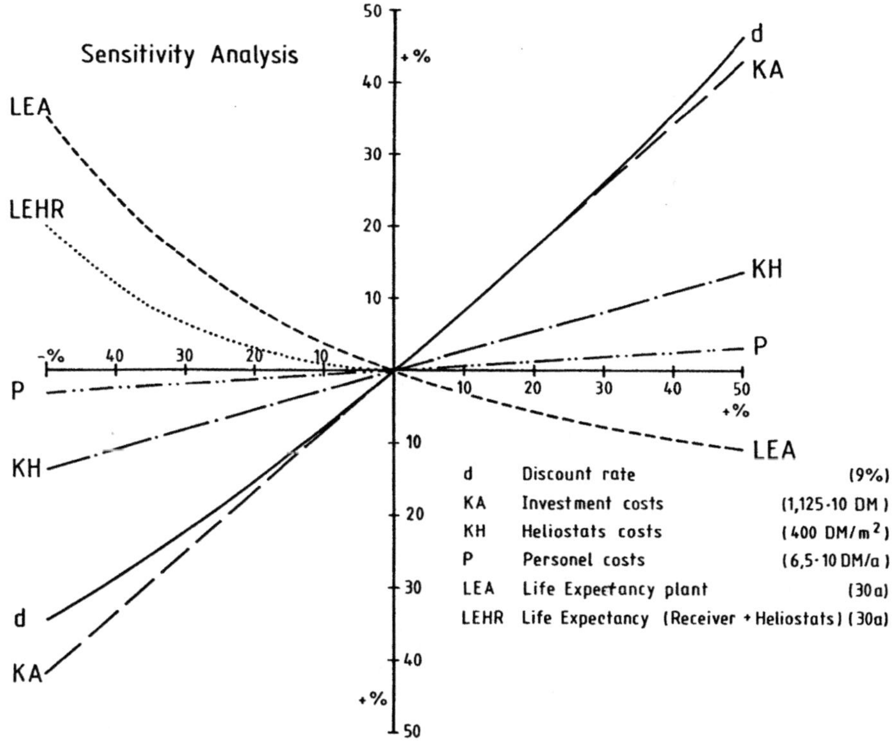

Fig. 4: Sensitivity analysis

Tab.: 4

Variation of Selected Parameters (Real Present Value Method)					
	Discount factor [d] %	Heliostat costs HEC [DM/kW]	Life expectancy plant LE [a]	Life expectancy Receiver + heliostats LERH [a]	Costs Pfg./kWh
Reference run	9	400	30	30	33,3
Variation 1	6*	400	30	30	25,1
Variation 2	9	900*	30	30	44,6
Variation 3	9	600*	30	30	37,8
Variation 4	9	400	20*	20*	39,1
Variation 5	9	400	30	15*	40,0
Variation 6	9	900*	20*	15*	58,4

Remarks: *) Changed parameters.
Reference: Eigene Berechnungen

Tab.: 5

Spec. Prices and Life Expectancy of Selected Power Plants		
	Size	Costs (DM/kW)
Oil and coal power plants (20 - 25 a)	up to 1 MW	5000 - 7000
	1 - 4 MW	4000 - 5000
	4 - 10 MW	3000 - 4000
	10 - 30 MW	2500 - 3000
	30 - 100 MW	1200 - 2500
	100 - 1000 MW	800 - 2000
Diesel power plants (5 - 15 a)	50 - 200 kW	2000 - 3000
	200 - 1000 kW	2000 - 2400
	1 - 10 MW	1600 - 2200
Gas turbine (10 a)	5 - 30 MW	600 - 1000
	30 - 80 MW	400 - 800
Geothermic power plants (20 a)	5 - 20 MW	4000 - 8000
Hydropower plants (30 - 50 a)	for all sizes	200 - 1200
Wind power plants	5 - 200 kW	5000 - 20 000
Solar energy	1 - 10 000 kW	30 000 - 60 000
	10 - 100 MW	10 000 - 30 000

References:

Kreditanstalt für Wiederaufbau, Energieprobleme der Entwicklungsländer, Heft 20/83, Frankfurt a.M. 1983.
World Bank, Renewable Energy Resources in the Developing Countries, Washington, D.C., November 1980.
Eigene Berechnungen.

Tab.: 6

Specific Investment Costs ($DM/kW_{el, net}$)				
Type of plant	1971	1975	1977	1982
Nuclear (1300 MW)	660	1325	1675	3000
Coal (bituminous)	485	945	1075	1650
Coal (lignite)	585	1015	1250	2000

Reference:

U. Hansen, Die Entwicklung der Wirtschaftlichkeit der Kernenergie, atw, Mai 1984.

10 MWe SOLAR THERMAL CENTRAL RECEIVER PILOT PLANT TOTAL CAPITAL COST

H. F. NORRIS
(presented by A. C. SKINROOD)
Sandia National Laboratories
Livermore, California USA

Summary

A cost analysis of the 10MWe Solar One Thermal Central Receiver Plant near Barstow, California, is presented to help predict costs of future solar thermal central receiver plants. In this paper, the Solar One Pilot Plant's total capital cost ($141,200,000) has been broken down into four different structures: (1) project cost, (2) plant system cost, (3) elements of work cost, and (4) recurring and nonrecurring cost. These cost breakdown structures have been correlated to show the interaction and the assignment of costs for specific areas. Since the Solar One Pilot Plant is a first-of-a-kind facility, many of its costs will be greater than those of future plants with similar designs. However, as the world's largest solar central receiver plant, Solar One offers a unique opportunity to examine actual design and construction costs.

1. INTRODUCTION

The purpose of analyzing the costs of Solar One is to provide a clear, detailed breakdown of costs in the form of a data set that can be used to plan new central receiver plants, to analyze alternate energy sources and technologies, and to reduce uncertainties in cost estimates for potential investors. The data also provide a checklist of types of costs that can be incurred in building a solar plant. Furthermore, the breakdown of costs can be applied to projections of costs of different sizes of plants.

The costs associated with designing, building and operating this first-of-a-kind plant are not typical of a commercial power plant. However, as the world's largest solar central receiver plant, Solar One offers a unique opportunity to examine actual design and construction costs. The costs of energy from this plant will be greater than those of either the next 10 MWe power plant of a similar design or a larger power plant that uses the same technology. Solar One was designed to be a scale model of a 100 MWe plant but the configuration was not optimized for the 100 MWe size.

The conceptual design of the pilot plant was completed in July 1977. The plant design was started in May 1979. Plant design and construction were completed in April 1982. The pilot plant total capital cost includes charges that can be considered final design, and perhaps preliminary design or research and development, as well as the construction costs. However, the plant was not totally complete in April 1982 and further costs have been incurred since that date.

2. TOTAL CAPITAL REQUIREMENTS

Although the total capital costs of Solar One can be identified, they cannot be understood unless they are examined in some detail from several different viewpoints. The total capital requirement can be categorized in several ways. The proper separation and detail can provide confidence in the cost data and the completeness of the overall charges. The total capital requirement ($141,200,000) remains the same for all types of categories.

In this article, total capital requirements are categorized four ways. These ways are: (1) project cost breakdown structure, (2) plant system cost breakdown structure, (3) elements of work cost breakdown structure, and (4) recurring and nonrecurring cost breakdown structure. The reason for providing the cost detail in each of these ways varies, as explained below.

2.1 Project Costs

The project costs include R&D costs, design costs, factory costs of engineered equipment, construction costs, and start-up costs. Costs categorized in this way are useful in examinations of costs of future plants, since similar parts, such as heliostats, may be used. The new plant may not have to pay for R&D costs since they have already been paid, or the engineering design costs may be greatly reduced.

The total capital costs broken down by project category are graphically shown in Figure 1. These costs are also listed below.

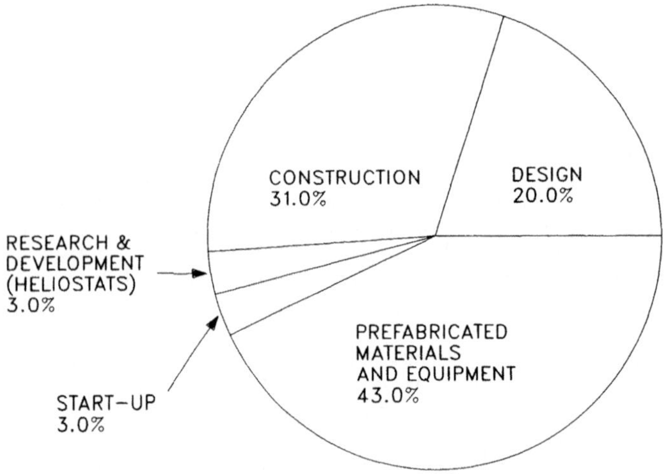

Figure 1. Cost Breakdown by Project Category

Project Costs

Research and Development		$ 4,868,145
Design--Final and Preliminary		28,804,112
Factory Costs		60,095,479
Construction Costs		43,011,283
Construction Packages	$27,506,758	
Installation of Heliostats	2,198,071	
Construction Management (Townsend and Bottum)	5,178,266	
Construction Management (Southern California Edison)	3,316,667	
Construction Package Support	208,679	
Program Management (Solar Facilities Design Integrator)	4,602,842	
Start-up Costs		4,384,368
Round-up, Miscellaneous Round-off		36,613
Total Capital Costs		$141,200,000

2.2 Plant System Costs

The plant system cost breakdown structure includes charges assigned to major parts of the plant. These areas consist of land, structures and improvements, the solar thermal portion of the plant, the turbine plant system, the plant electrical system, miscellaneous plant equipment, and plant-level costs. The solar thermal portion can be further divided into systems, including collector, receiver, thermal transport, and thermal storage. Plant-level costs are those costs that are difficult to assign to individual plant systems. These may include program and construction management charges and architectural and engineering fees. Although total costs are known, the distribution is unknown; no attempt to allocate these costs was made.

The breakdown by plant systems is useful if analysis is to be performed that considers only parts of the plant. Separating costs into these plant systems allows an understanding of system costs and the possibility of comparing the costs of other technologies with those of the solar thermal central receiver concept. The comparisons can be made with wind, photovoltaic, nuclear, and fossil technologies if enough detail is available. This breakdown is also needed if scaling of different-sized plants is attempted.

A total capital cost breakdown by the plant systems is shown in Figure 2. Subtotals for the solar plant portion and conventional plant portions are also provided.

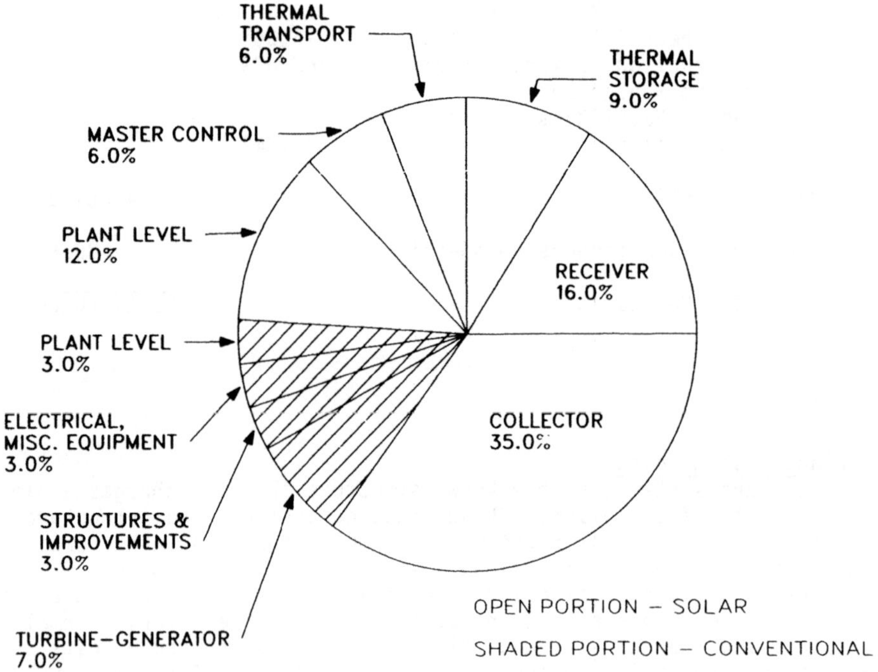

Figure 2. Cost Breakdown by Plant System

Plant System Costs--Solar Plant Portion

Collector System		$ 49,211,297
Installed Heliostat Field	$46,246,253	
Beam Characterization System	1,566,643	
Special Heliostat Instrumentation and Meteorological Measurements System	1,398,401	
Receiver System		22,570,587
Receiver	20,479,404	
Tower	2,091,183	
Thermal Storage System		13,176,982
Thermal Transport System		7,517,434
Master Control Subsystem		8,525,934
Plant Level Costs (Solar Portion)		18,198,701
Construction Management	5,178,266	
Program Management	4,602,842	
Construction Package Support	208,679	
Design Integration	4,241,322	
Checkout/start-up	3,112,617	
Start-up	854,975	
Solar Plant Portion Subtotal		$119,200,935

Plant System Costs--Conventional Plant Portion

Land	$ 0
Turbine-Generator Plant System	10,140,783
Electrical Plant System	2,829,554
Structures and Improvements	4,686,315
Miscellaneous Equipment	989,134
Plant-level Cost (Conventional Portion)	$ 3,316,667
Conventional Plant Portion Subtotal	21,964,793
Round-up, Miscellaneous Round-off	$ 36,612
Total Capital Costs	$141,200,000

2.3 Elements of Work Costs

The elements of work cost breakdown structure consists of sitework/earthwork, concrete work, metal work, architectural work, process equipment, piping and electrical work. This cost breakdown is of interest to architectural and engineering (A&E) firms, since many of the construction subcontracts will be organized by this breakdown. It is also useful when scaling new plants in the early design phases.

The total capital cost segregated by elements of work categories is summarized in Figure 3. A listing of the costs follows.

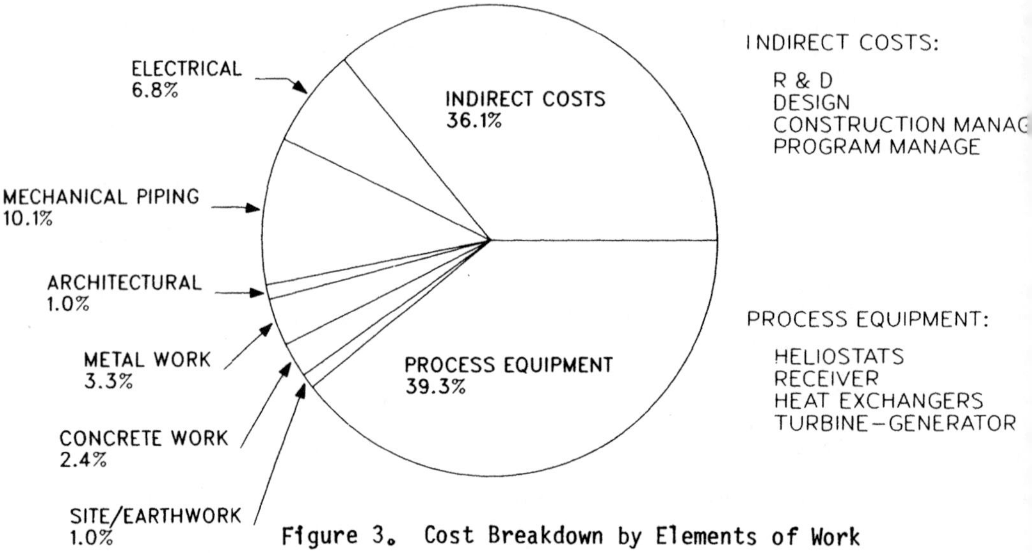

Figure 3. Cost Breakdown by Elements of Work

Element of Work Costs	
Sitework/Earthwork	$ 1,272,344
Concrete Work	3,417,007
Metal Work	4,641,180
Architectural Work	1,703,718
Process Equipment	55,454,738
Mechanical/Piping Work	14,199,637
Electrical Work	9,528,456
Indirect Cost Elements	50,946,303
Round-up, Miscellaneous Round-off	$ 36,617
Total Capital Costs	$141,200,000

2.4 Recurring and Nonrecurring Costs

The recurring costs include charges for off-the-shelf equipment that could be purchased from several sources and installed using standard practices. Nonrecurring costs at the pilot plant include charges for basic R&D, special pilot plant solar system instrumentation, data-recording systems, meteorological measurement systems, excessive factory and tooling amortization, unique engineering design, and extra program and construction management. In order to determine the cost of another plant like the pilot plant, the costs must be identified as either recurring or nonrecurring.

The previously described costs include recurring and nonrecurring costs. The nonrecurring costs would not be expected to be incurred in future plants.

The total capital costs can be separated into nonrecurring and recurring costs as follows.

	Nonrecurring	Recurring
Research and Development	$ 4,668,145	$ 200,000
Design--repeat 15% of design (except Visitors Center)	24,465,336	4,163,776
Pilot Plant Features		0
Visitors Center w/ design	583,403	
SHIMMS Factory & Construction	1,115,295	0
Data Acquisition System	1,085,695	
Factory Planning, Tooling	4,221,379	744,949
Other Factory, Construction	0	82,224,589
Start-up	3,726,713	657,655
Program Management	3,912,416	690,426
Construction Management	$ 3,397,973	$ 5,305,639
	$47,176,355	$ 93,987,034
Round-up, Miscellaneous Round-off	$ 36,611	
Total Capital Costs	$141,200,000	

3. CONCLUSIONS

The costs presented are summaries of more detailed costs that will soon to be published ("10 MWe Solar Thermal Central Receiver Pilot Plant Total Capital Cost," H. F. Norris, Jr., SAND83-8019). This report will provide a detailed cost data set for the pilot plant.

The cost of Solar One may seem high if compared to the construction of a commerical-scale, fossil-fuel, electrical generating plant; however, in reality, the cost of the Solar One experiment, whose primary purpose was to produce data, is relatively modest. Since the central receiver concept does not scale well to very low power levels, the Solar One experiment cost, relative to the energy produced, is expected to be high. However, the technology is making excellent progress toward the goal of $2000/KWe, and there is much to be learned from examining Solar One costs even though they are a factor of 7 higher.

REFERENCES

1. "10 MWe Solar Thermal Central Receiver Pilot Plant, Solar Facilities Design Integration, Pilot Plant Station Manual (RADL Item 2-1)," SAN/0499-57, MDC G8544, December 1980, revised September 1982.

2. "10 MWe Solar Thermal Central Receiver Pilot Plant, Solar Facilities Design Integration, Master Equipment List (RADL Item 2-19)," SAN/0499-84, MDC G9717, July 1982.

3. "Solar Thermal Central Receiver Cost Data Management System (CDMS), Software User's Guide," Polydyne, Inc., SAND83-8175, to be published.

4. "Solar Thermal Central Receiver Cost Data Management System (CDMS), Final Report," Polydyne,Inc., SAND83-8176, to be published.

5. "SCE Monthly Cost & Schedule Report," Southern California Edison, March 1982.

6. Personal conversations with personnel at pilot plant: D. Elliott (Dept. of Energy), T. Olson (Sterns-Roger), R. Gervais (McDonnell Douglas), and C. Lopez (Southern California Edison), March 1984.

7. Personal conversations with Martin Marietta personnel: R. Facchinello and M. Frohardt, February and March 1983, and January and February 1984.

8. Personal communication and conversations with D. Fellows and C. Winarski (Southern California Edison), March and December 1983, and February 1984.

ANALYSIS OF THEMIS EXPERIMENTAL PLANT COSTS

C. ETIEVANT

ECOLE CENTRALE DE PARIS

I) Construction Cost :

By the date of THEMIS dedication, 15 June 1983, the total construction cost of the facility was estimated to 300 Millions francs (in June 1983 currency). This cost differs noticeably from the estimated cost of 117.29 Millions francs initially evaluated when the construction started in September 1979.
The reasons for this increase of construction cost are partly due to inflation (+ 100.6 millions francs) and partly (+ 82 Millions francs) to other reasons :

. The mountain location of the work site (isolated area, winter works difficult, difficult access to the site, swampy land ...).

. The decision to instal in THEMIS plant a sophisticated prototype control system (automata, information transmission system, control room..).

. The large number of measuring equipments in connection with the experimental character of the plant.

. The repair of the heliostat field after the two violent storms of December 1980.

. The construction duration that was longer than expected (+ 1,5 year).

II) Cost of Operation, Maintenance and Experimentation :

Operation and Maintenance of THEMIS is ensured by Electricité de France Groupe de Production Thermique that has a team of 40 permanent agents on the site. In addition an evaluation team of 5 agents (3 belonging to Agence Française pour la Maîtrise de l'Energie and 2 belonging to Centre National de la Recherche Scientifique) is also working permanently on THEMIS site.
The total cost for Operation, Maintenance and Experimentation is estimated to 16,2 Millions francs for the year 1984.

III) Heliostat cost predictions :

The cost of production of heliostats for future plants has been analysed under several volume assumptions.
The results of this evaluation for low volume production have been published in Entropie 103 - 1982 page 39 .

In addition the production cost of a new heliostat design, the concrete heliostat, has been analysed by the SEMED company, a subsidiary of DUMEZ. This investigation has been done under a CNRS contact. The main results of this cost analysis for both low volume and mass production are indicated in page 8 and 9 of this present paper.

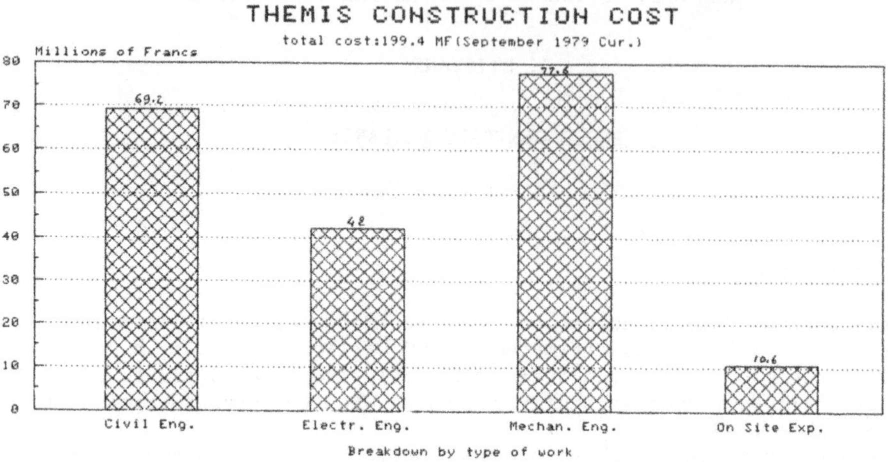

THEMIS Actual Cost

Comparison of actual cost versus estimated cost of THEMIS

	(Costs X 10^6F - Sept. 1979 currency) estimated cost	actual cost
Civil Engineering	31.96	69.2
Electrical Engineering	16.48	42
Mechanical Engineering	66.04	77.6
On Site Expenses	2.81	10.6
TOTAL	117.29	199.4

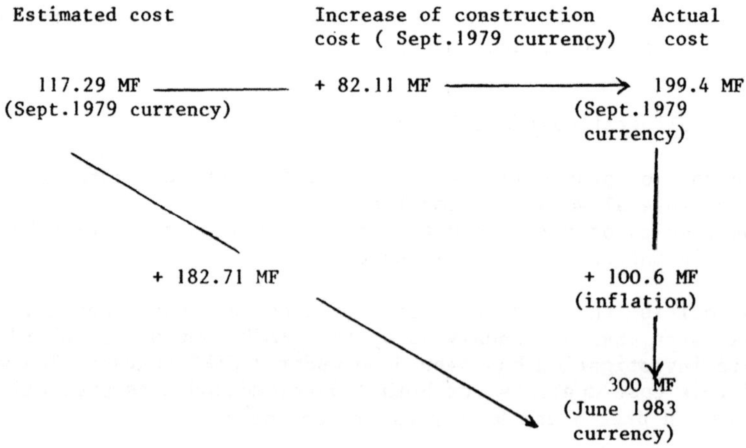

Breakdown of TEHMIS Construction Cost
by type of works

I) Civil Engineering :	Cost Estimate (Sept.1979 Francs)
. Site studies	450.000
. Access road	1.897.000
. Earth Works	4.060.000
. Civil engineering studies	1.240.000
. Architectural studies	970.000
. Buildings, Tower, Civil Engineering	6.240.000
. Structural steel works	5.500.000
. Control Room	200.000
. Office Building and Visitor Center	4.000.000
. Painting and finishing	500.000
. Handling equipments	100.000
. Contingency	6.800.000
Total cost for Civil Engineering	31.957.000 F

II) Electrical Engineering

. Electrical equipments	800.000
. Central computer	3.500.000
. Electronic bus	1.500.000
. Programmable automata	900.000
. Controls	900.000
. Measurements and instrumentation	1.500.000
. 20 KV Transformer station, 380 V equipments, converters, rectifiers, accumulators	1.070.000
. Auxiliary power unit	350.000
. Earth connection system	300.000
. Lighting system	800.000
. Grid connection	400.000
. Telephone	500.000
. Fire detection	100.000
. Meteo station	100.000
. Evaluation team computer	660.000
. Contigency	3.100.000
Total cost for Electrical Engineering	16.480.000 F

	Cost estimate (Sept. 1979 Francs)
III) Mechanical Engineering	
. Heliostats	32.400.000
. Solar Receiver	6.900.000
. Steam Generator	825.000
. Turbine Generator	4.500.000
. Storage Tanks	1.900.000
. Salt	2.000.000
. Salt handling equipments	770.000
. Salt pumps	750.000
. Tracing and heat insulation	1.000.000
. Salt valves	1.200.000
. Air Condenser	3.000.000
. Heat exchangers	620.000
. Conventional valves	800.000
. Fire protection	300.000
. Piping	2.400.000
. Tanks	500.000
. Pumps	450.000
. Vacuum System, Nitrogen loop, demineralized water	410.000
. Ventilation	160.000
. Elevators and travelling cranes	920.000
. Salt conditioning system and auxiliary heat source (THEK System)	2.840.000
. Contingency	1.400.000
Total cost for Mechanical Engineering	66.045.000 F
IV) On site expenses :	
. Supplies, Servicing, interim , telephone	100.000
. Site keeping	360.000
. Temporary buildings, offices, first aid post	150.000
. Equipments cleaning	300.000
. Various fluids	100.000
. Operators supplies	600.000
. Insurances	200.000
. Contingency	1.000.000
Total on site expenses	2.810.000 F

V) Increase of construction cost (Sept. 1979 currency) :

. Civil Engineering	37.000.000
. Electrical Engineering	25.800.000
. Mechanical Engineering	11.500.000
. on Site Expenses	7.810.000
Total increase of construction cost	82.110.000 F

THEMIS Experimental Plant
Cost of Operation, Maintenance and Experimentation
in 1984

		1984 Cost X 10^6 F
I) Salaries		
EDF (GRPT) Operation and Maintenance Team	40 agents	7.5
CNRS Evaluation Team	2 agents	0.5
AFME Evaluation Team	3 agents	0.5
Total Salaries		8.5
II) Operation and Maintenance		
EDF (GRPT) (Tax included, Salaries excluded)		5.15
III) Experimentation		
EDF (DER) (Tax included)		1.495
CNRS (Tax included)		1.038
Total 1984 Operation, Maintenance and Experimentation Expenses		16.183

EDF = Electricité de France
GRPT = Groupe de Production Thermique
DER = Direction des Etudes et Recherches
AFME = Agence Française pour la Maîtrise de l'Energie
CNRS = Centre National de la Recherche Scientifique

HELIOSTAT COST
Concrete Design SEMED Design

Low Volume Production (1981 Francs, Heliostat area 50 m^2)

Number of Heliostats	Heliostat cost $X\ 10^3$ F	Cost per unit of reflecting area F/m^2
1	952	19.000
4	495	9.900
200	131	2.520
2000	75	1.520

Mass Production (1981 Francs, Heliostat area 69.12 m^2)

Number of Heliostats yearly manufactured per factory	Heliostat cost $X\ 10^3$ F	Cost per unit of reflecting area F/m^2
868	78.3	1133
4340	62.7	908
13020	55.6	805

STRUCTURE OF HELIOSTAT COST
Concrete Heliostat, SEMED Design, Reflective Area 69.12 m^2
Mass Production - Costs in 1981 Francs

Annual Volume/Factory	868 units	4340 units	13020 units
Labor (manufacturing and installation)	13202	10206	8950
Supplies	13785	12334	11725
Tools (matrixes)	1500	1066	965
Reflective surface	7672	6808	6808
Drive and Hydraulic equipment	20497	16035	14008
Controls	8000	5600	3600
Factory	2160	1728	1080
Fuel, energy	1039	1010	1003
Transport	3745	2563	2520
Equipments for assembly on site	1397	1109	1028
Design and Prototype	208	21	7
Risks, Profit, Insurance, Taxes, Sundry	5134	4293	3950
Total installed cost :			
. per heliostat	78339	62773	55644
. per m^2	1133	908	805

CENTRAL RECEIVER COSTS FOR ELECTRIC POWER GENERATION

J. A. Dirks
Battelle Pacific Northwest Laboratory
(presented by J. B. Wright)
Sandia National Laboratories
Livermore, California USA

Summary

In the near future, solar electric power will become cost-competitive with fossil-fueled electric power generation. Several factors will lower the cost of solar generated power over time while the cost of electricity from fossil-fueled plants will tend to rise. In this paper, the cost of electricity from various generating technologies located in the southwest United States are compared using a consistent set of assumptions for utility owned plants. It is shown that while the capital cost of a solar central receiver electric plant will remain much higher than a fossil-fueled plant of the same size, the levelized energy cost from the solar plant will be the lowest cost alternative by the mid-1990's.

1. INTRODUCTION

In this paper the cost of solar electric power is compared to the cost of several fossil-fueled electricity generation technologies. There are two factors at work that will make solar power cost-competitive in the near future. First, the cost of solar electric generation technology is being reduced over time. This is due in a large part to the efforts of the national laboratories and industry to lower the cost through innovation and practical experience. Secondly, although solar power plants are capital intensive in comparison with fossil-fueled plants, they have no fuel input costs.

One of the major obstacles facing solar thermal electricity generating is maintaining continued reduction in cost. The perceived risk in building the first commercial plant is very high, and industry currently appears reluctant to accept those risks without significant financial incentives. The actual risks involved are probably considerably less than those perceived by industry. The possibility of total system failure, and hence a total loss of the investment, is indeed very low. Most of the actual risks and problems that are faced would not result in total failure but rather can be repaired.

Figure 1 shows the total capital requirements in first quarter 1984 dollars for plant start-ups between 1982 and 1995. Since, there is no inflation assumed in the analysis, the cost of the fossil-fueled plants are constant over this time period. However, the figure clearly shows the capital cost of solar electric power plants falling over time. This decrease in cost occurs because as a new technology is developed there is a reduction in the cost (in constant dollars) as improved versions of the new technology are subsequently built. This effect is often referred to as the learning curve.

Figure 2 shows the levelized electricity cost for the various power generation technologies as a function of the year of the plant start-up. All the fossil fuel plants have levelized costs that rise over time. This is due to the fact that fossil fuels are scarce resources in limited supply, and thus it is assumed that the input fuel price will rise in real terms over the life of the plant. At the same time it can be seen that the levelized cost of solar generated electricity will fall over time and will actually be the lowest cost alternative in the long term. The reason for this is three-fold:

1. Solar power generation does not require a scarce resource as an input.
2. The capital cost of solar power plants will fall as experience is gained and the technology moves down the learning curve.
3. The operating and maintenance cost will also fall over time due to experience and economies of scale.

2. ASSUMPTIONS FOR LEVELIZED COSTS

All cost assumptions for the conventional systems evaluated in this report were from Electric Power Research Institute/Technology Assessment Guide (EPRI/TAG) unless otherwise noted. The costs reported below for the conventional systems are given in December 1980 dollars. For the graphical comparison the values (levelized and capital) were updated to first quartre 1984 dollars by using the implicit GNP deflator. Specifically, 1980 costs were increased by 19.2%.

The discount rate used here is based on the capital structure given in TAG and is applied on an after-tax basis. It is assumed that the owner has a combined federal and state income tax rate of about 50%.

Type of Security	% of Total	Constant Dollar After Tax Return
Debt	50	.55
Preferred Stock	15	.40
Common Stock	35	2.20
		3.15

The assumptions and parameters listed below are used for all power generation plants in this paper.

- Location - Southwest United States
- Plant life is 30 years
- 10% investment tax credit
- No energy tax credits
- Property taxes are assumed to be 1% of the initial capital cost per annum
- No real escalation is assumed for capital, or operating and maintenance costs

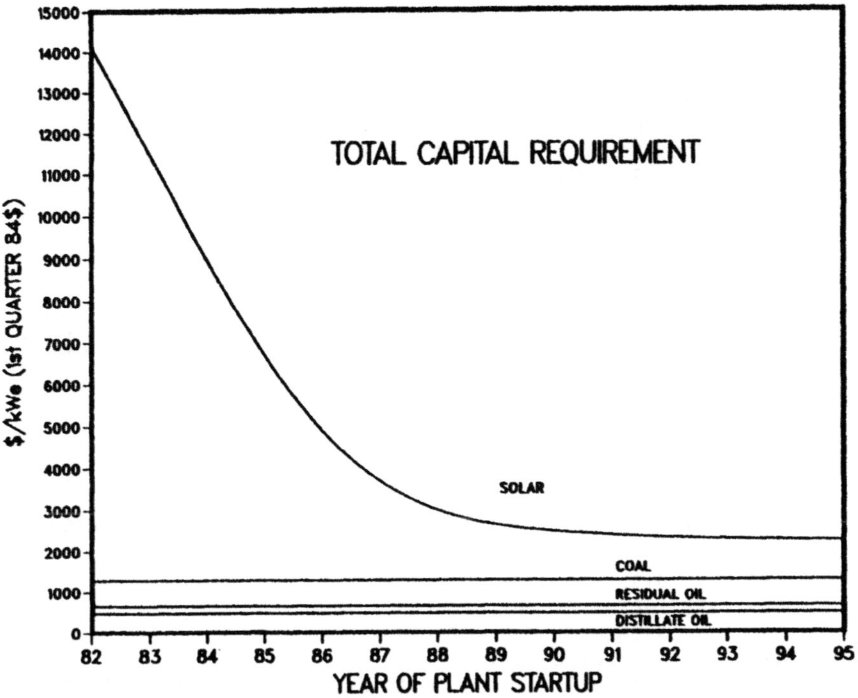

Figure 1.

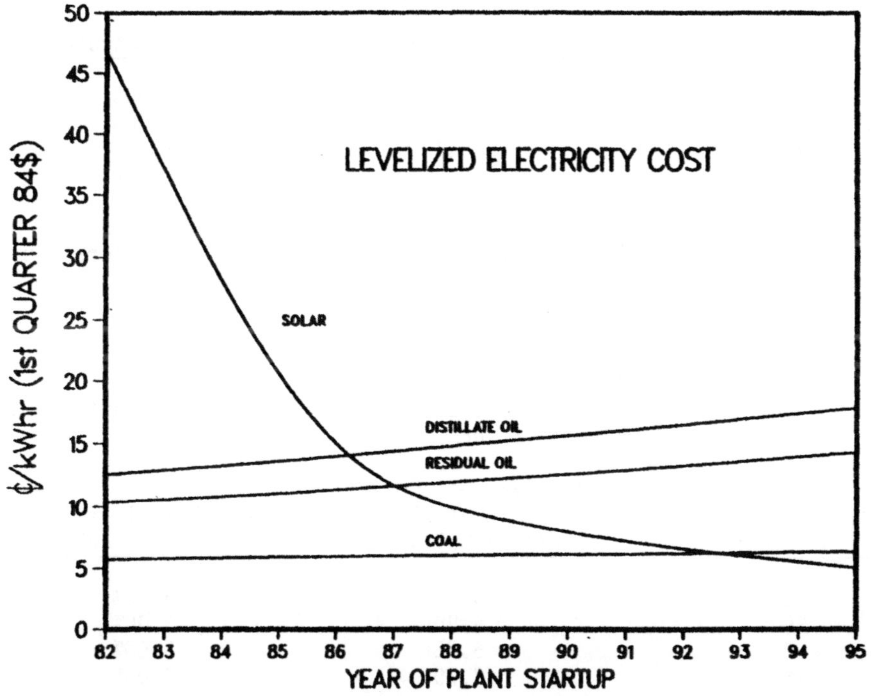

Figure 2.

For the intermediate load coal-fired plant, the following parameters are assumed:

- Two unit plant, subcritical, 500 MW unit size, using Bituminous coal
- Capacity factor of 42%
- Capital costs are $1100 per kilowatt
- 15-year ACRS depreciation
- Operating and maintenance costs are $23.33 per kilowatt per year
- Heat rate is 9970 Btu/kWhr

Year of Plant Start-Up	Coal Cost at Start of Plant ($/MBtu Delivered)	Annual Real Coal Price Escalation
82	1.78	2.14%
85	1.86	2.15%
90	1.96	2.26%
95	2.17	1.96%

For the distillate oil-fired plant, the following parameters are assumed:

- Two unit plant, 250 MW unit size
- Capacity factor of 28%
- Capital costs are $415 per kilowatt
- 15-year ACRS depreciation
- Operating and maintenance costs are $8.04 per kilowatt per year
- Heat rate is 8600 Btu/kWhr

Year of Plant Start-Up	Distillate Oil Cost at Start of Plant ($/MBtu Delivered)	Annual Real Distillate Oil Price Escalation
82	7.00	3.00%
85	7.65	3.00%
90	8.87	3.00%
95	10.29	3.00%

For the residual oil-fired plant, the following parameters are assumed:

- Two unit plant, 250 MW unit size
- Capacity factor of 28%
- Capital costs are $480 per kilowatt
- 15-year ACRS depreciation
- Operating and maintenance costs are $10.32 per kilowatt per year
- Heat rate is 8685 Btu/kWhr

Year of Plant Start-Up	Residual Oil Cost at Start of Plant ($/MBtu Delivered)	Annual Real Residual Oil Price Escalation
82	5.30	3.00%
85	5.80	3.00%
90	6.72	3.00%
95	7.79	3.00%

Three solar plants were considered in this paper: Barstow, Carrisa Plain, and a long-term cost estimate from the Solar Power Tower Design Guide. All costs for the solar plants are given in first quarter 1984 dollars.

Barstow:

- 10 MW prototype
- Capacity factor of 30%
- Capital costs are $14,160 per kilowatt
- 10-year ACRS depreciation
- Operating and maintenance costs are $333.00 per kilowatt per year
- Plant start-up in 1982

Carrisa Plain:

- 30 MW commercial project
- Capacity factor of 30%
- Capital costs are $4,923.33 per kilowatt
- 10-year ACRS depreciation
- Operating and maintenance costs are $90.00 per kilowatt per year
- Plant start-up in 1986

Solar Power Tower Design Guide:

- 100 MW commercial project
- Capacity factor of 42%
- Capital costs are $2280 per kilowatt
- 10-year ACRS depreciation
- Operating and maintenance costs are $45.00 per kilowatt per year
- Plant start-up in 1995

REFERENCES

1. ELECTRIC POWER RESEARCH INSTITUTE (May 1982). Technical Assessment Guide. EPRI P-2410-SR
2. BATTLESON, K. W. (April 1981). Solar Power Tower Design Guide: Solar Thermal Central Receiver Power Systems, A Source of Electricity and/or Process Heat. Sandia National Laboratories Livermore. SAND81-8005

SESSION III

LESSONS LEARNED AND ASSESSMENT OF THE TECHNOLOGY "HOW WOULD WE DESIGN TODAY A COMMERCIAL CRS SOLAR POWER PLANT"

Summary of the Session by the Rapporteur
A.C. SKINROOD, Sandia National Laboratories, Livermore, USA

Lessons learned : considerations for new design

SUMMARY OF THE SESSION BY THE RAPPORTEUR

A.C. SKINROOD
Sandia National Laboratories, Livermore, USA

1. INTRODUCTION

In Session IV, Mr. K.T. Cherian of the U.S. Department of Energy outlined the goals of the U.S. Solar Thermal Technology Program. Mr. Clif Selvage, IEA/SSPS Project, followed with an introductory presentation of the lessons which have been learned to date. Projects were then presented, and a lively discussion ensued.

2. CHERIAN

The goals of the U.S. Solar Thermal Technology Program are as follows:

1) develop high-performance and reliable systems and components resulting in cost competitive systems for electric and process heat applications;
2) transfer technology to the private sector; and
3) develop technology for solar enhanced fuels and chemicals production.

The strategy for achieving these goals is to foster increased development of innovative concepts and continue research and development on low-cost, high-performance components and subsystems. There is also an increase in efforts with industry to assure technology transfer and to evaluate high-risk, cost break-through opportunities. Furthermore, there is an emphasis on high temperature solar thermal technology for fuels and chemicals production applications.

Preliminary guidance for FY 85 funding (which begins October 1, 1984) is as follows:

ACTIVITY	BUDGET REQUEST
* Materials and Components Research	$ 8.42M
* Concentrator R&D	4.55M
* Central Receiver R&D	15.33M
* Planning and Assessment	1.60M
* Distributed Receiver R&D	7.10M
* Capital Equipment	0.50M
TOTAL	$37.50M

Mr. Cherian indicated that international cooperation continues to be an important part of the U.S. Solar Thermal Program.

3. LESSONS LEARNED

3.1 Selvage

Lessons have been learned in five critical areas:

(1) Meteorological measurements should be made at the actual site and carefully reviewed;

(2) Conventional equipment must be carefully selected to suit the solar application;

(3) Larger plants must be built because they are more cost effective;

(4) These plants should be designed for minimum operating staff; and

(5) There is a need to learn from the experiences of the six operating plants.

3.2 Project Presentations

Brief presentations were made on several of the operating central receiver projects. Some of the points raised by the speakers were as follows:

Barstow 10 MWe Pilot Plant (R. Gervais): The digital/distributed control system works well and increases plant on line time. Unique solar conditions, especially cyclic operation, affect the reliability of conventional equipment, and standard specifications may not be adequate. The plant has had minimum environmental impact.

Eurelios (J. Gretz): Start-up times need to be reduced since several of the projects have start-up times of one to two hours. Perhaps extra heliostats should be added to the west portion of the fields to increase the power available for start-up. Efficiency needs to be increased, particularly in the prime mover, since increased efficiency will result in a direct savings in the cost of solar equipment.

Themis (F. Pharabod): Two major components of a central receiver plant are new: the heliostat field and the receiver. The heliostat field has worked well as its performance has demonstrated. The receiver has also worked well and has accommodated rapid flux changes. It is important to reduce parasitic power.

Cesa-I (F. Sanchez): Local insolation is very important and broad-scale meteorlogical maps may be misleading. Foam core heliostats are probably not the best type of mirror construction because of mirror corrosion. Superheater location for a water/steam receiver is very critical and is sensitive to heat flux variations. The location therefore must be selected very carefully. Heat tracing of a molten salt system requires extra care because the reliability of the system is very important. Differences in the length of day between winter and summer cause some staffing problems because, while only one shift is necessary during winter, nearly three shifts are necessary during summer.

3.3 Discussion

A lively discussion took place with emphasis in three areas: 1) the amount of thermal storage which should be incorporated into a central receiver plant; 2) whether electrical energy or chemical energy is the best application of central receiver technology; and 3) what is the best heat transport fluid--sodium, molten salt, water/steam, or air. Some of the views expressed were as follows:

M. Fischer

In central Europe there is a shortage of insolation, so if solar energy is to be utilized the energy must be transported to Europe from either the Mediterranean region or Africa. The best means of transport has not been determined, but perhaps it will be in the form of chemical energy. The situation is different from the United States.

D. Gorman

Perhaps we should recognize that we are not going to totally replace fossil fuel. We should then use solar energy as it is collected rather than attempting to incorporate thermal storage. This type of system has been identified and built by Arco at Bakersfield, California.

C.J. Winter

Germany uses only 17% of their energy in the form of electricity, so we are interested in thermal applications.

C. Ortiz

From a technical point of view, sodium receivers seem to have higher efficiencies and the ability to withstand rapid flux changes. However, cost must also be considered; and from an overall standpoint, the fluid must be tailored to the application.

C. Etievant

The configuration of a plant to minimumize cost energy may not be the same as the configuration of a plant to maximize the rate of return to investors.

F. Blake

Development of trackers for photovoltaics is a major factor in the current low price quotations for heliostats.

D. Stahl

Annual energy production is more important than achieving high efficiencies in measuring the success of a central receiver plant.

Fricker

The solar multiple should be high and the storage capacity sufficient to operate plants twenty-four hours per day to avoid start-up and shut down problems.

LESSONS LEARNED: CONSIDERATIONS FOR NEW DESIGN

C. S. SELVAGE
IEA/SSPS - ITET
Almeria, Spain

Summary

Considerations for new design should take into account the following:

1. Learn as much as is humanly possible about the meteorology at the specific site chosen to construct a new plant.
2. Look very carefully at the conventional equipment from the standpoint of the needed thermal cycling capability.
3. Design as large a size as the availability of money permits, consistent with the constraints of tower heights, heliostat image size, and atmospheric attenuation.
4. Design to minimize the staffing.
5. Pay attention to prior efforts to avoid repeating mistakes.

1. INTRODUCTION

My approach to this subject is to first consider lessons learned in a broad general sense and then to see what this guides me to do if I am charged with designing a plant today. Consequently, I have listed the most generally observed "Lessons".

-METEOROLOGY
-CONVENTIONAL EQUIPMENT
-SIZE - MW/MWhrs.
-AUTOMATION/STAFFING
-EXPERIENCE

Certainly there are other lessons, most of which will become very specific, and I leave that to the true and solid scientists and engineers, some of whom are on this panel. So let me, in a general way, expand on each of the above areas.

2. METEOROLOGY

Yesterday, we heard presentations on design, construction, operational history, and performance of the six Central Receiver plants that have been operated. All have observed some significant effect on operational performance and/or serious damage brought on by the weather. It seems that the plant least surprised by some aspect of the weather was Solar One, although even at Solar One the overall performance was below expectations for 1982 and possibly 1983.

The location of Solar One at Daggett, California, was selected after very detailed analysis of meteorologic data acquired at the Daggett Airport, the nearby test station at China Lake, and Edwards Air

Force Base. Even then, years before construction started, a special meteorology station was installed and elements as detailed as sun shape (circumsolar) measurements were begun. The designers knew what to expect--still there were surprises such as freezing of water sight glasses and some pipes. Trace Heating had to be installed because nighttime temperatures were lower than expected.

For the SSPS project we made calculations based on somewhat limited global insolation data and wind data collected at the Almeria Airport, some 40 km away and with a mountain range in between. We used the usual calculation methods to determine expected insolation as a function of day of the year, then from optimistic arguments we moved the expected insolation curves upward because, "we move to a desert environment at a higher elevation and there will be less moisture or smoke in the air." I was wrong. The Operating Agent of the SSPS, and DFVLR, attempted to obtain better data between the time of site selection and construction, but the effort was hindered by limited funds and uncertainty about actual construction.

The SSPS plant has encountered serious lightning strikes twice, and in each case much damage resulted. The electronics in most of the heliostats were severely damaged; some damage occurred with the heliostat control computers and meteorological stations. Certainly we have not solved the problem of lightning susceptibility at the SSPS project.

The atmosphere near Tabernas has been surprising, not only due to the peak intensity of direct insolation, but also by the effect of the environment on our mirrors. Because of either dust, smoke, or other atmospheric aerosols, the rate of soiling or loss of reflectivity has been much higher than expected. We should have suspected something like this just from looking at what happens to an automobile left parked outside overnight. Winds are repeatedly stronger than had been anticipated. The peak insolation at summer solstice is lower than the 920 W/m^2 expected, and the peak at winter solstice is higher than expected. The atmosphere is clearer in winter than in summer. The SSPS best days are December, January, and February--a surprise.

CESA-UNO could have been affected in the same way as SSPS was because we are at the same location, but it seems they were a bit more conservative than SSPS and, by designing to a higher solar multiple, are able to compensate for lower than expected insolation. On the other hand, they are affected by the higher than expected winds and greater than expected soiling and loss of reflectivity.

Themis was surprised by turbulent winds in a very damaging way although excellent comprehensive meteorological data has been collected for some 20 years at nearby Odeillo. The winds were a surprise. Eurelios was sized using meteorological data from the Catania Airport meteorological station, many kilometers away, without detailed measurement of atmospheric effects on solar input--and of course, as with many of the plants, economic limitations caused a reduction in the number of heliostats installed. The combination has made performance less than expected.

All plants in operation today have been affected by unknowns in the specific meteorology of the place where the plant was built.

What Should This Tell Us?

First of all, when making a site selection, get all the data possible. After making a selection of the site, install a meteorological station while design proceeds, financing is cared for, and approvals are arranged. This should provide for at least one year of good specific data about that specific location. This in itself will not guarantee an elimination of surprises, but the designers might decide to move to another site.

3. CONVENTIONAL EQUIPMENT

All of the plants described and operated have reported reasonably good performance with the solar specific hardware. Certainly, in the sessions this morning on heliostats, receivers, and controls, those solar specific subsystems were discussed in detail, and their problems aired. Also, the hardware that is much needed to complete the system and allow operation--the so-called "conventional hardware"--was touched on. Many of the significant failures that have caused much loss of operational time have been brought about by failures of the valves, pumps, pipes, tanks, and electronics, the "conventional hardware."

In most of those failures it is possible to understand, in retrospect, that the hardware has been misapplied. It has been misapplied. It has been misapplied because of the cyclic nature of the energy source, the sun. Those valves, pumps, pipes, etc., were not designed to be cycled as much as is necessary for a solar thermal plant. So there have been many failures.

The Lesson - ?

Don't assume that equipment designed for a fossil-fired or nuclear thermal plant can be directly applied to a solar thermal plant. Again, check the experience gained.

4. SIZE - MW/MWhrs

Many of us have performed studies analyzing the influence of size of a Central Receiver plant as a function of cost of the plant in money (dollars, DM, francs, or whatever) and also in terms of cost per kilowatt hour. I believe that in all cases it has been obvious that these costs are very dramatically affected by the size of the plant.

I remember the results of one study, I believe performed by Martin-Marietta, that produced a curve similar to the one here. Note that I avoid putting numbers on the capital cost scale, mostly to avoid argument, but we recognize that with time, these numbers, and probably the shape of the curve, will change as we progress, learn more, and drive the cost of subsystems down. But, the curve surely represents that facts that cost per kW decreases rapidly in the range between one megawatt and one hundred MWe.

I show in this curve that the cost may--and I emphasize may--start back up again around 300 MWe because of tower heights, piping sizes, heliostat image size, and atmospheric attenuation. Certainly a very ripe area for discussion, and perhaps the session on Economics got into this area; Prof. Claude Etievant, who chaired the session, may give us some comment.

All of the six operating plants are much smaller than the size where the curve starts to flatten out. None of those who worked at justifying design and construction of the six operational plants were able to argue that it was essential to build a 100 MWe plant as a pilot or demonstration plant. Therefore, the plants are small, and are not cost-effective; it was never argued that they would be or could be. They are all either pilot plants or technology demonstration plants.

The Lesson - ?

Can I say big is beautiful? Keep in mind that just to pay for the people to run the plants, not to pay interest on the capital costs or pay for the capital costs, and at a very good price for the electricity, requires a very good location, hence lots of solar hours and very respectable sizes - certainly larger than most of our six plants. Therefore, the lesson is: <u>design a new CRS Plant as large as possible.</u>

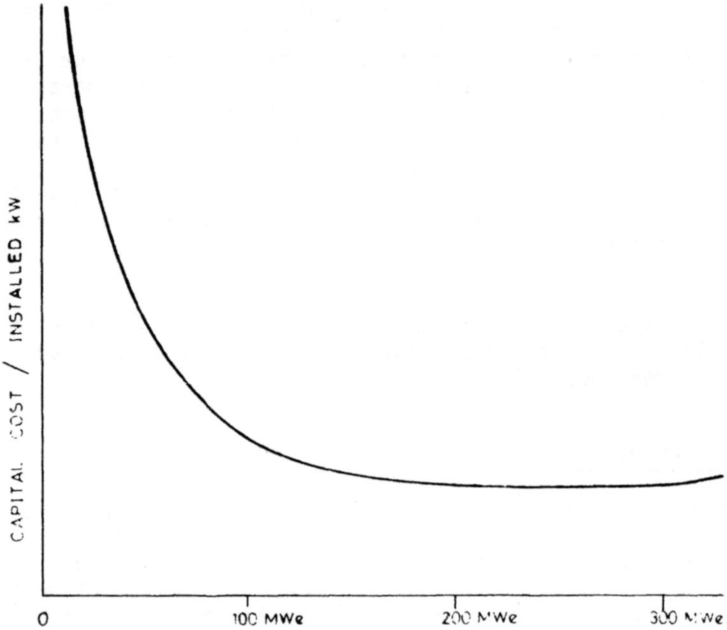

Figure. Capital Cost/Installed KW versus Plant Size (in MWe)

5. AUTOMATION/STAFFING

The above closing statement leads us to the subject of automation. Solar Thermal Central Receiver systems are rather complicated. They have a very large number of motors, pumps, valves, and electronic gadgets that must operate, or be operated, for the CRS to start-up at the right time, work, heating up the many thermal parts, and then operate a steam generator, turbine, etc. Any modern engineer knows that computers can do all this and automation or semi-automation of some of the pilot plants is becoming a fact. As we examine the early plans being developed for automation, we see that not all operating staff is eliminated, and certainly a reduction of maintenance and personnel can only come with high reliability of the hardware--not automation. As we learned at the first international CRS workshop at Clairmont in 1982, there are certain requirements in most countries, brought about either by social rules or Union requirements, that for every position identified, some minimum number of people (always greater than one) must be employed. Of course, this also makes very good management sense just to provide for sickness/vacation, holiday, etc., but in many cases it goes beyond what would be the number from a management point of view. Again, it all comes back to O + M costs and therefore the productive hours of operation, size of the plant, and finally, income. If complete unattended automatic operation is possible, it should be a major goal. I do not believe that all positions can be eliminated, but a maximum effort in that direction should be made.

The Lesson - ?
Design so that a minimum number of staff are needed.

6. EXPERIENCE

The world's solar community has built six Central Receiver plants--a spectrum of technologies, a variation of heat transfer options--salt, sodium--and four variations of water/steam. There are seven different heliostat designs in service and at least six different control systems for those heliostats. All the groups have collected large amounts of data and all have performed extensive analysis of the data in an effort to understand why the plants operated in the way that they have, and to attempt to improve the performance.

We are meeting this week to share our experiences, our discoveries, and to examine our failures. Those who are presently designing new, bigger, hopefully better central receiver plants have this information available to them as of today. I hope they are able and interested in using the results from these efforts, particularly the results of difficulties encountered, such as trace heating failures, and so many things with the conventional hardware. One problem I see is availability of the data, the evaluation, the reports, particularly of the failures which we will experience in the near future. We now have a status report from each--what about next month? next year?--how can we maximize the availability of information from these pilot or demonstration plants to the benefit of the designer? I don't know the answer, but I certainly hope that we can find a mutually acceptable way to achieve an easy and continuing information exchange.

I am reminded of a very pertinent saying by Talleyrand:

> Those who do not learn from the mistakes of history
> are bound to repeat those same mistakes.

Please let us learn from our experiences.

There is one particular experience with the SSPS that was not discussed in the sessions this morning that I would like to address here because I feel that it can be a serious problem with any storage coupled system. It is not an unsolvable problem if it is addressed during the design, accounted for during the design, and if actions are taken to compensate for it. We have called it thermal inertia.

Solar powered thermal systems must start each day from a lower temperature than that required for operation. It takes time for subsystems to reach rated temperature and neglecting 'thermal inertia' makes the performance of most thermal systems a disappointment when compared with the design goals.

A notable example is the thermal inertia associated with the energy storage tank in the SSPS sodium system. A steel tank has a large thermal capacity. This may be an advantage if the tank is already hot and is well insulated, but if it is allowed to cool, it may take a long time to bring it up to rated temperatures.

Serious consideration must be given to this fact when designing a sodium system. Inertia also affects operation strategies. When one considers all the interfaces--solar to thermal (the receiver), receiver to storage (the pipes and storage), and steam to water (the steam generator), it is apparent that careful consideration must be given to the thermal inertia of the components or else the overall daily performance will be poor.

We have published some papers on this problem area and we hope designers of new storage coupled systems are paying attention. Again, let us avoid making the same mistakes.

SESSION IV

BASIC ASPECTS ON SOLAR CHEMISTRY AND STORAGE

Summary of the Session by the Rapporteur
M. FISCHER, Institut für Technische Physik, DFVLR, Stuttgart, Federal Republic of Germany

Some basic aspects on solar chemistry and storage

Solar central receiver costs for electric power generation

Chemical storage of solar energy and solar fuels

Application of solar energy for chemical processes

A modularized sensible heat storage system for process heat generation

SUMMARY OF THE SESSION BY THE RAPPORTEUR

M. FISCHER
Institut für Technische Physik, DFVLR, Stuttgart
Federal Republic of Germany

1. INTRODUCTION

This session concentrated on solar thermal chemistry for storage and transportation. It was one of the highlights of the Second International Workshops on the Design, Construction, and Operation of Solar Central Receiver Projects, that, besides the generation of electricity, the development of a chemical energy carrier with a high energy density and which is easily transportable and storable on a large scale, without losses, receives more and more priority for solar energy conversion R & D.

In particular for European countries, where the need to store and transport solar energy on a very large scale over long distances is obvious, the conversion of solar energy at sites with high insolation in southern parts of Europe to solar fuels and chemicals is mandatory.

It has been shown that already now concentrated solar radiation is a relatively cheap "fuel." And it may become even cheaper, if the forecasted further reductions of heliostat cost will be realized. In various presentations of the workshop it became obvious that the actual cost reduction of heliostats indeed follows the trend of the cost reduction curve projected in the last years. Therefore, there is a good chance that further progress can be realized on the way to produce a chemical energy carrier economically--a chemical energy carrier which is compatible with our present energy technology and infrastructure. Such a large scale energy storage and transportation allows, among others, the design of the solar conversion system as well as the transport system according to the mean value demand, instead of its peak value, which is technically and economically attractive.

2. PROGRAMS

2.1 United States

The US solar fuels and chemicals program concentrates in the near-term on the commercialization of current technology and in the long-term on new applications for central receivers: industrial process heat and the production of fuels and chemicals. This is based on the corresponding huge US energy consumption in industry (17 Quad or 24.3% of total US consumption in 1982), respectively. The US solar fuels and chemicals program has two main activities:

(1) a technology project,
(2) system design studies.

The approach of the technology project is to develop the technologies of General Atomics thermochemical hydrogen process as a focus of the technology R & D efforts. The main activities are the development and test of high temperature and high pressure catalytic reactors, respectively. Moreover, work on ceramic heat exchangers and process design is part of the technology project.

The purpose of the system design studies is to establish conceptual designs for the production of fuels or chemicals using a solar central receiver as a major source of heat. In addition, a data base will be the development of various processes, designs, operating strategies, and economic predictions to guide future R & D in this field. Three design study projects with specific processes and products have been selected:

(1) Manufacture of activated carbon with gas and liquid fuels generated as by-products. The central receiver is cooled with carbonate salt as heat transport media.

(2) Manufacture of ammonia and nitric acid. Carbonate salt has been chosen too as heat transport media in the receiver. However, a direct flux absorption in a falling salt film in the cavity receiver is the important aspect in this design.

(3) Manufacture of ethylene. The important systems are MWt design point, 12 hour air rock storage, cogeneration of electric power, and an air-cooled, volumetric receiver type.

2.2 Switzerland

A basic laboratory experiment in Switzerland has been reported and concentrates on solar chemistry reactions in the high temperature range, preferably 1500K and above. Although suitable reactions for both energy storage and production of chemicals are known and under investigation, there exist no definitely favored processes. Therefore, the present laboratory experiments have the objective to explore possibilities to produce SiC, Si, and Al in a carbothermal reaction on the basis of direct coupling high concentration solar radiation into the process. First experimental results show the principal feasibility of these processes.

2.3 Germany

The present German R & D program on the application of solar energy for chemical processes puts special emphasis on steam reforming of methane and the refinement and purification of hydrocarbons, i.e. coal pyrolysis and heavy oil cracking. The availability of high temperature solar heat for endothermal chemical reactions is one important basis for the realization of the high development potential of the total solar system including large scale storage and transportation of solar energy via hydrogen-rich chemical energy carriers. Hydrogen and synthesis gas are the keys on which a whole energy system technology can be built up.

The main problems of high temperature chemical technologies based on solar heat are the discontinuity of solar energy inputs, transients, very high temperatures and corresponding material problems, heat storage, and on-site availability of base material needed for the processes. The objectives of the planned German experimental program on solar steam reforming of methane and heavy oil cracking are mainly concerned with the adaptation of the corresponding conventional processes to solar heat, i.e. development of receiver, vaporizer, and reactor components. An overview of development tasks are given in the papers presented in the session.

The specific problems and feasible solutions of modularized sensible heat storage systems for process heat generation were discussed in another paper of the session on Solar Chemistry. Thermal storage is most advantageous for buffering short term insolation shortages. Possible media for heat storage at temperatures above 110°C are solids and molten salts. Advantages of solid storage systems over molten salt are chemical stability, no corrosion, no toxicity, low price, availability of raw materials, simple technology, and broad know-how with cowper in iron casting industry. More complexity is involved in solid storage systems with respect to energy balance and control of the charge- and discharge-rates. The design characteristics and dynamic behavior of such storage systems have been studied with the computer simulation program PROSES (Parametric Representation of Solar Energy Systems). The hydrogen production by steam reforming of methane has been taken as an example for chemical processes on solar basis to illustrate the potential of storage modules that the solar energy input can be perfectly decoupled from the chemical processes.

3. CONCLUSION

The session Solar Chemistry indicated that large scale use of solar energy may also become possible in the non-electric sector if some of the proposed (and possibly other) ways to convert solar energy into fuels, chemicals, and process heat can be successfully developed.

SOME BASIC ASPECTS ON SOLAR CHEMISTRY AND STORAGE

P. Kesselring

Swiss Federal Institute for Reactor Research
CH-5303 Würenlingen

Summary

It is shown that concentrated solar radiation is relatively cheap at power levels of 10 MW_{th} and higher. This forms the basis to look into solar thermal chemistry for storage and transportation. R + D-areas of interest are new reactions at higher temperatures (e.g. above 1200 K), high concentrations (including secondary mirrors), and chemical reactors. It is felt that international cooperation is highly desirable.

1. INTRODUCTION

This day is devoted to the discussion of other solar thermal application than electricity generation. Emphasis is put on solar thermal chemistry. In discussing some basic aspects of these applications I will start by showing that concentrated solar radiation is a relatively cheap form of energy. This important fact is one of the sources for the interest, currently shown in our topic.

Another fact is the need to store and transport solar energy in order to make it a universally useful energy source. Chemical storage is the method of choice to achieve this goal. This will be the second part of my talk. The last part contains a few remarks concerning the different ways of doing solar chemistry, deals briefly with the efficiencies that can be expected for solar thermal processes and ends with discussing the elements of a possible Solar Thermal Applications Program.

There is no discussion of specific processes and applications. This will be left to the next speaker.

2. THE COST OF CONCENTRATED SOLAR RADIATION

Solar radiation is a high quality, low density kinetic form of energy that is abundantly available. Although it is basically free of charge, the low density leads to high collecting cost and large collector fields. Thus to give solar energy utilization a chance, both from the point of view of money- and energy economy, the need for cheap collectors, using a minimum of material is mandatory.

In order to achieve these goals, different routes have been followed. The one we have discussed these days is to concentrate sunlight by means of mirrors. A mirror is in some way an ideal collector, as it needs in principle not more than a few grams of e.g. aluminum/m^2 collector area. Of course cost and the amount of construction materials needed are determined by the structure which is necessary to protect and direct the active surface properly. By the way a similar situation prevails e.g. also in photovoltaics.

The result of concentration, e.g. at the receiver of a CRS plant, is radiation energy confined in a relatively small volume and at densities comparable to those generated by fossil fuel burners. In contrast to natural sunlight this high quality "solar fuel" is now not free of charge, but has its price.

Taking actual prices for installed heliostats of 230 to 250 $/$m^2$ and an energy at the receiver of 1200 to 1500 kWh/m^2a, we ask how much we have to invest in order to earn a kWh of concentrated radiation per year. The answer is

0.15 to 0.2 $a/kWh.

Assuming an annuity between 10 and 20 %/a this is equivalent to a kWh-cost of

0.015 to 0.04 $/kWh.

Thus, concentrated solar radiation is a relatively cheap fuel already now. It may become even cheaper, if the forecasted further reductions of heliostat prices become true.

By the way, we note that this kWh cost is about an order of magnitude lower than that for electricity from CRS-plants. This shows how much we loose in the conversion of radiation into electricity (mainly carnotic-, storage-, heating up- and cooling down-losses). It is an intriguing and - in connection with the topic of the present talk especially - relevant question, whether we have already found the best way to make use of concentrated solar radiation. (Of course this does not mean, however, that we should neglect the further development of solar thermal electricity generation, as this technology has the chance of becoming economically competitive in the future).

3. SOLAR ENERGY AND STORAGE

Substitution of oil is one of the most generally accepted goals of solar energy r & d. What makes it so difficult to reach this goal are inter alia the almost ideal properties of oil as an energy carrier, namely its high energy density, easy transportability and storage. Thus, we are able to "move around" energy easily in space and time. This fact allows to design the production system as well as the transportation system according to the mean value of demand (instead of its peak value), which is not only simpler but - most important - cheap.

Solar energy on the other hand is difficult to store and to transport over large distances. The basic reason for this difference is simple: radiation is kinetic, the chemical energy of oil is potential energy. Kinetic energy is difficult, potential energy is easy to store. We mention as an example the storage of mechanical kinetic energy in a flywheel, which is limited because of friction losses. We compare it to a hydro reservoir in the mountains that has practically no losses. The same difference remains true on a molecular level. Heat is the kinetic energy of the stochastic molecular movement and is difficult to store because of the irreversible heat conduction losses. They make the time constant of any thermal storage short compared to e.g. the storage in a chemical (as e.g. oil). The chemical bond, however, is the prototype of a potential energy "storage device" on a molecular level.

The task thus is to convert the kinetic solar energy into a storable form of potential energy. Which forms are available?

To any potential belongs a force and thus we have to ask what forces in nature might be suitable to do the job. We easily see in the table below that their number is very limited.

Type of bond	Relative bond strenght (Order of magnitude)
Van der Waals	10^{-3}
Hydrogen bond	10^{-1}
Covalent bond *)	1
Ionic bond	1
Nuclear forces	10^{6}

*) Typically a covalent bond is equivalent to about 418 kJ/mol or 4.35 eV.

In fact, Van der Waals forces and hydrogen bonds are too weak for storage purposes, leading to very large storage volumes. Nuclear forces on the other hand are much too strong. Consequently, we are left with chemical bonds essentially. Thus, solar chemistry is the method of choice to store solar energy for long periods and to transport it over long distances.

4. DIFFERENT KINDS OF SOLAR CHEMISTRY

Three basic disciplines come into play if we deal with solar chemistry:

- Fotochemistry
- Electrochemistry
- High temperature chemistry

Fotochemistry is probably the most direct and elegant way to use solar energy. It takes advantage of the quantum nature of light and is nearest to the principles used in nature (photosynthesis). Fotochemistry is in a state of basic research and it is not safe to make any predictions as to the point in time, when practical applications for the utilization of solar energy will become feasible.

Electrochemistry may be used to convert solar generated electricity into chemical energy. The best known example for such a process is the combination of solar cells with water electrolysis.

Solar high temperature chemistry is the topic that is of highest interest to us here. It differs from ordinary high temperature chemistry mainly in three ways:

- The useful temperature range attainable with solar radiation goes up to 2000 K and more, whereas the conventional upper limit lies in the order of 1500 K. Potentially, additional reactions and processes become feasible.
- The method of coupling solar energy into the chemical reactor may differ considerably from that known in conventional processes.
- The discontinuous and stochastic nature of the solar energy source leads to problems not known in ordinary process heat applications.

In many other respects, however, the large experience available from conventional high temperature chemistry in industry is applicable to the solar case, too. This is a major advantage.

5. INDUSTRIAL PROCESS HEAT APPLICATIONS

Though the conventional high temperature process heat demand constitutes only a few percent of the end energy used in industrialized countries, it may well be that carefully selected processes from this field may offer a good chance to promote solar high temperature applications. A typical temperature distribution for industrial process heat demand of today is shown in the picture below /1/.

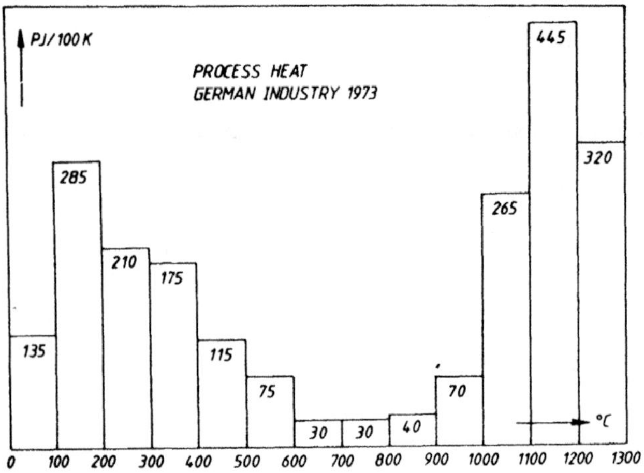

The high temperature peak is due to applications in the steel, cement, ceramic and glass industry and in brick production. Strictly speaking, this has little to do with solar fuels and chemicals. However, even if only a few processes can be modified to solar or solar hybrid operation and perhaps in experimental plants only, industry know-how would be involved directly and much could be learned concerning general problems of solar thermal chemistry such as coupling the radiation into the reactor.

6. EFFICIENCY ESTIMATES OF SOLAR THERMAL PROCESSES

At the beginning we have asked ourselves whether solar high temperature chemistry might make better use of concentrated solar radiation than e.g.

thermal electricity production. One aspect of this question is process efficiency.

Neglecting all losses but reradiation from a black body receiver surface of temperature T, the fraction of useful solar radiation is given approximately by

$$\eta_r = \left(1 - \frac{\sigma T^4}{I}\right) \qquad \begin{array}{l}\sigma = 5.67 \cdot 10^{-8}\ W/m^2\,K^4 \\ I = \text{Intensity of concentrated radiation } (W/m^2)\end{array}$$

We have also deliberately neglected effects such as e.g. reradiation solid angles differing from 2π.

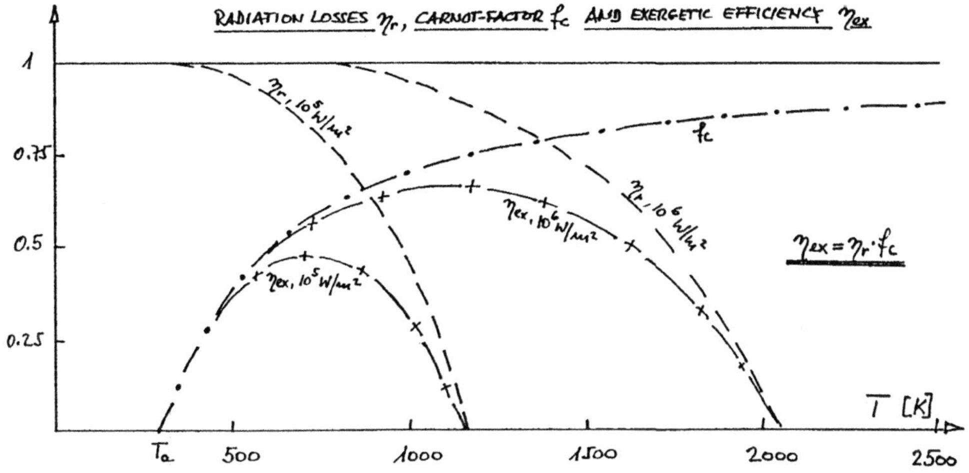

The figure shows this function for $I_1 = 10^5\ W/m^2$ and $I_2 = 10^6\ W/m^2$, i.e. for concentrations of approximately 100 and 1000 suns.

A second interesting function is the exergy fraction contained in the useful, not reradiated energy. It is given approximately by the usual carnot factor

$$f_c = (1 - T_a/T) \qquad \begin{array}{l}\text{T: Receiver temperature} \\ T_a\text{: Ambient temperature}\end{array}$$

neglecting that the exergy content of direct solar radiation is not 100 %.

Combining the two functions, we get the fraction of radiation available as exergy in a receiver at temperature T

$$\eta_{ex} = \eta_r \cdot f_c = (1 - \frac{\sigma T^4}{I}) \cdot (1 - T_a/T)$$

It has the well known shape with a maximum at approx. 700 and 1100 K.

The corresponding exergetic efficiencies are around 50 % and more than 60 % respectively. Of course these values will be considerably lower in reality because of the neglects made. However, they basically show that concentrated solar heat really is a high quality fuel because of its ability to generate high temperatures.

As an example we mention an experimentally measured energy efficiency (which should not be mixed up with exergy efficiency) of about 40 % at 1260 K and about 300 kW/m^2 radiation intensity. It has been reported for the gasification of biomass in a small experimental setup using direct insolation of the reactants /2/. The efficiency refers to the fraction of the incident solar radiation stored as chemical energy (enthalpy) in the reaction products. Care had been taken to make use of the sensible heat of the products.

7. ELEMENTS OF A SOLAR THERMAL APPLICATIONS PROGRAM

Until now R + D-work on solar chemistry has been done mainly in France (Odeillo) and the US. This workshop could be the occasion to stimulate or initiate activities in other countries. In this context it is important to come to a consensus with respect to the key elements of a solar thermal R + D-Program. According to my judgement there might be three main areas of work:

- Work on reactions, preferably in the temperature range of 1500 K and above. Although suitable reactions both for energy storage and production of chemicals are known and under investigation, there exists no definitely favourized process. It would be of much help, if the efforts in the different countries could be focused on a common R + D-subject as it has been previously the case with solar thermal electricity. The concentration of existing and the mobilisation of additional resources would become much easier, if a few reactions or processes could be identified and commonly accepted as being of highest priority.
- Work on reaching high concentrations with large heliostat fields (e.g. 10 MW_{th}).
  This may include R + D on secondary reflection methods, bringing us into the vicinity of the solar furnace community (e.g. Odeillo).
- Work on reactors, stressing the importance of solar specific aspects as e.g. the necessity of frequent start up's and shut down's which should not cause unbearable drops in the mean efficiency. Although strictly speaking reactors have to be tailored to the needs of each process individually, there exist a few reactor types which are especially interesting for solar chemistry. One example is the direct heated fluidized bed reactor. General research could concentrate e.g. on making these reactors fast thermally (low heat capacity) and "chemically" (direct irradiation of reactants, feed stock flow following the solar intensity). Another topic might be hybrid operation of reactors for continuous operation.

8. CONCLUSIONS

I hope you can follow me in drawing the following conclusions from my presentation:

- Concentrated solar energy is a "high quality fuel" with unique properties. It is already now relatively cheap.
- The basic economic and technical problem is to find applications for this fuel, such that the cost of the application may be kept low.
- Solar thermal electricity generation is such an application and has proven its viability.
- Long term storage and transportation is of prime importance for solar energy utilisation, if we want to give it a chance to take over functions now fulfilled by fossil fuels (oil).
- Solar chemistry is the method of choice to accomplish this task.
- Three different routes can be envisaged
  - Fotochemistry
  - Electrochemistry (+ solar electricity)
  - Solar thermal chemistry

  It is worthwhile to follow all three routes.
- The prospects for solar thermal chemistry are good technically (existing experience in France and USA. Relatively high efficiencies expected). It is not yet possible to give serious cost estimates of solar chemical plants (except cost of "fuel", i.e. of heliostat field). However - and I add here a very personal opinion - , the chance exists that selected industrial process heat applications might be economically competitive at the turn of the century, provided a serious R + D-effort starts now.
- Except in France and the USA, solar thermal chemistry is in a "status nascendi". International cooperation is highly desirable.
- Main areas of internationally coordinated R + D-work could be
  - Chemical Reactions
  - High concentration (including secondary concentration)
  - Solar chemical Reactors.
- A first - not easy - step of cooperation and coordination should be the identification of a few key processes on which the main efforts will be concentrated.

SOLAR CENTRAL RECEIVER COSTS FOR ELECTRIC POWER GENERATION

J.B. WRIGHT
Sandia National Laboratories, Livermore, USA

In this paper the cost of solar electric power is compared to the cost of several fossil-fueled electricity generation technologies. There are two factors at work that will make solar power cost-competitive in the near future. First, the cost of solar electric generation technology is being reduced over time. This is due in a large part to the efforts of the national laboratories and industry to lower the cost through innovation and practical experience. Secondly, although solar power plants are capital intensive in comparison with fossil fueled plants, they have no fuel input costs.

One of the major obstacles facing solar thermal electricity generation is maintaining continued reduction in cost. The perceived risk in building the first commercial plant is very high, and industry currently appears reluctant to accept those risks without significant financial incentives. The actual risks involved are probably considerably less than those perceived by industry. The possiblity of total system failure, and hence a total loss of the investment, is indeed very low. Most of the actual risks and problems that are faced would not result in total failure but rather can be repaired.

Figure 1 shows the total capital requirements in first quarter 1984 dollars for plant start-ups between 1982 and 1990. Since, there is no inflation assumed in the analysis, the cost of the fossil-fueled plants are constant over this time period. However, the figure clearly shows the capitatl cost of solar electric power plants falling over time. This decrease in cost occurs because as a new technology is developed there is a reduction in the cost (in constant dollars) as improved versions of the new technology are subsequently built. This effect is often referred to as the learning curve.

Figure 2 shows the levelized electricity cost for the various power generation technologies as a function of the year of the plant start-up. All the fossil fuel plants have levelized costs that rise over time. This is due to the fact that fossil fuels are scarce resources in limited supply, and thus it is assumed that the input fuel price will rise in real terms over the life of the plant. At the same time it can be seen that the levelized cost of solar generated electricity will fall over time and will actually be the lowest cost alternative in the long term. The reason for this is three-fold :

1. Solar power generation does not require a scarce resource as an input.
2. The capital cost of solar power plants will fall as experience is gained and the technology moves down the learning curve.
3. The operating and maintenance cost will also fall over time due to experience and economies of scale.

APPENDIX

Assumptions for levelized costs:

All cost assumptions for the conventional systems evaluated in this report were from Electric Power Research Institute/Technology Assessment Guide (EPRI/TAG) unless otherwise noted. The costs reported below are given in December 1980 dollars. For the graphical comparison the values (levelized and capital) were updated to first quarter 1984 dollars by using the implicit GNP deflator. Specifically, 1980 costs were increased by 19.2%.

The discount rate used here is based on the capital structure given in TAG and is applied on an after-tax basis. It is assumed that the owner has a combined federal and state income tax rate of about 50%.

Type of Security	% of Total	Constant Dollar After Tax Return
Debt	50	.55
Preferred Stock	15	.40
Common Stock	35	2.20
		3.15

The assumptions and parameters listed below are used for all power generation plants in this paper.

- Location - Southwest United States
- Plant life is 30 years
- 10% investment tax credit
- No energy tax credits
- Property taxes are assumed to be 1% of the initial capital cost per annum
- No real escalation is assumed for capital, or operating and maintenance costs

For the intermediate load coal-fired plant, the following parameters are assumed:

- Two unit plant, subcritical, 500 MW unit size, using Bituminous coal
- Capacity factor of 42%
- Capital costs are $1100 per kilowatt
- 15-year ACRS depreciation
- Operating and maintenance costs are $23.33 per kilowatt per year
- Heat rate is 9970 Btu/kWhr

Year of Plant Start-Up	Coal Cost at Start of Plant ($/MBtu Delivered)	Annual Real Coal Price Escalation
82	1.78	2.14%
85	1.86	2.15%
90	1.96	2.26%

For the distillate oil-fired plant, the following parameters are assumed:

- Two unit plant, 250 MW unit size
- Capacity factor of 28%
- Capital costs are $415 per kilowatt
- 15-year ACRS depreciation
- Operating and maintenance costs are $8.04 per kilowatt per year
- Heat rate is 8600- Btu/kWhr

Year of Plant Start-Up	Distillate Oil Cost at Start of Plant ($/MBtu Delivered)	Annual Real Distillate Oil Price Excalation
82	7.00	3.00%
85	7.65	3.00%
90	8.87	3.00%

For the residual oil-fired plant, the following parameters are assumed:

- Two unit plant, 250 MW unit size
- Capacity factor of 28%
- Capital costs are $480 per kilowatt
- 15-year ACRS depreciation
- Operating and maintenance costs are $10.32 per kilowatt per year

Year of Plant Start-Up	Residual Oil Cost at Start of Plant ($/MBtu Delivered)	Annual Real Distillate Oil Price Excalation
82	5.30	3.00%
85	5.80	3.00%
90	6.72	3.00%

Three solar plants were considered in this paper: Barstow, Carrisa Plain, and a long-term cost estimate from the Solar Power Tower Design Guide. All costs for the solar plants are given in first quarter 1984 dollars.

ASSUMPTIONS

- PLANT LOCATION - SOUTHWEST UNITED STATES
- 30 YEAR PLANT LIFE
- CONVENTIONAL POWER PLANT ASSUMPTIONS FROM ELECTRIC POWER RESEARCH INSTITUTE (EPRI) TECHNOLOGY ASSESSMENT GUIDE
- UTILITY FINANCING ASSUMED
- NO REAL ESCALATION ASSUMED FOR CAPITAL, OR OPERATING AND MAINTENANCE COSTS

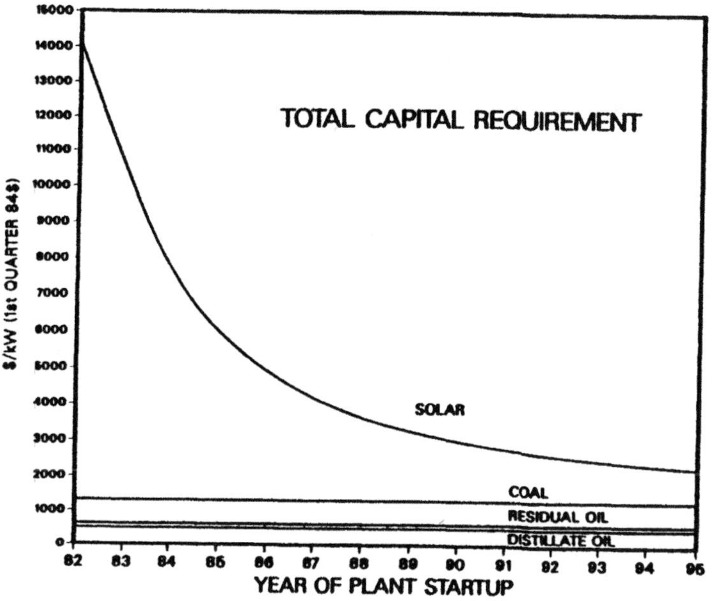

Figure 1.

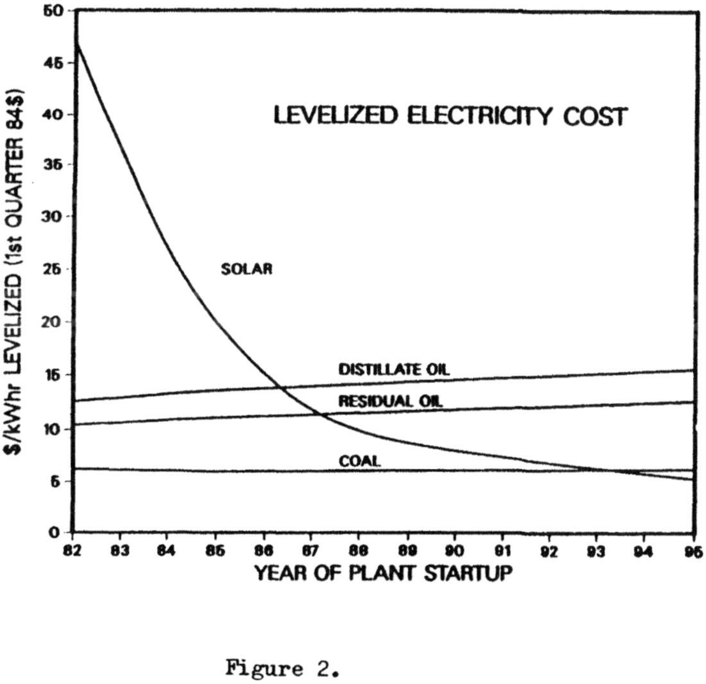

Figure 2.

Barstow:

- 10 MW prototype
- Capacity factor of 30%
- Capital costs are $14,160 per kilowatt
- 10-year ACRS depreciation
- Operating and maintenance costs are $333.00 per kilowatt per year

Carrisa Plain:

- 30 MW commercial project
- Capacity factor of 30%
- Capital costs are $4,923.33 per kilowatt
- 10-year ACRS depreciation
- Operating and maintenance costs are $90.00 per kilowatt per year

REFERENCES:

1. Electric Power Research Institue, "Technical Assessment Guide," EPRI P-2410-SR, May 1982

2. K. W. Battleson, "Solar Power Tower Design Guide: Solar Thermal Central Receiver Power Systems, A Source of Electricity and/or Process Heat," SAND81-8005, April 1981, Sandia National Laboratories, Livermore

CHEMICAL STORAGE OF SOLAR ENERGY AND SOLAR FUELS

Hp. Eicher, T. Weisskopf, R. Schlögel and H.J.Güntherodt
Institut of Physics, University of Basel
Klingelbergstrasse 82
4056 Basel Switzerland

SUMMARY

Chemical storage of energy will become more and more important with the increasing use of solar energy. This paper describes chemical storage of solar energy via carbothermic reduction of metal oxides. The resulting elemental metals fulfil the requirements for a chemical storage system to a great extent. The reactions are carried out in the solar high-temperature furnace "HELIOS 2000". Different analytical tools are used to characterize the resulting products. Metallic aluminium and silicon have been produced successfully.

Furthermore, we investigate the feasibilty of the production of industrial raw materials using the solar furnace as energy source. The preparation of beta SiC demonstrates, how a material which is accessible conventionally only with difficulties may be synthesized in the solar furnace.

1. INTRODUCTION

We started with the project on chemical storage of solar energy in sommer 1982. The stimulation was theoretical and experimental work in this field at the Swiss Federal Institute for Reactor Research (EIR) (1,2). In early 1983, we begann to construct our solar furnace with a thermal power of about 1kW in order to produce temperatures above 2500K. In may 1983 the solar furnace came into operation. Since then we carry out experimental work and set up the analytical tools to investigate the reaction products in situ and after reaction (3).

2. OBJECTIVES OF INVESTIGATION

Our aim is the investigation of selected chemical reactions in the high temperature range for chemical storage and transport of solar energy. The following reactions have been chosen.

$$Al_2O_3 + 3C \; - \; 2Al + 3CO$$
$$2Al_2O_3 + 3SiO_2 + 12C \; - \; 4Al + 3Si + 12CO$$
$$SiO_2 + 2C \; - \; Si + 2CO$$
$$C + H_2O \; - \; H_2 + CO$$

Furthermore a solar furnace could be used to carry out chemical reactions at high temperatures in order to replace conventional heat sources in the production of inorganic materials.

Therfore we are interested in the investigation of the preparation of inorganic materials like elemental Al and Si and Al-Si alloys (4). The preparation of element carbides such as SiC and CaC_2 may be carried out using solar energy.

The principal physical differences between pure thermal and photo-thermal energy sources may well lead to different reactions mechanism and therfore to different or new products. To establish the existence of this differences which we belive to be responsible for the selectivity of our beta SiC synthesis is another principal aim of our studies.

3. EXPERIMENTAL INSTALLATION "HELIOS 2000"

In order to investigate suitable reactions with the least expence we build a flexible installation, that had to fulfil the following demands: Achievement of temperatures greater than 2'500 K, reactions in vacuum or variable gases, measurments of pressure and temperature, analyses of gases and products, easy and quick manipulation

Figure 1 shows the principal design of HELIOS 2000

Fig.1

Solar furnace
HELIOS 2000

The required high energy concentration factor is obtained with two consecutive lenses. The first one is a fresnel-lens, the second one a glass concentrator. Behind the secondary concentrator the quarz receiver is located inside a cylindrical quarz vessel with a planar quarz front window.

4. ANALYSIS OF PRODUCTS

To analyse the untreated materials and the products of the reaction we use scanning electron microscopy, X-ray diffraction, gas chromatography and online mass spectrometry.

5. PRESENT RESULTS

-Carbothermic reduction of SiC:

Starting from a powdered homogeneous mixture of SiO_2 and C formation of SiC occured repeatedly. In contrast to the conventional process which yields the alpha SiC we always get beta SiC.

-Carbothermic reduction of Si:

Small amounts of elemental Si have been found in the reaction products of only one sample. The reproducibility and the conditions of formation of silicon in a solar furnace are beeing investigated during the summer of 1984.

-Carbothermic reduction of Al:

The reactions

$$Al_2O_3 + 3C \; - \; 2Al + 3CO$$

$$3SiO_2 + 2Al_2O_3 + 12C \; - \; 4Al + 3Si + 12CO$$

have been carefully investigated and may be summarized as in the diagramm shown below as Figure 2.

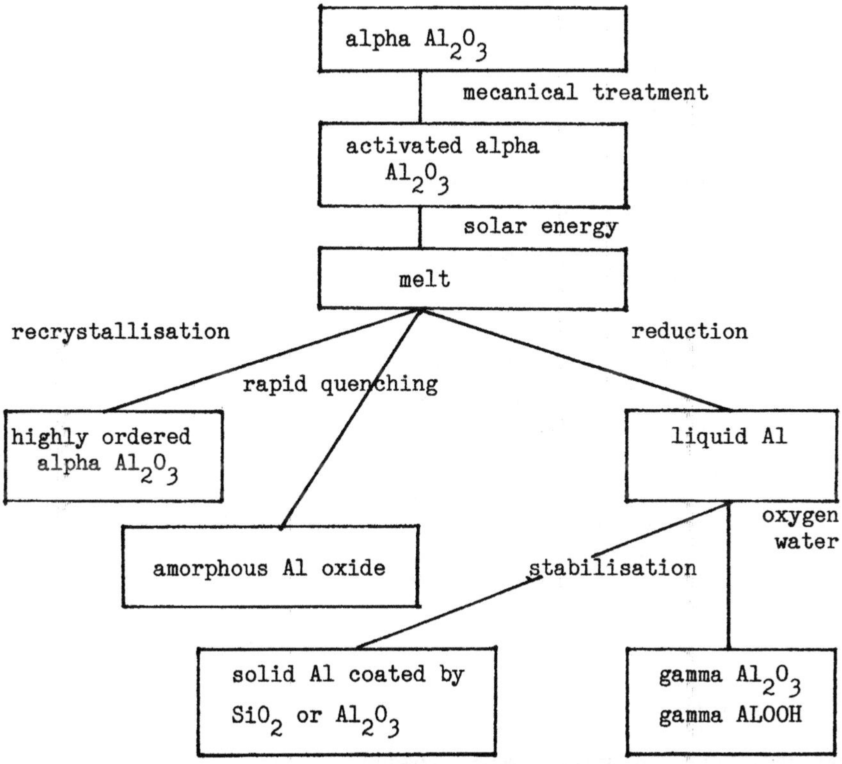

Fig.2

The untreated alpha Al_2O_3 is activated by milling. In the solar furnace the Al_2O_3 melts. One part of the Al_2O_3 melt directly recrystallizes to higly ordered alpha alumium oxide. In the major remainder of the melt carbon may dissolve. It can be observed that the melt then forms droplets of a diameter of about 0.2 to 1 mm. We note a strong influence of the heating rate and final temperature on the formation and size of the aluminium droplets. The evolution of gases (CO,CO_2) is clearly visible by the "boiling" of the material and the immediate increase of pressure. Reactions take place very rapidly in the center of the sun spot in the solar furnace. The resulting liquid of aluminium metal has the tendency to evaporate. Part of the aluminium is reoxydized from gaseous oxygen and water present in the receiver forming the new phase of gamma aluminium oxide. The rest of the liquid aluminium solidifies and is stabilized by a coating of silica or alumina. The detection of elemental aluminium by powder X-ray diffraction is shown in Figure 3.

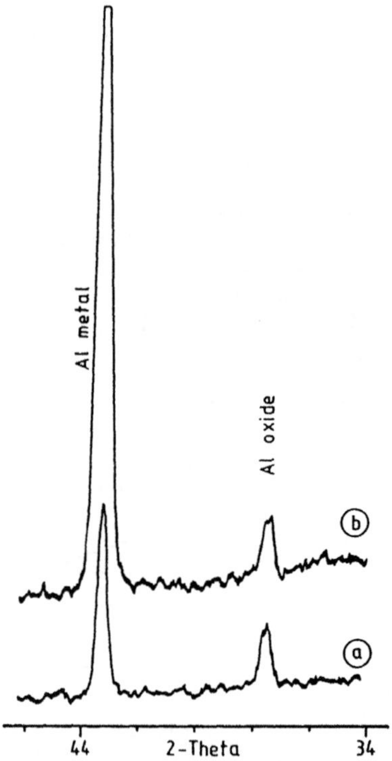

Figure 3

a Reaction product

b After addition of aluminium metal powder

6. CONCLUSIONS AND PROSPECTS

The formation of aluminium, silicon and silicon carbide in a solar furnace has been shown. This is the first step, in the assessment of the scientific feasibility of chemical storage of solar energy.

The use of other metal oxides and cheaper sources of carbon than graphitized carbon black are investigated now. Insight into the kinetics and mechanisms of the high temperature reactions is expected from controlling and monitoring the gas phase in the reaction zone.

Finally, we are aware of the fact that the sun is not only a source of thermal energy like any conventional furnace but also a source of photons of energy sufficently high to expect selective breaking and formation of chemical bonds as in photochemistry. The role of photochemical reactions in the chemical storage of solar energy is beeing investigated.

7. ACKNOWLEDGEMENTS

We wish to thank P. Kesselring and H. Weber from the EIR and V.Franzen from Lonza for helpfull discussions.

We greatly appreciate the cooperation with U. Scheidegger from Shell Switzerland. Financial support of the project is acknowledged to the NEFF (Nationaler Energie-Forschungs-Fonds).

8. REFERENCES

(1) Reaktionen die sich für die chem. Energiespeicherung eignen.
W. Durisch, O. Haas, H. Weber EIR Bericht TM-13-82-2

(2) Das System Bolex Helios 1500
Dr.P. Kesselring, EIR Würenlingen CH.

(3) Erzeugung chemischer Energieträger mit Sonnenenergie
Hp. Eicher, J. Bitterli, und H.J. Güntherodt
4.Symposium über Forschung und Entwicklung der Sonnenenergie in der Schweiz. Lausanne 17./18. Oktober 84

(4) Carbothermic reduction to form alluminium silicon alloys
AD Little ANL/OEPM-79-4

APPLICATION OF SOLAR ENERGY FOR CHEMICAL PROCESSES

M. Becker and R. Köhne
DFVLR, Köln and Stuttgart
R. Harth
KFA, Jülich

Summary

The objectives of using chemical processes in solar-thermal facilities are derived. The possible applications of these chemical processes are given under the boundary aspects of meteorology and operating conditions. Radiative and thermal energy transfer procedures are considered.

Special emphasis will be given to steam reforming of methane as initial process for the production of synthetic fuels and chemical feedstocks. Also, the refinement and purification processes of hydrocarbons, in essence coal pyrolysis and heavy oil cracking will be discussed in some detail.

Finally some practical conclusions will be drawn.

1. OBJECTIVES FOR THE USE OF CHEMICAL PROCESSES IN SOLAR-THERMAL FACILITIES

The solar-thermal power stations built and operated up to date yielded very valuable experiences. They clearly demonstrated the viability of the selected solutions. But also two major problem areas could be detected in most of the investigations:

- o The conversion of solar to electrical energy caused considerable troubles, surprisingly many were with conventional hardware. The solar-specific parts - heliostats and receivers -, however, had very positive and beneficial operating records.

  This situation motivates to consider a way which more directly applies concentrated sunlight respectively thermal energy to chemical processes, either parallel to the above mentioned power station philosophy or in a coupled manner as thermal-electrical plant.
- o The task of proper storage and efficient transport has been recognized since the start of development. A simple and entirely sufficient solution is not yet on sight.

  Then, the availability of solar-thermal heat input for endothermal chemical reactions suggests the idea of chemical storage and transport. This is, at least, a convincing idea that needs detailed studies.

Both conceptions lead to the employment of chemical reactors to function up to temperatures of 1000 °C as chemical production units as well as energy absorbers and releasers.

2. POSSIBLE APPLICATIONS OF CHEMICAL PROCESSES

A considerable restriction or boundary limitation is the variable availability of solar energy. Three methods of how to cope with this meteorological fact can be thought of:

- o A longtime storage must bridge the overnight gaps and occasionally even longer periods.
- o A hybrid operation of the solar plant with a fossil fired facility enables constant output of the chemical products.
- o Chemical processes can be selected towards the qualification for interruptions. Pyrolysis processes seem to be very advantageous in this respect.

The most favorable use of concentrated sunlight would be the immediate application of radiation at open energy transfer systems. This interesting but difficult topic, especially, needs basic research activity. At the test facilities in Odeillo (CNRS) and in Atlanta (Georgia Tech.) considerable efforts have been spent already, respectively are on the right track. Some more application-oriented activities are under definition (IEA-program, EIR and DFVLR).

The more conventional and straight-forward way that might be realized in shorter time would be the use of solar-thermal input at closed chemical reactors. Here, the otherwise fossil source is substituted by solar energy or the solar system is connected to a traditional system in a hybrid way. This procedure seems to be attractive if coupled operation and total energy use enable improved efficiency or if the fuel substitution/upgrading plays an important role. Favorite processes for the IEA program of thermal application are reforming and pyrolysis procedures. Also the thermal heat provision for high temperature vapor electrolysis has high esteem.

In the following lines a listing of some candidate processes is given.

Radiative heat transfer to solid particles:

- o Basic heat transfer investigations to sand material
- o Calcination of limestone
- o Fluidized bed technology for lignites

Thermal heat transfer through walls:

- o Phase changes and general heat application
- o Provision of thermal portion for high temperature vapor electrolysis at hydrogen production
- o Thermo-chemical conversions at
  - hydrogen production
  - steam reforming of methane
  - pyrolysis of coal or
of ehtylene from ethane
  - heavy oil cracking.

In the following two chapters, the steam reforming of methane process will be discussed as to its potential towards suitable application of solar energy in a larger total system and the coal pyrolysis and heavy oil cracking will be detailed as to an imagination of solar use.

3. STEAM REFORMING OF MEHTANE

3.1 General Background

The conversion of light hydrocarbons to different types of synthesis gases and to H_2 is one of the most fundamental procedures in the chemical industry today. In recent years industry turned nearly completely to natural gas as feedstock.

For the conversion of methane respectively natural gas to H_2- and CO-containing gases, the steam reforming process is used exclusively.

Steam reforming means the reaction of methane and steam through the absorption of heat in contact with a catalyst. By this way as well the methane as the water molecules are splitted following the equations (1) and (2)

(1) $CH_4 + H_2O + \text{heat} \rightleftharpoons 3\,H_2 + CO$

(2) $CH_4 + 2H_2O + \text{heat} \rightleftharpoons 4\,H_2 + CO_2$.

Because of the high temperatures necessary to achieve high conversion rates, the heat is provided to the steam reforming process by burning fuel directly. Thus, the steam reformers constitute the major fuel consuming step for the synthesis gas production. Generally, at present, natural gas is used both as feedstock and as fuel. Depending on the specific product approximately 30 to 40 % of the natural gas is used as fuel and 60 to 70 % as feedstock.

Due to the large scale application of the steam reforming process there is an advantageous field for the introduction of solar heat to save fossil fuel.

Beyond this, in the endeavour to make solar energy storable and transportable, the steam reforming process is an excellent candidate for solar heat conversion. The endothermic chemical reaction can be used to convert heat into enthalpy of formation. By this means, sensible heat is bound to a suitable substance which can be stored without considerable loss or can be used as feed material for various subsequent processes.

3.2 Technical Status

In principle, steam reforming of methane is a very convenient process with respect to the technological elements involved:

- o gaseous fluid, no phase change
- o straight tubes for chemical reaction
- o moderate heat fluxes
- o heat transfer by radiation and convection
- o no change in selectivity by temperature change.

Conventional steam reforming is carried out in industrial plants as sketched in Fig. 1. The prevailing conditions of reaction are as high as about 860 °C temperaure and up to 40 bar pressure. The standard dimensions of steam reforming tubes are:

- o heated length 8 - 12 m
- o inner diameter 80 - 130 mm
- o wall thickness 10 - 18 mm.

Because of high tube wall temperatures the reforming tubes are generally manufactured from austenitic alloys. They are usually arranged vertically in large furnace chambers. The heat is transferred to the tubes by radiation and convection. The reaction rate of the reforming process is promoted by use of catalysts.

The development of the chemical reactions is determined by thermodynamics and kinetics. As shown in Fig. 2, the main parameters to influence the thermodynamic equilibrium are temperature and pressure.

Various considerations, as to achieve a complete conversion, may contrarily influence such parameters. Therefore, the proper combinations are essential and affect the economics of the overall process, i.e. the quantity of external (solar) heat input as well as the secondary energy balance, i.e. the consumption of electricity for the overall process.

3.3 Solar Specific Features

Because of the high temperatures between 750 and 860 °C, necessary for the steam reforming reaction, only a central receiver system would be capable to serve as solar heat source. This holds even in case of a two stage concept.

There are two competing systems which introduce the solar-thermal energy into the chemical process as shown in Fig. 3:

a) direct solar radiation of the chemical reactor tubing,

b) indirect solar heating of the chemical reactor by means of a transport fluid which in term had been heated in a solar receiver.

The integration of receiver and reformer (a) means the arrangement of reformer tubes inside a receiver cavity. The reformer tubes themselves will be heated by solar radiation. This would have the advantage that only one heat transfer step from the wall to the process gas is necessary and that material temperatures and also mean cavity temperatures could be lower. On the other hand the high transients of solar radiation will influence in the integrated case the reactor immediately.

The separation of both components (b) requires a mechanism to transport the heat collected by the receiver to the steam reformer. In this heat transfer loop a circulating medium takes over the heat at a suitable temperature in the receiver and transfers it without significant heat losses to the reformer unit where it releases the heat needed for the endothermic chemical process. After that the transfer medium has to be pumped back to the receiver.

The main problem at introducing solar heat into the steam reforming process will arise from very high temperatures and the respective transients.

In order to achieve only moderate stresses at the reforming tubes, two stage reformer concepts are under discussion. They are outlined in Fig. 4. The first low temperature stage will be solar heated. The high temperature stage can be designed as a standard fossil heated reformer with the available experience, especially in attaining homogeneous heat flux distributions and in avoiding high temperature transient rates.

Today's advanced status of conventional steam reforming and both knowledges and experiences gained from the KFA-NFE-project (EVA/ADAM-facilities) and from the GAST-technology program suggests to study, design, and build such an experimental plant. The facilities at the Plataforma Solar near Almeria/Spain present themselves as a well qualified site for such a plant.

The aim of this experiment should be the production of solar process heat to operate the steam reforming of methane as a system linked with consecuting processes. The items to be investigated would be:

- o verification of design values of process and components
- o function of receiver and steam reformer

- collecting operational experience under solar conditions
- operation of a complete process chain when introducing solar heat with its diurnal changes.

4. COAL PYROLYSIS AND HEAVY OIL CRACKING

4.1 General Background

The discontinuously offered solar energy and succeeding changes in thermal power inputs to connected chemical processes present serious problems for the realization. In the long-term future, a commercial application of solar-thermal energy in chemical processing plants seems to be favorable if two conditions can be realized:

- Thermal power interruptions or changes should be tolerated by process requirements.
- Base material needed for the process can be provided by the system itself or is cheaply available on site.

4.2 Chemical Processes

For the refinement or upgrading of solid or liquid heavy fossil feedstock three chemical processes look promising when coupled to solar-thermal plants:

a) Gasification of highly reactive fuels (e.g. lignite, biomass)
Superheated steam is fed into a fluidized bed reactor. The reaction which takes place at temperatures up to 800 °C delivers synthesis gas. It is described by the well known relation of equation (3)

(3) $C + H_2O + heat \rightleftharpoons CO + H_2$.

b) Pyrolysis of coal
This process is of special interest since load changes introduced by solar energy variations might be tolerated by the process under the provision of maintaining a minimum level of temperature. Working temperatures range from 550 °C to 650 °C. Final products are coke, liquid and gaseous products.

c) Heavy oil cracking
Because of the high temperature level near 500 °C also this process is presently under consideration; eventually in combination with hydration (up to 800 °C). Final products would be benzenes and light oils.

4.3 Experimental Device

The experiment to combine one of these processes with high temperature solar-thermal heat is proposed to be managed in three phases. In order to increase reliability, prevent too high costs, and thus reduce any risks a small experimental device is planned for the first and second phase:

Phase 1: Functional tests of the test bed concentrator and reactor at DFVLR, Stuttgart

Phase 2: Transfer of the device to the Plataforma Solar at Almeria and conduction of experiments together with the chemical processing loop system

Phase 3: Upscaled experiment with SSPS-CRS

The experimental device will consist of a membrane dish collector coated with thin glass segments (0,6 x 0,6 m). The total diameter of the mirror is 17 m. A receiver - probably of the cavity type - which may be designed as chemical reactor also, is movable along the optical axis. Thus, different radiant flux distributions can be obtained in the receiver aperture. The width of the aperture is optionally foreseen between 0,2 and 0,7 m. Maximum expected radiation power into the receiver is about 150 kW assuming a beam irradiance of 850 W/m^2. The temperatures range from about 500 to 850 ^oC, the pressure may extend up to 50 bars.

Depending on the finally selected process there are two alternative versions for the loop system (Fig. 5). The one on the left hand side is simple, however requires high temperature technology. It is closed, however not a closed cycle. The raw material (heavy oil products) is pumped into the receiver reactor where it can be heated up to 850 ^oC. Additional reactants (e.g. H_2O-steam or H_2) can be fed into the reactor. The final products are collected in a vessel and a chemical analysis will be performed.

The other loop system on the right hand side is a closed cycle where any heat transfer medium (preferably air for the experimental device) is heated in the receiver up to maximum temperature. The air delivers its thermal energy via a heat exchanger to the actual chemical processing system. This arrangement is especially suited for the chemical process (b) - the pyrolysis of coal. Fig. 6 shows a scheme of the coupling with a Lurgi reactor. An inert gas which is heated in the heat exchanger is fed into the lift pipe of the reactor transferring its thermal energy to the raw material.

The items to be investigated are comparable with those mentioned under 3.3.

5. CONCLUSIONS

In the present situation, substantial hardware decisions are not yet required. It is, however, very important to initiate careful analyses and solid experimental research activities. Special regard should be given to the operational system to which the solar input has to be accommodated.

Questions on requirements and load changes have to be answered. Design alternatives concerning the coupling of solar receivers with chemical reactors must be assessed thoroughly. Finally, for a complete evaluation of future possibilities, some pilot experiments have to be prepared, built, and experienced.

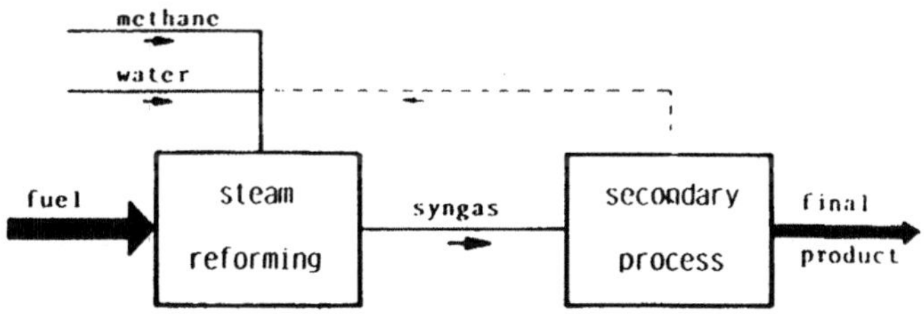

Steam Reforming Standard Parameters in Conventional Plants:

Methanol	850 ^oC	20 bar
Ammonia	780 ^oC	35 bar
Hydrogen	850 ^oC	25 bar

Fig. 1
General Principles of Steam Reforming Plants

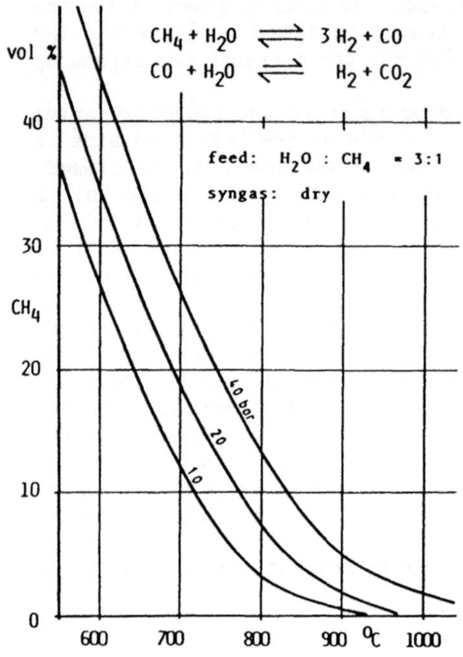

Fig. 2
Conditions of Thermodynamic Equilibrium

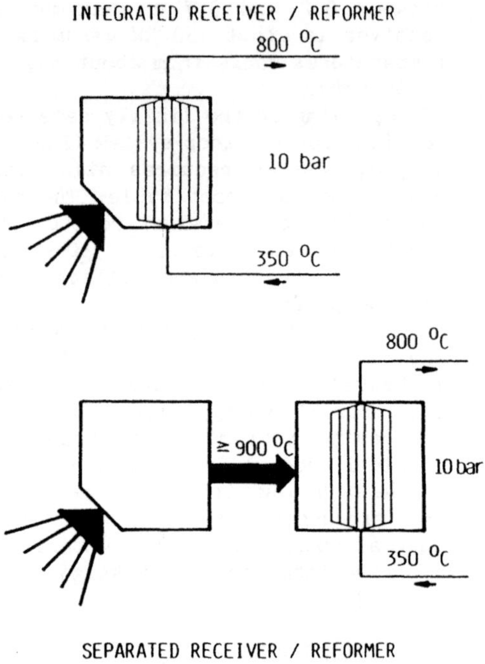

Fig. 3
Receiver/Reformer Arrangements

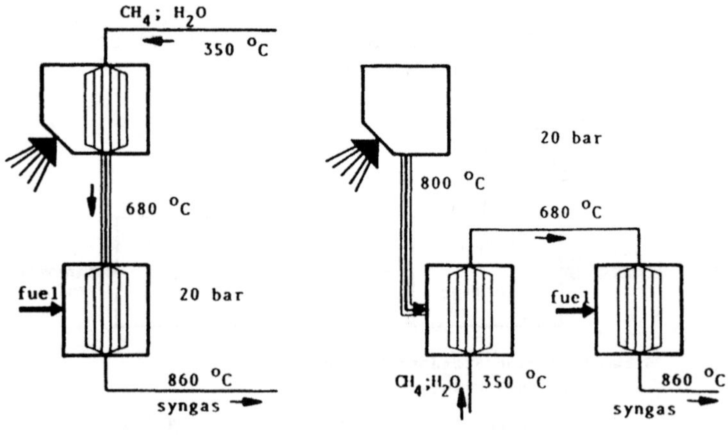

Fig. 4
Alternative Concepts for Solar Steam Reforming

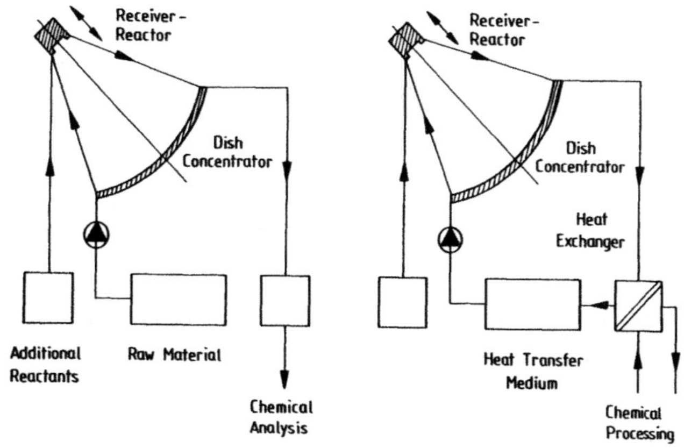

Fig. 5
Block Diagram for Alternative Experimental Arrangement

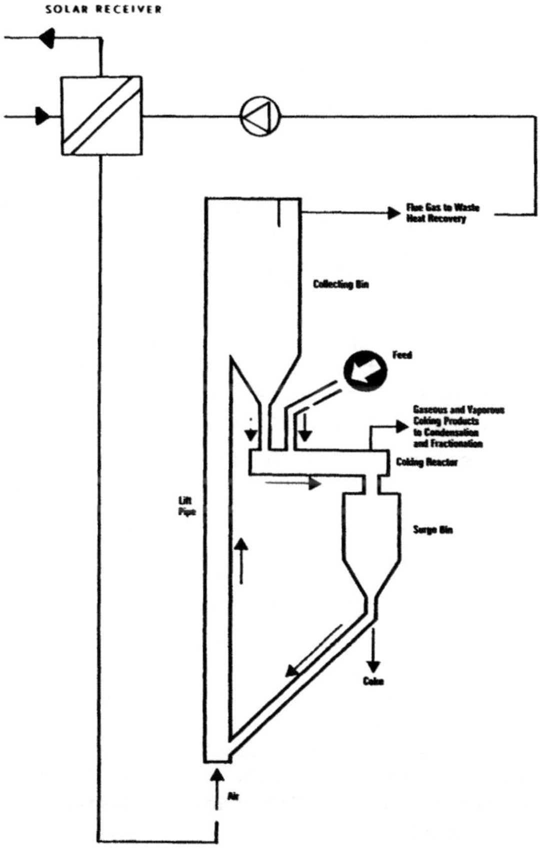

Fig. 6
Adaptation of Heat Exchanger
to Test Bed Loop and
Lurgi Reactor

A MODULARIZED SENSIBLE HEAT STORAGE SYSTEM FOR PROCESS HEAT GENERATION

M.GEYER
Institute for Technical Physics
DFVLR - Stuttgart

Summary

The storage of thermal energy is the prior condition for substituting conventional power generation systems by solar thermal plants. Energy storage is necessary in order to synchronize the diverging enrgy supply and demand functions. In this paper the operational possibilities of sensible heat storage modules are studied for the case of high temperature process applications.

1. INTRODUCTION

The energy density of thermal storage systems is 1 - 2 orders of magnitude smaller than that of the liquid or solid carbon hydrates; the heat capacity of 1 m^3 of MgO bricks at an ΔT = 400°C roughly amounts to 300 kWh/m^3 and is comparable to the energy content of 1 m^3 of gaseous hydrogen at 100 bar. Therefore thermal storage may not be suitable for storing energy longer than a couple of hours or days, but it is most advantageous for buffering short term insolation shortages and thus providing heat for a continuous thermal process. Possible media for heat storage at temperatures above 700° C are solids and molten salts.

Solid storage systems using ceramic bricks are wellknown as "cowper " in the iron casting industry and present a number of advantages over molten salts, e.g. chemical stability at high number of cycles, no corrosion of containment, not toxic, low priced, abundant occurence of raw materials, simple technology allowing their production and operation in developing regions. However, energy balance and control of the charge- and discharge cycle are much more complex for solid systems than for two-tank molten salt systems for the following reasons:

- Whereas for the two-tank system the utilizable energy capacity is given directly the content of the hot tank, the energy content of the solid system must be determined by its actual temperature profile which depends on the preceding charge|discharge cycles.
- While charging/discharging the solid system, the energy exchange rate between solid and the charging medium declines as the solid approaches the medium's inlet temperature.
- For extracting a required amount of energy from the solid system, the controlling system has to match continuously the massflow of the descharging medium with the declining discharge temperature.

Whereas the cowper of the iron casting industry are operated at constant charging and discharging massflow, the integration of such systems into solar thermal plants is accompanied by highly varying charge/discharge rates due to the irregular irradiation conditions.

For these purposes, this work presents a modular and standardized sensible heat storage system, which has been optimized with the computer simulation model PROSES (Parametric Representation of Solar Energy Systems) and for which the dynamic behaviour is studied.

2. MODEL OF A PASSIVE COWPER TYPE STORAGE SYSTEM

A solid storage system for sensible heat consists of the following three elements: The **storage mass** which absorbs or releases thermal energy by changing its temperature, the **heat exchanger** where the energy is transferred from the storage mass to the energy carrying medium and viceversa and the **containment** which isolates the storage mass against ambient temperature and which provides the supporting structure.

The heat exchange between storage system and energy carrying medium becomes simple, if the medium is gaseous. In this case, the exchange can be accomplished by the direct contact of the gas with the solid. For this purpose, gas leading channels can be incorporated into the structure of the storage mass; the surface of these channels represents the heat exchange area.

2.1 Thermodynamics of the Storage Module

Assuming a cylindrical storage module traced by a set of parallel, coaxial, rectangular gas channels, a linear thermodynamical model has been developed with the following approximations:

- Axial heat conduction is neglected (which can be realized by incorporating airgaps between the set of cylindrical storage slices)
- Radial heat conduction is initially assumed to be ideal and is later corrected by an empirical function.
- The isolation of the containment is assumed to be ideal; heat losses and boundary effects have been neglected.

By that means, the heat transport in axial direction of the storage module is restricted to the forced convection of the gas through the channels and to the heat exchange between the gas and the storage mass. For purposes of computation, the storage cylinder is divided into infinitesimal slices, each having the mass dM_S. During the time period dt an infinitesimal volume element of gas with the mass dM_G exchanges heat through the exchanger area dA_S. The energy balance is then given by the equations:

heat delivery of the gas: $dQ_G = -c_{p,G} \cdot dM_G \cdot dT_G$

heat exchange: $dQ_X = \alpha \cdot (T_G - T_S) \cdot dA_S \cdot dt$

heat absorption of the solid: $dQ_S = c_S \cdot dM_S \cdot dT_S$

In thermal equilibrium these quantities become equal and the following system of coupled, partial differential equations can be deducted:

$$\left[\frac{\partial T_G}{\partial x}\right]_t = -\frac{\alpha A_S}{L\,\dot{m}_G c_p}(T_G - T_S)_{x,t}$$

$$\left[\frac{\partial T_S}{\partial t}\right]_x = \frac{\alpha A_S}{M_S\, c_S}(T_G - T_S)_{x,t}$$

An exact solution with modified Bessel functions is given by (1) for the case of constant massflow of gas and special boundary conditions. For varying massflows and arbitrary initial temperature profiles a finite difference iteration algorithm has been implemented in **PROSES**.

In order to make the different storage states comparable, characteristic storage coefficients have been introduced following (2), which quantify the storage utilization and the utilization of the charging medium:

Storage utilization factor: $$\omega(t) = \frac{\Delta Q_S(t)}{\Delta Q_{S,\infty}}$$

Gas utilization factor: $$\gamma(t) = \frac{\Delta Q_G(t)}{\Delta Q_{G,max}}$$

where $\Delta Q_S(t)$ is the actual energy content of the module, $\Delta Q_{S,\infty}$ the thermal capacity of the module for the given temperature interval, $\Delta Q_G(t)$ the heat of the gas that is absorbed by the module and $\Delta Q_{G,max}$ the total heat content of the charging gas corresponding to the given temperature interval. These quantities are determined by the following integrals:

$$\Delta Q_S(t) = \frac{M_S\ c_S}{L_S} \int_0^t \int_0^{L_S} T_S(\xi,\tau) - T_o \quad d\xi\ d\tau$$

$$\Delta Q_{S,\infty} = M_S\ c_S\ \Delta T_G$$

$$\Delta Q_G(t) = c_{P,G} \cdot \int_0^t \int_0^{L_S} \dot{m}\ (\tau) \quad T_G\ (\xi,\tau) - T_o \quad d\xi\ d\tau$$

$$\Delta Q_{G,max} = c_{P,G} \cdot \Delta T_G \cdot \int_0^t \dot{m}\ (\tau)\ d\tau$$

where M_S means module's total mass, c_S its specific heat and L_S its axial length; $\dot{m}(t)$ is the massflow of the medium and $c_{p,G}$ its specific heat.

3. OPTIMIZATION OF THE STORAGE MODULE

The goal of the technical design of the storage module lies in maximizing the storage utilization at minimum costs. Although the operation of the storage system has to be evaluated in conjunction with the solar plant, the following items have been optimized separately by simulation: **containment size, rockbed geometry** and **storage material.**

3.1 Containment Size

Since the storage modules have to resist the plant's closed thermal cycle pressure (10 to 40 bar depending on the process), a cylindrical containment was chosen. Due to on field fabrication constraints (3), the diameter of the modules was restricted to 6m. The capability to absorb the heat of the charging gas then increases with growing module length. At a module length $L_S = 35$m, a satisfactory gas utilization γ can be achieved. Beyond this length, the further gain of γ becomes comparatively small.

3.2 Rockbed Geometry

The storage model of **PROSES** simulates a rectangular structure of bricks and gas channels (Fig.1). The variable parameters are the thickness **s** of the bricks, their width **b** and the width **d** of the gas channels. At the determination of these parameters, two countercurrent objectives have to be balanced: On the one hand, one wants to minimize the containment volume by reducing **d** as much as possible; on the other hand the heat exchanging area determined by **s** and **d** has to be as large as possible in order to provide a good heat transfer between the bricks and the gas. For purposes of fabrication a minimum brick thickness must be kept. Under these circumstances parameter variations lead to the conclusion that a quadrangular brick renders high brick densities as well as large channel surfaces. For the reference module an edge length **s** = **b** = 5cm and a channel width **d** = 2.1cm was chosen, leading to a fraction of 70.4% brick to total volume and to a volume specific heat exchanger area of 40 m^2/m^3.

3.3 Brick Materials

For the performance of the storage module three material dependant parameters are of major importance: the **specific heat** c_S, the **density** ρ_S, and the **thermal conductivity** λ_S. The first two parameters determine the thermal capacity of the module and, hence, the required volume. The thermal conductivity is taken into account by an empirical function. Heat capacity and thermal conductivity have to be matched in order to achieve a satisfactory gas utilization factor γ.

Among the materials that have been studied, magnesium oxide (MgO) and cast iron yield high thermal capacity, satisfactory heat conductivity and are well suited for high temperatures. For economical reasons, MgO bricks have been selected for the reference module having a density of ρ_S = 3000kg/m^3, a specific heat c_S = 1.26kJ/kgK and a thermal conductivity λ_S = 0.0105kJ/msK. Tab.I summarizes the parameters and deducted quantities of the presented reference storage module.

3.4 Dynamic Behaviour of the Reference Module

The dynamic behaviour of the storage module over the daily and yearly cycle has been studied with the **PROSES** model of a gas cooled CRS plant in order to understand the impact of the **form**, **rate** and **amount** of charging and discharging, respectively.

It could be shown that the final storage profile at the end of the charging/discharging period is mostly invariant with respect to variations of form and rate of the charging/discharging function and is to first order only depending on the initial profile and the integral amount of the charged/discharged energy.After cold start an equilibrium discharge profile is obtained within 10 day-cycles. This equilibrium profile experiences a slight fluctuation throughout the seasons of a year. Fig.2 shows the charge and discharge profiles for the 15th of the twelve months (1=JAN...12=DEC).

4. SOLAR PROCESS HEAT APPLICATIONS

In order to illustrate the quality (temperature behaviour and amount) of process heat that could be produced by a solar plant with a storage module and its implications on a chemical process, the hydrogen production by the steam reforming of methane has been taken as an example, since data about its temperature dependance where available. According to (4) this process accounts for a third part of the world production of hydrogen.

In current reformer designs (5,6) the steam/methane mixture (with a 3:1 molar $H_2O:CH_4$ ratio) enters catalyst packed reformer tubes under a pressure of 10 bar at a temperature of 450°C and is heated there in countercurrent by a heat carrying gas to a temperature of about 700 - 850 °C. In this process the endothermic reaction of steam and methane yields primarily:

$$CH_4 + H_2O + 206 \text{ kJ/mol} \rightarrow 3H_2 + CO$$

$$CH_4 + 2H_2O + 163 \text{ kJ/mol} \rightarrow 4H_2 + CO_2$$

The composition of the product gases is temperature and pressure dependant. Although basic research has yet to be done to understand the details of the chemical process under non steady conditions, its temperature dependance as it is computed by (7) is taken for first (and simplified) storage system studies. For these purposes, the reformer itself is treated as a black box, for which a rated power of 20 MW has been chosen; at the chosen design temperatures (850°C inlet, 450°C constant outlet) this would require a massflow of 44.6kg/s to be provided by the solar plant, if air was used as the heat carrying medium. The air cooled CRS plant of **PROSES** therefore was layed out for 57 MW thermal power at noon of December 21st (at a latitude of 37° north), leading to 6.5 hours of direct operation at 20MW rated power and providing additionally 151 MWh for the charging of one storage module. Hereby, the following design point (noon of 21.12) efficiencies have been assumed in **PROSES**: Heliostat field (123000m²) 73.8%, Receiver 82.1%, Solar multiple 2.85. On June 21st, the model yields 69.1 MW at noon; 358MWh are available for charging storage. (see also Fig.3)

To this solar plant, one of the presented storage modules is connected in parallel between receiver and reformer allowing the following modes:

1 REFORMER DIRECT	All energy flows from receiver to reformer.
2 REFORMER DIRECT AND CHARGING	If the thermal power collected by the receiver exceeds the reformer design power, the excess is used for charging the storage module.
3 STORAGE BOOSTED	If the receiver output falls below the reformer design power, the difference is extracted from storage.
4 DISCHARGING ONLY	Thermal power is extracted from storage only.
5 CHARGING ONLY	Receiver output is used for charging only.

In modes **1**,**2**,**5** the receiver outlet temperature is held constant at 850°C by controlling the air massflow; in modes **3**,**4** the temperature of the air which enters the reformer is determined by the storage discharge temperature. In order to reveal the impact of the decreasing storage outlet temperature during discharging, the reformer outlet temperature of the heat carrier air has been kept constantly at 450°C.

For the 400°C difference between charging and discharging the theoretical module capacity $\Delta Q_{S,\infty}$ amounts to 293MWh. However, its usable capacity is restricted by the required minimum discharge temperature. The impact of this restriction is analyzed for the following thresholds of discharge temperature: 700°C **(A)**, 750°C **(B)** and 800°C **(C)**.

The equilibrium discharge profiles then contain a non utilizable fraction of the 293MWh total capacity, which grows with increasing discharge temperature threshold: While case (A) is binding 20% (60MWh), this fraction increases to 27% (78MWh) in case (B) and to 37% (109MWh) in case (C). If a charging gas utilization $\gamma > 85\%$ is to be sustained, the utilizable net capacity reduces to 175MWh (A), 157MWh (B) and 126 MWh (C). To load them, requires 203MWh (A), 183MWh (B) and 148MWh (C) of charging energy.

If the so charged module is discharged by the design massflow (44.6kg/s) the discharge temperature threshold is reached after a period of 10h (A), 8.7h (B) and 6.8h (C), yielding during this time a mean discharge rate of 17.5MW (A), 17.9MW(B) and 18.5MW (C). The results of the simulations are summarized in Tab.II.

The energy flows of the **PROSES** reference plant on December 21st are shown in Fig.3. From 9 to 15 hours, the reformer is fed directly by the receiver and a surplus of 152MWh is passed through the storage module, where 134.5MWh (A), 124.4MWh (B) and 115.9 MWh (C) are stored (see Tab.III). Starting at about 15 hours, the module is discharged with the design massflow. Rated power (20MW) is kept until about 17 hours, when the discharge temperature begins to decrease. Fig.4 shows the effect on the reforming process: The fraction of unreformed methane increases while the production of hydrogen is decreasing. Fig.5 shows the ordered load hours for Dec.21st: Case (A) allows for the longest total operation time and provides with 277.8MWh the highest process heat output. Although cases (B) and (C) provide rated power (20MW) for a longer period, their total energy output is less than for case (A). The temperature dependance of the reforming process does not balance this difference; case (A) yields the highest hydrogen output (23t) but requires a higher specific throughput of steam/methane. (For purposes of storage analysis an ideal part load elasticity of the reformer has been assumed).

On June 21st, the reference plant delivers 358MWh to the storage module. This amount exceeds the net capacity of the modules for all cases by a factor 2 - 3 (a second module could be charged, which is not studied here). The utilization of the storage module is therefore $\omega > 96\%$ in all 3 cases. So at beginning of discharge, the module's content is almost the same for (A), (B) and (C) resulting in the same period of 19.5 hours (Fig.6) of rated power. Only (A) renders a continuous 24 hour operation; again it provides the highest amount of process heat (461MWh) and the highest hydrogen output (38t).

5. CONCLUSIONS

By means of sensible heat storage modules the solar energy input can be perfectly decoupled from the chemical process. Startup phases of the solar plant and of the chemical reformer become independant. The massflows and energy rates can be controlled and optimized independantly for the solar plant and the connected process. Multi receiver concepts feeding a central storage system become possible. Breakdowns of the solar plant and of the process can be buffered by the storage modules yielding a higher system reliability. For the future, the influence of bad weather periods will be analyzed and a cost analysis will be done.

6. REFERENCES

(1) B.I.Kitaev et al.: Heat Exchange in Shaft Furnaces, Pergamon Press, Oxford 1967

(2) Traustel: Auslegung von Regeneratoren, BWK (24), 1972

(3) BOEING: Closed Cycle High Temperature Central Receiver Concepts for Solar Electric Power, Vol.I-III ER629, Final report 1979

(4) IEA: Assessment of the potential future markets for the production of hydrogen from water, 1979

(5) Sandia National Labs.: Steam Reforming of Methane, Phase 1 Study submitted to the IEA SSPS Project, Livermore, February 1984

(6) U.Boltendahl, R.Harth: Wärmetransport auf kaltem Wege, Bild der Wissenschaft, 4, 1980

(7) Computer programs "**GLEICH**" and "**ENTH**", KFA Jülich, NFE Project, 1984

Characteristic Parameters of the Reference Module:	
length of containment	L_s = 35 m
diameter of containment	D_s = 6 m
thickness of containment wall	W_s = 0.07 m
thickness of bricks	s = 0.05 m
width of bricks	b = 0.05 m
width of gas channels	d = 0.021 m
brick material	MgO
density	ρ_s = 3000 kg/m³
specific heat capacity	c_s = 1.26 kJ/kg K
thermal conductivity	λ_s = 0.0105 kJ/ms K
channel cross section to brick cross section ratio	ε = 0.429
brick volume to total volume ratio	v = 0.704
channel cross section to total cross section ratio	f = 0.696
specific heat exchanger surface	a = 40 m²/m³
total volume of containment	V_c = 989.6 m³
total volume of bricks	V_s = 696.9 m³
total mass of containment	M_c = 350 t
total mass of bricks	M_s = 2091 t
total cross section area	F_s = 28.3 m²
total gas channel cross section area	F_a = 8.5 m²
total heat exchanger surface	A_s = 39584 m²
total heat capacity per degree K	$M_s \cdot c_s$ = 0.727 MWh/K

Table I: Summarized data of the reference storage module

ITEM	UNIT	CASE A	CASE B	CASE C
Discharge temperature threshold	°K	973	1023	1073
Charging temperature	°K	1123	1123	1123
Discharging temperature	°K	723	723	723
Maximum capacity	MWh	292.7	292.7	292.7
Bound capacity	MWh	59.1	77.9	108.8
Non untilizable fraction	%	20.2	26.6	37.2
Net utilizable design capacity	MWh	175.2	156.6	125.6
Net utilizable fraction	%	59.9	53.5	42.9
Required charging energy	MWh	203.0	183.0	148.0
Gas utilization γ	%	86.4	85.5	84.9
Discharge time at 44.6 kg/s	h	10.0	8.7	6.8
Mean discharge rate	MW	17.5	17.9	18.5

Table II: Utilizable capacity and required charging energy for different minimum discharge temperatures

LIN	ITEM	UNITS	CASE (A)		CASE (B)		CASE (C)	
			DEC 21	JUN 21	DEC 21	JUN 21	DEC 21	JUN 21
1	Direct solar radiation	MWh	629.6	1277.0	629.6	1277.0	629.6	1277.0
2	Receiver output	MWh	294.9	597.0	294.9	597.0	294.9	597.0
3	Reformer direct	MWh	143.3	239.0	143.3	239.0	143.3	239.0
4	Fraction of receiver output 2	%	49.0	40.0	49.0	40.0	49.0	40.0
5	Total charging energy	MWh	151.6	358.0	151.6	358.0	151.6	358.0
6	Fraction of receiver output 2	%	51.0	60.0	51.0	60.0	51.0	60.0
7	Utilized charging energy	MWh	134.5	222.0	124.4	212.0	115.9	188.0
8	Fraction of 5	%	88.7	62.0	82.1	59.2	76.5	52.5
9	Fraction of 2	%	45.6	37.2	42.2	35.5	39.3	31.5
10	Storage utilization (ω)	%	65.5	96.2	68.1	97.8	75.7	98.7
11	Charging losses	MWh	17.1	136.0	27.2	146.0	35.7	170.0
12	Fraction of 2	%	5.8	22.8	9.2	24.5	12.1	28.5
13	Total process heat 3+7	MWh	277.8	461.0	267.7	451.0	259.2	427.0
14	Fraction of 2	%	94.2	77.2	90.8	75.5	87.9	71.5
15	CH_4 input (450°C, 10 bar)	t	61.4	98.6	57.2	95.1	54.3	89.1
16	H_2O input (450°C, 10 bar)	t	207.1	332.9	193.2	321.1	183.3	300.8
17	CH_4 output	t	6.8	7.8	4.9	7.0	3.8	5.8
18	H_2O output	t	123.3	195.4	113.7	188.1	107.2	175.5
19	H_2 output	t	23.0	38.0	21.9	36.8	21.1	34.8
20	CO output	t	60.5	104.1	59.5	101.5	58.4	96.9
21	CO_2 output	t	54.8	86.2	50.4	82.8	47.2	76.9
22	Reformed fraction of CH_4	%	88.9	92.1	91.4	92.6	93.0	93.5

Table III: Summarized results of the PROSES plant simulations for the steam reforming of methane

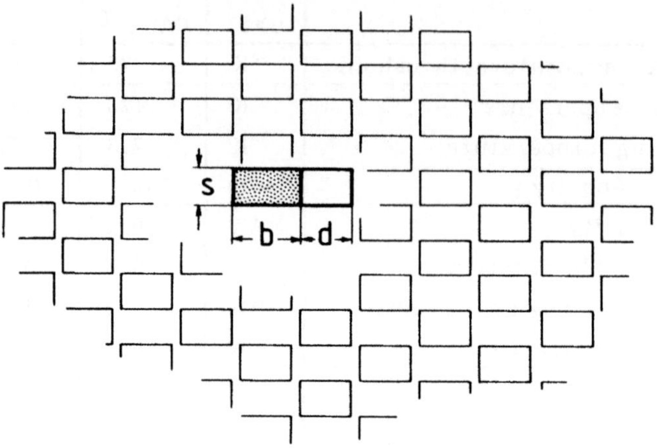

Figure 1: Brick geometry of the **PROSES** storage module. The dotted area represents one brick with thickness s and width b. d is the width of the gas channels.

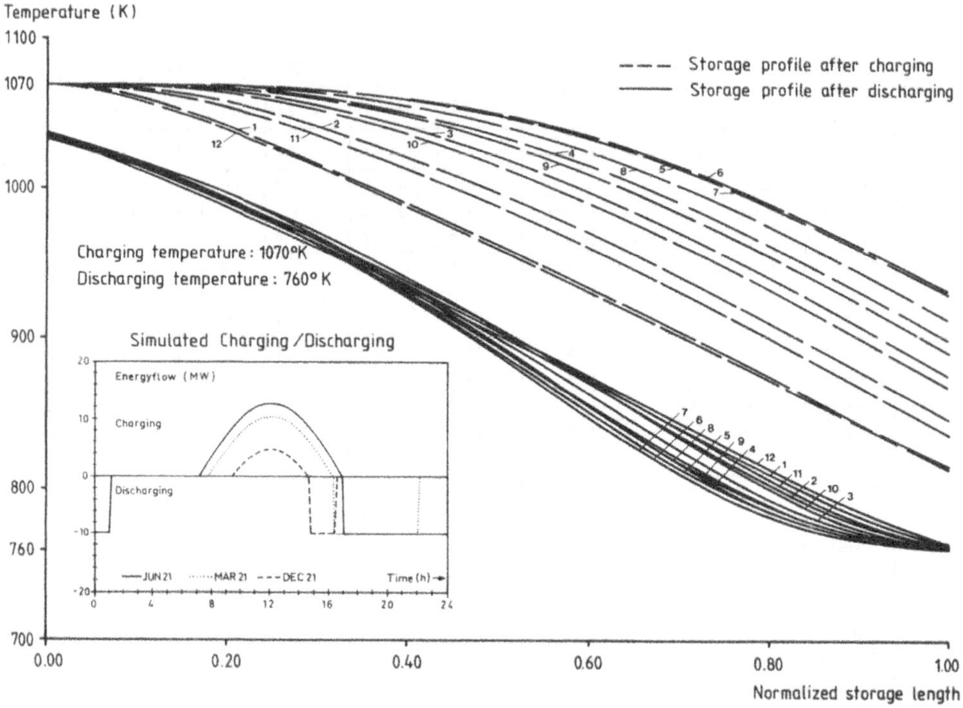

Figure 2: Charging and discharging profiles at the 15th of each month (1 = JAN,..., 12 = DEC). The charging temperature has been 1070°K, the discharging temperature 760°K. The energy flow corresponds to a solar multiple of 2.5. On the 21.6. total charging was 80MWh.

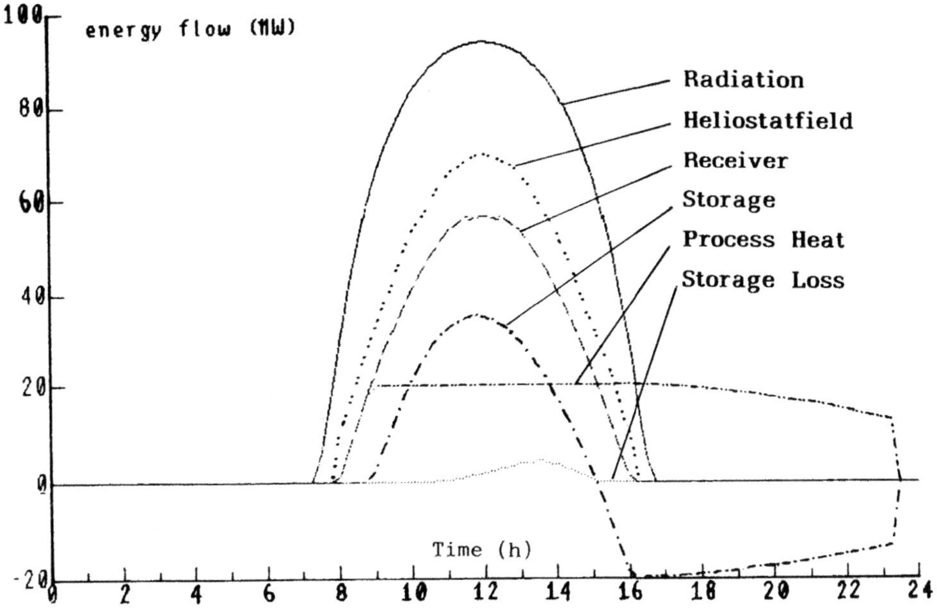

Figure 3: Energy flows of the reference plant on December 21. Negative energy flows refer to storage discharging.

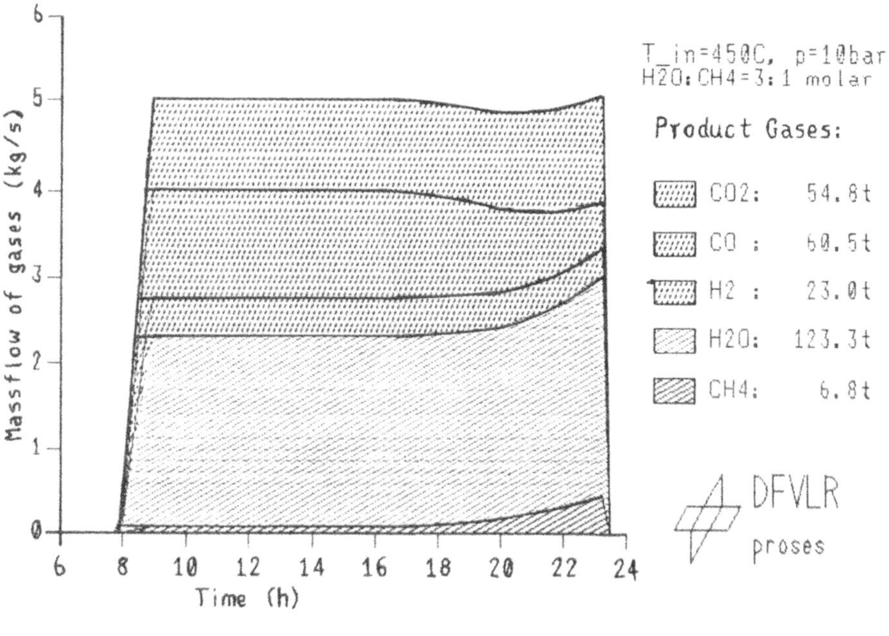

Figure 4: Output of methane/steam reformer on December 21. The temperature dependance of the process has been computed with the programs "GLEICH" and "END" of KFA Jülich, NFE project.

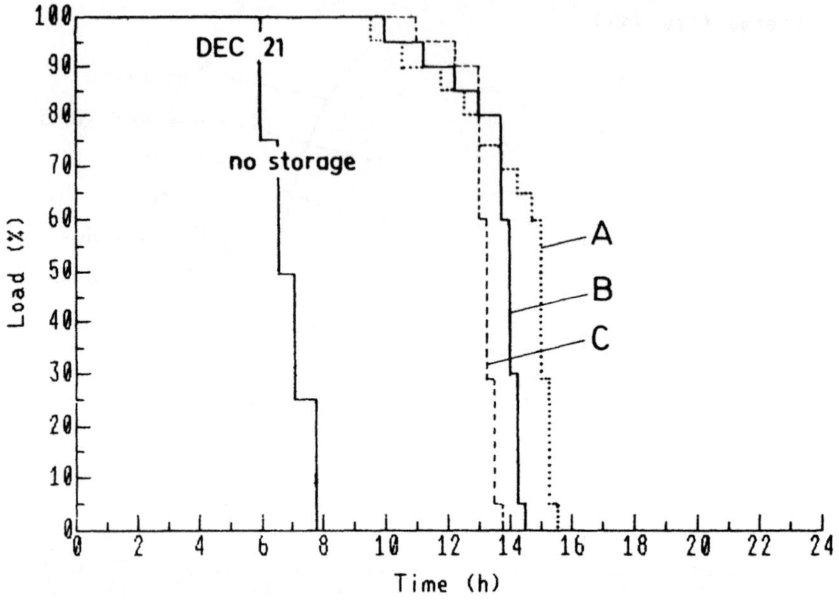

Figure 5: Ordered hours of thermal load for December 21st for the cases (A), (B) and (C). 100% load means operation at rated power (20MW).

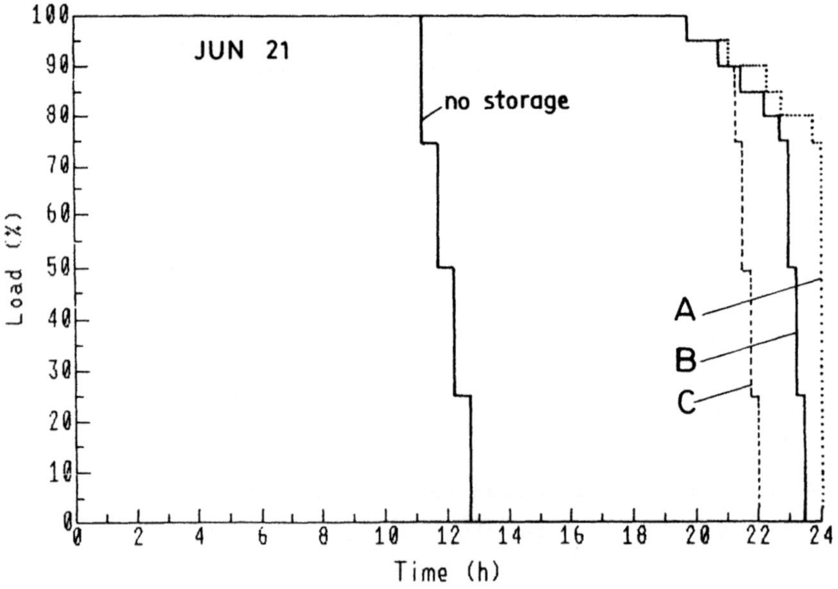

Figure 6: Ordered hours of thermal load for June 21st. All cases provide 19.5 hours of rated power (20MW). Case (A) allows continuous 24 hour operation.

SESSION V - PART 1

PROCESS HEAT GENERATION

Summary of the Session by the Rapporteur
E. BURCK, JRC, Ispra, Italy

ARCO central receiver solar thermal enhanced oil recovery project

SUMMARY OF THE SESSION BY THE RAPPORTEUR

E. BURCK
Commission of the European Communities, Joint Research Centre
Ispra, Italy

1. INTRODUCTION

In the session, the presented paper was:

"ARCO Central Receiver Solar Thermal Enhanced Oil Recovery Project," F. A. Blake, D. N. Gorman, J. H. McDowell.

Mr. F. A. Blake presented this paper illustrating the historical, economical, and technical details of this rather original application of Central Solar Receiver Projects.

2. COMMENTARY

The questions arising after the rather detailed presentation aimed essentially at getting further technical and economic information on this project. The answers can be summarized as follows:

The vapor necessary to extract the remaining oil consumes about 25 to 30% of the extracted oil; this technique is actually the only one to extract the remaining oil. Limitations due to pollution problems and the engagement of ARCO in the CRS business were the reasons for the choice of ARCO which is considered economically valid. The size of the heliostat mirrors is 52 m^2, vapor temperature outlet is 550^{0}C with a maximum pressure of 1000 psi. The current requirements are of vapor pressure of about 400 psi, which allows the injection of vapor until 800 feet depth. The heliostat mirrors are designed to operate with wind velocities of up to 30 miles/hour, and a special emergency power supply system guarantees safe operation even in special black-out situations which happened already during operation which started automatically in May 1983. Stand-by situations of the heliostat field due to excessive wind velocities have been registered only two or three times during the operation period.

ARCO CENTRAL RECEIVER SOLAR THERMAL ENHANCED OIL RECOVERY PROJECT

F.A. Blake, D.N. Gorman, J.H. McDowell

ARCO Power Systems
Littleton, Colorado

I. Introduction

A pilot module Solar Thermal Enhanced Oil Recovery (STEOR) system installed at ARCO's Fairfield Lease in Kern County, California is well into its second year of operation. Completed in November of 1982, the 30 heliostat central receiver system augments the injection steam produced on the site by natural gas fired steam generators. The project was designed to have an output at or above the 1 MWt (3.4 x 10^6 btu/hr) level for the central six hours of clear days throughout the year.

The pilot plant is an element in ARCO Power System's program to develop solar central receiver systems, component equipment, and markets. While significant hardware and system development is present in the STEOR pilot plant, the major objective of the project is attainment of operational and maintenance experience in the oil field environment. Upon its completion, the plant was given a short checkout and system performance evaluation, and put into daily steam generation service the first week of 1983. A six month engineering development was conducted without impacting operational availability between January and June 1983. The major objective of this period was the attainment of computerized control of the full system together with establishment of the O&M procedures.

Major visible features of the Fairfield STEOR project are illustrated in the graphics art of Figure 1. The computer controlled heliostats are positioned in a 90 degree sector north of the tower which supports the Struthers Wells receiver. Support equipment for the receiver, secondary loop steam generation and water treatment equipment are installed at ground level west of the tower. The system is a 1/10 scale module of a commercial STEOR system. Heliostats are full scale in size but reduced in number. The heat transfer/steam generation system is scaled down in size while preserving functional features of the commercial scale unit.

II. Description

The ARCO Solar Thermal Enhanced Oil Recovery (STEOR) system provides the capability for generating reservoir injection steam by the use of concentrated solar energy as an alternative to combustion of fossil fuels. It uses the central receiver concept for solar thermal energy collection, concentration, and conversion. A field of 30 tracking heliostats reflect and focus radiant energy into saturated steam. Rated receiver output is 4180 pounds of steam per hour at 598 degrees F and 1515 PSIA.

Steam Generation

The receiver is a natural circulation boiler with a waterwall and steam drum configuration closely resembling elements of conventional fired boilers. The receiver operates with a closed water/steam loop except for make up water to replace blow down, as shown in the schematic of Figure 2. Energy exchange from the receiver discharge steam starts with pre-heating of the boiler feedwater which reduces quality to the 72-76 percent range. The balance of the extractable heat is used to generate 3435 pounds per hour of 1000 PSIA, 545 degree F, 80% quality injection steam in the condenser/boiler.

Two levels of water treatment are used. A "primary water treatment" consisting of softening and chemical additives for oxygen and pH control is given all supply water and is the only treatment for water to be used in the injection circuit. De-ionization polishing of the make-up water is performed.

Collector

Site of ARCO's STEOR development module is the ARCO Oil and Gas Company's Fairfield Lease, 45 miles West and South of Bakersfield, California. A nearly level 3-4 acre site adjacent to the lease headquarters and the producing field was supplied for the installation as shown in Figure 3. The 30 heliostat collector subsystem is shown in operation. Control of the heliostat tracking is accomplished by a microcomputer in the trailer located along the West side of the field.

Layout of the heliostats in five circular arcs is illustrated in Figure 4. Spacing in the front row was limited by corner to corner clearance of the heliostats. Spacing in the back rows is the result of a shading and blocking analysis directed toward minimum use of land while maintaining maximum unobstructed geometric performance.

Heliostat

Heliostats used in the STEOR module represent the development status attained during 1981. The area of each is 568 square feet ($52.8M^2$), the same produced under the DOE/Sandia sponsored "Second Generation Heliostat Development" program completed in early 1981. Cost reduction developments from ARCO's ongoing R&D program in the foundation, structure, drive and electronics subsystems were incorporated in the STEOR heliostats with the result that they represent a major advancement over the "second generation" heliostats. The front view of the heliostats during check out operation is shown in Figure 5. The reflecting surface consists of 24 mirror facets 4 feet x 6 feet mounted in pairs on 12 mirror module substrates. These are identical to the "second generation" mirror modules except for a sharper cant angle due to the short slant range of the field. The facets are mounted at varied cant angles to effectively achieve a single spherical curve across the heliostats.

Working equipment of the heliostat is shown in Figure 6. The foundation pedestal is a single 2 foot diameter pipe with 15 feet below grade and 12 feet above grade. Concrete grouting locks the pedestal to ground. A dual axis gear drive by Winsmith mounted at the pedestal top supports the rack structure halves and converts drive motor movement into accurate sun tracking motion through 18400:1 gear reduction. D.C. drive motors controlled by a microprocessor electronics package position the drive in response to communications from the master computer located in the control trailer.

Receiver

The receiver unit is a natural circulation steam generator configured to efficiently accept and convert the radiant energy projected from the collector field. Figure 7 illustrates the primary features of the receivers. The absorbing surface is a 10 foot x 10 foot wall of steam generation tubes. The steam/water mixture collects in an upper header and flows to the drum through a centrally located riser. Steam separates in the drum and exits the receiver while water recirculates through the downcomer and lower header to the tube wall. The receiver is shown venting steam during check out operation in Figure 8.

Tower

The tower installation is also shown in Figure 8. The boiler platform is 65 feet above ground level. A heliostat beam characterization target is mounted on the tower face centered at the 50 foot level. This enabled operational check out of the heliostats ahead of receiver installation. Installation of the receiver, support equipment, piping, instrumentation, and tower wiring was the final construction activity.

A separate boiler control console, shown in Figure 9 was designed for the EOR installation. Control, temperature and pressure transducers are located at their positions in the system schematics on the console face for ease of monitoring and operator response.

III. Construction

Construction was initiated with the start of 1982. January's accomplishments included grading of the site, on-site pedestal and rack fabrication, trailer set-up and pedestal installation. February and March activities centered around the electrical installation. Trenching and wiring for the power and data bus wiring to the heliostats and trenching and wiring for control and data acquisition of the steam generators was installed. Box mounted heliostat controllers were installed and communication with the computers was checked out. Assembly of the mirror modules on the rack structure was initiated. Figure 10 shows the construction site at the end of March when the first drives arrived allowing the final assembly of heliostats to begin.

Heliostat final assembly was accomplished during April and early May with the activity being paced by delivery of drives. Tracking at the standby spot, shown in Figure 11, during the day and cycling operation during the night provided an accelerated check out and endurance test of the collector system during the months of June, July, and August, 1982.

The tower installation was done in late May allowing tracking accuracy tests to be performed as part of the collector check out.

Steam generation equipment was received in early September and installation of it was completed by the first week of November. The secondary steam generation equipment used to generate "injection steam" into the oil recovery steam system is shown during check out venting in Figure 12. A view of the collector field heliostats from the tower is shown in Figure 13. An aerial view of the mid July configuration is shown in Figure 14. One heliostat is tracking the target and the balance are tracking the "stand by point" East of the receiver.

IV. Computer Control System

The STEOR system includes two distinct control packages, each with its own hardware and software, but both packages function under the overall supervision of a master control program. The reason for this approach is that the collector field control package (having been developed for other similar collector field applications) was largely already in existence, whereas the new and unique water/steam system required the development of a totally new control package.

The collector field control hardware includes a Hewlett Packard 9825 desktop computer, five line-driver circuits (one for each radial row of heliostats) and a microprocessor control package mounted to each heliostat pedestal together with azimuth/elevation drive-motors and limit switches on each heliostat. The 9825 continually updates time and sun position so that it can repetitively calculate new azimuth/elevation pointing data for each heliostat operating in a tracking mode. The HP 9825 also transmits command data (either from keyboard input or from master control computer) to the heliostats and obtains status data from heliostats. The line drivers serve to condition the command signals and transmit data between the computer and the individual heliostats. The heliostat microprocessor deciphers the command data and controls power to individual 1/8 horsepower D.C. motors for azimuth and elevation positioning. Torque from each motor is transmitted through a combined planetary/worm gear drive with an 18,400:1 reduction ratio to effect heliostat motion. Position status is continuously updated by monitoring motor revolutions. Extreme heliostat azimuth and elevation limits are established by microswitches which stop the motor when actuated. The limit switches are also utilized to check and reset absolute azimuth and elevation positions once a day.

Control package for the water/steam system (shown schematically in Figure 15) consists of a Hewlett Packard 9826 desktop computer, a Hewlett Packard 3497A Data Acquisition Control Unit, a custom operations console and the valves, pumps, sensors and transmitters in the system. The 9826 computer also serves as the system master control computer, sending commands to the collector field 9825 computer as conditions dictate.

The control system includes a third computer, also a Hewlett Packard 9825 desktop, which is used only to drive two video graphics monitors to display system status data. Figure 16 is a photograph showing the computer control station. From left to right on the desk are the collector system 9825, the data graphics 9825 and the master control 9826 with printer. The data graphics monitors are on the shelf above the computers.

The water/steam system operations console is shown in Figure 17. The console contains all necessary controllers for valves, pumps and the receiver door, as well as two multi-point temperature recorders and a video monitor which displays a pressure gage and water level indicator mounted on the receiver steam drum. Elements of the water/steam system can be controlled in any of three modes:

1. Manually, using controllers on the console,

2. Local automatic, using set-point control on the console,

3. Computer automatic, whereby the 9826 computer supplies all command data to the console via the 3497A DAC unit.

The STEOR system has been routinely operating under automatic computer control since May, 1983. A system operator is always onsite, but rarely intervenes in direct control of the system. His major activities include periodic monitoring of system operating parameters, performing a once-a-day water chemistry check, and tending to routine maintenance functions.

The 9826 computer software is a combination of several subprograms to perform the various functions of water/steam system control, overall system master control, data processing, data logging and maintaining up-to-date status information. The system control function is all-inclusive, from start-up from cold or hot standby conditions, to maintaining normal operation through varying sun conditions through the day, to shutting the system down due to insufficient power availability. Throughut the operating cycle, the control program monitors system pressures, flows, temperatures and water levels, issuing a warning if normal operating ranges are exceeded, and performing an orderly system shutdown if pre-defined critical limits are reached. In addition, there are built-in protective features to insure that certain parts of the start-up sequence have been completed before a subsequent operation is initiated. For instance, heliostats will not be commanded to focus on the boiler panel unless a "receiver door open" signal is sensed, and a feedwater pump will not be started unless its recirculating bypass valve is open. This approach will suspend the start-up operation in the event of malfunction of a critical component.

The DAC unit continuously scans approximately 100 data signals, including 60 temperatures, 6 flow rates, 6 pressures, 3 water levels and the set points and positions for all control valves. The 9826 computer regularly samples the data, converts it to appropriate engineering units, applies calibration factors and transmits selected processed data to the graphics 9825 computer for immediate display on the video monitors. Data updates are accomplished about 3 times per minute. At longer time intervals (currently every 6 minutes) all data are stored on a floppy disc for later retrieval and evaluation.

V. Operational Data

The STEOR system has demonstrated a reliability during start-up and its first year of operation which exceeds expectations. The system was operated safely during periods when the response times and other subtle characteristics were still being determined. The heliostat field has performed almost perfectly with downtime limited to individual heliostats for fractions of a day. Total time lost in the heliostat field was 6 1/2 heliostat-days. The heliostats were cycled at night for about 120 nights with each night's cycling equivalent to a month of actual operation. The steam generation system was available for operation 345 days of the year with downtime allocated to these activities: Fifteen days total were lost trying to find a good gasket material to eliminate leaking in the condenser/boiler tube and shell heat exchanger. Four days were lost repairing a pin hole leak in the receiver downcomer. Finally, 1/2 a day was lost changing the makeup water pump to a higher output pump. Other downtime was due to maintenance on the oil and gas steam lines and management decisions which resulted in the site being unmanned for much of November and December.

The annual output of the system is described by Table I. This table provides a monthly and annual summary of the injection period and the total steam injected. Figure 18 is a stairstep plot of daily totals of the steam actually flowing into the oil and gas steam system. These values are calculated by summing the product of the injection flow rate F(2) and interval of time since last data was read, usually about 20 sec. An annual stairstep of daily injection period is presented in Figure 19. These values are a measure of time from when the value connecting the system to the field steam piping is opened until it is closed at shutdown. The bar graph in Figure 20 shows the daily totals of the steam injected on a daily graph instead of the annual stairstep in Figure 18. The injection time period is also presented in this form in Figure 21. The graphs make the long unbroken production runs more clear. The period of June, July and August when the system ran almost everyday is evident by the solid black base. The lower values of total steam injected denote the partly cloudy summer days. Figure 19 provides daily peak insolation values. These values are the peak insolation during the period when steam was being injected into the oil field.

A more detailed look at a good run day is presented in Figures 20, 21, 22. In Figure 20, the two primary flows, F(2) the injection flow rate and F(7) the receiver steam flow rate out of the steam drum are shown. The steam rate on the receiver side is a result of maintaining a constant pressure, 950 psig, on the steam drum and the flow fluctuates with variations in insolation and feedwater makeup to the drum. The injection flow rate is controlled to be 90 percent of the receiver steam rate by the master control computer. This ratio has been determined to provide wet steam out to the field.

The primary system pressures are shown in Figure 21. The highest pressure P(6) is the pressure in the receiver steam drum which is ramped up to operating pressure. The master control computer monitors drum surface temperatures to maintain top to bottom temperature differential of less than 200 degrees F which keeps thermal stresses below allowables during ramp up. The injection pressure P(3) is brought up to operating pressure, 600 psig, at start-up and the flow rate F(2) is kept to a minimum, 1000 lbs/hr, until the drum reaches operating pressure. The lowest pressure on the graph is the pressure in the field piping. When injection starts, that pressure rises to equilibrium with the rest of the field piping. The pressure in this piping will vary from day to day depending on the field injection activities, but it is typically 300 to 500 psig.

The last graph shows how clear it was this day. The insolation level was above 700 W/M^2 for 8 hours, but the peak level never exceeded 900 W/M^2. The system was designed for typical daily insolation, derived from insolation models based on data from earlier years, of 950 W/M^2.

High atmospheric ash due to unusual volcanic activity in the Northern hemisphere has resulted in the lower recorded values. Early data in 1984 have spawned hope that this affect may be attenuating and we look forward to a record breaking year.

Conclusion

The design, fabrication, and installation of the central receiver solar thermal enhanced oil recovery system proceeded in a generally satisfactory manner. The full cycle was performed in the 1982-83 calendar years. The overall project pacing item was the steam generation system.

Operation with minimal O&M expenditure was demonstrated from the outset and the ability to operate virtually unattended has been demonstrated.

The project has demonstrated that all elements of a solar thermal steam generation system are ready for a full scale application such as the enhanced oil recovery application of this project or similar process heat or desalination applications.

TABLE 1

STEOR Operational Data 1983

Month	Actual Hrs Injected	Total Stm Injected (lbs)
Jan	34	78535
Feb	51	119950
Mar	76	161210
Apr	78	190960
May	91	207235
Jun	198	385415
Jul	196	424865
Aug	102	207540
Sep	163	346815
Oct	137	278145
Nov	18	39860
Dec	24	39900
Total Annual	1168	2480430

Figure 1. Graphic of Fairfield Lease EOR Project

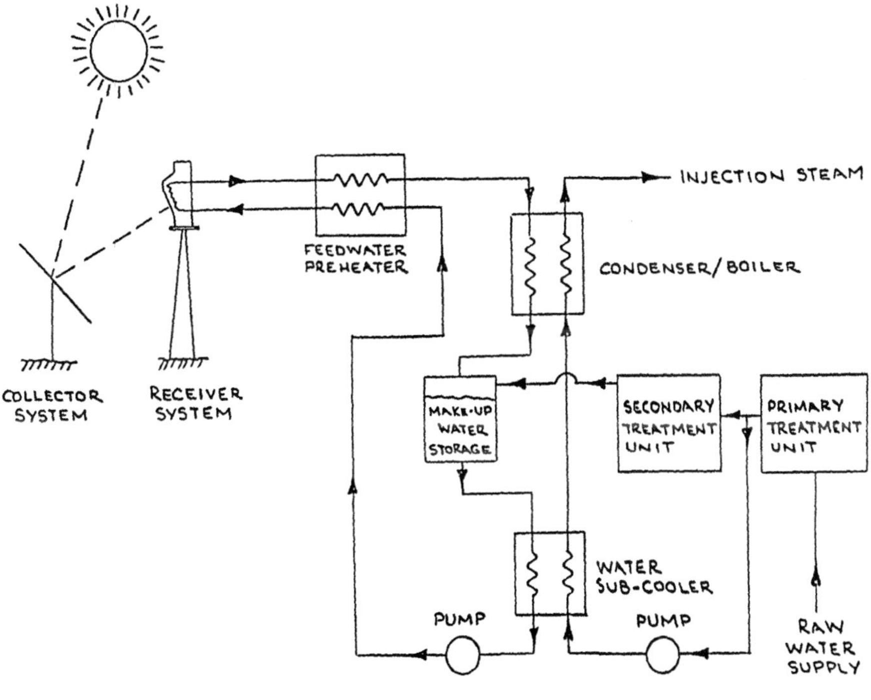

Figure 2. Solar Thermal Enhanced Oil Recovery Process Schematic

Figure 3. Fairfield Lease STEOR Pilot Plant In Operation

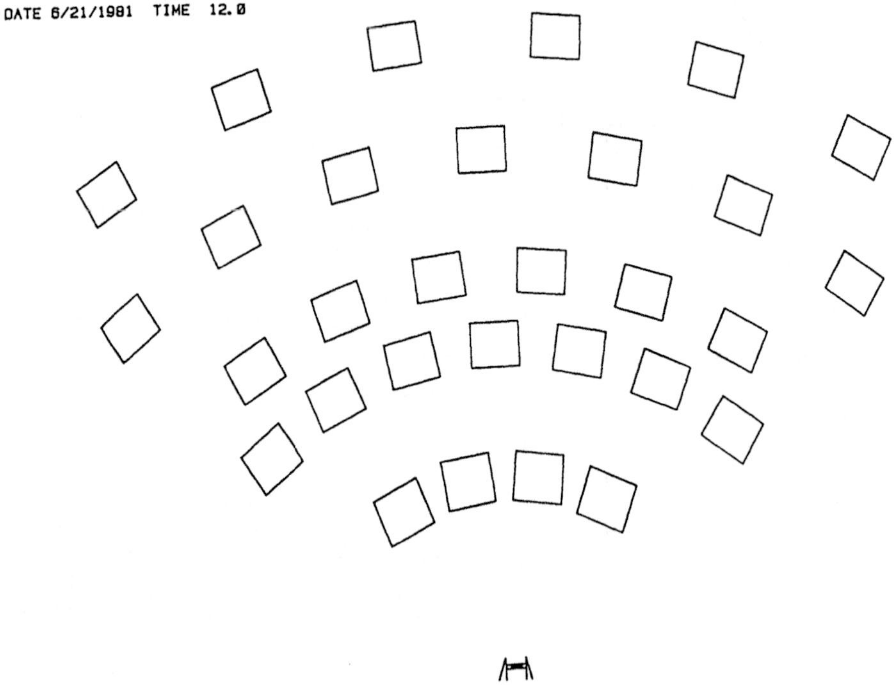

Figure 4. Heliostat Layout For ARCO 'Fairfield Lease' Project

Figure 5. Heliostats, Mirror Side, During Checkout Operation

Figure 6. Heliostats, Equipment/Structure Side, In Operation

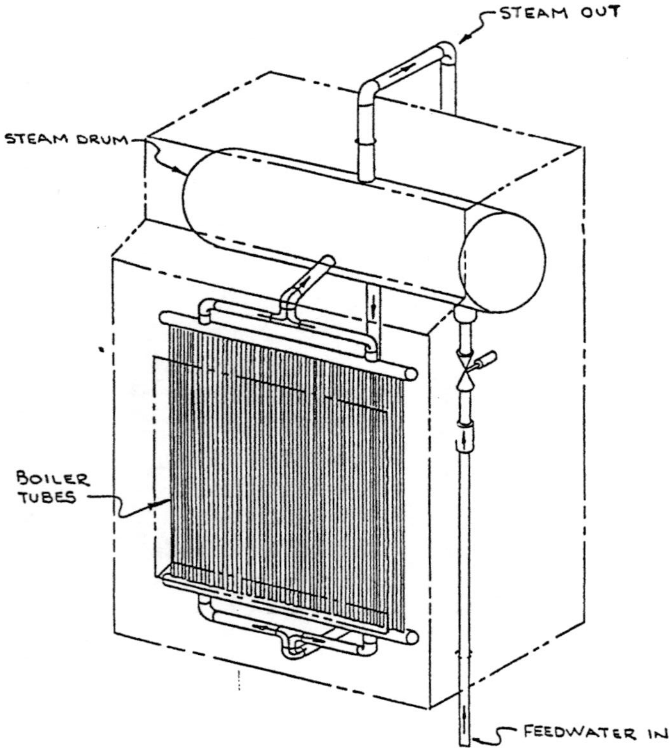

Figure 7. Solar Steam Generator Arrangement

Figure 8. Tower/Steam Generator Installation
Steam Venting During Checkout Operation

Figure 9. Steam Generation Control Console

Figure 10. 'Fairfield Lease' construction Site
March 31, 1983, Prior to Final Heliostat Assembly

Figure 11. 'Tracking at the Standby Position

Figure 12. Injection Steam Generator,
Venting Steam During Checkout Operation

Figure 13. Fairfield Collector 'from the Tower'

Figure 14. Aerial View of Fairfield Lease Project, July 1982

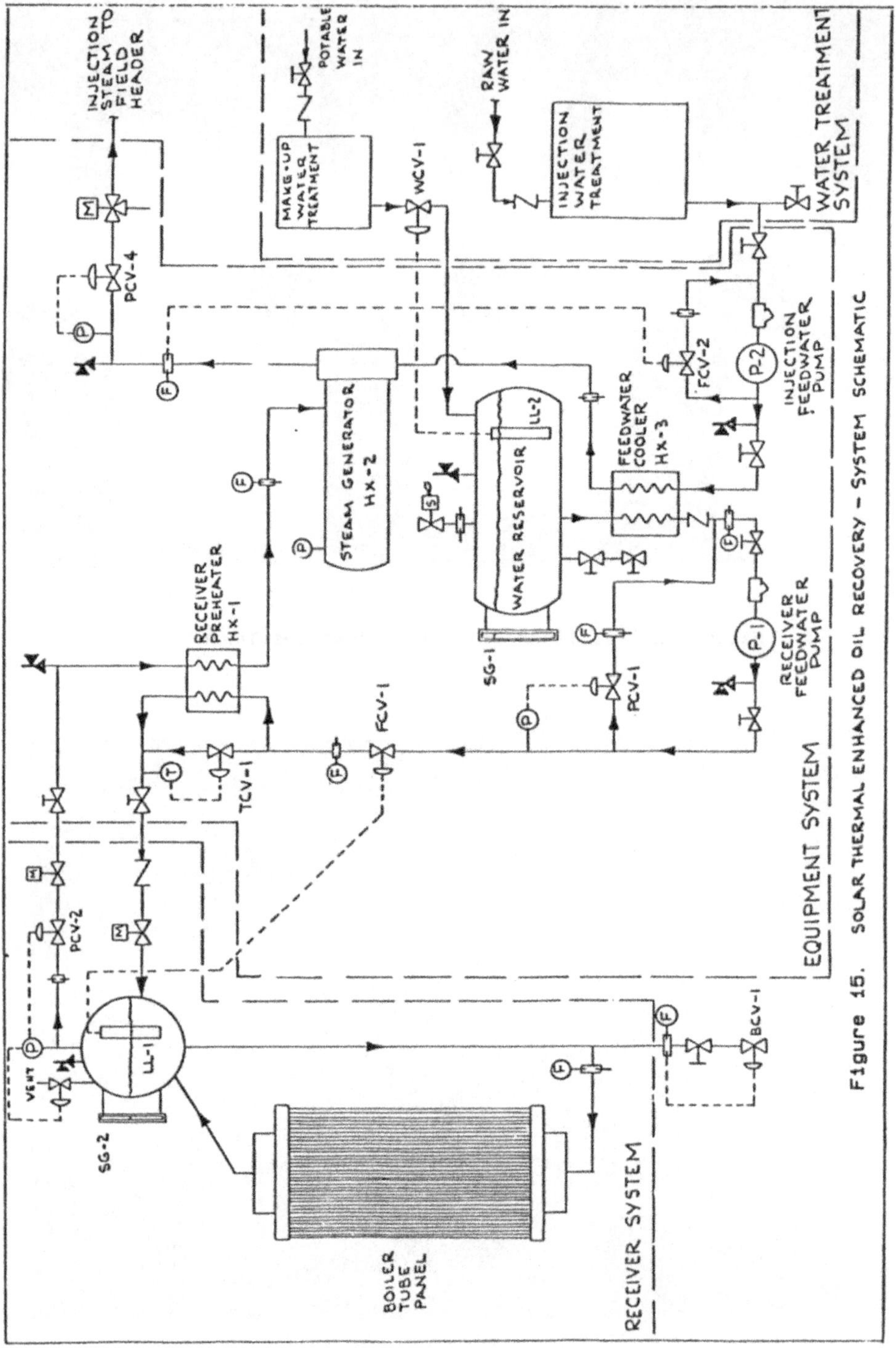

Figure 15. SOLAR THERMAL ENHANCED OIL RECOVERY - SYSTEM SCHEMATIC

Figure 16. Heliostat Control, Graphics-Data Acquisition, & Steam Generation Control Computers (HP9825, HP9825, HP9826)

Figure 17. Steam Generation Control Console, Fairfield Lease Project

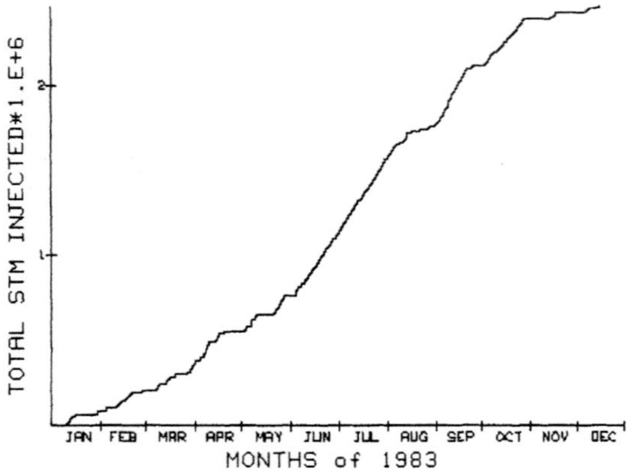

Figure 18. 1983 Total Steam Injection, Millions of Pounds

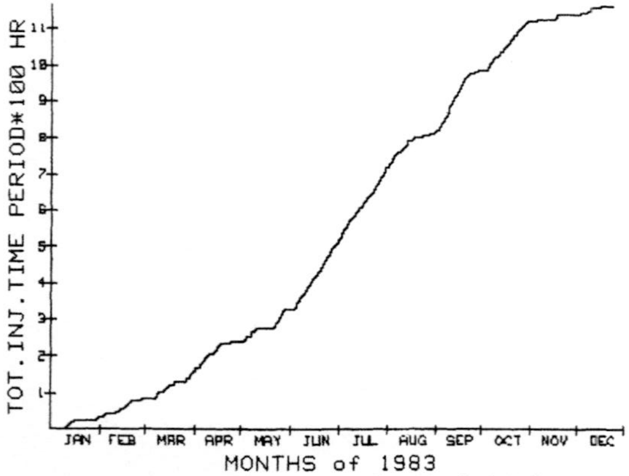

Figure 19. 1983 Total Injection Mode Operation Period, Hours

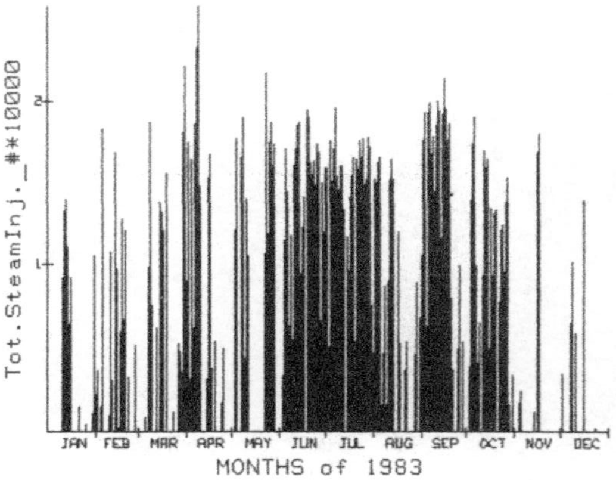

Figure 20. 1983 Daily Steam Injection

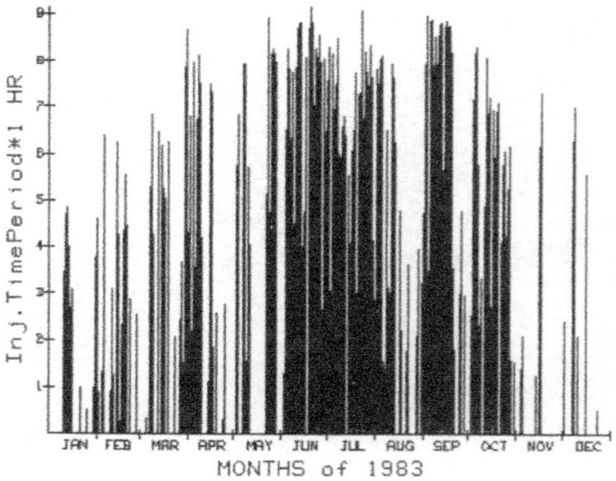

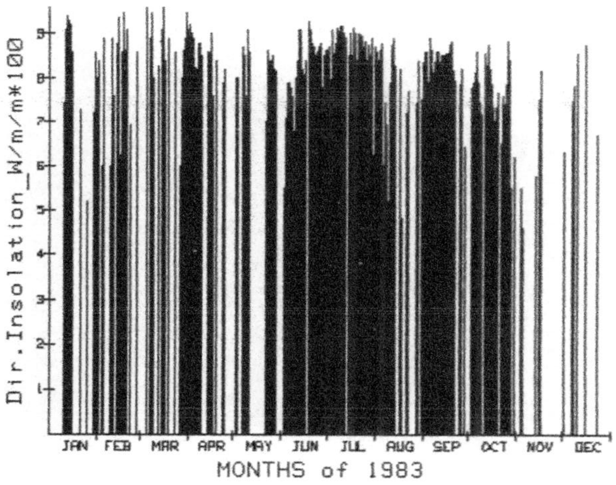

Figure 22. 1983 Daily Peak Insolation

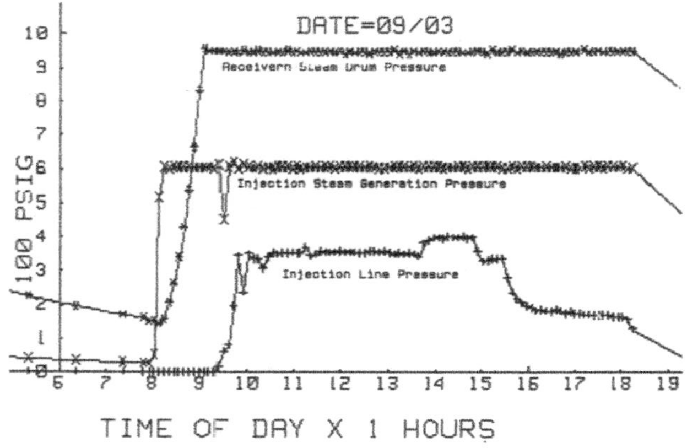

Figure 24. Single Day Pressure Pattern Profile, September 3, 1983

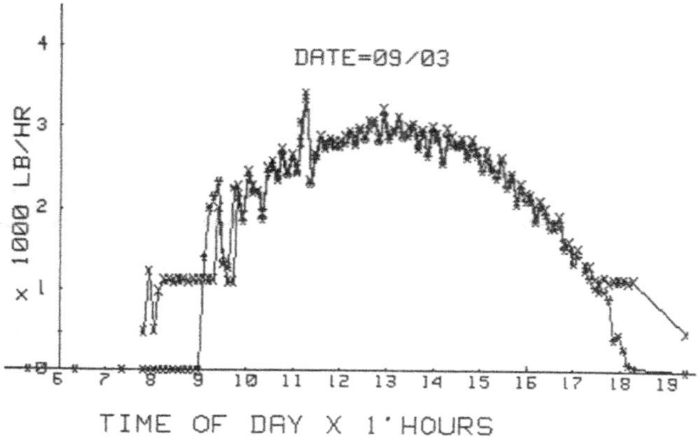

Figure 23. Single Day Flow Pattern Profile, September 3, 1983

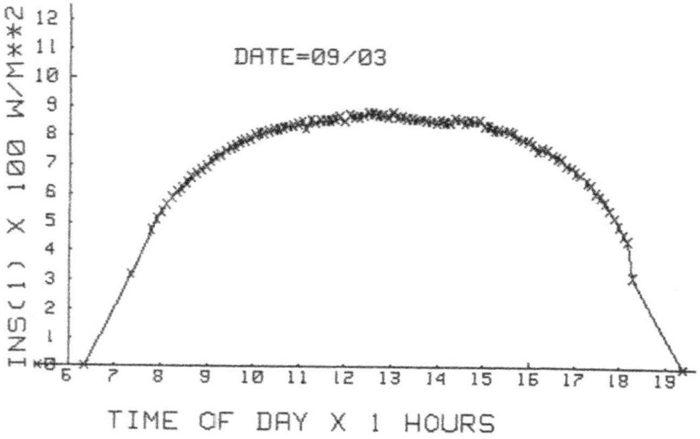

Figure 25. Single Day Insolation Profile, September 3, 1984

SESSION V - PART 2

NEW TECHNOLOGIES

Summary of the Session by the Rapporteur
B.P. GUPTA, SERI, Golden, USA

Conceptional design of solar thermal electric power plant with optical fibers and channels

Low tension stress design for a glass receiver concept

The university of Houston solar research program

SUMMARY OF THE SESSION BY THE RAPPORTEUR

B.P. GUPTA
SERI, Golden, USA

1. INTRODUCTION

The operation of the Central Receiver Solar Thermal Power Plant has shown the feasibility of this technology for obtaining thermal energy over a wide temperature range (450°C - 650°C). We have also learned about the integration of various subsystems to form the entire power plant as well as the use of thermal energy for electrical output.

The experience to date clearly points out the need for development of new technologies to either reduce the cost of subsystems, or the operating reliability and efficiency of individual components, or increasing the applicability of solar thermal systems for other uses, e.g. IPH and F&C by operating at temperatures required by the end use. Generally, this end use would drive the development for high temperature systems.

Significant among the need for new technologies are lower cost designs for heliostats, low loss receivers, low loss energy transport and storage, high temperature receivers, high temperature transport and storage, efficient heat extraction, and coupling of solar flux to a potential application. A more exhaustive list is included in Table 1, presented by the session chairman Mr. M. Becker of DFVLR.

2. PAPERS

In this session, four papers were presented. The titles and names of the authors are given below:

1. Conceptual Design of Solar Thermal Electric Power Plant with Optical Fibers and Channels; N. Ikeda (NEDO), T. Tani (ETL), T. Horigome (NEDO), Japan.
2. Low Tension Stress Design for a Glass Receiver Concept; B. Karrais, R. Kochendörfer, DFVLR, Stuttgart, Germany.
3. High Temperature Research at SERI, Direct Absorption Receiver Concept; B. P. Gupta and R. Copeland, SERI, Colorado, USA.
4. Research at the University of Houston; L. Vant-Hull, University of Houston, Texas, USA.

The first paper presented by Mr. Ikeda proposed a concept of optical fibers to transport energy from the point of collection to the point of conversion. He compared this optical system to the current experience from Nio and Solar One Power Plants and derived

Table I. New Technologies.

System

* Concepts
* Analysis
* Computation
  (Thermal, Stress, Design)
* Power Range
* Application (e.g. MHD)

Components

* Heliostats
  (Dome, Membrane, etc.)
* Chemical Reactors
  (Separated, Integrated, Application)
* Storages
  (Sodium, Molten Salt, Sensible Heat, Methane, Hydrogen, etc.)
* Transport
  (Electrical, Chemical, Application)

Heat Transfer Medium

* Sodium, Salt
  (Development of Salt Components at High Temperatures)
* Steam
* Gases
  (Air, Helium)

Material

* High Temperature
* Alloys
* Ceramics
* Glass, Quartz

the conclusion that the optical system may have a potential benefit in transporting energy from the receiver to a central electrical generating station. He felt that the system using optical energy transport could be applied to both dish and central receiver systems. This system is being proposed to address the apparently high cost and high thermal losses experienced in the transport and storage systems at the Nio plant. The system described in the paper suggests that development of optical fiber bundles for energy transport would be required and that light weight low cost plastic fibers would be needed to make the proposed system cost effective. In response to a question during discussion, the speaker said the Japanese manufacturers of optical fibers feel they can achieve the required optical efficiency of greater than 80%. But the question remains whether such efficiencies are achievable over the broad band for energy transmission as opposed to light transmission in narrow wavelength band.

The second paper proposed a concept where solar radiation is absorbed in a multiple reflection/absorption process as the working gas passes through a complex glass receiver. The design proposed addresses problems typically experienced in high flux/high temperature solar receivers. Specifically, the tension stresses are reduced or eliminated, the thermal stresses are reduced by use of glass, and the heat losses are reduced by arrangement of gas flow and by designing the tubes with specific transparency of the glass tubes. Special attention was given to the need for developing a selective coating for window material, keeping the aperture at low temperature, and more uniform heating of the tubes.

The principle of fully absorbing the radiation in the working fluid once the radiation is trapped inside the receiver is important for more efficient use of solar energy in the receiver. The paper also presented the future plan for verification tests using a bundle of glass tubes at the focal point of a dish collector. During the discussion, questions were raised regarding the weight of this receiver. Answers revealed that flux of $1 MW/m^2$ could be used in this concept. In response to the weight question, it was estimated that the glass filled approximately 20 percent of the volume of the receiver which would give an approximate weight for a specific power output of the receiver.

The third paper presented by Mr. Gupta summarized the research thrusts of the U.S. Department of Energy's Solar Thermal Research Program, managed by SERI, and described one area of research in detail, namely the High Temperature Absorption Receiver concept. The overall research emphasis described included research to establish the technical feasibility of new concepts, providing a basis for next generation technology, as well as exploration of new and innovative ideas that may lead to cost reduction, performance improvements, and unique or beneficial use of concentrated solar flux. The presentation of the specific high temperature Direct Absorption Receiver and Thermal Storage Concept (DARTS) included the present laboratory research ongoing, and the description of future plans for testing the concept using the solar facility at Georgia Institute of Technology. During the discussion and in response to questions, the specific benefits of secondary concentrator at the aperture, the elimination of tubes in the receiver, and the use of liquid as the working media were emphasized.

The fourth paper, presented by Mr. Vant-Hull, outlined the research activities of the University of Houston. He briefly described the ongoing research in materials behavior in concentrated solar flux, chemical reactor operations under cyclic loads for methane reforming, and other fuels and chemicals related research being conducted at the University under Department of Energy and SERI sponsorship.

CONCEPTIONAL DESIGN OF SOLAR THERMAL ELECTRIC POWER PLANT WITH OPTICAL FIBERS AND CHANNELS

N.Ikeda (NEDO),T.Tani (Electrotechnical Lab.) and T.Horigome (NEDO)

Summary

Conceptional design of central receiver type solar thermal electric power plant with optical fibers and channels is reported. There are important problems with the heat transportation between components in present systems of solar thermal electric power plants. It is required for the plant with central receiver tower to concentrate sun light with high accuracy. On the other hand, it is required for the plant with distributed collector system to decrease the heat loss from the heat transportation pipes. Therefore, the conditions required for advanced solar thermal electric power system are as follows;

1) Tracking method of collector system is a simple one that the sun tracks directly.
2) Energy transports with light as long as possible.
3) System is to consist of many small units.

From the above conditions, parabolic dish collectors or fresnel lens and optical channels are chosen and optical fibers are required as energy transport ways with flexiibility. It is possible for this system to lay down the receiver on ground. 10 MWe plant is designed following the proposed system. The plant has 909 parabolic collectors or 4030 fresnel lens collectors. It is found that the system has several technical and economical merits compared with present plants.

1. INTRODUCTION

The conceptional design of solar thermal electric power plant with optical fibers and optical channels are presented. Operation of Nio plant brings us the comprehension of solar thermal electric system. It also lead us to the development of the next generation system for sloar thermal electric generation which we want to call. It is found from the experiences of operation of Nio plant that heat loss on the way of gathering solar heat energy is very important,though it is necessary to develop the components of the system.

2. CONDITIONS FOR IMPROVED SYSTEMS

Two systems at Nio are compared with each other from the aspect of the energy transportation as shown in Tab.1. In a central receiver system, collector system needs the high accuracy of sun tracking within 1 degree and the receiver has to locate at the top of tower. Energy transportation between collector system and receiver are performed by light. High temperature heat generated in the receiver transports to thermal storage system and then it is sent to turbine and electric generator system in the same energy form. On the other hand, in the distributed collector system, collector system which consists of almost parabolic trough, is easy in sun tracking. The receiver system is able to separate in many units. The distance between these collector and receiver systems is, of couse,very short.

Solar energy changed into heat has to be gathered from the wide area through long distance to thermal storage system. After then, heat is sent to turbine and electric generator system.

From the comparision with two solar systems, following items might be needed as the conditions for improved system:

1) Direct sun tracking system.
2) Energy tranportation by light as long as possible.
3) Many standard units for consisting of system.

Item 1) brings system easy tracking like the parabolic trough system, not like heliostat system. and it contributes the increase of the availability of mirror and cost down. Item 2) makes minimum in energy loss on the way of gathering heat. Item 3) brings us standardization of components and cost down by mass production.

3.PROPOSED SYSTEM FOR THE NEXT GENERATION

The system for the next generation which satisfies above conditions are presented. If a direct tracking system for collectig system are taken, it will be parabolic dish system or fresnel lens system. When the direct tracking are used, optical path with flexibility are needed for energy transportation by light. This is the reason that the system with optical fibers are proposed. However, there is yet no optical fibers for energy transportation at present and it is very expensive to use the optical fiber for communication.

In the proposed system, energy gathering and transportation are similar to tower receiver system,because larger turbine has better efficiency. Solar energy is gathered by parabolic dish mirrors or fresnel lenses. Energy transport in long distance are performed by light with optical fibers and channels. receiver system is able to locate on the ground. This contributes freedom of design,decrease of heat loss and cost down by lack of tower. The conceptional view of the proposed solar thermal electric power system with optical fibers and channels is shown in Fig.1.

Some preliminary designs are performed for getting the quantative characteristics of the proposed system. That is:

CASE 1 10 MWe solar thermal electric power plant with large parabolic dish collectors are designed. 10 MWe electricity needs 50 MW thermal output in the efficiency 20 %. The diameter of the parabolic dish is 10 m which is minimum cost by R.L.pones'paper. The area of the dish is 78.5 m^2 and the collecting energy is about 55 kW/unit when solar insolation assumes 0.7 kW/m^2. If solar multiple factor is 2, the system needs 1818 units of the parabolic dish collectors. If area needs a didh 225 m^2 (15m x 15m),total field is about 409,000 m^2. It is necessary to optimize to minimum area. In comparison with Solar one plant with solar multiple factor 1.25, the collector field of the plant is 71,447 m^2. Receiver takes the same shape and size of Solar one in this design, The diameter is 7 m and its hight is 13.7 m. This receiver is housed in the room with thermal insulation wall. This housing room has 30 m in diameter and half circular dome. If 70 cm width is taken as an optical channel, the room give us 135 channels. If a focusing of the parabolic dish needs about 100 cm^2 cross sectionarea of optical fiber, the nergy density is about 550 W/cm^2. It is no problem in present optical fibers.

CASE 2 10 MWe solar thermal electric power plant with small size fresnel lens collectors are designed. Thermal output needs the same value as CASE 1. Diameter of fresnel lens is 150 cm which is the largest fresnel lens in

Japan. The area of the lens is 1.77 m^2 and the collecting energy is 1.24 kW/unit. The plant with solar multiple factor 2 needs 80,645 units. If area of a sun tracking lens system needs about 4 m^2, total field is about 322,600 m^2.

CASE 3 If plant efficiency increases to 30 %, CASE 3 is as follows in comparison with CASE 2. Needed thermal output decreases to 33.3 MWt. The same quantities with fresnel lens in CASE 2 are used. The plant at the solar multiple factor 2 needs 53,763 units and total field is about 215,000 m^2.

CASE 4 Plants in this case and next case are examined in comparison with 1000 kWe Nio plant. Plant efficiency is taken 13.8 % which is the same value of Nio plant. The same quantities with fresnel lens are taken. The plant with solar multiple factor 1 needs 5,840 units and 23,360 m^2 as total field. On the thoer hand, Nio plant is about 25,400 m^2.

CASE 5 Expecting to decrease of heat loss in the system, above 20 % in efficiency will be attained. In this case, plant needs 4,033 units and 16,132 m^2. Receiver takes the equivalent one as that of Nio plant except corn part. It has 2 m in diameter and 11 m in hight. That is, the equivalent area is 69 m^2. Both diameter and hight of it are taken the same values. Diameter of the housing room is 9 m. The plant has 40 channels in the 70 cm width by one channel. Therefore, one channel has 101 lens units.

Parameters of plants examined in case study are arrenged and shown in Tab. 2.

4. PROBLEMS TO BE SOLVED

The conception of main components and its problems have to consider in detail.

Collecting system

There are three type of point focusing system: a) parabolic dish, b) deep parabolic dish, c) fresnel lens. At present, fresnel lens system is better than others. Therefore, there remaine the life problem about the lens and the cost.

Optical fiber system

There is no problem with the limit of energy density such a low value of 1 kW/cm^2 in this designs. It seems to be at present above several tens GW/cm^2. There still some problems to apply the optical fiber for energy transportation. One is selectivity of light wave length and other is very expensive. Therefore, light guided tube with cheap and high efficiency must be developed. In this tube, optically thick and transparent liquid must be enclosed shown in Fig.2. And also plastic optical fiber which its shape of cross-section is not circle,but hexagon for increasing efficiency,has to be examined from its cheap cost shown in Fig.3.

Optical channel

Optical channels are another way to cost down. An example of connecting part of fiber and mirror is shown in Fig.4. The mirror is a kind of parabolic characteristics.

Receiver system

The same receiver on the top of tower in usual plants might be used with slight improvements. It is possible to install the door with thermal insulation between the housing room and the channel. Heat loss in night is able to reduced to minimum. The receiver system results in high efficiency and relieving thermal stress of receiver tubes caused by heat cycles.

Storage system

The storage system in this plant is able to located in the receiver tubes. this arrengement gets the minimum length of pipe to transport the heat.

Turbine and generator system

This system must be developed along the "solar specification" that the efficiency of the turbine and generator do not decrease, but keep the maximum value with wide range of steam condition.

5. CONCLUSIONS

Solar thermal electric power plant with optical fibers and channels has some features above mentioned. It must be examined from technical aspects and at the same time, must be evaluated from economical efficiency. This system might be decreased 20 or 30 % of morror area due to rise up of the availability of mirror and saves the cost of tower which is 15 or 20 % in usual tower system. It is possible to apply the mass production system even in one plant in case 2,case 3 and case 4. If some auxiliary heat sources are installed in receiver house, the start up time and transient characteristics of operation will be considerably improved. This receiver system is very compact. Therefore, it will be easy to design the system with sodium loop or molton salt loop.

Tab.1 Comparison With The Problems Of Two Systems And The Proposed System

	Central Receiver System Energy Transport	Distributed Collector System Energy Transport	Light Concentrating System Energy Transport
Collector System	High Accuracy Tracking	Almost Trough Easy Tracking	Parabolic Dish Fresnel Lens
	Light, Long Distance	Light, Short Distance	Light, Optical Fiber & Channel
Receiver System	Tower Top Location	Many Separating	Central Ground Location
	Thermal Heat, Medium Distance	Thermal Heat, Long Distance	Thermal Heat, Short Distance
Thermal Storage System			
	Thermal Heat	Thermal Heat	Thermal Heat
T / G			

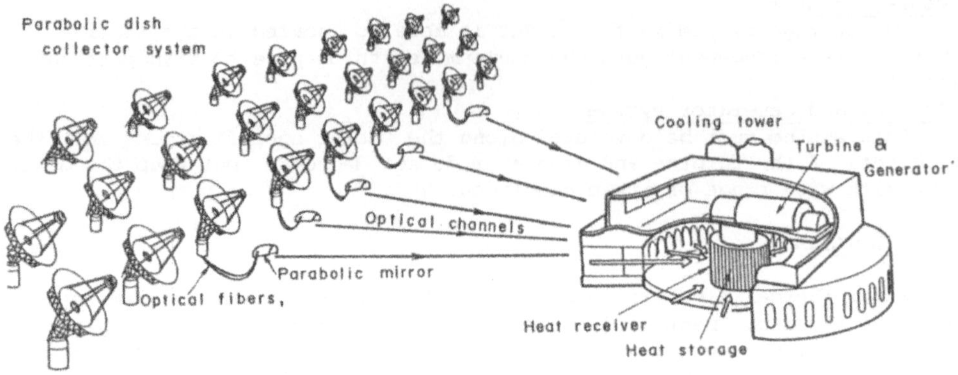

Fig.1 CONCEPTIONAL VIEW OF SOLAR THERMAL POWER GENERATION SYSTEM WITH OPTICAL FIBERS AND OPTICAL CHANNELS

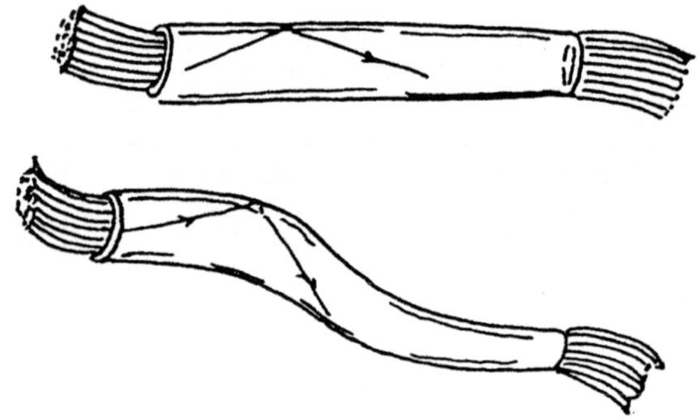

Fig.2 Proposed Optical tubes

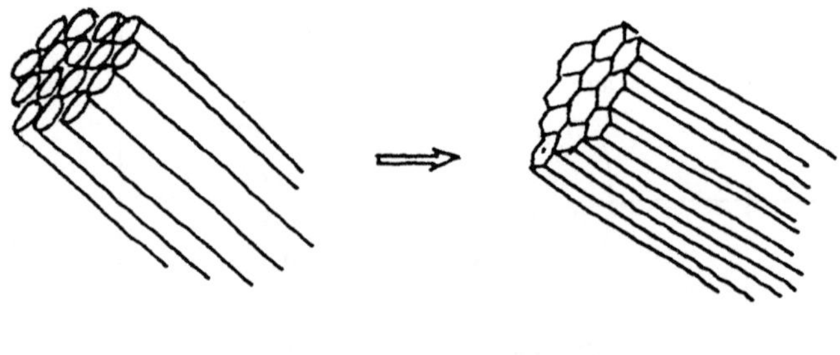

Fig.3 Improved Optical fiber with Hexagonal cross-sections

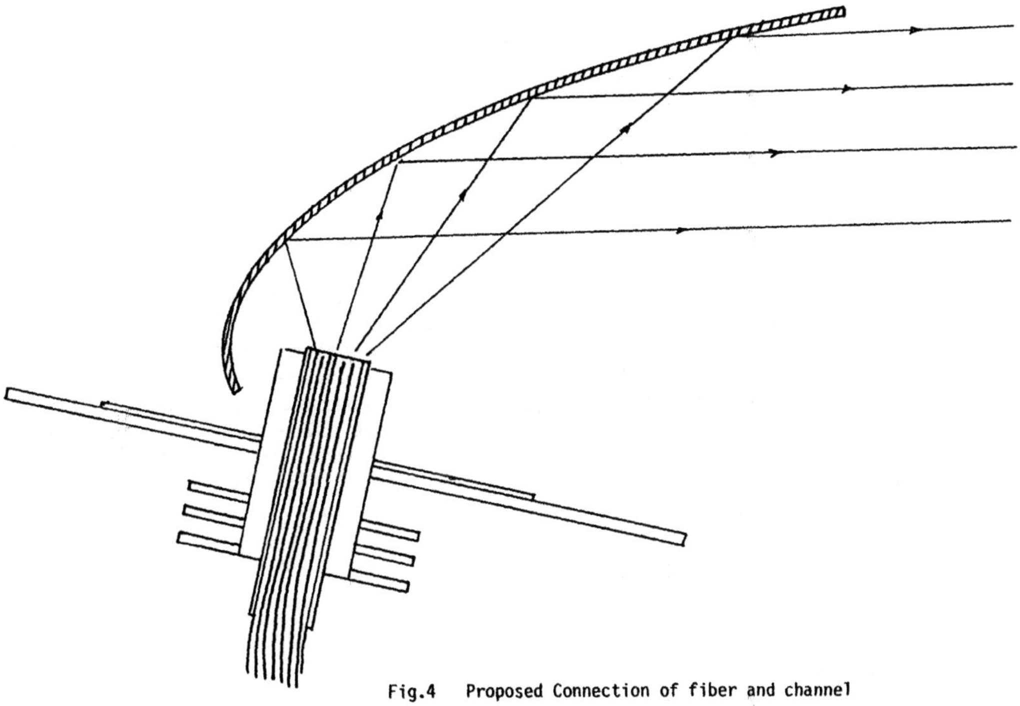

Fig.4 Proposed Connection of fiber and channel

Tab.2 Parameters of The Proposed Solar Thermal Electric Power Plants

		CASE I	CASE II	CASE III	CASE IV	CASE V
Electric Output	(MWe)	10	10	10	1	1
Thermal Output	(MWt)	50	50	33.3	7.24	5
Plant Efficiency	(%)	20	20	30	13.8	20
Collector System		Parabolic Dish	Fresnel Lens	←	←	←
Diameter	(m)	10	1.5	←	←	←
Collect.Energy	(kW/unit)	55	1.24	←	←	←
Solar Multi.Factor		2	2	2	1	1
No. of Units		1,818	80,645	53,763	5,840	4,033
Field Aera	(m^2)	409,000	322,580	483,867	23,360	16,132
Receiver System						
Receiver	(m)	$7^o,13.7^h$			$4.7^o,4.7^h$	
Housing Room	(m)	30°	⇐	⇐	9°	⇐
No. of Channels		135			40	
Energy Density of Optical Fiber	(W/cm^2)	550			177	

LOW TENSION STRESS DESIGN FOR A GLASS RECEIVER CONCEPT

B. Karrais, R. Kochendörfer
DFVLR Stuttgart
Pfaffenwaldring 38-40, 7000 Stuttgart 80

Abstract

In the proposed concept, solar radiation energy is absorbed in a multiple reflection/absorption process as the working gas passes through a complex glass receiver. By the receiver design the tension stresses usually resulting from the pressurized working gas are eliminated or reduced. The extremely low thermal expansion of glass minimizes thermal stresses. A clever gas flow concept uses the working gas at low temperatures as a cooling system for a translucent aperture window, thus reducing reradiation losses. A modular concept allows even large scale aperture windows, taking advantage of low cost serial production.

1. DESIGN GOALS

It is well known, that thermal cycles become more and more efficient with increasing temperatures. Combined with a solar receiver, the overall efficiency is affected to a large extent by reradiation losses of the receiver, because these losses increase with the 4th power of temperature. Our design goal is to reduce the reradiation losses of the receiver, thus increasing the overall efficiency.

To realize this goal, a number of design principles were applied:

1. Selective window material.
   The aperture is closed with a window, which has a selective effect like a greenhouse, namely sunlight passes through the glass, while the infrared radiation from the inside is reduced. So reradiation losses will be diminished.

2. Aperture at low temperature.
   The window temperature is kept as low as possible to reduce the emission of the window, which is part of the radiation losses.

3. Low stress design.
   The thermal and mechanical stresses are held to a low level. Especially when using a brittle material like glass, the tension stresses must be low. To avoid stress peaks in the attachment points, the loads are distributed onto large areas.

4. "Smooth absorption".
   The absorption of the radiation is distributed uniformly on a large surface to minimize temperature gradients, which may cause thermal stresses.

2. SOME PROBLEMS WITH EXISTING CONCEPTS

In the literature (1, 2, 3), there are a few receiver concepts with aperture windows (fig. 1). In some the window is slightly cooled by the inlet fluid. However, if the window is pressure loaded, thick and expensive windows are required, or the size of the aperture is limited. The different thermal expansion between the window and the housing material always causes difficulties as far as stresses and tightness are concerned.

In receiver designs incorporating heat exchanger tubes, the solar radiation usually hits the tubes at one side (fig. 2). The different temperatures between the exposed and the shadowed areas result in a bending of the tubes and consequently in stresses due to restraints, especially at the tube attachment points. Additional tensile stresses are caused by the pressurized medium.

3. ABSORPTION AND HEAT TRANSFER MECHANISM

In our low stress design approach, the tube axis is directed to the center of the heliostat field (fig. 3). Then the beams enter the tube symmetrically and there is no temperature gradient in circumferential direction. Assuming this tube is of reflective material, the solar radiation will be reflected again and again until it reaches the tube end. Because no material is a perfect mirror, a small part of the beam energy will be absorbed at each reflexion. This leads to a slightly higher temperature at the heliostat facing tube end than at the other one, where the remaining solar radiation exits the tube and hits a "black" wall. This wall is heated to a high temperature and emits intensive infrared radiation.

This infrared radiation enters the tube also symmetrically and passes through the tube by multiple reflexion but in the opposite direction. Using glass as tube material the infrared radiation is absorbed much better than the solar one, resulting in a temperature profile over the tube length which is much higher at the wall facing side than at the other one. Together with the temperature profile caused by solar radiation, this leads to tube wall temperatures which are constant in circumferential direction and slightly increase to the black wall in axial direction.

By this mechanism, a very smooth absorption and a relatively low temperature at the aperture can be realized. A suitable gas flow can serve to further lower this temperature and to receive the heat absorbed by the tube, as is shown later.

In reality a glass tube does not offer as good a reflectivity as is necessary to conduct the radiation through the tube, because if a beam hits a glass tube wall, only a part of the beam is reflected, a small part is absorbed and the rest will leave the tube. But this leaving part is not lost, if the tube is part of a tube bundle. Then each beam entering a tube has an antimetric beam that enters the neighbouring tube. This antimetric beam will hit the wall at the same place, with the same intensity, but at the antimetric angle. So that part of a beam leaving one tube is exactly replaced by the beam leaving the neighbouring tube.

This mechanism is valid up to the outside surface of the bundle, where a reflective coating is necessary to minimize heat losses.

4. GLASS RECEIVER MODULE

To couple the absorbed energy into a gaseous medium, a pressurization of the gas flow is advantageous. This requires that the tubes be closed by transparent windows at least at the heliostat facing ends.

Our favourite way to realize this concept is shown in figures 4 and 5. A square shaped glass tube is closed with a window at one end. Within this square shaped tube, a central inlet tube and a number of outlet tubes are positioned in such a way, that a gap - that means as a gas flow path - remains between the window and the tube ends. This combination of glass tubes builds up a module (fig. 4).

Solar radiation coming from the window side is absorbed within the module, heating the walls. Cold gas is blown into the central tube and flows to the window, where it is reversed to flow back in the total volume betwen the square shaped tube and the inlet tube. The function of the outlet tubes is just to increase the surface for absorption and heat transfer. The gap between the tube ends and the window is designed to accelerate the gas velocity, thereby increasing the heat transfer coefficient at the window. This keeps the window at a low temperature.

5. STRUCTURAL ASPECTS

Several modules can be stacked to a block (fig. 5) whose cross section is as large as the focal spot of the heliostat field. The stacked modules are sealed by welding lines at the open ends of the square shaped glass tubes. This results in a compact block, which can be seen as a very thick window with integrated cooling channels.

The next design goal was to realize a distributed load introduction to hold this thick window against the internal pressure load. The ideal load introduction to balance the resulting force of the axial pressure load would be by shear forces, distributed uniformly over the entire side surfaces of the block. These shear forces can be produced by friction, if a jacket is pressed against the sides of the block. It is favourable, that the jacket pressure is in the same order of magnitude as the internal pressure. Then the outer walls of the modules at the periphery of the block will not deform.

In reality the outside pressure load can be applied by a double walled pressure vessel (fig. 5). The inner wall of the vessel, the jacket, is a thin membrane of steel, which is welded leaktight to the outer wall. For insulation purposes, this steel membrane is lined with ceramic fiber boards of small bending stiffness. The glass block is embedded in this insulation, the two halves of the vessel are closed and the volume between the vessel walls is pressurized. Now the block lies comfortably clamped between air cushions, leaktight and insulated.

The outlet tubes are fixed within the square shaped tubes, the rear wall is positioned to the vessel, forming a hot gas collector volume between the block and the rear wall. The insulation of this wall acts as a black body. The inlet tubes are inserted and held by a perforated plate. After the end cap is closed, the receiver is ready for operation.

Cold gas is blown into the cap inlet, distributed to the modules, flows to the windows, where it is reversed and exits as hot gas between the rear wall and the glass block through a number of outlets.

6. RECEIVER DESIGN PROSPECTS

Using this concept, a number of advantages can be expected:

1. Small reradiation losses due to low window temperature.
2. Large heat exchanger surface in a small volume, because the high number of inserted tubes offers a big surface.
3. The chemical resistance and the transparency of glass allow the receiver to be used as a chemical reactor.
4. A long service life can be expected due to the low stress level.
5. A modular design offers cost saving and adaptability to different heliostat field sizes.
6. Using quartz glass as material, the receiver will be resistent to thermal shocks and to temperatures up to 1000 ^{o}C.

7. VERIFICATION TESTS

To test whether these advantages can be realized, we plan to build a small receiver atop a parabolic dish collector (fig. 6), with a diameter of 5 meters and a maximum concentration factor of about 3000. By varying the distance from the collector to the receiver, the concentration at the aperture can be reduced.

The receiver consists of four modules of quartz glass (fig. 7) each 55 x 55 mm in cross section, with walls 2.5 mm thick. The thickness of the window is 8 mm. The length of the block is 1.9 m. The sides of the block are gold plated, to avoid superheating at the periphery. Into each module 9 tubes are inserted.

At an internal pressure of 8 bar, the maximum tension stress for the glass modules was calculated to be 9 N/mm^2 in the middle of each window. Due to material and design, thermal stresses are estimated to be low. The negative effects due to different thermal expansion of the glass modules and the stainless steel vessel are eliminated by relative motions within the fibrous insulation layer.

So we hope, that the receiver will withstand a long testing period, to answer the questions on material and gas temperatures, mechanical behaviour and the efficiency.

REFERENCES

1. M.A.N., Interatom, Dornier: Abschlussbericht GAST Phase 1
2. Kühn, T.H. Dr. Innovative Receiver Design for a Parabolic Trough and Paraboloidal Dish Collector. Dept. of Mechanical Engineering, University of Minnesota, Minneapolis
3. Proceedings of the Solar Thermal Research Workshop, 7.-8. Sept. 83, Roster

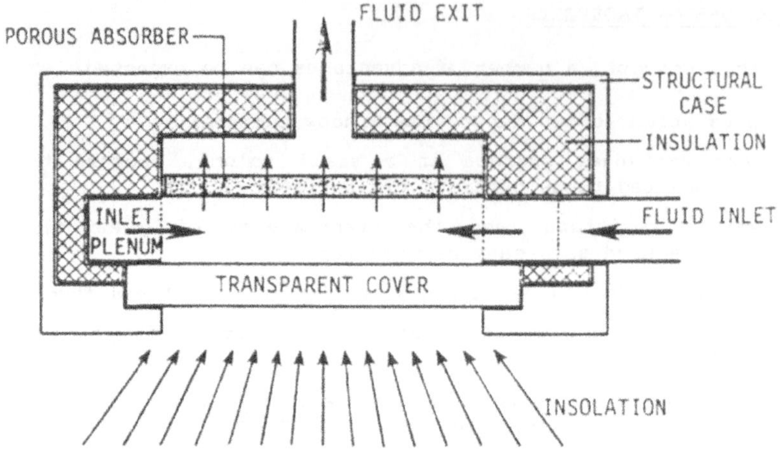

fig. 1: Receiver with an aperture window /1/

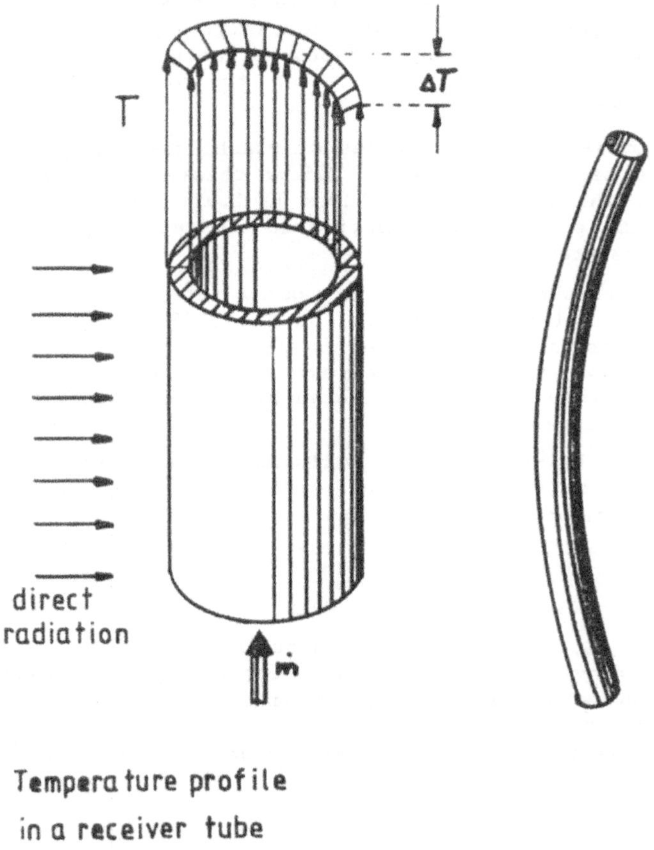

fig. 2: Receiver tube bent by temperature gradient induced stresses.

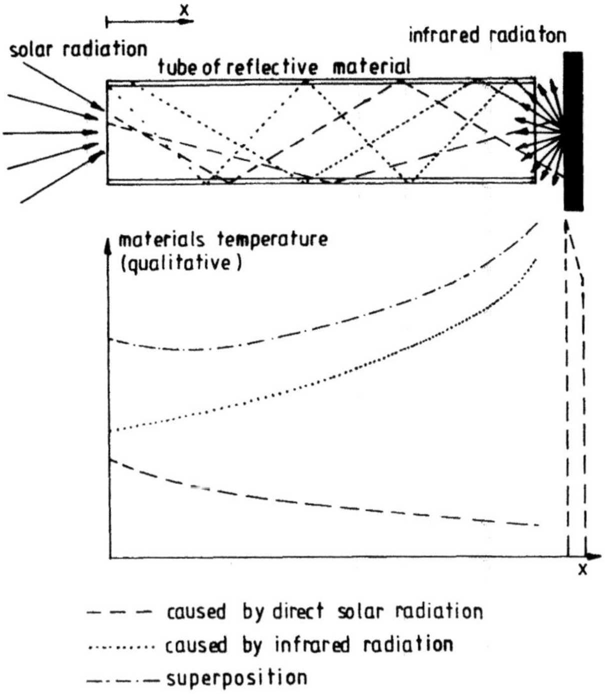

fig. 3: Schematic temperature profiles over the tube length

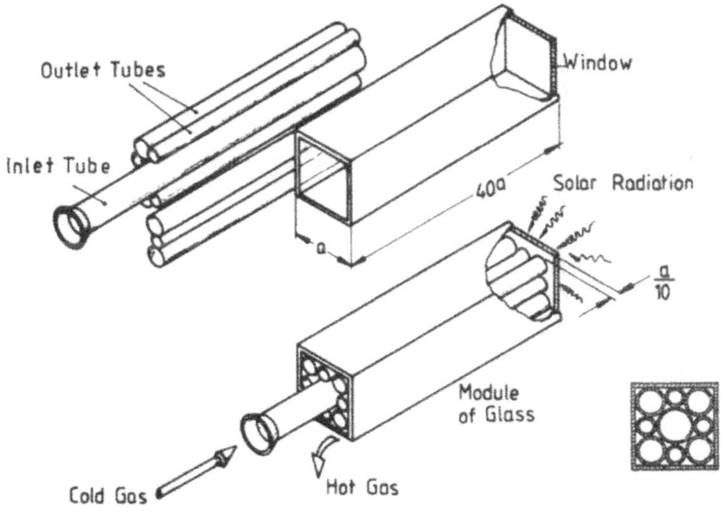

fig. 4: Glass receiver module

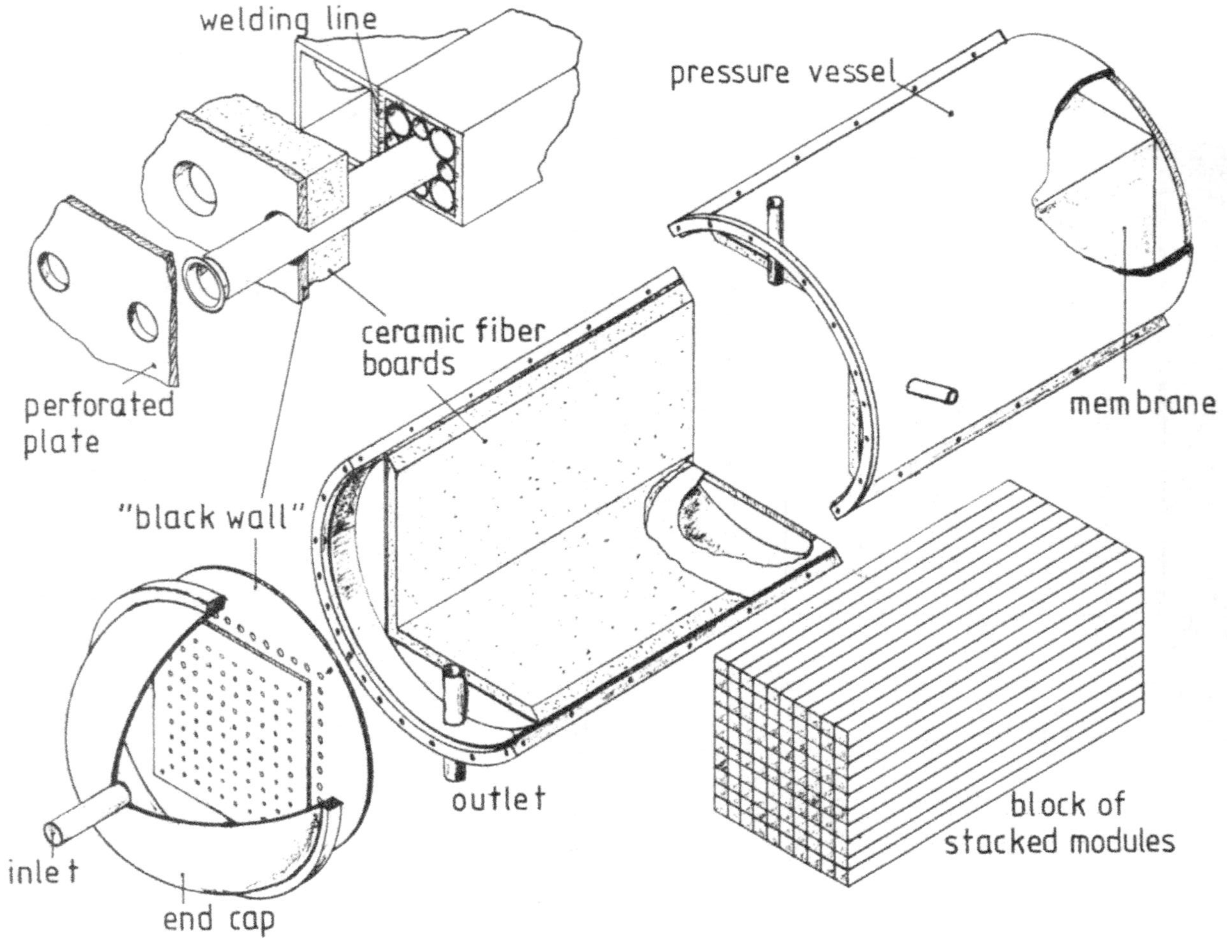

fig. 5: Assembly of the glass receiver

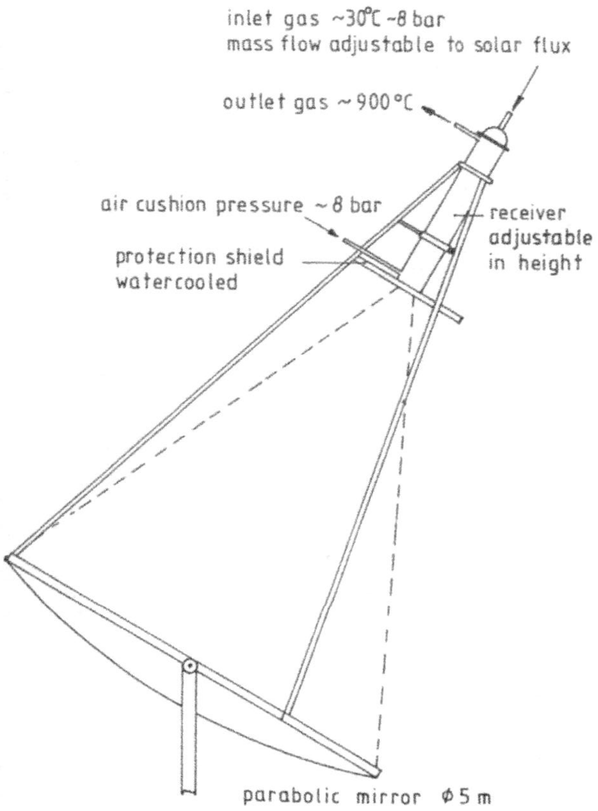

fig. 6: Receiver test facility

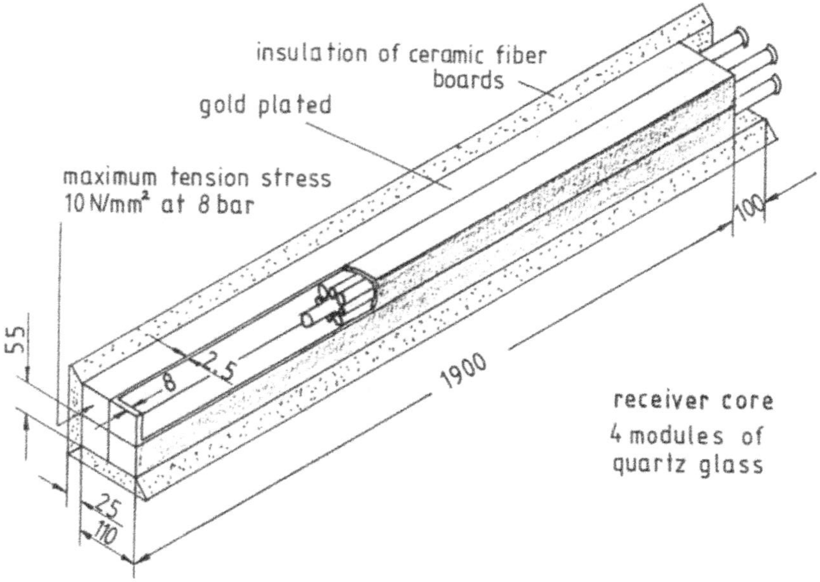

fig. 7: Four module test receiver

The University of Houston Solar Research Program

L. L. Vant-Hull
University of Houston
Houston, Texas, USA 77004

Summary

The University of Houston has been active in the Solar Central Receiver area since 1970 and has had an active program in chemical energy storage and chemical energy transport since 1976. Since 1981 we have operated a Solar Thermal Advanced Research Center for the United States Department of Energy, funding up to seven discrete research projects related to the Solar Thermal Program. The areas of emphasis are analytic model development, high temperature materials and receiver, and fuels and chemicals processes. In this paper I will present only a very brief survey of each task, interested parties can contact the named principal investigators for more detailed information or see reference 1 and subsequent annual reports from the STARC project.†

1. Analytic Model Development

1) Heliostat Field Analysis (Vant-Hull, Energy Lab)

A series of very general computer codes have been developed, applied to numerous design studies, documented and delivered to several users. For a central receiver system these codes provide optical performance and design data, cost effective annual performance optimization, heliostat field layout and performance calculations for flat or multiplane fields coupled to any receiver configuration. Flux distributions on external receivers, aperture planes or cavity interiors can be provided and a two band cavity radiation interchange program is operation.

Our most recent study on high loss receivers was reported at the Georgia Institute of Technology and Stuttgart workshops.

2) Passive Latent Heat Storage (Shamsundar, Mechanical Engineering)

Finite element and finite difference methods have been used to study the freezing and melting of phase change materials on heat exchanger tubes. From this work simple analytic models have been developed which are suitable for solution on a small

† (DOE/SF/11557-1) Solar Thermal Advanced Research Center, University of Houston Annual Report, December, 1982. NTIS, PC A03/MFA01, order number DE 84005709.

computer. These models can handle rapid changes of coolant flow rates and allow effective design of the minimal PLHS unit for a given task, such as buffer storage in the receiver - generator of a dish electric system, etc. Documentation of the computer codes for heat recovery is in progress and the heat storage code is nearly complete.

2. High Temperature Materials and Receivers

1) High Flux Photoenhanced Process (Ignatiev, Physics)

Compared to oven heated degradation processes, many materials are affected by the presence of a solar flux of 1000 to 2000 suns. Surface atoms or absorbants may be desorbed or activated by the solar flux. Under high vacuum conditions clean chromium or iron showed no photoeffects while uncleaned samples desorbed CO_2. Iron heated in air by solar simulated radiation showed enhanced oxidation (photoactivation) over oven heated samples, decreasing from 30% enhancement at 410° C to 8% enhancement at 615° C. In contrast, aluminum shows 30% less oxidation at 1300 suns radiation than does an oven heated sample. This result suggests an enhanced surface (1-3 atom layers) temperature due to the solar flux. These results were reported at the Yugoslavian and SERI materials workshops in 1984. This prediction will be checked by LEED temperature analysis of the surface. Studies of surface catalytic effects will also be initiated in the near future, as will work on other materials used in high flux environments.

2) High Flux Liquid Jet Cooled Receiver (Lienhard, Mechanical Engineering)

A boiling-liquid jet can cool a disc very effectively, removing heat at flux levels of 10 MW/m^2 or more. This project has generated very successful correlations to existing high flux cooling data obtained by the Japanese and others. These correlation studies will soon appear in the Journal of Heat Transfer. Using these correlations, a high pressure water jet apparatus has been designed and constructed to study multijet systems at 100 KW. Cooperative research with the University of Pisa has been established. Extensions to higher temperatures using liquid metals is envisioned for the future.

3. Fuels and Chemicals Processes

1) Design of Receiver Reactors for Solar Chemical-Heat Pipe Applications (Richardson, Chemical Engineering)

A methane-steam reformer (endothermic part of the EVA-ADAM energy transport system) has been designed, constructed and operated in a steady state mode. After validating the mathematical model this one standard liter per second process development unit is now operated in a cyclic mode under computer control to simulate diurnal solar cycles, cloud passages, etc. 15-20 KW of solar simulated radiation is provided to the catalytic reactor bed by IR heating of a 2.3 m long, 5 cm diameter sodium heat pipe. This heat pipe acts as a flux transformer from the 200 KW/m^2 "solar" input to ~ 50 KW/m^2 delivery to the catalytic bed, delivering heat only to the cooled reaction zone. It also serves as a thermal diode, transmitting heat into the bed but not out.

Catalyst stability and coking (deactivation) will be studied during the current cyclic operation tests in which detailed reaction studies will be followed by inlet and outlet gas chromatography using a tracer gas for calibration.

4. Chemical Storage of Solar Energy with the AHS Cycle

Decomposition of Ammonium Hydrogen Sulfate (AHS) produces NH_4OH and SO_3 (or SO_2), which can be stored as liquids at near ambient conditions. The chemicals are cheap and readily available, storage density is high and the energy can be recovered at 500 to 600° C.

1) AHS Decomposition (Wentworth, Chemistry)

Previous work identified a two-step process involving a metal pyrosulfate as the limiting high temperature step for AHS decomposition. A preferred route using $ZnSO_4$ decomposition, which is a common intermediate step in water splitting reactions, has been identified. To avoid the technically difficult decomposition of solid $ZnSO_4$ at 900° C, a molten eutectic was sought. A 50% mixture of $ZnSO_4/LiSO_4$ melts easily and reacts to form ZnO and SO_3. When catalyzed with V_2O_5, the reaction is greatly accelerated. Use of a ceramic rather than metal reaction boat leads to much smoother reactions. The catalyzed reactions occurs at about 750° C with considerable SO_2 evolution. Simultaneous SO_2 and SO_3 quantitative analysis are carried out by reacting 100% of the SO_3 quantiatively with CaC_2O_4 at 325° C to form $CaSO_4 + CO_2 + CO$. Simultaneous analysis for CO_2 and SO_2 is easy and provides a new understanding of the reaction products.

2) Recovery of Energy Stored in the AHS Cycle (Prengle, Chemical Engineering)

For operation in a cogeneration facility requiring energy at several temperature levels, ~ 100% of the stored energy can be recovered via a system employing a counter current flow of liquid sodium at 600° C. The reactor design is critical. First gaseous H_2O and SO_3 are combined in an adiabatic reaction zone to produce the desired 600° C. As the reactants move into the heat exchanger zone, the temperature falls and the reaction goes to completion producing liquid H_2SO_4. In the second reactor AHS (to prevent side reactions) and NH_4OH are introduced into teh H_2SO_4

stream in a second adiabatic zone and again heat is removed as the reaction goes to completion. The detailed differential equations and boundary conditions for each zone have been developed and solutions obtained on a digital computer. A series of optimization runs have led to a substantially reduced size for the tubular reactors. A bundle of 90 reactor tubes surrounded by counterflowing sodium in a heat exchanger shell would be adequate to handle the energy of a 10 MW pilot plant.

stream in a second adiabatic zone and again heat is removed as the reaction goes to completion. The detailed differential equations and boundary conditions for each zone have been developed and solutions obtained on a digital computer. A series of optimization runs have led to a substantially reduced size for the tubular reactors. A bundle of 70 reaction tubes surrounded by counter-flowing coolant in a heat exchanger shell would be adequate to handle the needs of a 1/2 MW pilot plant.

SESSION VI

ROUND TABLE DISCUSSION ON THE "FUTURE OF HIGH TEMPERATURE SOLAR THERMAL CONVERSION"

Summary of the Session by the Rapporteur
F.A. BLAKE, ARCO Power Systems, Littleton, USA

Solid particles solar thermal loop for production of heat at 1000°C

Aspects of future high temperature solar thermal conversion

SUMMARY OF THE SESSION BY THE RAPPORTEUR

F.A. BLAKE
ARCO Power Systems, Littleton, USA

1. INTRODUCTION

Introductory papers by Claude Royere discussing the CNRS experiment to heat sodium salt particles to 1000°C and by Manfred Becker discussing the "State-of-the-Art" and constraining influences opened the session. Mr. Royere described the experiment schematically and physically as it is currently installed at the "Four Solaire" in Odeillo, France. Results to date include a production run test achieving a 6000 kg/hr rate of particle with the 50 KW_t solar powered experiment (a small fraction of the facility power).

Mr. Becker referred to the progress since 1973 when only two solar facilities San Ilario, Italy and Odeillo, France were in existence. He also highlighted that a major role of failures is the provision of fuel for analysis to enable development to progress. He briefly covered the comparative basic characteristics of solar thermal and solar photovoltaic systems which are increasingly being compared at the present time.

2. ROUND TABLE DISCUSSION

The latter part of the session was a round table discussion featuring summary talks by C. J. Winter (DFVLR), N. Ikeda (NEDO), C. Ortiz (INITEC), K.T. Cherian (DOE), F. Pharabod (AFME), J. Gretz (JRC), and C. Etievant (GESER).

2.1 C. J. Winter

Both project and R & D activities supported by the German government and in cooperation with other governments have been under way for over 7 years and will be strongly continued. Project expansion at Almeria and component developments (heliostats and receivers) are near term emphasis. Longer term is focused on storage and portability of stored energy. In-house DFVLR research has continuing activities in innovative receiver development, receiver loss distribution definition, heliostat field optimization, and secondary energy applications (such as LH_2 automobile).

2.2 N. Ikeda

There are no new plant activites at present in Japan. Research emphasis is shifting toward photovoltaic electric generation and industrial process heat. Enthusiasm continues due to the non-polluting character of solar energy.

2.3 C. Ortiz

Spain is now involved in SSPS, CESA-1, and GAST. The drive in current projects is to increase their KWh/yr productivity. R & D activities on heliostats and receivers continues strong. Technical/economical analyses focus near term applications with a perception that re-powering will be selected. Efforts to stimulate private sector participation are under way and international cooperation will continue. The "Solar Platform" at Almeria is essential to a healthy solar program.

2.4 K.T. Cherian

The U.S. near term work will feature continued support of the 10 MW_e pilot plant near Barstow and of the Molten Salt Electric Experiments and emphasis on automatic controls development.

Near term areas of research interest include Design of Advanced Electric Systems, IPH application, Fuels and Chemicals, and Innovative Component Development. The latter includes receivers heating particles and carbonate salts and stretched membrane heliostats.

2.5 F. Pharabod

Intense testing with the THEMIS Plant to achieve improved performance and generate an experience base is the basic major thrust of the French program. The need exists for a full scale 50-100 MW conceptual design to be built with international support.

2.6 J. Gretz

Eurelios is deficient in performance for a variety of reasons and will be modified to correct performance. The program will feature summer operation. JRC planning includes a PV experiment conceptual design.

2.7 C. Etievant

"Do not Abandon the Solar Thermal Electric Goal!" We currently see a wide range of component cost reduction successes. We must initiate a full scale power plant supported by cooperating nations.

SOLID PARTICLES SOLAR THERMAL LOOP FOR PRODUCTION OF HEAT AT 1000°C

C. ROYERE
Solar Energetics Laboratory
CNRS (ODEILLO)

Summary

The experiment presented concerns the evaluation of solid particle materials as a medium for direct conversion of focused solar radiation into heat in the range of 1000°C .

INTRODUCTION

Applications of high and very high temperatures solar heat have been demonstrated with the 1 MW_t solar furnace .

Solar plants presently in operation and under evaluation for power generation through conversion of solar thermal energy use fluids at medium and medium high temperatures .

For high temperatures in the range of 1000°C industrial applications of solar thermal energy solid particles of appropriate materials and the use of direct coupling the radiant energy to this type of heat carrier show very attractive features for evaluation of possible future uses .

DESCRIPTION OF THE SOLAR THERMAL LOOP

The main subsystems of the experiment are the following : (fig.3).

- receiver : mixed bed rotary kiln
- storage : 2 tanks (one hot, one cold)
- heat exchanger : multistage fluidized beds (4 stages)
- transport : upward by pneumatic conveyor
  downward by gravity

The solid particles material used presently in the loop is ordinary sand (particles size : between 400 and 800 µm).

The medium for unloading the heat from the loop in the heat exchanger is air .

Design characteristics of the loop have been established as follows :

- temperatures : - solid particles : 1000°C
  - air outlet from heat exchanger : 800°C
- power : 50 KW (nominal design of the heat exchanger)
- storages capacities : - hot tank : 700 kg
  - cold tank : 1000 kg

PRELIMINARY EXPERIMENTAL DATA

The thermal loop in under evaluation. It is currently operated over long periods up to 8 hours in a row matching the available clear days sunshine hours .

The main results concerning the different subsystems of the experiment are as follow :

- receiver :
  . steady outlet temperature around 950°C
  . efficiency of the receiver is to be determined in the next phase of operation
- storage tank :
  . efficiency for constant level continuous operation : above 95%
  . thermal losses after loading : 5 to 10°C per hour
- heat exchanger
  . air outlet temperature of 750°C is achieved for steady state operation
  . stairsteps diagramme of temperatures (air,solid particles on the different stages) is in good agreement with prediction
  . efficiencies of 85% and above are achieved at power levels between 50 and 75 kw.

PROGRAMME OBJECTIVES

The general objective is to investigate the feasibility and conditions of direct conversion of radiant energy to thermal energy at temperatures level in the range of 1000°C for high temperature applications of solar energy .

The short term specific objectives for this experiment are :

- optimization of operating conditions (steady state) at maximum output (temperatures : sand 950°C, air 800°C and power : up to 150 KW on the present heat exchanger)
- determination of the loop performances
- investigation of the loop under transcient operating conditions to design automatic control
- learning to use of the storage as a buffer and getting data for designing larger storage .

REFERENCES

1 . C.ROYERE , Solar Chemical Engineering research and test activities using the 1 MWth Solar Furnace at CNRS , Solar Thermal Test Facilities Users Association, Facility operators and Experimenters Workshop, May 3-4,1979 , Albuquerque NM, Proceedings pp. 24-29 .

2 . A.HASSAN,C.ROYERE,JL.TUHAULT, LES CNRS/C.LAGUERIE, M.PEYMAN IGC TOULOUSE, Evaluation of multistage fluidized bed heat exchanger to transfer thermal energy at high temperature to a gas from a granular solid storage heated by concentrated solar radiation, Proceedings of annual meeting STTFUA, April 15-17, 1980, Las Cruces NM, pp.145-156 .

Fig.1 General view of the $1MW_t$ Solar Furnace

Fig.2 The receiver in operation at the focus (viewed from the concentrator)

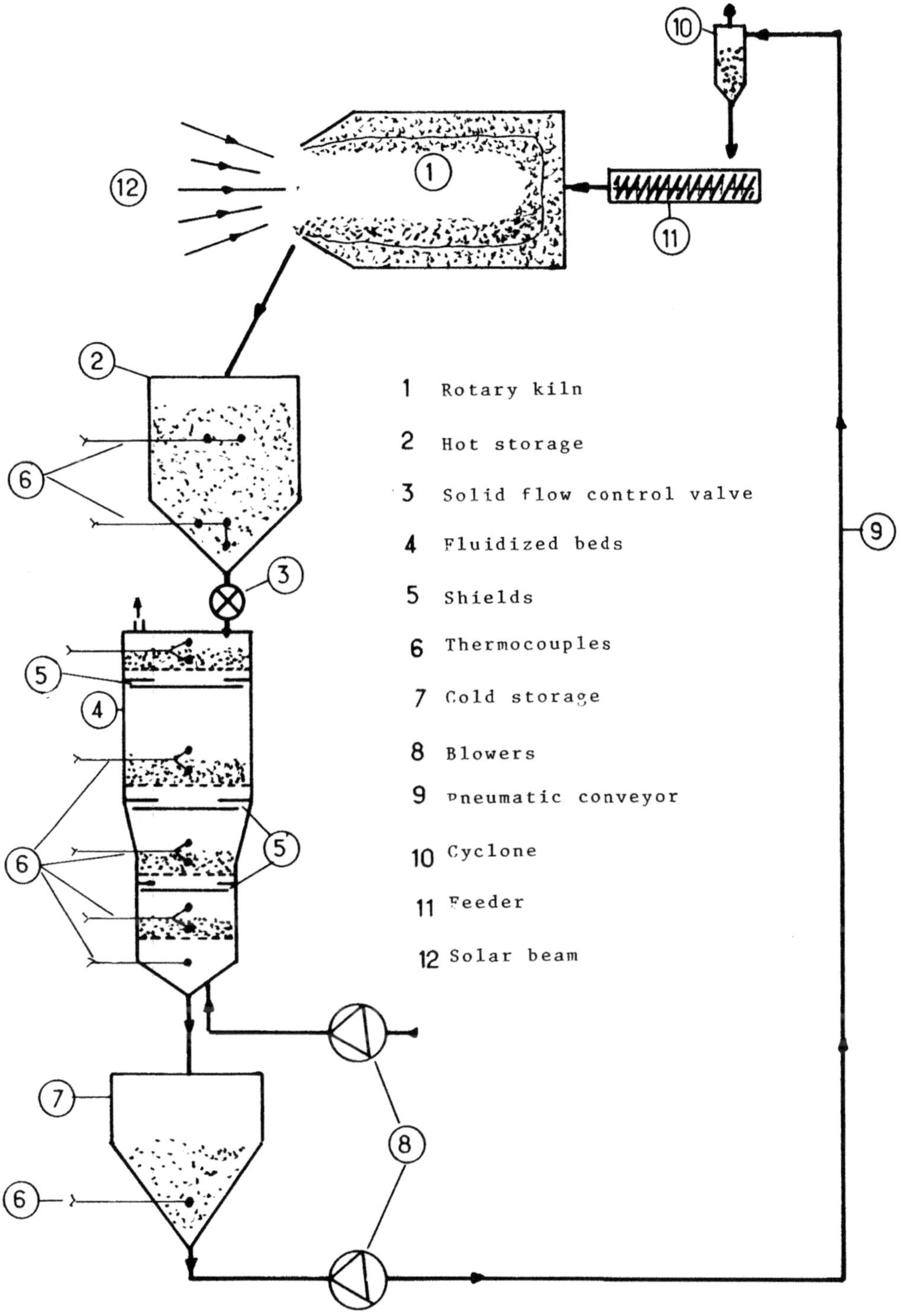

Fig. 3 Schematic diagramme of the loop .

Fig.4 General view of the loop

ASPECTS OF FUTURE HIGH TEMPERATURE SOLAR THERMAL CONVERSION

C.-J. Winter, M.Becker
DFVLR - Stuttgart

Summary

Some short descriptions on status and achievements of the solar-thermal high-temperature projects are given. The idea of using in alternative to heat and electricity a chemical carrier for energy storage and transport purposes is developed. Also solar-thermal characteristics are compared with photovoltaic features. Possible heat applications are mentioned and outlined concerning the use of radiative and thermal energy. As example the solar EVA process, the steam reforming of methane is described and possible end products as steam, synthesis gas, hydrogen, methanol, and ammonia are specified.

1. STATUS AND ACHIEVEMENTS

Almost ten years ago, after the energy crisis had questioned our economic structures, activities for far reaching use of solar energy were started. Beside research investigations important projects took shape. Except for the pioneer facilities at San Ilario, Italy and Odeillo, France, they all contributed as pilot power stations to the development of further electricity supply. The present Ispra-Workshop as well as its predecessor at Claremont in 1982 were organized to give a summary of the status and an account of the achievements.

Table 1 presents a survey on the central receiver power stations in the world. Only the main characteristics are given. More details can be drawn from the special publications of the workshop. The data of the Russian plant were taken over from "Entropie", No. 103, 1982.

Encouraging results towards commercialization are known especially from larger facilities. Difficulties and failures were reported on various subjects, surprisingly on many conventional items.

The situation must not be disturbing; modification and progress shall be the answer. All facilities had been built to learn more about them. Therefore, problems and errors do inhibit an extremely positive view. They stimulate thorough solutions or reconsiderations of alternatives. At any time, they shall result in evident improvements, both technically and economically.

These elements should be employed to define the trends of future research and development in high-temperature,solar-thermal conversion as outlined below:

A) The power conversion from thermal to electrical energy caused substantial delays and is still very cost-intensive. The consequence must be to look firmly and carefully for better solutions and/or to use in a new attempt the radiative/thermal energy in a more direct way for endothermal chemical processes.

B) The storage problem is still an open issue. The technical solutions do not yet fulfill the requirements that originate from meteorological conditions and operational necessities.

Chemical storage in connection with suitable reversible processes could be combined with the idea of chemical energy transport purposes over short and/or long distances.

2. IDEA OF CHEMICAL CARRIER

In order to find out whether and why a chemical energy carrier is needed, let us pursue the following line of argu-- ments:

- o From the very beginning, solar energy research and development was entirely devoted to solar heat production, solar thermal electricity generation, and photovoltaic electricity generation.
- o Industry has begun to commercially install solar heat production facilities and solar-electric power stations.World-wide are around 100 MW (together with wind energy converters some 100 MW) in operation.
- o Almost in any case the generated heat is used in a nearby facility (cooling process, industrial production heat a.o.); the electricity is fed into an existing grid which serves as frequency constanter, storage device and energy transport vehicle.
- o Statistics for industrially needed process heat provides us with a temperature curve of a "camel back" type with distinct peaks at 100-200 °C and 900-1200 °C respectively; the lower peak excellently meets the temperature (and pressure) and unit capacity specifications of one axis tracking longitudinal paraboloids with typical maximum temperatures of 300-400 °C and maximum unit capacities of less than 1 MW_{th}.

  A system analysis, by the way, of the mediterranean countries on both the European and African side (without France) shows that substantially more than 50 % of the overall primary energy demand is process heat below 300 °C and that 90 % of the heat production facilities have a unit capacity of less than 1 MW_{th}. On the other hand the high-temperature domaine exhibits a much smaller number of units with, however, significantly higher capacities (10-100 MW_{th} each) which only can be met by the specifications of point focussing solar tower type converters. This is a reasonable solution on thermal terms , only if the solar converter and the industrial heat user are in an acceptable vicinity to each other. Otherwise the heat losses become too high.
- o For even more elevated temperatures around 1000 °C point focussing parabolic dishes with fairly small unit capacities of 150 kW_{th} or assemblies of them are available up today only in experimental versions. The technology is not mature, yet; the potential of the highest solar-electric efficiency of 30 % (measured) of all solar energy converters, however, fully justifies further intensive R+D measures.
- o Centralized facilities of medium to large capacities as needed for the energy supply of industry complexes can only be point focussing solar towers. Multi-tower arrangements for modularized high capacity facilities are also viable.
- o Generally, these facilities are too large as to expect that the transport of the generated secondary energy carrier "heat" or "electricity" could be economic. What is needed in this case could either be a solar fuel which is longtime stable and efficiently transportable over long distances, or it could be a solarchemical, directly manufactured on site with the solar irradiance coupled into the endothermic chemical reaction.

3. SOLAR-THERMAL OR PHOTOVOLTAIC

On the question of solar-thermal (STEC) or photovoltaic (PV) energy conversion the following considerations should be taken into account:

- o STEC needs direct radiation for conversion; hence STEC is not utilizable in world regions with high diffuse light fractions.
  STEC has the potential for both the production of heat and electricity in heat/power-units.
  STEC has the potential of direct conversion of solar irradiance into chemicals and fuels.
- o PV uses without preference direct as well as diffuse solar radiation. Especially in world regions with traditionally temperate climates, PV is operable when because of lack of direct light, there is no chance for STEC.
  PV directly converts light into D.C. electricity by means of semiconductor cells. There is no need for heat exchangers, turbines and rotating electric generators.
  PV's overall conversion efficiency is less than half of STEC's, both with today's technology (for example: Solar Tower 10-25 %, Parabolic Dish = 30 %, PV 8-15 %).
- o For process heat production within the whole industrially applicable temperature range there is a distinct preference for STEC, since STEC's first (out of two) conversional step is the conversion of irradiance into sensible heat.-The same is true for solar-chemical conversion in one step,namely the direct conversion of solar irradiance within a chemical reactor in order to obtain and maintain a chemical reaction.
- o For solar electricity generation both STEC and PV are applicable. PV is still capable to produce electricity even under conditions when only diffuse light is available; this holds although PV's conversion efficiency is low and the investment cost is high. STEC - on the other hand - yields electricity only if there is concentrable direct light of a certain intensity level available; the conversion efficiency is significantly higher and the cost for the conversion facility is lower than in the PV-case.
- o Since the electricty market offers - regionally depending - higher or lower diurnal and seasonal market prices for the kWh, the question is which conversion system yields more marketable kWhs, especially at high price time periods. It is not easily predictable whether this is STEC or PV since such dissimilar calculation factors like solar offer over time,conversion efficiency, cost of investment and maintenance a.o. do significantly influence the cost-calculation of the kWh offered on the electricity market.
- o Both secondary energy carriers of STEC and PV - heat and electricity - should preferable be consumed on site. If long storage time periods and long transportation distances between power station and consumer facility are to be overcome, another - chemical - energy carrier is needed.
- o PV - and STEC - electricity can electrolytically be converted into hydrogen. The advantages of this conversion are threefold:

- The substrate (water) is ubiquitous and plentiful.
- The driving force (sun) is unlimited, however, intermittantly available.
- The product (H_2) is chemically stable and non-polluting; it stems from water and recombines into water when combusted.

o STEC furthermore offers the possibility to directly convert solar irradiance in a chemical reactor into storable and transportable chemical seondary energy carriers. A solarized methane steam reforming process is imaginable; also thermo-chemical hydrogen generation with concentrated sunlight can be used as continuous energy source.

The conversion of electricity into hydrogen by means of electrolysis is not really new, the electrolytically produced hydrogen does not care for the origin the electric current stems from.- However, two things must be thoroughly taken care of:

- the dynamics of the combined process of the solar electricity converter/electrolytic cell must be matched
- the detail efficiencies of the various steps of the multi-step energy conversion system must be high, the cost of the various elements needed must be low, in order to come up with an agreeable overall cost for the stored and transported kWh.

o Coupling of concentrated solar irradiance into a continuous chemical reaction, however, is - to our knowledge - nowhere in the world off laboratory status. Questions arising in this context are:

- How can the principally different dynamical behaviours of solar radiation and endothermic chemical processes be matched? Is it realistic to think of a single loop process? Or is it imperative to transform the solar energy into a sensible heat transfer medium in a first step and consequently in a second step use the heated transfer medium in the chemical reactor?
- Is direct coupling of solar energy into a solid transfer medium in a reactor without transmittant walls a promising solution? Of course, in such a case we would not need walls of high temperature resistant materials; but we do, consequently have to face atmospheric processes, which some times result in bulky geometries of the respective reactor designs.

4. POSSIBLE HEAT APPLICATIONS

The role of solar energy shall be discussed here in concern to the radiative and thermal character, not in respect to biosynthesis or photolysis . So the most favorable use of concentrated sunlight would be the immediate application of radiation. Quite a few investigations deal with this interesting and at the same time difficult topic. Also and probably in a more straight-forward way a widespread employment of thermal heat inputs is viable for well known events and processes.In any case, the time dependent availability of solar radiation has to be taken care of either by hybrid operations or by the choice of interruptable processes.

A few of the possibilities mentioned are outlined below.

Radiative heat (heat transfer to solid particles):

- o Sand as basic heat and mass transfer material
- o Calcination of limestone
- o Fluidized bed technology for lignite.

Thermal heat (heat transfer through walls):

- o Phase changes
- o Provision of thermal portion to high temperature vapor electrolysis for hydrogen production
- o Thermo-chemical conversions for hydrogen production reactions, steam reforming of methane, pyrolysis of coal or ethylene from ethane, heavy oil cracking and others.

Finally the purpose to produce solar fuels and chemicals can serve various intentions, for instance as means for

- o Propellants (hydrogen)
- o Storage and feedstocks (hydrogen, methanol, ammonia)
- o Transport (hydrogen, EVA+ADAM-system).

5. SOLAR EVA PROCESS:AN EXAMPLE

As an example for an attractive process the solar application to the KFA/Rheinbraun EVA+ADAM process-system shall be mentioned.The original idea is to verify the process conditions as described in Tab. 2 in a closed system. The EVA-part absorbs the solar radiation by the endothermic process of steam reforming of methane (look up Fig.1). A mixture of hydrogen and carbon-dioxide can be transported in cold state to the location of thermal energy consumption. Here, in the ADAM part, the exothermic methanization process takes place. Methane and water vapor - again in cold state - can be transported back to the reformer.

Presently, the solar energy induced steam reforming process and its application to the production of useful chemical feedstock are under intensive investigation.Figs.2 to 4 shall serve as examples of how to organize a system with natural gas and water to produce methanol and hydrogen (Fig.2), hydrogen and steam (Fig. 3), and ammonia (Fig. 4).

6. THINGS TO BE DONE

Since we are at the begin of a development many ideas are at stake. So are the ways for realization. An analysis of what has to be done in future and a strategy to achieve valuable results in a given time under certain restrictions seems to be mandatory. For that reason a few thoughts shall be described:

- o The international solar-thermal energy R+D community needs the perpetuation of continued exchange of views, results, thoughts, and information. What has begun with Claremont/Cal. and was set forward with Ispra/Italy is highly worthwhile to be further pursued.

- o The existing facilities should be improved. Problems should be solved in a thorough way; subsystems could be oriented towards the know-how of mass fabrication (heliostats) or optimized due to highest material standards and longtime experience (receivers). There are many rewarding tasks.
- o Within PV- and STEC-R+D-efforts during the past 10 years, there is one key element still lacking which also was by far not as intensively taken care of as it should have been; solar-chemical energy conversion as a means for making solar secondary energy carriers longtime storable and transportable over longer distances.
- o Chemical endothermic reactions need energy. If solar radiative energy could be directly coupled into a chemical reaction the results were of extraordinary importance for all those regions of the world where indeed sufficient solar radiation over time and adequate land areas are available, however, no conventional energy infrastructure. Again, this field of solar-chemical energy conversion,i.e. the field of "solar chemicals", as that of the a.m."solar fuels" is by far not as thoroughly researched as it needs to be before we are in a position to give a competent judgement on solar fuels and chemicals production capabilities.
- o Since the scientific results of the appropriate solar-thermal/solar-chemical research work is only conditionally scalable, analytical and laboratory work needs verification under field test specifications. Thus, solarthermal/solarchemical test facilities are to be maintained in order to provide researchers and industry engineers with equipment necessary to test and evaluate solar components under realistic conditions.

  IEA-SSPS-Almeria could be one of these facilities needed, versatile in lay-out and measuring equipment and since a couple of years successfully managed and operated for and on behalf of a truly international clientel.
- o High temperature applications is one of the main issues of future solar thermal conversion. Like pursued in the German-Spanish-GAST-Project (Gas cooled solar tower project) (see Table 3) there is a wider R+D-committment to high temperature investigations necessary.
- o As long as the technology is far from being marktable the endeavour should be internationally pursued, because of the obvious and trivial reasons of cost sharing, information exchange and the avoidance of duplicated work.

Name Location Country	SSPS (CRS) Tabernas/Alm. Spain	SUNSHINE (CRS Nio/Shikoku Japan	EURELIOS Adrano/Sicily Italy	CESA-1 Tabernas/Alm. Spain	THEMIS Targasonne/ Pyrenees France	CES 5 Lenino/Crimea USSR	SOLAR 1 Barstow/Calif USA
Sponsor Designer	IEA INTERATOM MMC; SAIT	AIST EPDC	EEC ANSALDO, ENEL CETHEL, MBB	CEE CASA, SENER EISA, INITEC, TR	AFME EdF CNRS	MEE	DOE, SCE MDAC, MMC T+B, SCE
Performance Net El. Power MW	0.5	0.8	1.0	1.0	2.3	5.0	10.0
Start of Design End of Construction Start of Operation	Mar. 1978 Sep. 1981	Jul. 1977 Sep. 1981	Nov. 1978 Jun. 1981	Jun. 1977 Jun. 1983	Jul. 1979 Aug. 1982	1985 ?	Mar. 1979 Apr. 1982
Heliostat Field Manufacturer Heliostats Single Area m^2 Total Area m^2	MMC 93 39,3 3 655	MITSUBISHI 807 16 12 912	CETHEL/MBB 70/112 52/23 6 216	CASA, SENER 300 39,6 11 880	CETHEL 201 53,7 10 740	1 600 25 40 000	MMC 1 818 39,3 71 447
Tower Height Receiver Centre	43,5	60	55	60	106	89	80
Receiver Manufacturer Type Fluid Outlet Temperat. °C Outlet Pressure bar	SULZER/FT Cavity/External Sodium 530 3	MITSUBISHI Cavity Water/Steam 249 40	ANSALDO Cavity Water/Steam 512 62	Cavity Water/Steam 525 108	CNIM Cavity Hitec Salt 450 1-2	External Water/Steam 250 40	ROCKETDYNE External Water/Steam 516 105
Storage Fluid Time h	Sodium 2	Press. Water 3	MoltenSalt/Water 0.5	Hitec Salt 4	Hitec Salt 5	Press. Water 3	Caloria HT43 4

Table 1

Main Data of Solar-Thermal Power Plants

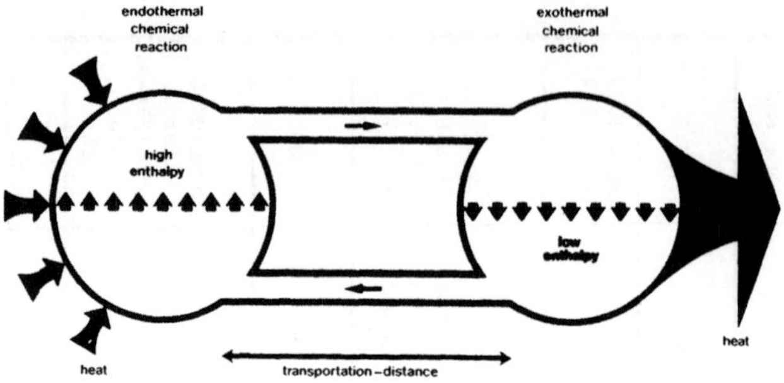

Fig. 1
Principle of Energy Transport by a Cyclic Thermo-Chemical Process (EVA and ADAM)
Source: R. Harth, KfA, 1983

$CH_4 + H_2O \rightleftharpoons 3H_2 + CO \ \Delta H^\circ = 206$ kJ/mol

- Parallel reaction:
  $CH_4 + 2H_2O \rightleftharpoons 4H_2 + CO_2 \ \Delta H = 165$ kJ/mol
- Process conditions:
  - endothermal process:
    feed gas $H_2O/CH_4 \geq 2/1$ (molar)
    $\triangleq$ H/C/O = 8/1/2
    temperature range: $450 \leq \vartheta \leq 850\,^\circ C$
    pressure range: $1 \leq P \leq 50$ bar
  - exothermal process:
    temperature range: $250 \leq \vartheta \leq 650\,^\circ C$
    pressure range: $18 \leq P \leq 80$ bar
- Properties of reactands:
  - unlimited available
  - no corrosion problems
  - no toxicity problems
  - gaseous and liquid

Table 2
Process Conditions of EVA and ADAM Process
Source: R. Harth, KfA, 1983

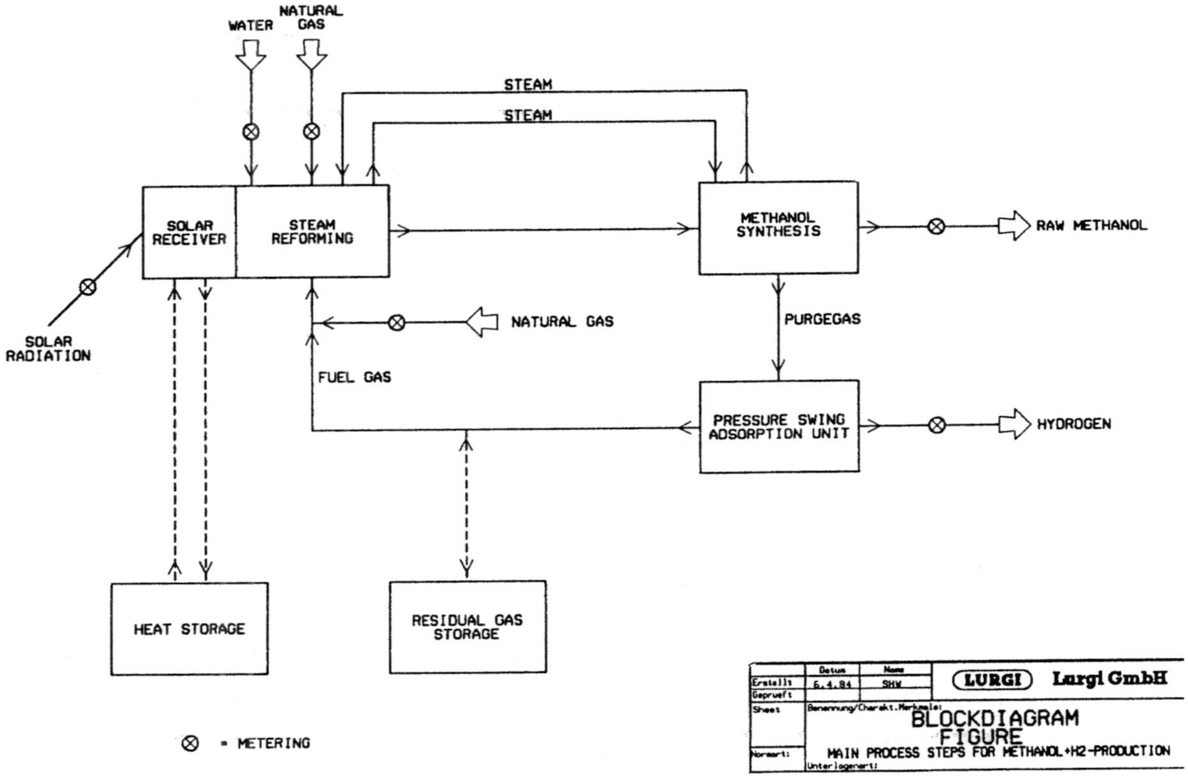

Figure 2

Block Diagram of Solar Steam Reforming Process with Methanol and Hydrogen Production
Source: Lurgi Co., 1984

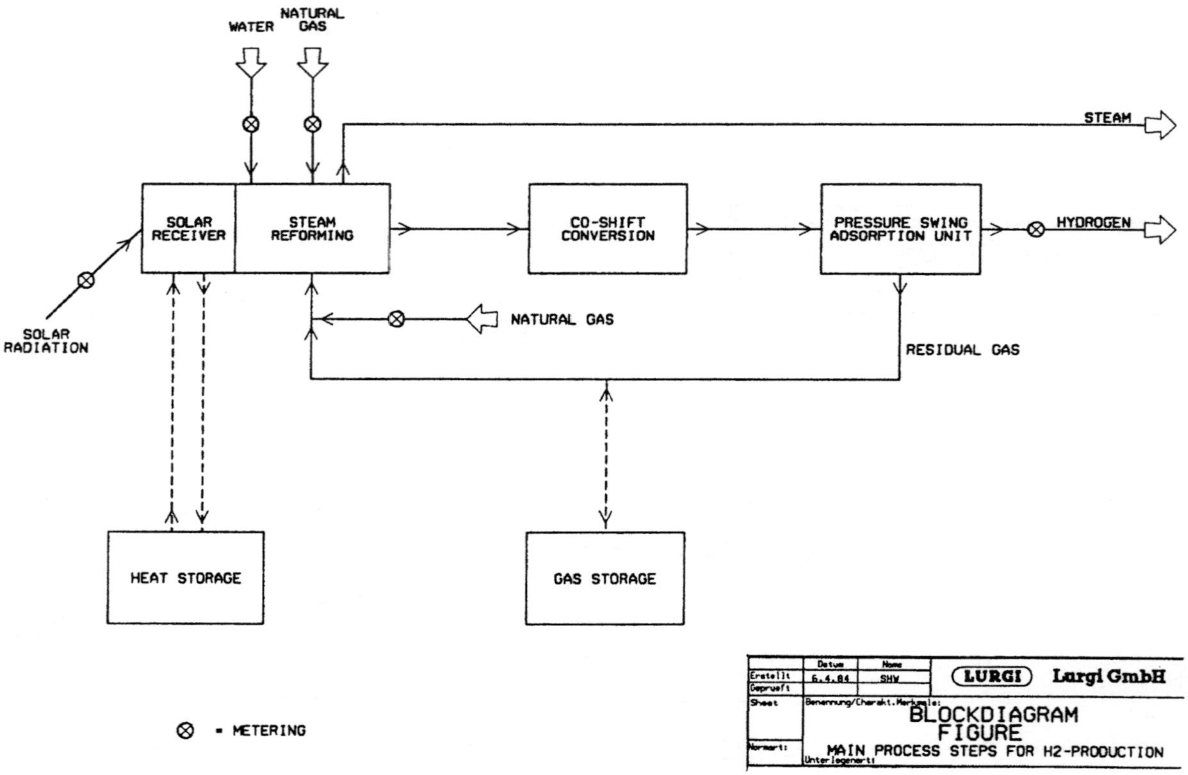

Figure 3

Block Diagram of Solar Steam Reforming Process with Hydrogen and Steam Production
Source: Lurgi Co., 1984

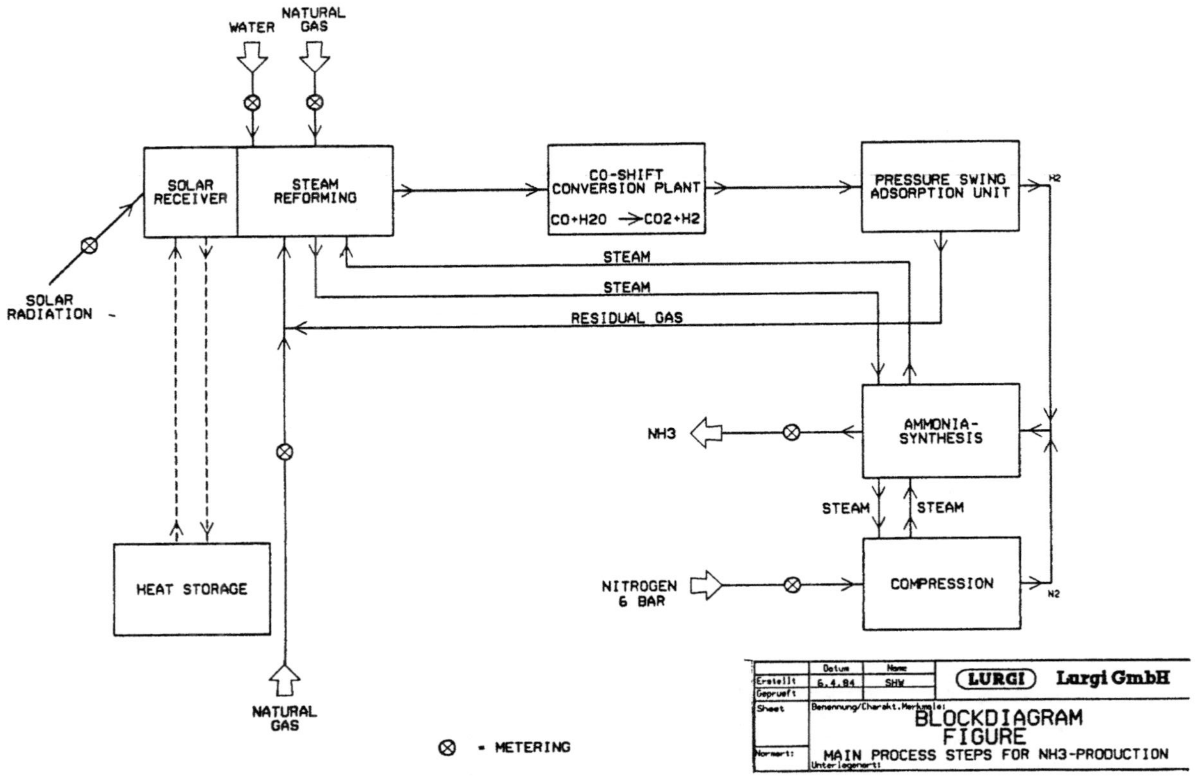

Figure 4

Block Diagram of Solar Steam Reforming Process
with Ammonia Production
Source: Lurgi Co., 1984

- o Consideration on System
  - Mixed Gas/Steam Cycle for Electricity Production
  - Medium Gas (Air, Helium)
    Hybrid Operation with Fossil Facility
  - Repowering and Cogeneration (Process Heat)
  - Optimal High Temperature Range
    Radiation/Thermal Cycle Interaction
    Chemical Process Requirements
  - Adaptation of Chemical Reaction to Solar Thermal Conditions
  - Commercialization

- o Optimization of Heliostats
  - Size and Concentration (Image)
  - Stiffness and Elasticity
  - Economoy and Mass Production

- o Selection of Receivers
  - Gas Leads to Cavity Version and High Concentration/
    High Cost Heliostat Requirements
  - Sodium/Salt Results in External Type and Low
    Concentration/Low Cost Heliostat Requirements
  - Configuration Optimization
    Tubing, Heat, Pipe, Dome, Volumetrics, Particles
  - Materials
    Metals, Ceramics

Tab. 3
Future High Temperature Aspects
(GAST - Technology Prospects)

SESSION VII

PRESENTATION OF NEW CRS PROJECTS

Summary of the Session by the Rapporteur
L.L. VANT-HULL, University of Houston, USA

The gas-cooled solar tower project "GAST"

Central receiver research facility at the Weizmann Institute of Science

Molten salt electric experiment

Solar thermal central receiver pilot plant overview - Part II : A utility perspective

SUMMARY OF THE SESSION BY THE RAPPORTEUR

L. VANT-HULL
University of Houston, USA

1. INTRODUCTION

At this session, papers were presented which described five of the nine active solar central receiver projects (besides the six currently operating pilot plants, the 5MWt test facility at Albuquerque, the 1 MWt solar furnace at Odeillo and the numerous smaller test facilities). These projects vary widely in size, objective, project phase, working fluid cost, funding status, etc. In Table I, we list these nine plants in order of current design maturity (as perceived by the Rapporteur), along with the main parameters describing each. However, it must be recognized that important design parameters are subject to change until a final design is under way. Important information is currently being developed by the various operating CR systems, by special heliostat and receiver tests, by component tests, and by computational modeling. Thus, reduced heliostat costs or more cost optimal designs can lead to reduced tower heights; the surprisingly high loss figures and high costs derived for cavity receivers can lead to the adoption of external receivers, either flat plates or cylinders (as used effectively in Solar One); etc.

2. STATUS OF PROJECTS

The most obvious problem facing most of the projects (3 through 9) at the moment is the inability to show economic feasibility. These demonstration projects, using pre mass-production facilities and having significant engineering costs and allowance for contingencies, will cost 30 to 50% more than their value. Numerous approaches to covering this excess cost have been evaluated, but the economics do not seem to be there. Tax credits are inadequate or too short lived, venture capital requires an inordinate return on investment, no single company has a strong enough patent position to gamble on the future market, utilities must protect their stockholders, and utility commisions will protect the ratepayers. Consequently, a period of stagnation seems imminent. Several governments seemed inclined to develop advanced programs and to take action designed to reduce future contingency allowances or to reduce the perceived technical risks. To date, however, there seems to be little tendency of any government to take the more effective route of cost sharing, sharing the excess costs of the demos, or of offering no--or low--interest loans to support the development of the alternative energy source.

Table I

#	Main Participant	Name	Site	Status 6/1/84	Objective	Power (MW) Thermal/elec	Receiver Configuration	Focal Height	Peak Flux Limit MW/m^2	Storage Mat./Hrs
1	DOE, EPRI Utilities & Solar Ind.	Molten Salt Electric Expt	5 MW CRTF Albuquerque, USA	Operational System Test in progress.	Verify advanced receiver concept. Provide US operating exper. with molten working fluid.	5/.75	Cavity	65 m	.6	salt/3
2	Weizman Institute	Weizman Institute CR Research Facility	Institute Campus Rehovat, Israel	Advanced Design Start constr. Fall 84, Project funded	National Research Facility in High T Chemistry and High Intensity Photochemistry	3/NA	Test Areas	45-38 m	3	NA
3	Pacific Gas and Electric ESG of Rockwell	Solar Two "Carissa Plains"	100 Km North of Santa Barbara, CA	Final Design 50% complete. Technology available.	Demo for commercial sodium advanced receiver system.	118/30	Flat Plate	122 m	1.2	Sodium 1.1
4	So. Cal. Ed, McDonnell Douglas	Solar 100	Lucern Valley 50 Km south of Solar One.	Preliminary design complete (12/83) Component development (salt receiver) in progress (test in 85)	Demo for commercial molten salt advanced receiver system	320/100	Semi-cavity	217 m	.85	Salt/.5 (2nd field-6)
5	DFVLR Interatom MAN, MBB	GAST	Solar Platforma Almeria, Spain	Preliminary sys redesign in progress. Component devel.	Develop High temp gas receiver technology. Test metal (800C) panel til 6/85 and ceramic (1000C) panel til 6/86.	75/20	NE and NW facing cavities	(200 m)	.1	None. Hybrid gas combustor in parallel
6	Arizona Public Serv. Martin-Marietta Babcock & Wilcox	Seguaro*	Arizona 43 Km NW of Tucson, Az	Preliminary design. Complete 9/83	Demo for commercial salt advanced receiver system	190/66	C shaped cavity	165 m	.53	Salt/4
7	El Paso Light and Power. Martin-Marietta	Newman*	24 Km NE of El Paso, TX.	Preliminary design. Complete 12/83	Demo for commercial advanced water-steam receiver system	112/41	External hemi-cylinder	152 m	.37	None (Hybred)
8	Cambrian Eng. Babcock & Wilcox	20 MWe CR* Plant	South Jamaica	Pre-feasibility study. Complete 6/84.	Demonstrate small molten salt. Peak shaver to reduce oil imports.	83/20	Cavity	107 m	.53	Salt
9		Russian* CR system	Crimea (?)	Conceptual design. Complete 1977.	Demonstrate CR technology	50-5/?	external?	?	?	?

*Not represented at workshop.

Projects 1 and 2, which are primarily designed as test and research projects, are currently funded and well under way. Solar One and Themis are currently scheduled for 2-3 years of systems operation and evaluation while Cesa Uno, Nio, IEA, and SSPS will be operated primarily for special tests and evaluations for a year or two. Government funding is available for continued work and component development on projects 3 and 5, while projects 4, 6, and 7 are essentially on hold until rational financial arrangements can be made. Project 8 is in a very preliminary phase, while the status of Project 9 is difficult to determine--conflicting reports read from "nearly started" to "nearly completed."

3. CONCLUSION

Prospects for continuation of the work of the last 12 years to successful commercialization are good if one or more projects in the 100+ MWt range are funded during 1984. If no final design and construction work is authorized in 1984, it is likely that most of the currently existing expert design teams will disintegrate, severely affecting the future of the entire central receiver program. Continued governmental support for component development, risk reduction, and cost reduction programs offers hope that in a few years, when a need for alternative energy systems again rises, the central receiver technology will be considered a low risk option.

THE GAS-COOLED SOLAR TOWER PROJECT 'GAST'

P. Wehowsky, D. Stahl
INTERATOM, Internationale Atomreaktorbau GmbH

J. de Marcos, L. Crespo
ASINEL, Asociación de Investigación Industrial Eléctrica

Summary

A survey of the 20 MWel GAS-cooled Solar Tower power plant concept and the Technology Programme "GAST" at the time being worked out in bilateral German-Spanish co-operation is given.

1. INTRODUCTION

In October 1978 the gas-cooled solar tower project 'GAST' was started in Germany. The German Federal Minister for Research and Technology (BMFT) was promoting the work for design and development of the pilot plant with an electrical power rating of appr. 20 MW and for testing of its solar-specific components.

Within the first 3 years, the work was a joint venture formed by the German industrial partners INTERATOM (as project leader), M.A.N., MBB and Dornier-System for

- project definition and preliminary design of the 20 MWel pilot plant
- development of methods and computer codes for large heliostat field design and overall system optimization/simulation
- alternative design studies and selection of reference concepts for heliostat, receiver, cycle/hot gas piping, tower, master control/process computer system
- begin of detailed engineering and design of the plant's systems and components
- definition and start of test work required for the solar-specific components

assuming a construction site at Almeria/Spain.

Since summer 1981 the project has been continued as a Technology Programme in bilateral German-Spanish co-operation by INTERATOM and the Spanish research association ASINEL in Madrid, and further participation of the a. m. German companies and Spanish subcontractors, e. g. CASA for Spanish heliostat development.

Up to the end of 1985 solar-specific GAST components have to be developed, qualified and tested under realistic solar conditions using the existing solar tower plants at the "Plataforma Solar de Almeria" - the Spanish CESA-1 and the IEA-SSPS/CRS - as test facilities.

On German side the DFVLR (German Aerospace Research Establishment) is acting as the official project attendee supervising the industrial project work on behalf of the BMFT.

2. "GAST" PILOT PLANT CONCEPT

The 20 MWel GAST pilot plant utilizes a field of approx. 1900 heliostats of two types with a reflective mirror surface of 52 m² and 55 m² respectively. The heliostats concentrate the incident solar radiation into two identical solar receivers mounted on top of a tower of about 200 m height. Fig. 1 shows schematically the arrangement of the plant's systems.

Air has been selected as heat transfer medium which is heated within the receivers to more than 800 °C for high-efficient Brayton cyle demonstration. An open gas turbine cycle with a bottoming steam turbine circuit is being provided for optimized energy conversion. In place of a passive energy- storage system, this plant concept contains a supplementary fossil-fuel fired heating system allowing electric power generation even during periods with insufficient or no incident solar radiation.

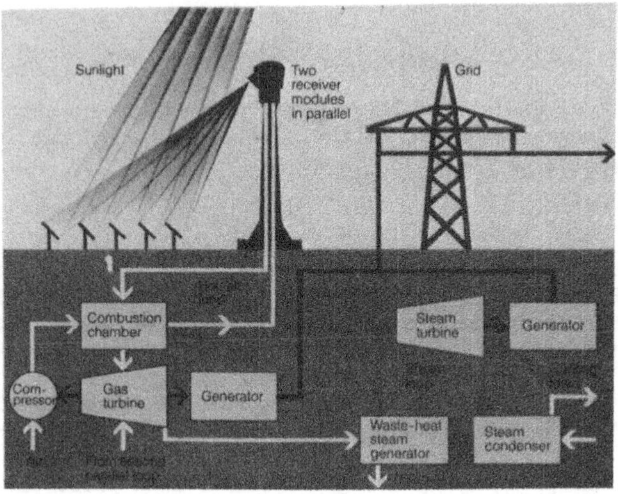

Fig. 1 GAST Block Diagram

For the tower a reinforced concrete design has been chosen. In a turret below the receiver modules the components of the KWU gas-turbine generator set with 2 directly flanged fossil-fired combustion chambers as well as the waste heat steam generator are arranged. Thus, only short length of highly complicated and costly hot gas piping are necessary. The steam-turbine section of the energy conversion system is located at the ground level of the tower.

The heliostat field of appr. 99,000 m² reflective surface is - roughly kidney-shaped - horizontally arranged around the tower and divided symmetrically into 2 identical segments by the north/south axis. The required land area for the plant amounts to about 307,000 m².

In accordance with the components' test site in Almeria (37 °N latitude) for the design point on June 21 at the solar noon the plant concept design data are as follows:

- direct insolation: 960 W/m²
- ambient conditions: 28 °C air temperature, 45 % rel. humidity, 1 bar atm. pressure
- cooling water temperature: 22 °C
- electric net power: appr. 20 MW

- modes of operation: pure solar operation over the day acc. to solar radiation as well as 24 hours continuous 20 MWel power production day and night by utilzing fossil-fuel firing whenever necessary.

This ensures that the plant concept is adaptable to different future requirements of gas-cooled solar power stations being operated in grid connection as well as for an insulated net-work.

A revision of this concept is to be expected at the end of the presently running Technonogy Programme considering specifially process heat applications.

3. TECHNOLOGY PROGRAMME "GAST"

On the basis of the first 3 years' preliminary design work for a 20 MWel GAST pilot plant concept within the actual interims Technology Programme all key questions concerning system design shall be clarified and the essential soft- and hardware technologies for a following plant's construction phase have to be developed. At present the programme has entered the major experimental phase with the preparation of tests of receiver modules, hot gas piping and a small heliostat test field at the research center in Almeria, Spain. With respect to programme organization the activities are structured as follows:

3.1 Heliostat

The aim is the development of a cost-effective, mass-production oriented heliostat suitable for slant ranges up to 600 m and solar concentration factors $\doteq$ 1000 in large fields including the construction of German and Spanish heliostat prototypes and small test-series fields and their qualification and testing under realistic conditions.

On the German side MBB has started the development on the basis of their EURELIOS heliostat know-how. After extensive conceptual studies and component pre-tests prototypes of 40 and 52 m^2 mirror surface were built, qualified and tested at MBB in Ottobrunn/Munich as well as in Almeria.

For the relatively small aperture size of the cavity receiver (appr. 30 m^2 each module) an admissible deviation of the reflected beam of appr. 2 mrad was required originally. Because of improved test and analysis results with the first 52 m^2 prototype heliostat the high image accuracy requirements could be reduced by increasing the total error budget to appr. 3 mrad for a second unit. Regarding the reduced design requirements concerning acceptable wind velocities for survival and operation limits, cost and weight reductions (appr. 30 %) could be realized. The latest Almeria test results have already proven the full accordance with these requirements und realistic conditions (measured error budget appr. 1.8 mrad.).

On the basis of the qualified second prototype the design of the small test-series of 30 heliostats has been finalized by MBB. Since Sep. 1983 the manufacturing is being performed in Germany and Spain, the installation at the SSPS-CRS is planned to be started in July 1984.

On the Spanish side a global approach from the heliostat field system point of view was performed by ASINEL. The interest of a field design consisting of two different types of heliostats was demonstrated. The two ways, either more accurate heliostats for the rear part or heliostats with lower image quality for the nearest tower area were analysed, resulting the second one more cost effective. Therefore, on the basis of 3.9 mrad total error budget two Spanish prototypes of 55 m^2 with different features were specified, manufactured (by CASA and INTECSA) and finally satisfactorely tested in Almeria.

The specification for the 30 Spanish heliostats is under preparation, and the best solution of the two a. m. prototypes will be chosen using the results of the further theoretical calculations. In this way a bigger heliostat (appr. 65 m²) making use of the favourable features of the rectangular type will be obtained.

3.2 Receiver

The aim is the development of a gas-cooled solar receiver for air temperatures of min. 800 °C including test of heat exchanger tube panels in real scale using metallic tubes for a test temperature of 800 °C (M.A.N.) resp. ceramic tubes for 1000 °C (Dornier-System) in Almeria.

With respect to this new high-temperature receiver application complicated thermo-dynamic and stress analysis work has already been performed using specially developed computer codes. In addition extensive material and component laboratory tests in newly erected test facilities have been realized like:

- LCF tests on Incoloy 800 H tube material up to 830 °C, 10 bar inner pressure and axial loading
- infra-red tests on metallic as well as ceramic (SiSiC) tubes and on insulation material for the cavity walls simulating solar radiation and gas flow conditions
- life time tests on a hot metallic tube collector of a real size heat exchanger bundle under realistic operational load conditions esp. thermal cycling
- qualification of manufacturing and welding procedures for the receiver components.

An ASINEL test facility has been built using a high-concentrating LAJET parabolic dish to test receiver cavity materials, panel supporting and insulating structure.

The technology development on ceramic material at Dornier-System concerning feasibility of mass production of at least 6 m long SiSiC tubes and connecting "ceramic-metal" pieces was concentrated

- on the qualification and test of these components at 10 bar inner pressure and air temperatures up to 1350 °C as well as of the special soldering (ceramic-ceramic) and non-destructive testing techniques including elaboration of not yet existing material data
- on the fabrication of the SiSiC components.

After solving of extreme difficulties concerning the soldering techniques the feasibility of the ceramic panel tubing is ensured by now.

For the reference receiver design the metallic heat exchanging tubes were divided into a low and a high temperature part. Panel tests on the CESA-1 plant in Almeria will start with an original scale tube bundle of 1.5 m width and 8 m active tube height being tested at max. 800 °C air temperature. After finalization of the design for this test component on the basis of the reference receiver the manufacturing of the test panel is being performed since the end of 1983, workshop acceptance is planned for beginning of June 1984.

In summer 1985 the metallic tube bundle will be exchanged by a ceramic one containing all essential construction elements and tested with air temperatures up to 1000 °C.

3.3 Hot Gas Piping

For receiver air outlet temperatures of max. 800 °C the development and test of suitable hot gas pipe components have been performed at INTERATOM taking into account their experience from the high temperature nuclear reactor field (HTR).

On the basis of former development results, the design uses now a "cold" outside pressure pipe insulated inside e. g. with ceramic material, and an inner metallic liner for flow conduction ("Isoliner" concept). This concept was successfully tested within existing HTR-facilities at the nuclear test center KFA Jülich.

For the solar-specific load conditions (temperature/pressure cycling) also lab tests on metallic and insulation materials are performed at INTERATOM's test facilities.

All essential hot gas components (e. g. tee-connections, compensators, elbows) are under manufacturing and will be procured for a test rig to be installed in Sep. this year at the CESA-1 plant in Almeria in connection with the receiver test panel. Test under realistic conditions will start beginning of 1985.

3.4 Field Control and Data Acquisition

For the planned qualification of the heliostat test-series field of 2 x 30 German and Spanish heliostats two different control systems have been developed by Dornier-System and ASINEL. For the receiver test panels/hot gas piping components the necessary field control and data acquisition systems have been developed by Dornier-System. By the means of this, specific soft- and hardware technologies for the pilot plant will be developed and tested.

3.5 Almeria Test Facilities

In order to test the a. m. components a complete gas supply system has been designed. It will be placed at the free upper part of the CESA-1 tower using its heliostats for producing the required radiation flux profiles at the receiver test panels. A scheme can be seen in Figure 2. The other necessary components (recuperator, cooling system, piping etc.) are being manufactured in Germany and Spain. The instrumentation and control and the electrical systems will be supplied by ASINEL.

The tests will be conducted along the whole year 1985.

3.6 System Engineering and Optimization

Another major task of the Technology Programme GAST is comprehensive system analysis work and optimization support for design of test components, adaption and extension of Almeria test facilities, for definition, performance and evaluation of tests as well as for re-design of the GAST 20 MWel reference concept and the conceptual design of alternative plant concepts.

The following examples shall briefly describe the main activities:

- The newly developed computer codes for design optimization of the system "heliostat - heliostat field - tower - receiver - energy conversion" are applied for optimization studies. By varying the parameters heliostat size and error budget, receiver aperture size and position, tower height and cost, etc. Considering the part load characteristics of the receiver and gas/steam turbines the optimum configuration of heliostat field, tower and receiver has been elaborated for the nominal design conditions and for the criterion "maximum produced electric/thermal annual energy per heliostat".

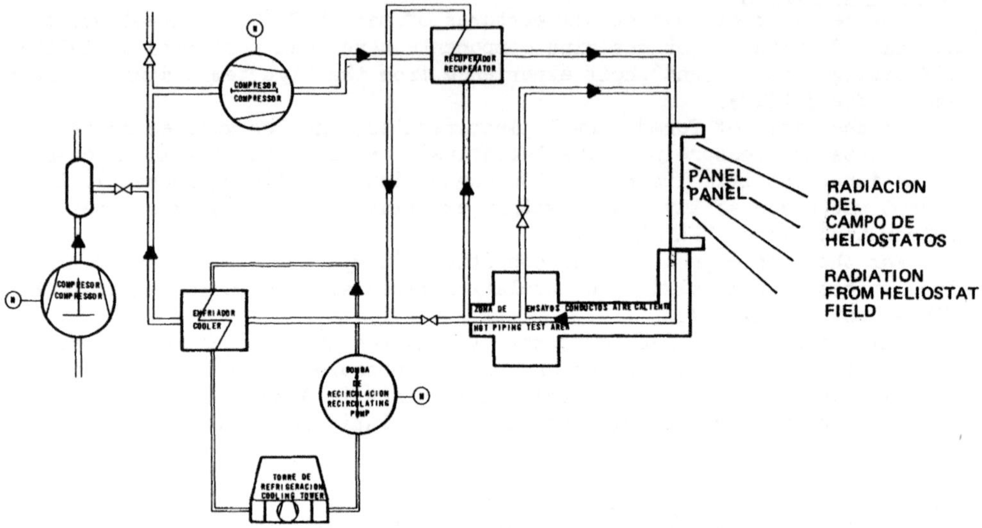

Fig. 2 Flow Scheme of Gas Supply System

- With a specially developed computer code (GASBIE) the GAST system and subsystems' operational behaviour and the plant's energy production have been simulated and analysed for realistic solar and solar/fossil fuel buffered operation with and without clouding, taking into account statistically evaluated annual meteorological conditions. Special dynamic cycle control problems are being studied by the help of a new code (DYNAG).
- The influence of higher receiver outlet temperatures up to 1200 °C on the total plant efficiency has been studied under consideration of available part load behaviour and design details of the concerned subsystems. The potential for a relative increase of appr. 20 - 25 % of the mean annual plant energy efficiency in comparison with the 800 °C reference concept could be found.
- Investigations concerning alternative GAST plant processes (e. g. high/low temperature process heat application, co-generation processes) have been started within the frame of system engineering activities.

Considering the evaluated test and optimization results of the investigated subsystems and components the reference pilot plant concept will be revised in 1985/86. Especially the utilization of the plant for high temperature process heat or synthetic fuel production will be applicable alternatives.

4. OUTLOOK

After successful termination of the Technology Programme in 1986 the continuation of the GAST pilot plant final design and subsequent construction work is expected but not yet defined. This concerns also the programme structure which could be re-organized according to future requirements. This could lead to an extension of the present form enabling the participation of interested parties/countries in the realization of the gas-cooled solar tower power plant GAST.

CENTRAL RECEIVER RESEARCH FACILITY AT THE WEIZMANN INSTITUTE OF SCIENCE

I. Dostrovsky and M. Epstein
The Weizmann Institute of Science
Rehovot, Israel

Summary

The design of a Solar Central Receiver facility aimed specifically for a chemicals and fuel programme is described. The heliostat field, with a total reflecting area of 3500 m^2 will deliver a maximum power of almost 3 MW(th). Several experimental stations are provided to give a flexible facility serving a variety of research projects. Some of these projects are outlined.

1. INTRODUCTION

The Center for Energy Research at the Weizmann Institute of Science supports a broad spectrum of projects in the energy field. Various aspects of solar energy are now receiving priority. To provide the tools for some new developments it was decided a few years ago to establish a central receiver facility on the campus.

Almost all the central receiver facilities operating today have been designed either as models of electric power generation system or for testing components.

Considering that electric power generation in the industrial nations accounts for about 30% of primary energy use, and in the developing nations only 16%, the importance of introducing solar energy into the larger non-electric sector is obvious.

Large scale use of solar energy in the non-electric sector may become possible if ways are found to convert this energy into fuels, chemical and process heat. The facility which is at an advanced stage of design (start of construction is scheduled for this fall), will be devoted entirely to research in high-temperature chemistry and high intensity photochemistry.

2. GENERAL

The facility will consist of a heliostat field, a tower and supporting laboratories and services. It will be constructed on the Institute's campus on the edge of the town of Rehovot (31°54' N, 34°45' E) at an altitude of 45 m. above sea level.

The Central Receiver will be regarded as a national facility open to all research workers in the country. Experiments will be evaluated by a users committee and those elicited will be allocated a time slot. The facility will provide the solar beam according to specification, all utilities and heat removal service. The experimenter is expected to provide his own equipment, manpower and budget.

3. THE HELIOSTAT FIELD

Since heliostats are readily available it was decided to rely on commercial units rather than embark on in-house development and assembly.

A review of various projected experiments indicated that a maximum reflected power of about 3 MW(th) at equinox noon will be adequate. The beam quality should be such as to deliver at least 1 MW(th) into a 1 m^2 aperture for six hours on any bright day of the year.

A total reflecting area of some 3500 m^2 is needed to achieve this. In terms of second generation heliostats one requires 63 units. We used MIRVAL and DELSOL computer codes to design the field.

The heliostats will be arranged in a north field in six areas. Heliostats in each area will be focused and canted according to the slant range to the upper experimental level.

The individual facets will be spherical and also focused to the same target. Because of the situation of the field in the midst of a built area, considerable attention was given during its design to the elimination of any hazard from reflected sunlight under any conditions. This aim was achieved by a suitable configuration of the terrain around the field and of the tower and laboratory buildings. A special computer code was developed to aid this design tasks and a modified MIRVAL code was used to check the safety of the final configuration.

4. THE TOWER

Since the facility will have to serve a wide variety of experiments, maximum flexibility was sought in the design of the tower. The entire volume of the tower can be used as a working space where experiments can be mounted. The roof can be used also for experimental equipment. The tower rises 51 meters above ground level. Its cross section is rectangular with dimensions of 10 m (east-west) by 15 m (north-south). Three main experimental areas have been designated in the upper part of the tower. An area for high intensity optical work (at the 45 m level, see below), an area for high temperature chemistry (at the 38 m level) and an area for component and sub-system testing. This last area will also be used as a beam dump, to raise steam for the Institute's supply system when the heliostats are not otherwise needed.

5. SUPPORTING LABORATORIES AND SERVICES

The tower rises up from a building which stretches from east to west. This is a 3 story building which serves also as physical protection of adjacent structures. The building contains four main wings: (a) Machine and Control Rooms. (b) Service Laboratories, Offices, Stores and Workshops, and Assembly space. (c) General Laboratories. (d) Entrance Hall. The main entrance to the building is from the south side. In addition to the usual services (electricity, de-ionized water, air) the tower will be fitted with a heat rejection system capable of handling the full design power.

The control room will house both the heliostat control equipment and the experimental areas' controls. Data acquisition facilities and beam characterisation systems will also be located there.

6. EXPERIMENTAL PROGRAMME

As was indicated above, this facility is meant to serve a variety of projects, most of which are probably not known precisely today. In what follows we outline some of the experiments which are among the first in line.

At the present time there are several experiments in various stages of development aimed at ultimate installation in one or other of the target areas. Some of these are still in the laboratory stage and some are already in the engineering test phase.

As was mentioned in the introduction, we regard research in solar fuels and chemicals and in solar process heat as high priority item. One of the main problems in the provision of large amounts of high temperature solar process heat is its storage and transmission over long distances. The first arises, of course, because of the intermittent nature solar energy while the second arises from the discrepancy between the distribution of high insolation areas and the concentrations of industry. Only if satisfactory solutions to these problems can be found is there a chance for a major contribution of solar energy to the industrial non-electric energy sector.

One of the first priority items, therefore, in the research programme at the new facility will be the study of the conversion of high temperature solar heat to chemical energy. There are many facets to this research programme, extending from the study of chemical systems and catalysts to the study and development of solar heated reactors. A key in this connection is Reforming, i.e. the reaction of hydrocarbons with steam or carbon dioxide to give a mixture of hydrogen and carbon monoxide. Steam reforming of methane or naphtha is a standard petrochemical procedure for producing hydrogen.

However, the conditions of a solar heat source are very different from those existing in the conventional, fossil fuel heated, process. The adaptation to solar heat is expected to be difficult but not impossible. Much of the initial experimental programme will be devoted to this problem.

The high temperature achievable in central receivers can also be used to produce hydrogen by thermochemical cycles and some work in this direction is also being planned.

Another field which is being developed involves solar driven lasers. The basic idea here is to avoid converting photons into heat, but rather use them directly to drive chemical reactions. A group at the Weizmann Institute is developing special lasers based on interhalogen compounds, suitable for excitation by solar energy. The laser radiation can them be manipulated, by techniques which are becoming well known, to obtain high fluxes of photons of exactly the right energy to drive the desired chemical reaction.

MOLTEN SALT ELECTRIC EXPERIMENT

W. R. Delameter
Sandia National Laboratories
Livermore, California USA

SUMMARY

The design, construction, and checkout of the Molten Salt Electric Experiment has recently been completed at the Central Receiver Test Facility in Albuquerque, New Mexico. This project is a full system demonstration of the use of molten nitrate salts as the receiver heat transfer fluid and also the thermal storage medium in a solar central receiver. Steam is produced to drive a 750 kWe turbine-generator. The Molten Salt Electric Experiment is now entering a 6-month test period of system performance characterization and component reliability verification. During this time, teams of design engineers and power plant operators from the private companies and public utilities who are project sponsors will be trained on-site in the operation of the system. Their feedback will be valuable in the design of future systems. Plans are presently being formulated for future testing and operation of the system which will optimize the technical value of this project.

1. INTRODUCTION

A major goal in the solar central receiver technology program is to develop systems which are both economically competitive with conventional sources of energy and are also reliable and operationally flexible enough to be attractive to energy users. Economic analyses performed several years ago indicated that central receivers using molten nitrate salts as a primary heat transfer fluid and as a thermal storage medium have a considerable cost advantage over other central receiver concepts. Such systems have the added advantage of simplified controls because the thermal storage buffers the end-use from temporary solar transients such as clouds.

To develop molten salt technology for central receivers, system designs have been performed to determine subsystem and component design requirements. Commercial scale molten salt receivers and steam generators have been designed. A 5 MW_t molten salt receiver was subsequently built and tested at the CRTF in 1979, followed by the testing of a two-tank thermal storage unit in 1980. In 1982 the Molten Salt Electric Experiment (MSEE) was begun, using the previously-tested receiver and thermal storage unit as part of this full-system demonstration of molten salt central receiver technology.

A consortium consisting of industries with solar technology experience, interested utilities, and the Electric Power Research Institute, was formed to help fund, construct, and operate the experiment. The consortium supplied half of the estimated funding in the form of both cash contributions and cost-shared engineering services. The Department of Energy supplied the other half of the funding, plus project management and on-site construction and operations through Sandia. DOE also made available for the experiment the CRTF with its existing heliostat field and receiver tower.

The MSEE has three phases. In Phase I, design and construction were completed in summer, 1983, and checkout and startup is nearly complete. In June, 1984, the MSEE enters Phase II, a six-month test period when system performance and reliability will be assessed, and when utility operators will obtain hands-on training and operating experience with the system. An optional Phase III is presently under consideration. This phase will most likely include continued system performance and component reliability assessment.

2. PROJECT DESCRIPTION

The MSEE is a full-system demonstration of an advanced central receiver system using molten salt. The overall goals of the project are to: (1) verify the capability, flexibility, and simplicity of an advanced central receiver concept, (2) provide performance information and operating experience on molten salt systems and components for utilities, system designers, component suppliers, and financial institutions, and (3) establish a test bed for component development and advanced controls.

The MSEE has five major subsystems as shown in Figure 1: the receiver, the thermal storage unit, the steam generator, the electric power generator, and the master controller. The MSEE also makes use of the existing CRTF facilities: the heliostat field, the tower, the data acquisition system, and the control room.

The receiver heats the molten salt from 590°F to 1050°F. Solar flux is concentrated by the heliostat field into the receiver cavity and onto the 13-foot by 18-foot absorber panel. Salt is pumped from the cold storage tank up the tower to the receiver where it flows through 18 vertical passes of tubes, each pass consisting of 16 0.75-inch Incoloy 800 tubes. A salt flow of 97,000 lb/hr is required for the full rated capacity of 5 MW_t.

The thermal storage unit consists of two large tanks, one to store hot salt at 1050°F and one to contain cold salt at 590°F. The hot tank is a special design to accommodate the corrosive nature of the salt and the high temperature. A corrugated liner, 0.050-inches thick and made of Incoloy 800, contains the salt. The structural load, however, is carried by an outer shell of 0.25-inch carbon steel. To insulate the carbon steel from the high temperature, there is a 13-inch layer of fire brick between the liner and the outer shell. The hot tank is 30 feet tall and 10 feet in diameter. The cold tank, a conventional design with a carbon steel shell and external insulation, is 15 feet tall and 12 feet in diameter. The full salt inventory is 176,000 lb or 11,000 gal.

The steam generator has three major components: a superheater, an evaporator, and a steam drum. Salt from the hot tank is pumped to the superheater where saturated steam at 567°F and 1200 psi from the steam drum is superheated to 1000°F. Salt leaving the superheater is attemporated with cold salt before entering the evaporator. This allows the use of low-alloy, chrome-molydenum steel in the evaporator whereas the superheater requires more expensive 304 stainless steel to withstand salt at the required temperatures. The evaporator has forced-circulation with a recirculation ratio of about 3.7:1 to maintain a high heat transfer coefficient. Both heat exchangers are U-tube within U-shell design to accommodate differential thermal expansion, with the high-pressure water/steam in the tubes and the low pressure molten salt in the shell. Outlet steam from the superheater is attemporated with saturated steam to 950°F and 1100 psi. The design output of steam is 11,600 lb/hr, and the thermal rating is 3.1 MW_t.

The electric power generator converts the steam enthalpy into electric power. The turbine-generator accepts 7800 lb/hr of steam at 940°f and 1050 psi and produces 750 kWe at 460 volts, 3 phase alternating current. The power is fed to the local power distribution grid. The steam condenses at 5 inches Hg and 133°F, and the waste heat is rejected through dry cooling towers.

The master controller for the MSEE is a distributed digital process controller manufactured by EMC Company. This system controls all flow rates, temperatures, and pressures, based on operator commands and process data. An operator interfaces with the system through video displays and a keyboard in the control room. Heliostat controls, which are not part of the MSEE controls, are located adjacent to the MSEE controls; the heliostat

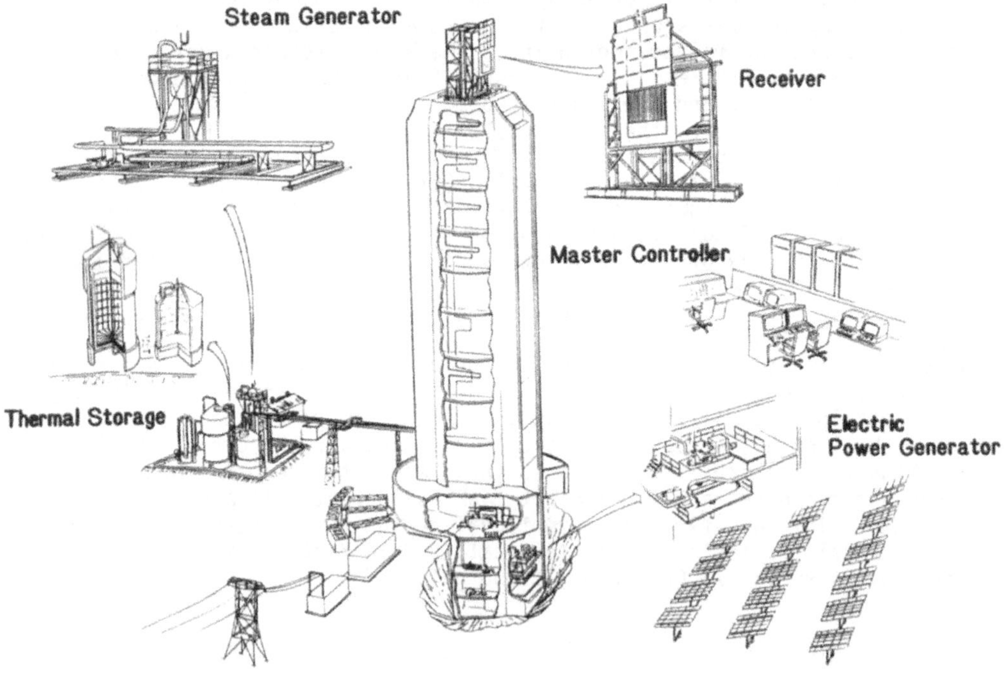

Figure 1. Subsystems of the Molten Salt Electric Experiment

operator and the MSEE operator communicate with each other verbally. The steam generator is controlled separately by a digital controller manufactured by Bailey Controls. The Bailey controller receives and carries out commands from the EMC controller, and it also passes on process information to the master controller. The MSEE controls are supported by a hard-wired relay logic system which automatically provides for the safe shutdown of the system in the event that the control computers, critical hardware component, or electrical power fails.

3. ACCOMPLISHMENTS AND STATUS

System design for the MSEE was started by Martin Marietta in early 1982. Black and Veatch was contracted to perform molten salt piping design between the top of the tower and the thermal storage area. The Black and Veatch contract was later expanded to include the design of the water/steam interface piping and the electric power generation subsystem. In August, DOE gave official approval for the project, and an organizational meeting of the MSEE sponsors was held the following month. At this time, the management structure was agreed to, and Martin Marietta was chosen as the System Integrator through system checkout. McDonnell Douglas agreed to plan and implement the system performance evaluation and utility training and operation phase. Also, in September, Babcock and Wilcox was contracted to supply the steam generator for the system. The requirements for steam conditions and flow rate were matched to a surplus naval turbine-generator which was procured by Sandia.

The turbine-generator was delivered to the CRTF in April, 1983, after a thorough inspection. The steam generator was delivered to the site in May and its installation and checkout was begun. The receiver, which had been undergoing refurbishment by Martin Marietta for 6 months, was raised to the top of the tower in June. All design activities were completed in July, and major construction, which had been closely following the design effort, was completed in August.

Construction and checkout had originally been scheduled for completion in October, 1983. It was found that the time allotted for checkout was unreasonably short and that this schedule could not be met. The schedule slippage was compounded by a series of equipment failures which slowed the checkout while repairs were made or replacement components obtained. Some of the problems encountered are specific to the molten salt solar technology and represent valuable lessons learned; these lessons will be discussed later in this report. Other problems, while no less frustrating, were the result of mistakes or were in no way related to solar central receiver or molten salt technologies, and will not be discussed here.

In October, 1983, the receiver produced rated hot salt (97,000 lb/hr at 1050°F) from solar energy for the first time. In December, rated steam was produced by the steam generator, and the turbine-generator was spun up to its rated speed of 17,000 rpm. No electricity was produced at this time, however.

Checkout progressed well until the middle of January, 1984, when the 70 kW recirculation heater which had been purchased by Sandia for the steam generator failed for the third time in six months. A new heater was required and the system was essentially shut down for 2 months before a new one was installed.

Synchronization of the turbine-generator was accomplished on April 13, 1984 and the full system was operated simultaneously for the first time on May 9. This milestone represented the first time in this country that electric power has been produced from solar energy using a primary work fluid other than water/steam.

The status of the MSEE at this time is: all subsystems are operating and are nearly checked out (only checkout of the Electric Power Generation Subsystem and the full system testing remain to be completed), the full system has been operated simultaneously, with electrical power produced from solar energy being fed to the power distribution grid; the responsibility of on-site test conductor has been successfully transferred from Martin Marietta to McDonnell Douglas; the current commitments from the government and utility/industry sponsors allow for an expanded scope for Phase II (additional system performance evaluation has been added) and the plans now go through November 1984; and finally, plans for the MSEE after the completion of Phase II are under consideration.

The plans for Phase II will be presented in the next section of this report and will be followed by a discussion of the pertinent lessons learned to date from this project.

4. PHASE II PLANS

Following completion of checkout and start-up testing in Phase I, the MSEE will be operated for approximately six months in a Phase II program. This program has the objectives to: (1) generate system and subsystem data to support development of utility power plants utilizing molten nitrate salts and (2) demonstrate operation with utility operators and obtain their feedback.

The Phase II tests are grouped into four categories. Part A will generate the test data to be used to evaluate performance and functional capability of the MSEE. Part B contains training tests which will be used to train utility operator teams. Part C contains those tests that will be performed to evaluate the operation of MSEE as a utility power plant. They will be used to produce extended performance information using the optimum operating strategy developed in Part A tests. These tests will be conducted by the utility operator teams. Part D tests will be conducted throughout Phase II by Olin to obtain data on salt properties, stability and corrosion.

Operation of the MSEE by teams of utility operators will be a major activity during Phase II. The objectives are to: (1) provide utilities with experience in operating a molten salt solar power plant with digital controls, and (2) obtain evaluation of the plant by the utilities. Four teams of utility plant operators will be scheduled.

A three-week training and operating period is planned for each team. During the first week, one-half day classroom sessions will be combined with one-half day console operation using simulation of the MSEE system. The second week will be devoted to training tests, first operating each individual MSEE subsystem alone and then in combination with other subsystems. The third week will be devoted to operation of the full MSEE system in a mode simulating routine utility power plant operation.

5. LESSONS LEARNED TO DATE

Several problems have arisen during the construction and checkout of the system hardware which have either identified new areas of design concern for solar molten salt technology or underscored the importance of pipes and valves.

However, it is important to recognize that there have been some positive lessons learned, also. Balanced against the few components that have failed are hundreds of components that function as they were designed to with no problems at all. It is noteworthy that the two previously tested subsystems -- the receiver and thermal storage -- have been shown to operate successfully as parts of a full system. It is also important to note that in the two previously tested molten salt subsystems and the new one -- the steam generator -- no major problems with the technology have been encountered. There have been minor problems, to be sure, but basically these major components do work. Most importantly, the concept of a system which uses the primary heat transfer fluid as a storage medium and buffers the end-use from solar transients has been shown to work extremely well.

5.1 Molten Salt Valves

Internal leakage has occurred through globe and rotary molten salt valves. All valves currently used in molten salt applications require metal seals, and these all have a very small leak rate. It is recommended that the system be designed to accommodate this leakage. Valves and their actuators are difficult to effectively insulate and trace-heat without interfering with the valve operation. It is recommended that molten salt valves be designed with extended yokes to allow actuators, position indicators, and limit switches to be placed further from the insulation and that these yokes be heavily insulated to reduce heat loss. External valve actuator seals are also an area of concern. Conventional packing materials are unsuitable for high temperature molten salt application. Packings which are not violently attacked by the salt have been shown to leak badly. Bellows seals have been used in the MSEE with only a few problems. Careful attention must be paid, however, to bellows design and materials, and to the operating procedures for the valves. Bellows seals may not be appropriate for full-scale applications (8-inch valves and larger) and other design solutions should be sought.

5.2 Molten Salt Pumps

The molten salt pumps used in the MSEE are all cantilever centrifugal pumps where the motor, seals, and bearings are all out of the aggressive environment of the salt. Problems occurred in the cold salt boost pump when salt splashed onto the area where the impeller shaft enters the packing at the roof of the pump sump and froze when the pump was turned off. The pump is too massive to trace-heat, and torches had to be used on the frozen area at several occasions to thaw the salt. Later, it became part of the operating procedures to switch the pump back on for a few turns during shutdown to prevent a layer of solid salt from forming. Finally, the pump was removed and disassembled so that modifications could be made to prevent salt from depositing in this critical area. The lesson is that close attention must be paid to the design of pumps because (1) they operate intermittently and (2) they cannot be trace-heated.

5.3 Insulation and Trace-Heating

Insulation is required to reduce heat losses to an acceptable level and to function along with the trace-heating to warm up the pipes and valves to a temperature above the freezing point of the salt (430°F) before salt flow is initiated from a drained condition. The trace-heating for the MSEE was designed to operate passively: the heat input to the piping just balances the heat lost to the environment through the insulation while piping temperature is maintained safely above the freezing point of salt but below the temperature limitations of the piping material. This is a very difficult design task, however, when the variation of heat loss to an ambient temperature of from 20°F to 100°F is considered along with winds that can vary from zero to 40 or 50 mph at the top of the tower. The design problem is compounded for complex shapes such as valves with long protruding actuators. The passive trace-heat design was reasonably effective in enclosed areas such as inside the tower. Severe problems were encountered, however, in exposed areas such as on top of the tower. Many test days were lost because the drain valves on the receiver were below temperature and salt flow could not be initiated. The solution was to put critical trace-heat circuits under active control (on at 600°F, off at 700°F). This led to a problem of failures of the trace-heating cabling and connectors due to thermal cycling. For the MSEE, wind shielding is being added to reduce heat losses from valves and piping in exposed areas. Trace-heat spares are being ordered. It is now recognized that trace-heating design and implementation needs to be given significant attention to avoid the problems encountered in the MSEE.

6. CONCLUSIONS

The MSEE is an important step in the development of solar central receiver technology. It is also unique in that its management and funding is provided by both the government and the private sector. Valuable lessons have been learned during the construction and checkout phase which will greatly benefit the designers, builders, and operators of future molten salt systems. A six-month test and evaluation phase is about to begin which will determine system performance characteristics, establish a measure of operating flexibility, and allow utility operators the opportunity to gain hands-on experience with the technology and to provide feedback to the designers.

SOLAR THERMAL CENTRAL RECEIVER PILOT PLANT OVERVIEW
PART II: A UTILITY PERSPECTIVE

J. N. REEVES
Southern California Edison Company
Rosemead, California USA

Summary

The Solar One Pilot Plant has been connected to the grid since April 1982 and has been operated and maintained by personnel of the Southern California Edison Company. The plant's operation has demonstrated the technical viability of the central receiver concept. Required staffing levels, environmental impacts, operating and maintenance costs, weather effects, mirror soiling rates, and auxiliary power consumptions rates have all been documented over a two year test and evaluation period. The results from this period generally have met or exceeded expectations. The plant now commences a three year power production phase to evaluate the plant's performance as a utility generation resource.

1. INTRODUCTION

The Solar One Pilot Plant is a joint venture by the U.S. Department of Energy, Southern California Edison Company, and the Los Angeles Department of Water and Power. Sandia National Laboratories serves as the technical manager of the test and evaluation program. Edison personnel have responsibility for the operation and maintenance of the facility and share project test direction with DOE.

The plant commenced operation on April 12, 1982 when the turbine was first synchronized to the Edison utility grid. As shown on Figure 1, by July 1984 the detailed test and evaluation phase is completed and the plant enters the power production test phase.

During the first two years of operation the plant has provided substantial information on each of the key utility objectives shown in Figure 2. Perhaps the most important objective, the technical viability of the central receiver concept, has been clearly demonstrated. The pilot plant is operated every day, 7-days per week, year around to simulate normal utility operation. Operating and maintenance experience is documented in monthly operating reports.

2. PLANT STAFFING

The plant is staffed by experienced Edison utility operating and maintenance personnel who were selected from the neighboring Cool Water Generating Station. Initially, a full staff of 40 people was provided as shown in Figure 3. Each shift included a foreman, three operators and a security officer. The maintenance personnel worked mainly day shifts to support the requirements of the test program. Over the first two years of operation, as the operators became more familiar with the plant and automation of the control system was completed, the staffing was adjusted

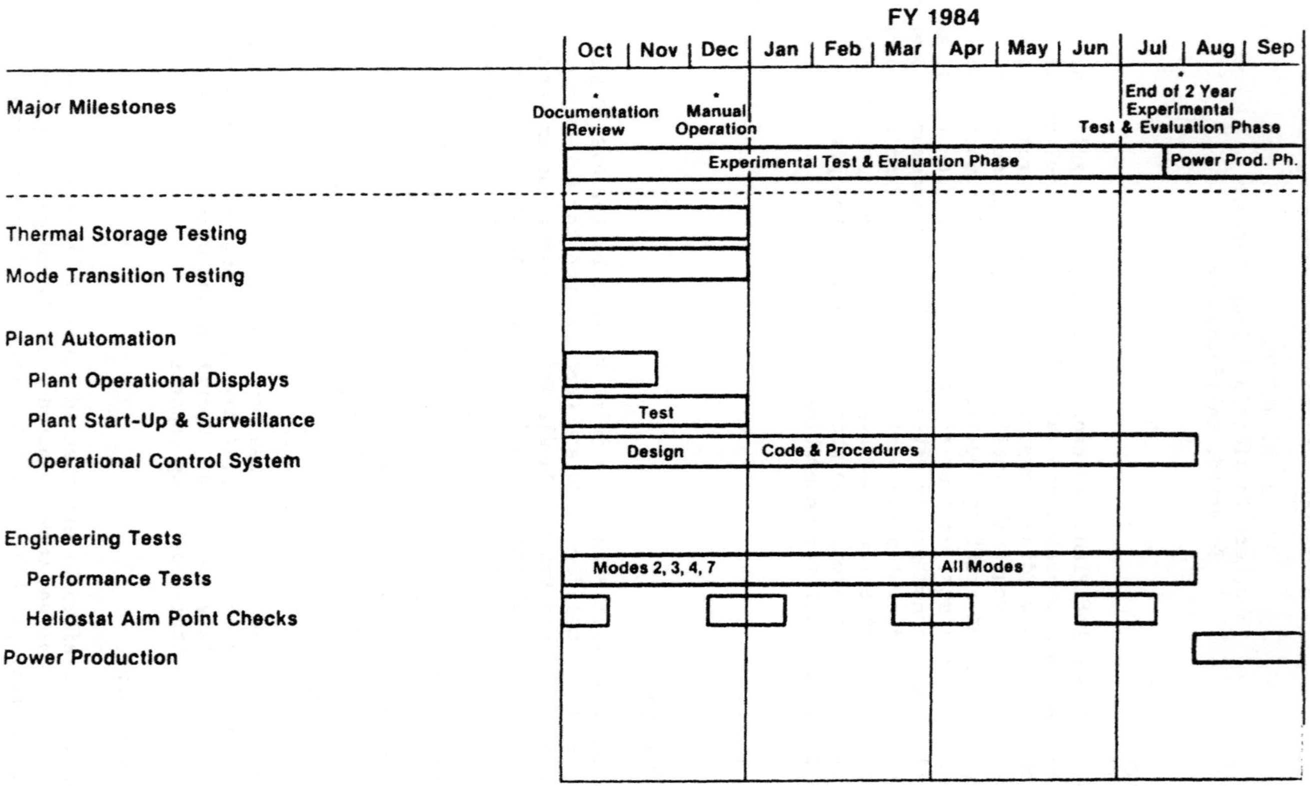

Figure 1. 10 MWe Pilot Plant Test Schedule

- **Demonstrate the technical viability of the central receiver concept.**
- **Examine overall plant reliability.**
- **Gain operating and maintenance experience with staffing levels and costs.**
- **Demonstrate plant performance as a utility resource.**
- **Demonstrate feasibility of all operating modes.**
- **Examine impact of weather on O&M and plant performance.**
- **Examine the environmental impacts.**
- **Determine mirror soiling rates.**
- **Reduce auxiliary energy consumption and optimize net energy to the grid.**
- **Upgrade the control systems.**

Figure 2. Solar One Key Objectives

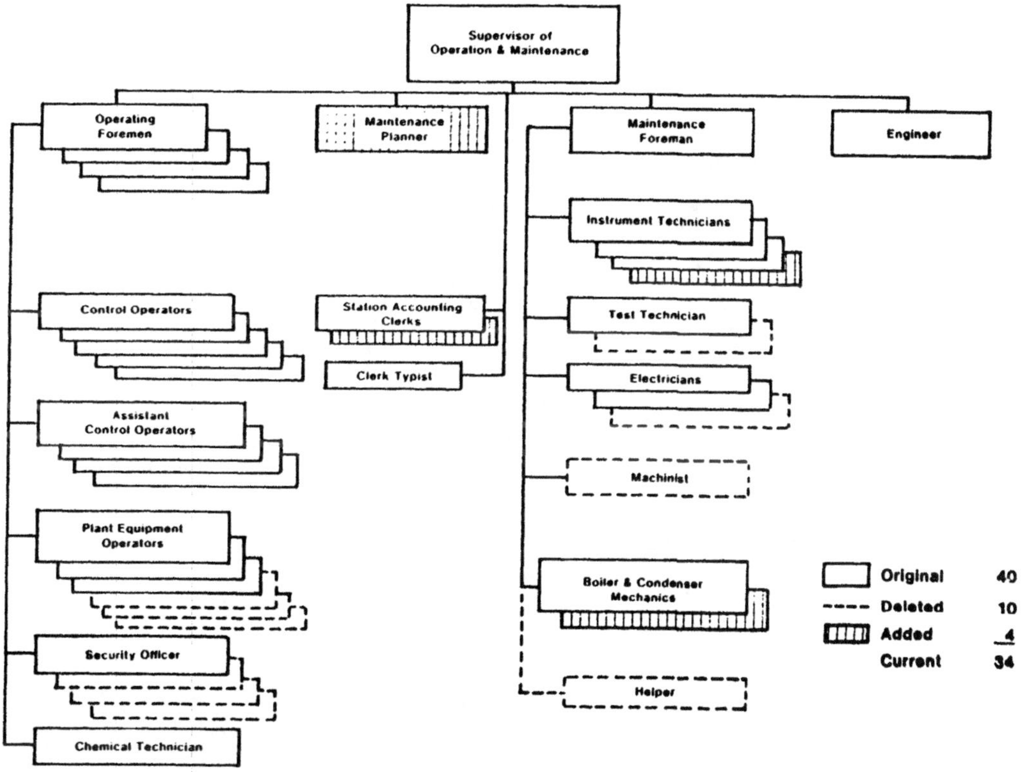

Figure 3. Solar One Operation and Maintenance Staff

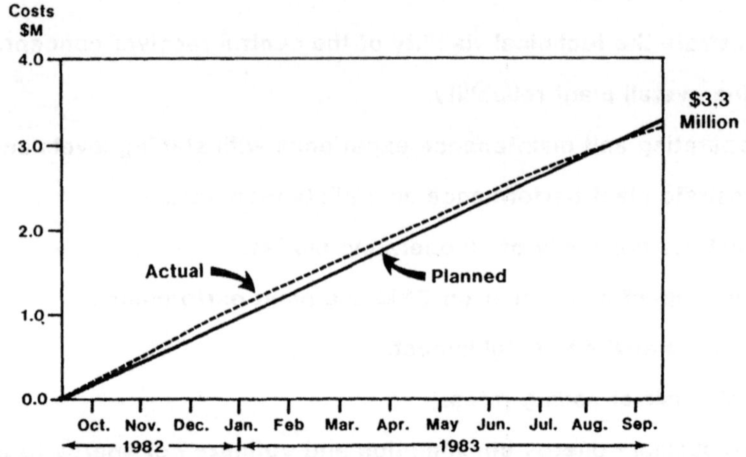

Costs $K													
Planned	275	275	275	275	275	275	275	275	275	275	275	275	3.300
Actual	290	381	301	233	267	246	244	322	276	211	254	213	3.238
Variance	(15)	(106)	(26)	42	8	29	31	(47)	(1)	64	21	62	62
Cumulative	(15)	(121)	(147)	(105)	(97)	(68)	(37)	(84)	(85)	(21)	—	62	62

Figure 4. Solar One FY83 Operation and Maintenance Budget

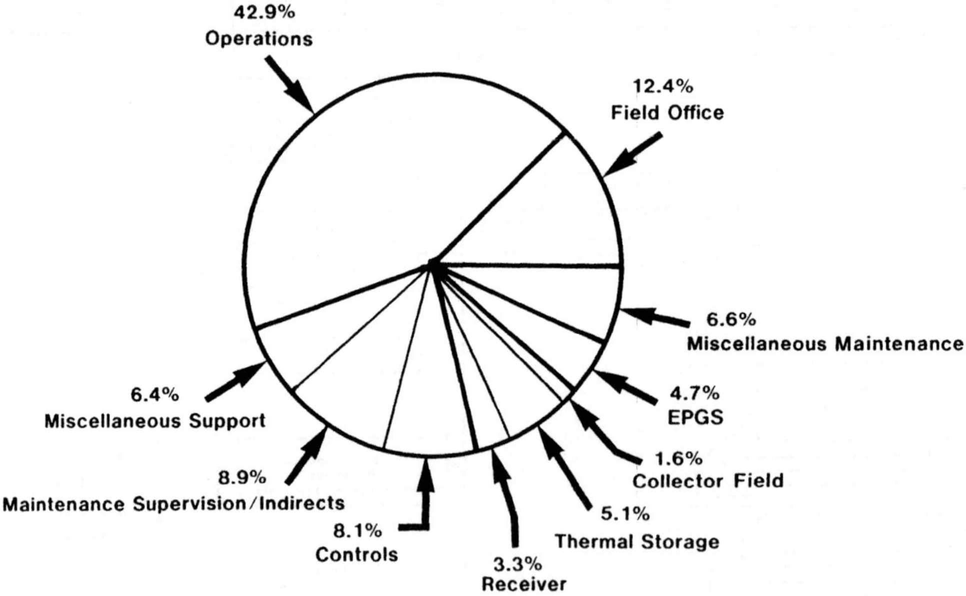

Figure 5. Solar One FY83 Operation and Maintenance Cost Breakdown

as shown by deleting 10 of the original positions and adding 4 new positions. The current staff of 34 is considered to be a prudent minimum stand-alone staff. During the power production phase additional reductions may be possible by consolidation with the Cool Water plant staff. These further reductions would probably not be possible with a stand-alone commercial plant that was not adjacent to an existing generating station.

3. COST OF OPERATION AND MAINTENANCE

The operating and maintenance costs for the pilot plant in Fiscal Year 1983 (October 1982–September 1983) are shown in Figure 4. The total O&M cost was $3,238,000. The variance between monthly planned and actual expenditures are also shown. Figure 5 shows that operations and maintenance expenditures are about equal. The biggest surprise in maintenance has been in the collector field which has required only 1.6 percent of the total O&M budget. Receiver maintenance, although initially high, has stabilized at a less-than-expected level. Conversely, non-solar equipment maintenance has been high, attributable primarily to repairs traceable to solar-induced cyclic operation.

The O&M budget for FY1984 is shown in Figure 6 at $3,286,000. If the effects of inflation are removed, the estimated expenditures in FY84 would be approximately 6 percent less than FY83. This has resulted primarily from the reductions in staffing.

4. ENERGY PRODUCTION

From a utility perspective, O&M cost and energy production are the two most important measures of plant performance. However, the first two years of operation have been devoted primarily to testing of the plant and not to energy production. Therefore, it must be recognized that any data on energy production to date is not representative of the plant's optimum energy output.

Figure 7 presents the monthly energy production data. Net transmitted is equal to gross generated minus station auxiliary power. Evident in the figure is the high auxiliary power requirements in the early months before the Thermal Storage System was fully operational. Much of the load during this period was due to the electric boiler used to produce auxiliary steam. Through 1983 and into 1984 the auxiliary power requirements have steadily reduced, reflecting the intense efforts by the operating personnel to reduce auxiliary load.

As seen from Figure 8, the plant has demonstrated its ability to produce positive power when not restricted by plant outages or bad weather. During a 23-day power generation test in May 1983, the plant produced 729,900 kWhrs net. Figure 6 illustrates the daily energy production experienced on four consecutive days during this period. Once data from the 3-year power production phase are available the energy production statistics for the plant should be much improved.

5. ENVIRONMENTAL IMPACT

The environmental influences of the pilot plant have been studied extensively by the University of California and the Los Angeles County Museum. Outside of the immediate plant site, no significant environmental impact from the pilot plant has been identified. One potential impact which was studied extensively was the possible effect of

Operation, Maintenance, Field Office	**$1,368,000**
Material, Equipment, and Services	**864,500**
Labor Overheads	**591,900**
Administrative and General	**462,400**
Total Budget	**$3,286,800**

Figure 6. Solar One FY84 Operation and Maintenance Budget

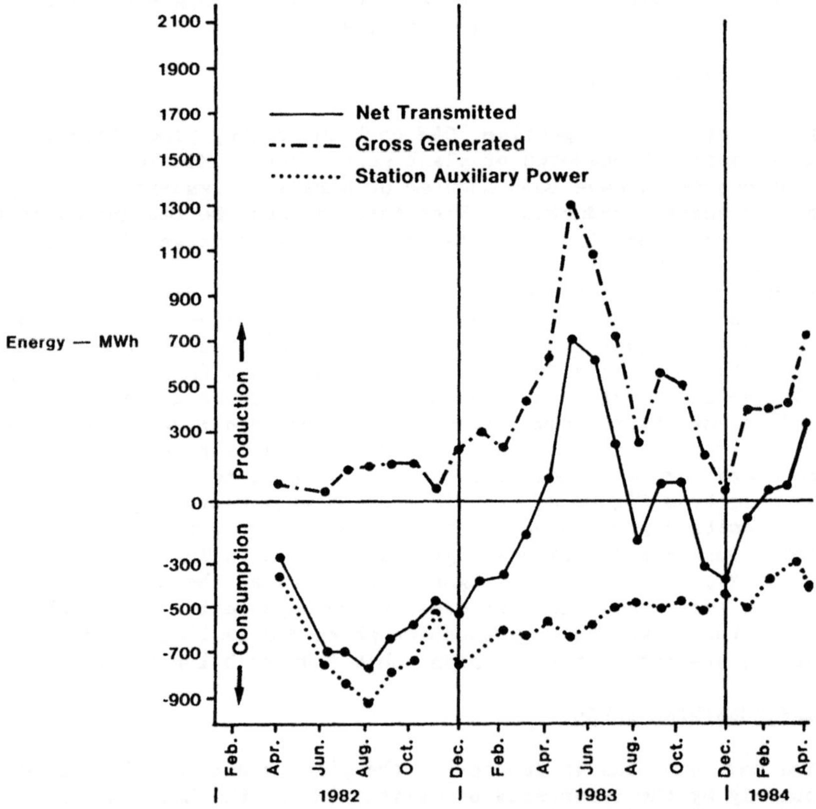

Figure 7. Solar One Monthly Energy Production

bird mortality due to heliostat beam incineration. Figure 9 shows the results. Considering the fact that over 100 species of birds occur in the vicinity of the plant and that 22,000 individual birds were counted in 102 days of observations, the number of fatalities are insignificant.

6. MIRROR CLEANLINESS

The Solar One experience with mirror cleanliness and rainfall is summarized in Figures 10 and 11. On the average, a heavy rainstorm is the best way to keep the mirrors clean. Measurements during the two years of operation indicate that rainfall can return the mirror surface to within 3-4 percent of original clean reflectivity. In the absence of sufficient rainfall, the Edison operators utilize a transmission line insulator wash truck (pressurized tank with high pressure nozzel) and deluge wash the mirror surfaces with demineralized water. This method will usually return the mirrors to within 5-6 percent of clean reflectivity. Lately the operators have come to notice an apparent non-soluble film gradually building up on the mirrors. It is readily removed by manually wiping the glass surface but not by the demineralized water spray. It is possible that some kind of manual or mechanical wiping of the mirrors may be necessary once every two years or so to remove this film buildup. Sandia Laboratories has designed a truck-mounted mechanical scrubbing device which may prove suitable.

7. LESSONS LEARNED

Some of the most significant lessons learned at Solar One are summarized in Figure 12. It is significant to note that many of the original areas of concern during the plant design have not been problems, specifically: flowstability in the once-through receiver, thermal losses from the external receiver, and receiver tube orifice blockage. It is also significant that thermal cycling has been more of a problem than originally anticipated, especially in the conventional components (heat exchangers, valves, flanges, pumps, etc.)

8. CONCLUSIONS

The results of the first two years of operation of Solar One generally have met or exceeded expectations and the test program is considered to have been very successful. Results from Solar One are currently being applied in the development of future commercial Solar Central Receiver Systems such as the Solar 100 (100MWe) project being planned for Southern California.

Date	Useable * Energy Hours	Hours On Line	Heliostats Available	Useable ** Energy ($kwhrs/m^2$)	Electrical Energy (MWHRS) Gross	Net	Aux.
5/25/84	11.9	10.0	1709	9.19	69.1	52.0	17.1
5/26/84	11.5	10.6	1706	8.76	69.1	51.8	17.3
5/27/84	11.4	10.8	1706	8.44	73.0	57.1	15.9
5/28/84	11.3	10.5	1699	8.48	71.4	53.2	18.2

* Hours with solar insolation greater than 450 $watts/m^2$
** Integrated energy above 450 $watts/m^2$

Notes: (1) Peak insolation approximately 930 $watts/m^2$
(2) Heliostats 90% clean; up to 64 out of service for washing

Figure 8. Solar One Sample Operation

- **Bird Mortalities**
  - **Spring 5.6/month**
  - **Fall 4.3/month**
  - **Winter none**

- **Insect Incinerations**
  - **1-5,034/hour**
  - **$\bar{x}$ = 2,145/hour**

Figure 9. Summary of Solar One Biological Effects

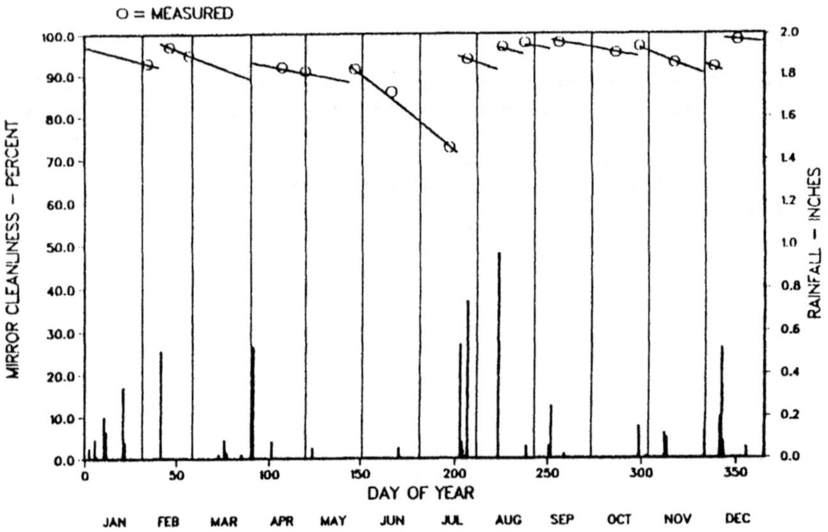

Figure 10. Solar One Mirror Cleanliness and Rainfall: 1982

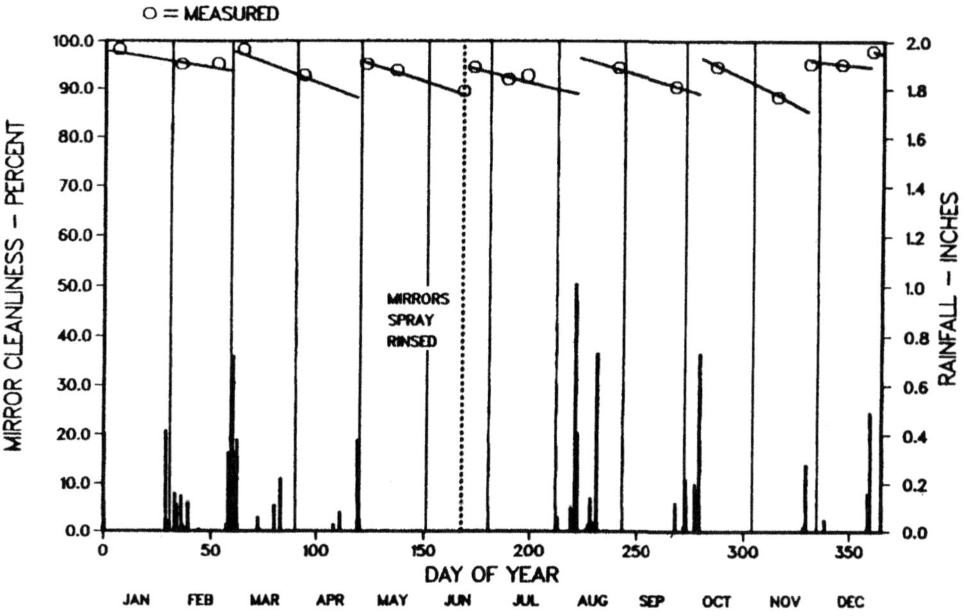

Figure 11. Solar One Mirror Cleanliness and Rainfall: 1983

- SYSTEM AND COMPONENT PERFORMANCE APPROXIMATES THEORETICAL PRÉDICTIONS

- MANY OF THE ORIGINAL AREAS OF CONCERN HAVE NOT BEEN PROBLEMS:

  ONCE-THROUGH RECEIVER FLOW STABILITY, EXTERNAL RECEIVER THERMAL LOSSES, RECEIVER TUBE BLOCKAGE, STORAGE MANIFOLD PLUGGING

- THERMAL CYCLING OF CONVENTIONAL AS WELL AS SOLAR COMPONENTS MUST BE THOROUGHLY ANALYZED

- DIGITAL DISPLAYS AND CONTROLS HAVE MANY ADVANTAGES

- THERMAL STORAGE SIGNIFICANTLY INCREASES OPERATING FLEXIBILITY

- WAYS TO REDUCE OPERATING AND MAINTENANCE COSTS NEEDED

Figure 12. Some Lessons Learned

WORKSHOP CONCLUSIONS

J. GRETZ, C. SELVAGE, A. SKINROOD

During the first two days of this 2nd International Workshop on the Design, Construction and Operation of Solar Central Receiver Projects, the operation and performance of the six existing solar power plants in Europe, the U.S. and Japan were presented and discussed.

With the exception of the U.S. plant in Barstow, California, well designed and operating under exceptionally good climatic conditions in the Mojawe desert, the operation of several other solar power plants is suffering from rather poor insolation, in quantity as well as in quality. Their "Solar Utilisation", if defined as the ratio of the actual to the site-specifically foreseen electricity production is in the neighbourhood of some 50%, and less, avaraged over the year.

Not only bad insolation is the reason for these rather poor performances, however, but also failures and bad operation of many conventional components.

The lessons learned in view of the design of future solar power plants are to:

- undertake scrupulous site selection, based on a several years' measurements campaign and bearing in mind that microclima is the decisive parameter;
- avoid the classical "new technology mistake", i.e. put all the software and hardware efforts on the solar specific parts and neglect the conventional engineering;
- design the plant in view of energy effectiveness rather than of rating efficiency by adequate circuit design (short start-up time, quick insolation responses);
- choose carefully the prime mover in view of flexible start-up operational capabilities and of good efficiencies (an increase in prime mover efficiency corresponds to an appr. 3 fold efficiency increase of the receiver and/or mirror reflectivity).

As for the economics of the existing plants, one should be very careful, if not avoid, to establish any costs of investment and of operation. All these plants are experimental or pilot plants with not representative performances. Numerically forwarded cost figures which are off competitivity by orders of magnitudes don't tell anything but are dangerous because of likely misinterpretation by the outsider public.

A review of some interesting potential possibilities for high temperature chemistry was given, by displaying the many possibilities with chemical processes powered with Central Receiver Plants. The industrial process heat application made by ARCO-Solar deserves specific mention as the first real application of CR in process heat and the first all industrially funded application of Central Receiver technology. This process heat project was reported as although small, it is cost effective and provided a step towards lowering the cost of heliostats.

In the session on New Technologies we heard of efforts in Japan, Germany, France and the U.S. on systems improvements, such as application of fiber optics, solid particles, and gases for heat transfer fluids, direct absorption in sand and salts as well as several other new heliostat approaches and receiver cooling methods. Storage, a very much reviewed but underdeveloped area, was touched upon but with a not too satisfactory outcome.

Finally, a round table discussion on the future of high temperature Solar Central Receiver technology showed that all countries or owners of the existing power plants expressed their will and desire to proceed operation of their plants with more or less extensive R&D programmes.

Some general trends and ideas from the conference are clear:

1) significant progress has been made in demonstrating the technical feasibility of the central receiver concept largely through the efforts of the six major central receiver projects;
2) although significant progress has been made in reducing the cost of energy delivered by central receiver systems, more reductions are needed if the technology is to compete with fossil energy sources;
3) future central receiver development should be a balance of support to completing the near term development of electrical generating systems and the long term development of process heat, fuels and chemicals high temperature technology;
4) thermal energy storage is an advantage of thermal conversion and this potential should be taken advantage of;

5) continued international information exchange on central receiver technology is absolutely necessary for the success of the technology;
6) some specific technical conclusions are as follows:
   a) more emphasis is needed on energy storage development;
   b) the use of molten salts has the advantage of having identical medium for storage and heat transfer. These mediums allow quicker start-up and easier plant control than water/steam systems. Gas is a good coolant for high temperature systems
   c) no particular receiver design (cavity or external) or receiver fluid seems to be definitely superior;
   d) increased effort is needed in the measurement and calculation of annual energy output ;
   e) temperature cycling of non-solar as well as solar components is a very important design consideration for CRS.

From a technology point of view, the future for Solar Central Systems is promising.

Discussions of the future of high temperature process heat shows the potential market to be even larger than previous studies have shown. In any case, there is consensus that the implementation of central receiver technology will depend to a large degree on the cost and availability of other fuel sources. The forecasting of this is not very good. We cannot wait until fuel shortage occur and prices become prohibitively high to begin development of alternatives. We should do this now, with our conventional energy supply system still operable, in order to be technically ready if potential circumstances require a new energy concept.

LIST OF PARTICIPANTS

AIELLO, F.
ENEL - Centro Ricerca Termica e Nucleare - Pisa
Via Cesare Battisti 69
I - 56100 PISA

ALTDORFER, F.
Science Policy Office
8, rue de la Science
B - 1040 BRUSSELS

AVELLANER LACAL, J.A.
INITEC, S.A.
Agustin de Foxa, 29
E - MADRID 16

BAKER, A.F.
SANDIA LABS
P.O. Box 969
USA - LIVERMORE, CA 94550

BALANZA ALONSO, R.
Engineer
Universidad de Sevilla
Cesa 1 P.O. Box 22
E - TABERNAS (Almeria)

BARTHELMESS, W.
MBB
Postfach 80 11 69
D - 8000 MUENCHEN 80

BECKER, M.
DFVLR - PT/ET
Linder Hoehe
D - 5000 KOELN 90

BEDOGNI, V.
Fi TOSI
Piazza Monumento 12
I - LEGUANO

BEER, G.
Phoebus SpA
Via Leopardi 148
I - CATANIA

BEZIAN, J.J.
G.E.S.T.
Centrale Themis
BP 27
F - 66120 FONT-ROMEU

BLAKE, F.A.
Arco Power Systems
7061 So University
Suite 307
USA - LITTLETON, Colo.

BOUDUELLE, B.
CNRS - GEST
Centrale Thémis
B.P. 27
F - 66120 FONT ROMEU

BOURDET, J.P.
Agence Française pour la Maitrise de l'Energie
Route des Lucioles
Sophia - Antipolis
F - 06565 VALBONNE

BOUTIN, V.
Themis (GEST)
B.P. 20
F - 66120 FONT ROMEU

BRATKE, U.
Deutsche Forschungs- und Versuchsanstalt für Luft- und Raumfahrt
Linder Höhe
D - 5000 KOELN

BROCKEL, D.
Deutsche Babcock Werke A.G.
Duisburger Str. 375
D - 4200 OBERHAUSEN

CACHET, P.
Electricité de France Région d'Equipement Alpes-Marseille
140, ave Viton
B.P. 560
F - 13401 MARSEILLE CEDEX 9

CARMONA, R.
IEA - SSPS - OTET
Apdo 14
E - TABERNAS (Almeria)

CIVITANO, L.
ENEL - CRTN
Via Cesare Battisti 69
I - PISA

CORVI, C.
ENEL
Via G.B. Martini 3
I - 00138 ROMA

CRESPO, L.
Asociacion de Investigacion
Industrial Electrica
(ASINEL)
P. de la Castellana, 143
E - MADRID 16

DE BENEDETTI, A.
AGIP SpA
AFOR/INCE
Viale Brenta 27
I - 20139 MILANO

DE MARCOS, J.
ASINEL
Gast Project
Francisco Gervàs 3
E - MADRID 20

DECELLE, O.
Agence Française pour la Maitrise
de l'Energie
27, rue Louis Vicat
F - 75015 PARIS

DELAMETER, W.R.
Sandia National Laboratories
USA - LIVERMORE, CA 94550

DI MEGLIO, A.
Franco Tosi Ind.
Piazza Monumento 12
I - 20025 LEGNANO (MI)

DIETRICH, B.
RWE Hauptverwalfung, Abt. AT
Kruppstr. 5
D - 4300 ESSEN 1

DINELLI, G.
ENEL - CRTN
Via C. Battisti, 69
I - 56100 PISA

DORON, U.
SIGOLEV Ltd
Weizmann Inst. of Science
14, Daniel Moritz
IL - 69345 TEL-AVIV

DOSTROVSKY, I.
Weizmann Institute of Science
14, Daniel Moritz
IL - 69345 TEL-AVIV

DURAND, T.
Ecole Centrale, Paris
F - 92290 CHATENAY MALABRY

DURISCH, W.
EIR
CH - 5303 WUERENLINGEN

EICHER, H.
Inst. für Physik
Universität Basel
Klingelbergstrasse 82
CH - 4056 BASEL

EPSTEIN, M.
Weizmann Inst. of Science
Energy Res. CTR
IL - REHOVOT

ETIEVANT, C.
Ecole Centrale
Grande Voie des Vignes
F - 92290 CHATENAY MALABRY

FANINGER, G.
Austrian Solar and Space Agency
ASSA
Garnisongasse 7
A - 1090 VIENNA

FINNSTROEM, B.
The Energy Research Commission
Box 43020
S - 10072 STOCKHOLM

FISCHER, M.
Institute for Technical Physics
DFVLR
Pfaffenwaldring 38
D - 7000 STUTTGART

FRANZKE, G.
Interatom
Fried Ebert Str.
D - 5060 BERG GLADBACH 1

FRICKER, H.W.
Sulzer Ltd
Ko. Str. 0348
CH - 8401 WINTERTHUR

FUHRMANN, H.
M.A.N. - Advanced Technology
Dachauer Str. 667
D - 8000 MUNICH 50

FUJINO, M.
Energy Research Laboratory
Hitachi Ltd
1168 Street
Hitachi-City
Japan - MORIYAMA - CHO

GAZZOLO, F.
ENEL
Compartimento di Firenze
Lungarno Colombo 54
I - FIRENZE

GEHRKE, H.
Dornier System
Postfach 1360
D - 7990 FRIEDRICHSHAFEN

GENIER, R.
EDF
6, quai Wattier
F - 78400 CHATON

GENONI, J.F.
Sandia National Labs
USA - LIVERMORE, CA 94550

GERVAIS, R.L.
McDonnell Douglas
5301 Bolsa Ave.
USA - HUNTINGTON BEACH, CA 92647

GEYER, M.
DFVLR
Pfaffenwaldring 38-40
D - 7000 STUTTGART 80

GINORIS MARTIN, J.
IEA/ Small Solar Power Systems
Int. Test and Evaluation Team
Apartado 14
E - TABERNAS (Almeria)

GOMEZ DOMINGUEZ, V.
Construcciones Aeronauticas S.A.
Departamento de Energias
Renovables
Avenida de John Lennon S/N
E - GETAFE, Madrid

GORMAN, D.N.
Arco power Systems
7061 S. University Boulevard
Suite 307
USA - LITTLETON, CO 80122

GRASSE, W.
DFVLR - SSPS
Apartado 649
E - ALMERIA

GRETZ, J.
Commission of the European
Communities
Joint Research Centre
I - ISPRA (VA)

GUEDEL, R.
A. Guedel AG
Antriebstechnik und
Maschinenbau
Industrie Nord
CH - 4900 LANGENTHAL

GUPTA, B.P.
Solar Energy Research Institute
1617 Cole Boulevard
USA - GOLDEN, CO

GUTIERREZ, L.
Sandia Nat'L Lab.
East Ave
USA - LIVERMORE, CA 94550

HARTH, R.E.
Nuclear Research Center Juelich
D - 5170 JUELICH

HOHL, F.
Interatom
Nittelgtr. 5
D - 4018 LANGENFELD

HOLMBERG, J.
Hugo Theorells
Consulting Engineers
Box 1261
S - 17124 SOLNA

HORIGOME, T.
New Energy Development
organization (NEDO)
Sunshine 60 Buld. 28F 170
3-1-1, Higashi-Ikebukuro
Japan - TOSHIMA, Tokyo

IKEDA, N.
New Energy Development
Organization (NEDO)
Solar Energy Dept.
Sunshine 60 Buld 28F 170
3-1-1, Higashi-Ikebukuro
Japan - TOSHIMA, Tokyo

IZYGON, M.
Ecole Centrale de Paris
Grande Voie des Vignes
F - 92290 CHATENAY MALABRY

JACOBS, H.
IEA/SSPS-ITET/Almeria
and Interatom (Germany)
Apartado 17
E - TABERNAS (Almeria)

KARRAIS, B.
DFVLR
Pfaffenwaldring 38-40
D - 7000 STUTTGART 80

KESSELRING, P.
Swiss Federal institute for
Reactor Research (EIR)
CH - 5303 WUERENLINGEN

KLAISS, H.
DFVLR
Pfaffenwaldring 38-40
D - 7000 STUTTGART

KLEINRATH, H.F.
Techn. Universität
Gusshausstr. 25
A - 1040 WIEN

KOCHENDORFER, R.
DFVLR
Pfaffenwaldring 38/40
D - 7000 STUTTGART

KOEHNE, R.
DFVLR
Pfaffenwaldring 38-40
D - 7000 STUTTGART

MAFFEZZONI, C.
ENEL - Centro Ricerca
di Automatica
Via A Volta 1
E - 20093 COLOGNO MONZESE (MI)

MAGNANI, G.
Studio di Informatico
Via Winkelmann 1
I - MILANO

MARTINEZ CORCOLES, F.
Central Solar CESA 1
CIA Sevillana de Electricidad, SA
Apartado 22
E - TABERNAS (Almeria)

MELCHIOR, E.
M.A.N.
Dachauer Strasse 667
D - 8000 MUENCHEN

MICHEL, A.
Belgonucleaire
25, rue du Champ de Mars
B - 1050 BRUSSELS

MINDER, R.
Electrowatt Engineering
Service Ltd
Bellerivestr. 36
P.O. Box
CH - 8022 ZURICH

MURO, M.A.
AS̱INEL
P° Castellana 143
E - MADRID 20

NICOLAY, D.
Commission of the European
Communities - DG Information
Market and Innovation
L - 2920 LUXEMBOURG

ORTIZ ROSES, C.
INITEC/IER
P.O. Box 22
E - TAVERNAS (Almeria)

PALZ, W.
Commission of the European
Communities - DG Science, Research
and Development
200, rue de la Loi
B - 1049 BRUXELLES

PERRONE, P.
ENEL - Centrale Eurelios
I - 95031 ADRANO (Catania)

PHARABOD, F.
Agence Française pour la
Maîtrise de l'Energie (AFME)
27, rue louis Vicat
F - 75015 PARIS

RAMOS, J.
Cia Sevillana de Electricidad
SSPS Project
Apdo. 18
E - TABERNAS (Almeria)

RASCH, W.
DFVLR
Linder Höhe
D - 5000 KOELN 90

REEVES, J.
Southern California
Edison Co.
2244 Walnut Grove Ave
USA - ROSEMEAD, CA 92667

RIVOIRE, B.
CNRS - GEST
Centrale Themis
Groupe Gest
F - 66120 FONT-ROMEU

ROMAN RAMOS, J.E.
Universidad de Sevilla
CESA-I
P.O. Box 22
E - TABERNAS (Almeria)

ROYERE, C.
CNRS
Solar Energetics Laboratory
P.O. Box 5
Odeillo
F - 66120 FONTROMEU

RUIZ MUNOZ, F.
Cia Sevillana
Plataforma Solar
Apartado 21
E - TABERNAS (Almeria)

RUIZ, V.
Universidad de Sevilla
Avda Reina Mercedes S/N
E - SEVILLA 12

SALA, C.
AGIP SpA Dep. AFOR/INGE
Viale Brenta, 27
I - 20139 MILANO

SANCHEZ SUNDON, F.
INITEC - CESA - I
Central Solar de Almeria
P.O. Box 22
E - TABERNAS (Almeria)

SANCHEZ, C.
Instituto de Energias Renovables
Junta de Energia Nuclear
Ciudad Universitaria
E - MADRID 3

SCHIEL, W.
DIVLR
Pfaffenwaldring 38-40
D - 7000 STUTTGART 80

SELVAGE, C.
Sandia National Labs
IEA - SSPS - ITET
Apartado 19
E - TABERNAS (Almeria)

SKINROOD, A.C.
Sandia National Laboratories
USA - LIVERMORE, CA 94550

SMADJA, P.
EDF
6, quai Watier
F - 78400 CHATOU

SPITERI, P.
EDF
Les Renardières
F - 77250 ECUELLES

SPRENGEL, U.W.
DFVLR/INSOLAR
Pfaffenwaldring 38-40
D - 7000 STUTTGART 80

STAHL, D.
Interatom Internationale
Aron Reactorbau GmbH
Friedrich Ebert Str.
D - 5060 BERGISCH GLADBACH 1

STRUB, A.
Director
Commission of the European
Communities, DG Science, Research
and Development
200, rue de la Loi
B - 1049 BRUXELLES

TAKEUCHI, H.
Engineering Development Dept.
Electric Power Development Co. Ltd
1-8 Marunouchi
Japan - TOKYO

UNGER, E.
Internationale Atomreaktorbau GmbH
INTERATOM
Friedrich Ebertstrasse
D - BERGISCH-GLADBACH

VANT-HULL, L.
University of Houston
Energy Laboratory
4800 Calhoun
USA - HOUSTON, TX 77004

WATTIEZ, P.
SSPS-ITET/SPPS (B)
Apartado 649
E - ALMERIA

WEMOWSKY, P.
INTERATOM GmbH
Internationale Atomreaktorbau GmbH
Fr.-Ebert-Str.
D - 5060 BERGISCH-GLADBACH 1

WETTERMARK, G.
COR
Västerlanggatan 31
S - 11129 STOCKHOLM

WINTER, C.J.
DFLLR
Pfaffenwaldring 38-40
D - 7000 STUTTGART 80

WRIGHT, J.B.
Sandia National Laboratories
USA - LIVERMORE, CA 94550

ZOSCHAK, R.J.
Foster Wheeler Solar
Development Corp.
12 Peach Tree Hill Road
USA - LIVINGSTON, NJ 07070

INDEX OF AUTHORS

AGATONOVIC, P., 175
AIELLO, F., 133
AVELLANER, J., 73

BALANZA, R., 221
BARANDAO, D., 253
BECKER, M., 344, 417
BEDOGNI, V., 155
BELLAGAMBA, B., 133
BEZIAN, J.J., 106
BISHOP, G.R., 4
BLAKE, F.A., 365, 410
BONDUELLE, B., 84, 106
BUEHLMANN, M., 253
BURCK, E., 364

CARMONA, R., 140
CAVALLERI, P., 155
CHILLE', A., 231
CIVITANO, L., 228, 231
CORVI, C., 10
CRESPO, L., 112, 433

DE BENEDETTI, A., 200
DE MARCOS, J., 112, 433
DELAMETER, W.R., 442
DI MEGLIO, A., 155
DINELLI, G., 133, 231
DIRKS, J.A., 307
DOSTROVSKY, I., 439
DURAND, T., 263
DURISCH, W., 253

EICHER, H., 339
EPSTEIN, M., 439
ETIEVANT, C., 263, 301

FANINGER, G., 248
FISCHER, N., 324
FUHRMANN, H., 175

GEHRKE, H., 181
GENIER, M.R., 130
GERVAIS, R.L., 190
GEYER, M., 281, 352
GORMAN, D.N., 365
GRASSE, W., 21
GRETZ, J., 3, 459
GÜNTHERODT, 339
GUPTA., B.P., 386

HORIGOME, T., 52, 390

IBRAHIMI, S., 263
IKEDA, N., 52, 86, 194, 390

JACOBS, H., 149

KARRAIS, B., 396
KESSELRING, P., 253, 327
KLAISS, H., 281
KOCHENDÖRFER, R., 396
KÖHNE, R., 344

MAFFEZZONI, C., 215
MAINI, M., 209
MARTIN, J.G., 140
MARTINEZ, F., 73, 238
MASCLE, V., 62
MAVIS, C.L., 124
McDOWELL, J.H., 365

MINDER, R., 275
MUKAI, T., 52

NORRIS, H.F., 293

ORTIZ, C., 73

PANDINI, G., 209
PERRONE, P., 231
PHAM HI, Duc, 263
PHARABOD, F., 8, 62

RAMOS, J., 92
REEVES, J.N., 449
RIVOIRE, B., 62, 162
ROMAN, J., 221
ROOS, D., 175
ROYERE, C., 412
RUIZ, F., 21

SAKAMOTO, T., 52
SALA, C., 200
SANCHEZ, F., 73, 120
SCHIEL, W., 100
SCHLÖGEL, R., 339
SELVAGE, C.S., 149, 317, 459
SKINROOD, A., 35, 314, 459
SPITERI, P., 169
STAHL, D., 433
STRUB, A., 5

TANI, T., 390
TOGGWEILER, P., 275

VANT-HULL, L., 48, 404, 430

WATTIEZ, P., 92
WEHOWSKY, P., 433
WEISSKOPF, T., 339
WINTER, C.J., 417
WRIGHT, J.B., 334

YOSHIKAWA, H., 86, 194

ZONI, U., 209

GPSR Compliance

The European Union's (EU) General Product Safety Regulation (GPSR) is a set of rules that requires consumer products to be safe and our obligations to ensure this.

If you have any concerns about our products, you can contact us on ProductSafety@springernature.com

In case Publisher is established outside the EU, the EU authorized representative is:

Springer Nature Customer Service Center GmbH
Europaplatz 3
69115 Heidelberg, Germany

Batch number: 10755284

Printed by Printforce, the Netherlands